CRM Series in Mathematical Physics

T0191619

Marc Thiriet

Biology and Mechanics
of Blood Flows

Part II: Mechanics and Medical Aspects

 Springer

Marc Thiriet
Project-team INRIA-UPMC-CNRS REO
Laboratoire Jacques-Louis Lions, CNRS UMR 7598
Université Pierre et Marie Curie
Place Jussieu 4
75252 Paris Cedex 05
France

ISBN: 978-1-4419-2576-3   e-ISBN: 978-0-387-74849-8
DOI:10.1007/978-0-387-74849-8

Series Preface

The Centre de Recherches Mathématiques (CRM) was created in 1968 by the Université de Montréal to promote research in the mathematical sciences. It is now a national institute that hosts several groups and holds special theme years, summer schools, workshops, and a postdoctoral program. The focus of its scientific activities ranges from pure to applied mathematics and includes statistics, theoretical computer science, mathematical methods in biology and life sciences, and mathematical and theoretical physics. The CRM also promotes collaboration between mathematicians and industry. It is subsidised by the Natural Sciences and Engineering Research Council of Canada, the Fonds FCAR of the Province de Québec, and the Canadian Institute for Advanced Research and has private endowments. Current activities, fellowships, and annual reports can be found on the CRM Web page at www.CRM.UMontreal.CA.

The CRM Series in Mathematical Physics includes monographs, lecture notes, and proceedings based on research pursued and events held at the Centre de Recherches Mathématiques.

Montréal   *Yvan Saint-Aubin*

Contents

Introduction

Scientifically one thinks of truth as the historical rectification of a long error, one thinks of experience as rectification of a primary and common illusion. (G. Bachelard) [1]

As the title suggests, the present text is aimed at providing basic knowledge and the state of the art on the biology and mechanics of blood flows. Due to its length, this book has been divided into two parts, the criterion being the size of the biological domain used to extract input data required to model the cardiovascular system (CVS). The subtitles of each part thus reflect this subdivision.[1] The present Part II, *Mechanics and Medical Aspects* refers to the macroscopic scale. It contains chapters on anatomy, physiology, and continuum mechanics, as well as pathology of the vasculature walls, including the heart, and their treatment. Methods of numerical simulations are given and illustrated in particular by wall diseases. The main information about cardiovascular biology is provided in Appendix A; biology of the cardiovascular system is given in Part I. Mechanical data and concepts are summarized in Appendix B, as well as major information on flows in deformable conduits. Basic information on numerical simulation methods are given in Appendix C, emphasizing the computational methods used by the author to simulate blood dynamics in large vessels.

The origin of this book is an invited contribution text, Biochemical and Biomechanical Aspects of Blood Flows, of the survey article collection, Modeling of Biological Materials.[2] The present review also answers a pedagog-

---

[1] Part I, *Biology* deals with the nanoscopic and microscopic scales. The nanoscale corresponds to the scale of biochemical reaction cascades involved in cell adaptation to mechanical stresses among other stimuli. The microscale is the scale of stress-induced tissue remodeling associated with acute or chronic loadings.

[2] The book *Modeling of Biological Materials* belongs to the Series Modeling and Simulation in Sciences, Engineering, and Technology, Bellomo N. (ed), Birkhauser.

ical objective of the project, Mathematical Modeling for Haemodynamics (HaeMOdel).[3] This book represents a much longer version of a draft given to the participants of the CRM–INRIA Spring School[4] on "Mini-invasive procedures in medicine and surgery: mathematical and numerical challenges," at the Centre de Recherches Mathématiques of the Université de Montréal in Québec, Canada.

Blood flow is widely explored because of blood's vital functions, flow-mediated regulation of vessel lumen caliber, and of wall structure (endothelial mechanotransduction), as well as stress-dependent focal wall pathologies. The endothelium is permanently exposed to biochemical and biomechanical stimuli that are sensed and transduced, leading to responses that involve various pathways. In particular, the endothelium is subjected to hemodynamic forces (pressure and friction) that can vary both in magnitude and direction during the cardiac cycle. The endothelium adapts to this mechanical environment using short- and long-term mechanisms. Among the quick reactions, it modulates vasomotor tone locally releasing vasoactive compounds. The endothelium also actively participates in inflammation and healing. Long-term adaptation leads to wall remodeling and vascular growth, with the formation of functional collaterals and vessel regression. Blood flow behavior then depends not only on cardiac pump quality and vasculature anatomy (Chap. 1), but also on its structure (Part I). Both the heart and blood vessels are controlled (Chap. 2). Furthermore, cardiac output interacts with blood circulation (pre- and afterload) and vice versa.

The understanding of a physiological system is based on identification of the features that determine its behavior. Whereas physicians and biologists are satisfied with enumeration of process components (data catalog) so they can appropriately react when experiencing unusual cases, physicists and applied mathematicians seek explanations according to few selected appropriate system features (reductionism).

The modeling of the cardiovascular system can have multiple goals: (1) prediction, (2) development of pedagogical and medical tools, (3) computations of quantities inaccessible to measurements, (4) control, and (5) optimization.

---

[3] HaeMOdel project corresponds to a Research Training Network of the fifth call of the framework program of the European Community. The main objectives of the HaeMOdel project were the development of numerical models, including three-dimensional reconstruction from medical imaging, Navier-Stokes solutions in deformable domains, and biochemical transport.

[4] The object of this combined school (short courses, May 16–20) and workshop (May 23–27, 2005) was to bring together several facets of mini-invasive procedures in medicine and surgery and identify issues, problems, trends, and mathematical and computational challenges. The school was structured around the following themes: medical image processing and geometrical modeling, fluid-structure interaction in biomedical problems, static/dynamic design and control of (implantable) medical devices, and finite element-based computer-aided design/manufacture.

The modeling of any biological system, in particular the cardiovascular system, entails several stages: (1) a definition stage during which the whole datum set related to the real system are collected and the problem and its goals are identified (concrete level), (2) a representation stage of the system, the system datum number being reduced (information filtering), data relevant to the problem being extracted to describe the model, simplified version of the reality, (3) a validation stage, which is the consequence of generality loss, comparisons with available observation data providing a meaning to the model, and (4) possible model improvements. Two main varieties of models include the representation and knowledge models. The former is associated with mathematical relations between input and output variables, the equation coefficients having no necessarily physical or physiological meaning. The latter is aimed at analyzing the mechanisms that produce the explored phenomena. Any model is developed in the framework of a theory, such as continuum mechanics, but the model is much more specific than the theory. The simplification degree of the model provides its application domain. Certain assumptions, which are not strongly justified, can be removed in iterative model refinement. Any model is characterized by: (1) its potential to improve a system's knowledge and underlying theory and the development of new concepts, and (2) its capacity to define new strategies.

A model must be easy to use and understand, reliable, robust, evaluating the system response to the extrema of the involved variables and parameters, adaptive to estimate the responses to various alternatives, and evolutive to respond to more complicated situations. The model is constituted of acting and interacting (system dynamical structure) components with its attributes, descriptive variables that define the system state, parameters and control factors; functional relations that describe the system behavior, using the variables and parameters; constraints due to the system and its environment, and functional criteria. The model with its behavioral equation set associated with hypotheses remains too complex to be solved by analytical methods. Numerical approximations are then required, based on approximations associated with a solving technique. Whatever the theory and its underlying assumptions, mathematical and numerical analyses of the associated equations are required.

Biomechanical research[5] can be defined by four main classes: (1) motion biomechanics,[6] (2) organ structures and rheology, as well as biomaterials, (3) biofluid flow and rheology, (4) heat and mass transfer, and (5) cell and

---

[5] Biomechanics investigates biological systems by means of mechanical laws and principles. Biomechanics links mechanics with biological disciplines as well as techniques used in various disciplines to: (1) find specific solutions to patient problems and (2) take into account the living tissues in a real environment.

[6] Biomechanics of motion deals with the locomotor system (bones, joints, and muscles) and its neural control, during normal activities or exercise, in normal conditions (gait, ergonomics, robotics, etc.) or in diseases (orthopedics, rehabilitation, etc.).

tissue engineering. All these topics are related to blood flows at manifold levels. For instance, postural changes, movements, and exercise require adaptation of the blood circulation.

Modeling adopts a reductionist strategy, selecting system properties suitable to the posed problem to define the model, and breaking the system into several parts at a given length scale to understand the behavior of each part. However, the top-down approach can neglect the interactions between the system's parts, the complete system being more than the sum of its parts, especially in the presence of non-linear interactions, with negative and positive feedbacks. Furthermore, the functions of the cardiovascular system depend on multiple parameters, local and global, and intrinsic and environmental.

Computational models of the cardiovascular system are now based on medical images (Chap. 3), because of high between-subject variability, to yield patient-specific data. The geometrical information is provided by four main imaging techniques, using X-rays, γ rays, ultrasound (US), or magnetic fields. Various image processing techniques give three-dimensional reconstruction. Furthermore, image-based experimental approaches, using stereolithography, or other rapid prototyping techniques are aimed at validating the numerical results.

The vasculature is constituted of prestressed, viscoelastic, more or less convergent (arteries) or divergent (veins), curved vessels with numerous branches or tributaries (Chap. 4). The wall has a nerve-controlled muscular layer that commands local blood input (Part I). Stresses applied by the flowing blood on the wetted surface of the vessel wall also regulate the local vessel bore (mechanotransduction). Pressure and flow variations along the arterial tree are associated with the propagation of their corresponding waves, with given phase lags. The waves run with a speed of 5 to 30 m/s and change their shape. The artery wall can be injured, with frequent lethal consequences, inducing either lumen dilations (aneurisms) or narrowings (stenoses).

The cardiovascular system has been studied from different complementary perspectives, solid and fluid mechanics, and command and control, most often without links between them. Most of the biomechanical works deal with arterial flow (Chap. 5). The arteries carry the blood from the cardiac pump to the tissue capillaries, loci of gas and substance exchanges. Intermittent cardiac output is transformed into an uninterrupted unsteady flow by the arteries.

However, the arteries only constitute a part of the cardiovascular system. Blood flow modelings also target heart functioning, to better describe the blood motion (Chap. 6). The sequence of depolarization spreading and repolarization depends on the intramural layout of myocardial fibers.[7] Genesis

---

[7] The elongated shape of myofibers accounts for myocardium anisotropy for electrical conductivity. The myofibers have different inclination angles in wall slices from the inner to the outer heart surface. Moreover, myofibers are not parallel to the heart surface (transmural obliquity). Myofiber orientation belongs to the input data both for direct and inverse problems.

and propagation of the electrochemical wave induce myocardium contraction. Vessel wall displacement during the cardiac cycle is induced by pressure wave propagation. These scientific fields thus deal with coupling problems, in particular heart electromechanical coupling, fluid-structure interaction (FSI) in the deformable heart and vessels, both equipped with valves and chemical coupling, i.e., substance transport in vasculature segments and within the vessel wall.

Hemodynamic quantities may cause or be cofactors in various pathologies of the cardiovascular system (Chap. 7).[8] Detailed investigation of local blood flows in pathological vessel segments is then required for a complete medical check-up, especially stress fields applied by the blood to the vessel wall as well as stress distribution within the wall. Moreover, blood flow modeling is helpful in therapy strategy; it can help to make the choice between surgical and interventional procedures. Conversely, heart and vessel diseases induce changes in hemodynamic variables that must be measured by functional testing. Biomechanical studies can also be aimed at designing surgical repairements and implantable medical devices[9] and predicting effects of implanted prostheses and devices (Chap. 8).

Modeling and numerical simulations of the functioning of the cardiovascular apparatus and its disturbances require a large scale range, from the molecule level (nanoscopic scale, nm, Part I), to the cell organelles associated with biochemical machinery (microscopic scale, μm), the whole cell connected to the adjoining cells and extracellular medium, the subdomain of the investigated tissue (mesoscopic scale, mm), and the entire organ (macroscopic scale, cm).

The complexity of the physiological systems and limited number of available values of implicated factors still lead to simplifications. Available computational techniques can only cope with limited problems and cannot accurately treat the coupling between the various involved scales and the whole set of biochemical and biophysical phenomena. Moreover, models are commonly aimed at describing phenomena (physiological processes and local biomechanical behaviors of solid organs under given loadings and/or biofluid flows) at the macroscopic level.

Nevertheless, investigations have started to deal with multiscale modeling to take into account the whole set of mechanisms involved in the functioning of blood circulation. Modeling remains sufficiently simple not only for computational efficiency but also for experimental set-up elaboration. (Measurements allow model validation when in vivo data cannot be acquired without great

---

[8] A research goal on pathophysiological mechanims underlying pathogenesis of cardiovascular diseases can be to improve diagnosis at a preclinical state, determining indices and enhancing signal processing.

[9] When the endothelium is damaged or has defective functioning, hemodynamical factors contribute to in-stent restenosis and vascular graft failure. Numerical tests of blood flow behavior during a whole cardiac beat in a geometrical model of the diseased vascular segment ensure better therapy after planning.

disturbances or/and tissue damage.) Moreover, modeling avoids large numbers of parameters that cannot be handled in the inverse problem solving, an interaction field between biomechanics and medical practice. An example is given by myocardial contraction, the source of blood circulation. Successive stages are considered, from the nanomotor level (actin–myosin binding) to the sarcome deformable structure, muscular fibers, and myocardium. Biochemistry is tightly linked to biomechanics.

The three-dimensional behavior of blood flow is commonly explored in subregions of the cardiovascular system, characterized by a short length in the streamwise direction. These domains usually undergo large spatial and temporal variations in stresses. Current investigations are aimed at coupling these three-dimensional models to simplified representations (one-dimensional flow, lumped parameter models) of blood flows to model the whole vasculature. In the recent years, coupling between detailed models and one-dimensional models suitable for wave propoagation and lumped parameter models for the main branches of the arterial bed has been investigated both mathematically and numerically.

There are still many challenges in cardiovascular biomechanics. The first challenge is the rapid reconstruction of the entire network of large vessels responsible for the blood irrigation and drainage of any organ using hierarchical models to produce more realistic predictions (Sect. 6.2.4). The second deals with blood flow coupling with interactive processes at different levels, between blood cells and plasma, vessel wall and flowing blood, between vessels and organ parenchyma, and input waveforms and the flow dynamics. The third corresponds to the incorporation of microcirculation and organ perfusion to close the vasculature loop. Microcirculation models can be based on several complementary methods based on recent investigations, such as reduced-basis element models for small-bore branching circulatory beds, multiphysic homogenization of deformable conduit networks, deformable particle flow for zoomed regions of interest in association with peculiar physicochemical processes, and transport in poroviscoelastic media.

Associated with image processing, computer graphics (infographics) and medico-surgical robotics, the research targets medical and surgical real-time simulators to train for interventional medicine and mini-invasive surgery as well as develops computer tools for diagnosis, treatment planning, and guided gestures.

> Nothing is given. Everything is constructed
> (G. Bachelard) [2]

Acknowledgments

The author, a CRM (Centre de Recherches Mathématiques) member, wishes to acknowledge two french public research organizations, namely CNRS (Centre National de la Recherche Scientifique), as he is a researcher from this center, and INRIA (Institut National de la Recherche en Informatique et Automatique), for supporting his activities via the project team REO localized in both Laboratoire Jacques-Louis Lions, UMR CNRS 7598, of University Pierre et Marie Curie (Paris VI University) and INRIA Paris Research Center, and via INRIA Associate Team "CFT," as well as ERCIM office (European Consortium of public Research Instituts) in helping him manage the working group IM2IM.

This book would never appear without numerous fruitful collaborations, especially in Canada (Y. Bourgault, M. Delfour, A. Fortin, A. Garon), in Europe (M. Bonis, P. Brugières, C. Delpuech, S. Deparis, D. Doorly, C. Fetita, P. Frey, A. Gaston, J.M.R. Graham, F. Hecht, L. Formaggia, Y. Maday, S. Naili, J. Peiro, O. Pironneau, F. Prêteux, A. Quarteroni, C. Ribreau, L. Soler, G. Szekely, J. Treiber, K. Wolf, etc. - those missing will have to pardon the author -, as well as Asclepios, Bang, Macs, Gamma, Reo [A. Blouza, L. Boudin, M. Boulakia, L. Dumas, M. Fernandez, J.F. Gerbeau, C. Grandmont, I. Vignon-Clementel], and Sisyphe team members), in Hong Kong (H.H.S. Ip), and in Taiwan (T.W.H. Sheu), and the help of A. Montpetit. The author also acknowledges the patience of his family (Anne, Maud, Alrik, and Damien).

1

Anatomy of the Cardiovascular System

> *On ne pourra bien dessiner le simple qu'après une étude approfondie du complexe. [One will well design the simple (model) only after a complete study of the complex (reality).]* (G. Bachelard) [1]

The cardiovascular system is mainly composed of the cardiac pump and circulatory network. The heart is made of two synchronized pumps in parallel, composed of two chambers. The left heart propels blood through the *systemic circulation*, the right heart through the *pulmonary circulation* (Fig. 1.1). Blood circulation is characterized by several space and time scales associated with vascular wall and caliber and cardiac cycle and control phenomena (Table 1.1).

1.1 Heart

The heart is located within the *mediastinum*,[1] behind and slightly to the left of the sternum[2] [3, 4]. The base of the heart is formed by vessels and atria and the apex of the heart by the ventricles. The heart rests on the diaphragm. The heart has four cavities, upper left (LA) and right (RA) *atria*, and lower left (LV) and right (RV) *ventricles*. The left ventricle is located posteriorly and leftward from the right ventricle. Heart chamber size varies during the

---

[1] The mediastinum is the chest space between the sternum (front), the spine column (rear), and the lungs (sides). It contains the heart and its afferent and efferent vessels, thymus, trachea and main bronchi, esophagus, thoracic nerves and plexi, and lymph vessels and nodes.

[2] About two-thirds of the heart is left of the midline, with its long axis oriented from the left hypochondrium to the right shoulder in the usual situation, with minor changes during respiration. The cardiac position varies between subjects with possible mirror-image configuration. The right border is given by the right atrium, the inferior border mostly by the right ventricle, the lower and upper parts of the left border by the left ventricle and atrium, respectively.

Table 1.1. Blood circulation length and time scales.

Vasculature geometry	
0.1 μm	Endothelium cleft width
1 μm	Averaged capillary wall thickness
10 μm	Capillary lumen bore
2–25 μm	Blood cell size
10–80 μm	Endothelial cell size
1 mm	Large artery wall thickness
3–5 mm	Lumen bore of large artery and vein
1 cm	Ventricle wall thickness
1–3 cm	Lumen bore of aorta and vena cava
2–6 cm	Width of heart chambers
1–2 m	Body height

Blood circulation-related activities	
1 s	Cardiac cycle
10–100 s	Control
mn–hours	Adaptation
day–weeks	Remodeling

cardiac cycle due to myocardium activity (Table 1.2). The left ventricle is the

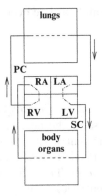

Figure 1.1. Pulmonary (PC) and systemic (SC) circulation. The blood is conveyed from the drainage veins of the systemic circulation to the right atrium (RA). It is then convected into the right ventricle (RV) and expelled into the arteries of the pulmonary circulation for oxygenation into the lungs. The oxygenated blood is sent to the left atrium (LA) via the pulmonary veins. It then enters into the left ventricle (LV) to be propelled into the arteries of the systemic circulation, which distribute blood to every body organ (including heart and lungs) for energy and nutrient supply and waste removal.

largest chamber with thickest wall[3] (Table 1.3). The septum separates the left and right hearts. The *pericardium* surrounds the heart and the roots of great blood vessels. It is attached by ligaments to the spine column, diaphragm, and other organs. The pericardium restricts excessive heart dilation, and thus limits ventricular filling.

Epicardial fat can be observed on human hearts during surgery (grafting) or by imaging procedures (CT, MRI, and echocardiography) [5]. In the adult heart, adipose tissue can be found in the atrioventricular and interventricular grooves. Epicardial fat can increase up to cover the entire epicardial surface.

Four valves at the exit of each heart cavity, between the atria and the ventricles, the *atrioventricular valves* (AVV), and between the ventricles and the efferent arteries, the *ventriculoarterial valves* (VAV, Fig. 1.2), regulate blood flow through the heart and allow bulk unidirectional motion through the closed vascular circuit.[4] The atrioventricular valves are inserted at the atrioventricular junctions, whereas the ventriculoarterial valves are hinged from semilunar

Table 1.2. Echographic measurements of cardiac kinetics. Estimated width (w) range and mean (mm). The volume of the ventricle, assumed to have the configuration of an ellipsoid, is evaluated by $V = \pi/6\,Lw^2$ (L: base-to-apex length of the cavity). The shortening fraction (%) is equal to the ratio of the difference between the end-diastolic width and the end-systolic one to the end-diastolic width.

	Range	Mean
LV width (diastole)	35–55	45
LV posterior thickness (diastole)	5–10	10
Interventricular septum thickness (diastole)	5–10	10
LA width	20–40	30
LV shortening fraction	35–45	36

Table 1.3. Estimates of heart wall thickness (mm).

RA, LA	1–3
RV	3–5
LV	10–15

---

[3] The right ventricle pushes blood into low-pressure pulmonary circulation; therefore, it has a moderately thick muscle layer. The left ventricle expels blood into high-pressure systemic circulation and thus has the thickest myocardium.

[4] On a frontal chest radiograph, the pulmonary valve is the upper one, above the mitral valve. The aortic valve is positioned at the south-west corner of the mitral valve, above the tricuspid valve. The right-heart valves, widely separated from each other on the roof of the right ventricle, are located anteriorly from the adjacent left-heart valves on the roof of the left ventricle.

Figure 1.2. Isolated pulmonary orifice axially cut at one commissure line to display the three retracted cusps with their curved insertion lines and the commissures. The ventriculoarterial valves are not connected to the cardiac wall by chordae tendineae.

insertions (without any ring attachment) at sinusal junctions.[5] The *tricuspid valve* (TrV) regulates blood flow between the right atrium and the right ventricle. It is composed of antero-superior, postero-inferior, and septal leaflets. The *pulmonary valve* (PuV) controls blood flow from the right ventricle into the pulmonary arteries, which carry blood to the lungs to pick up oxygen. The pulmonary valve is located at the end of the pulmonary infundibulum of the right ventricle. The *mitral valve* (MiV) lets oxygen-rich blood from pulmonary veins pass from the left atrium into the left ventricle. It consists of two soft thin cups attached to the atrioventricular fibrous ring (oblique position): a large anterior (aortic) and a small posterior (mural).[6] The *aortic valve* (AoV) guards the left ventricle exit; once opened, blood crosses it from the left ventricle into the aorta, where it is delivered to the body. Like the pulmonary valve, it consists of three semilunar cusps. Immediately downstream form the aortic orifice, the wall of the aorta root bulges to form the *Valsalva sinuses.* Orifice sizes are given in Table 1.4.

Table 1.4. Estimates of valvar orifice size (mm). Datum variability (Sources: [6, 7]).

Orifice	Perimeter	Caliber	Thickness
Tricuspid	110–130		
Mitral	90–110	20–25	
Pulmonary	75–85		0.4 ± 0.1
Aortic	70–80	32	0.6 ± 0.2
Coronary sinus		2–7	
Inferior vena cava		2–8	

---

[5] Semilunar hingelines extend from the sinotubular junction to a virtual ring joining the basal valvar insertion, thus crossing the junction between the ventricular infundibulum and the arterial wall.

[6] The large mural leaflet closes about two-thirds of the valvar orifice.

Trabeculae carnae, muscular columns, are observed in both ventricles, especially the left ventricle. *Papillary muscles*, ventricular muscular pillars, protrude into both ventricular lumina and point toward the atrioventricular valves. They are connected to *chordae tendineae*, narrow tendinous cords that are attached to the leaflets of the respective atrioventricular valves (Fig. 1.3).

1.1.1 Atria

The atria are composed of: (1) a body, quasi-absent in the adult right atrium, although clearly observed in the left atrium; (2) a venous chamber, with vein endings;[7] (3) a vestibule, leading to the atrioventricular valve; and (4) a

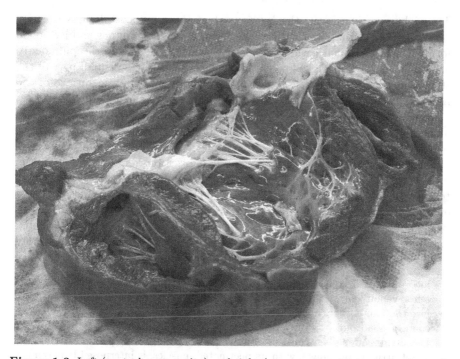

Figure 1.3. Left (central open cavity) and right (open cavity with two parts located at bottom left and top right (small) photograph corners) ventricles of a calf. Papillary muscles and chordae tendineae, associated with the atrioventricular valves (with a focus here on the mitral valve), are larger in the left than in the right ventricle. The apparatus composed of papillary muscles and chordae tendineae allows the atrioventricular valves to remain closed during the systole, the sheet made by contact leaflets taking a parachute-like shape, hence avoiding leakage into the atrium. In the upper right region, the entrance segment of the ascending thoracic aorta is displayed with removed aortic valve and the two coronary ostia in their respective Valsalva sinuses. These sinuses are separated by the valvar commissure.

---

[7] The venous chamber, or sinus venous, is the connection region of the terminal veins to the atria. Superior and inferior vena cavae and the coronary sinus are

muscular region, which consists of the atrial appendage and the pectinate muscle[8] [13]. The atrial septum is composed of the septum secundum,[9] the inferior edge of which is the limbus and fossa ovalis on the right atrium side, and the septum primum, which is the flap valve of the oval foramen on the left atrium side. The atrioventricular septum is associated with the atrioventricular valves. In the right atrium, the *Eustachian valve*, or valve of inferior vena cava (IVC), is an endocardial crescentic fold situated in front of the orifice of IVC.[10] The *coronary valve* is a semicircular fold that protects the orifice of the coronary sinus.

1.1.2 Ventricles

The right ventricle forms the larger part of the sternocostal surface and a small part of the diaphragmatic surface of the heart. Its upper and left angle forms a conical pouch (conus arteriosus), which give rise to the pulmonary artery. The left ventricle is longer than the right. It forms a small part of the sternocostal surface and the larger part of the diaphragmatic surface of the heart. It corresponds to the apex of the heart.

Both ventricles have three kinds of muscular columns: some are attached along their entire length on one side, certains are fixed at their extremities but free in the middle, others (papillary muscles) are anchored by their bases to the ventricle wall and their apices serve as insertions for chordae tendineae. Each ventricle possesses two papillary muscles connected to anterior and posterior walls of the ventricle. In the right ventricle, chordae tendineae connect either the anterior papillary muscle to the anterior and posterior cusps or the posterior papillary muscle to the posterior and medial cusps of the tricupid valve. In the left ventricle, chordae tendineae from each papillary muscle are connected to both cusps of the bicuspid mitral valve.

---

incorporated by the right atrium in the sinus venosus (venous chamber or systemic venous sinus of the right atrium), whereas the four pulmonary veins drain into the four corners of the left atrium posterior dome. These terminal veins can connect abnormally to the atria.

[8] Left atrium appendage constitutes a diverticulum containing pectinated muscles. Right atrium appendage, with its pectinated muscles, is interposed between the smooth-walled venous chamber and the vestibule. In the right atrium, the junction between the appendage and the venous chamber is internally defined by the prominent terminal crest and associated groove, the sulcus terminalis, on the external right atrium surface. The crista terminalis, a myocardium ridge, forms from the superior part of the right venous valve. The sinoatrial node resides near the crista terminalis. The flap valve of the fossa ovalis, a remnant of the oval foramen and its inferior rim are septal structures, whereas the other rims are infoldings enclosing fat [14].

[9] The septum secundum is an infolding of the atrial wall. The actual atrial wall is the flap valve of the foramen ovale.

[10] In the fetus this valve directs the blood from IVC to LA through the foramen ovale. In the adult, it occasionally persists.

1.1.3 Aortic Valve

The aortic valve consists in three quasi-equal semi-lunar cusps (right, left, and noncoronary leaflets; thickness 0.2–0.4 mm [8])[11] (Fig. 1.4). Dimensionless heights and bores are displayed in Fig. 1.6. Dimensions of the three cusps of the aortic valves are also given in Table 1.6 for porcine hearts, which provide orders of magnitude of cusp sizes[12] [9].

The three *Valsalva sinuses* of the aorta root match the three valve leaflets [10] (Figs. 1.7 and 1.8). The coronary arteries branch off from two of the sinuses (Fig. 1.5). The intersection regions between the aorta wall and the cusp free edges are called the *commissures*. The *nodulus of Arantius* is a large collagenous mass in the central part of the coaptation[13] region. The *lunula* is the region on each side of the nodule. When the aortic valve is closed, the free margins of the cup-like leaflets seal each other, defining two planes, a sealing vertical (the aorta root axis being vertical) and a bottom oblique plane from the insertion line to the channel axis, with an angle of nearly 20 degrees.[14] When the reference length is the channel bore at the insertion line bottom, the sinus height is equal to 0.87, the length between the insertion line bottom and the commissure 0.71, the coaptation length 0.17, and the maximum sinus half width 0.65–0.73; the aortic bore downstream from the sinuses is almost equal to 1 [8]. The open leaflets form a triangular orifice when the cusps are

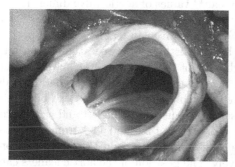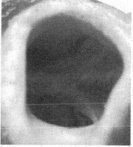

Figure 1.4. View from the arterial lumen of the aortic orifice. (**Left**) Two displayed cusps with their free edges and the lunulae; the commissure is at the lower left corner. (**Right**) Three close cusps; the insertion line of the three semilunar leaflets, from the cusp nadir to the commissures are hidden by the aorta wall.

---

[11] Higher thickness values up to 700 μm can be found.

[12] Between-species geometry variations might explain xenograft failure with porcine bioprostheses.

[13] The cusp coaptation region is the leaflet portion below the free edge, which comes into contact with the neighboring cusps.

[14] The more or less curved bottom cusp below the coaptation region, which is connected to the wall, is approximated by a straight line in a plane crossing the mid-cusp.

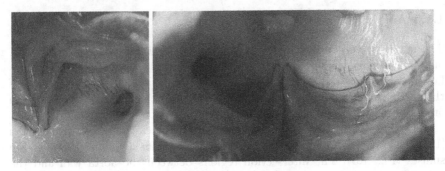

Figure 1.5. Valsalva sinuses of the aortic root with the coronary ostium. In the calf heart, the commissure reaches the sinotubular junction, whereas it remains below in humans. (**Left**) Coronary ostium with its smooth convergent. (**Right**) View of the sinotubular ridge.

not fully open. To limit the valve obstacle, the flow can flatten the cusp along the sinus cavity, the cusp end projecting slightly inside the cavity [11].

Certain dimensions of Valsalva sinuses and cusps of the aortic valve, left coronary cusp (LCC), right coronary cusp (RCC), and non-coronary cusp (NCC) have been measured [12]. The height of the cusps are assessed from the bottom of the Valsalva sinus to the cusp free margin at the middle point between the commissures. The mean[15] values of aortic valve dimensions (cusp height, lunula width and length, and intercommissural distance)[16] for the three cusps (LCC, RCC and NCC) are given in Table 1.5. The position of the ostium can be determined by the distances between the ostium and the commissures at the left and right side of the ostium on the one hand and the length between the ostium and bottom (nadir) of the corresponding Valsalva sinus. The authors have observed that both coronary ostia can be located in the left coronary sinus and be supracommissural. The left and right coronary ostia are located on average at a distance of 9.6 and 11.1 from the left commissure, 11.0 and 11.1 from the right commissure, and 13.3 and 14.8 from the Valsalva sinus bottom, respectively. The position of the aortic leaflets can also be defined with respect to the ventricular septum, using the distance between the septal extremity and RCC-LCC commissure, the septal end and the RCC-NCC commissure, and the septum and the NCC-LCC commissure. The mean values of these distances are 9.5 ($\pm$ 5.3), 5.7 ($\pm$ 4.2), and 19.2 ($\pm$ 2.9) mm. The mean aortic diameter is equal to 21.8 ($\pm$ 3.6) mm.

---

[15] One hundred healthy hearts of both sexes from people 9 to 86 years old have been studied.

[16] Semilunar leaflets have a thin free lunula with a dense nodule at the midpoint. The size of the lunulae are estimated by the width at the commissural level and the length of the free margins. The intercommissural distances, either straight (chord) or curved (arclength) along the aortic orifice circumference, have also been measured.

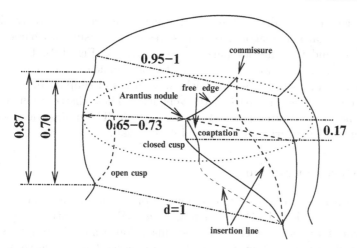

Figure 1.6. Schematic drawing of the aortic valve and dimensionless lengths (Source: [8]).

Table 1.5. Mean dimensions of human aortic leaflets (mm; LCC: left coronary cusp; RCC: right coronary cusp; NCC: non-coronary cusp; Source: [12]).

	LCC	RCC	NCC
Cusp height	15.2	15.0	14.9
Lunula width	4.6	4.4	4.3
Lunula length	30.6	30.1	30.0
Intercommissural distance	25.2	24.5	24.1
Intercommissural chord	19.8	19.2	20)

1.1.4 Heart Vascularization

The heart is perfused by the right (RCA) and left (LCA) *coronary arteries*, originating from the aorta just above the aortic valve.[17] LCA, which divides

Figure 1.7. Top view of a mesh of the aorta root with the three leaflets of the aortic valve and the Valsalva sinuses (from T. Cherigui).

[17] The lumens of the coronary arteries are not obstructed by open leaflets.

into the left anterior descending and the circumflex arteries giving birth to obtuse marginal and diagonal branches, respectively, supplies blood to the left atrium and the anterior part of the ventricles (Fig. 1.9). RCA, which branches successively into the conus, right ventricle, acute marginal, posterior descending, and posterior left ventricle branches; and supplies blood to right atrium and the posterior part of the ventricles. These distribution coronary arteries lie on the outer layer of the heart wall. These superficial arteries branch into smaller arteries that dive into the wall.

1.1.5 Heart Innervation

The heart is innervated by both components of the autonomic nervous system (ANS; Table 1.7, Fig. 1.11). *Parasympathetic* (pΣc) innervation originates in the *cardiac inhibitory center* in the medulla oblongata of the brain stem and is conveyed to the heart by way of the vagus nerve (cranial nerve X). *Sympathetic* (Σc) innervation comes from the *cardiac accelerating center* in the medulla and upper thoracic spinal cord. The cardiac sympathetic nerves extend from the sympathetic neurons in stellate ganglia, which reside bilateral to the vertebrae. Sympathetic nerve fibers project from the base of the heart

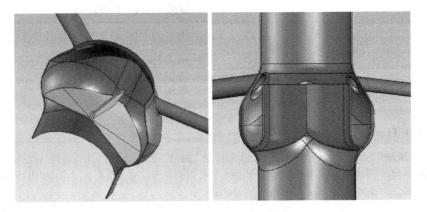

Figure 1.8. Model of the aorta root with aortic valve cusps in open and closed configurations (from T. Tran).

Table 1.6. Dimensions (mm; mean data of a set 10 postmortem porcine hearts) of right (RCC), left (LCC), and noncoronary (NCC) leaflets of porcine aortic valves [9].

	RCC	LCC	NCC
Width	13.3	13.9	13.7
Free edge length	33.0	31.5	32.7
Insertion line length	46.4	47.6	48.1
Perimeter	79.4	79.1	80.8

into the myocardium. They are located predominantly in the subepicardium in the ventricle. The central conduction system (sinoatrial node, atrioventricular node, and His bundle) is abundantly innervated with respect to the myocardium.

Normally, parasympathetic innervation represents the dominant neural influence on the heart. Maximal stimulation of vagal fibers can stop myocardium contractions. When stimulated, the sympathetic fibers release noradrenaline with several effects (Sect. 2.6.2).

The function of most of the ganglia in fat pads on the surface of the heart remains poorly understood.[18] Intraganglionic neuronal circuits are found within the sinoatrial ganglion defining an intrinsic cardiac nervous system [15].

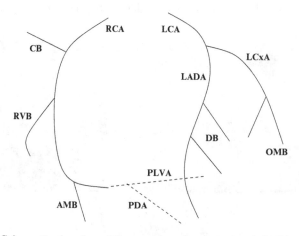

Figure 1.9. Schematic drawing of the coronary artery network (LCA, RCA: left and right coronary arteries; CB: conus branch; RVB: right ventricle branch; AMB: acute marginal branch; PLVA: posterior left ventricle artery; PDA: posterior descending artery; LADA: left anterior descending artery; LCxA: left circumflex artery; OMB: obtuse marginal branch; DB: diagonal branches).

Table 1.7. Mediastinal nerves control blood circulation and respiration that affects the venous return.

Nerve	Target
Sympathetic	Heart
	Blood vessels
	Tracheobronchial tree
Parasympathetic	Heart
	Blood vessels
	Bronchial smooth myocytes and mucous glands

---

[18] The interacting sinoatrial ganglion and posterior atrial ganglion affect heart rate (negative chronotropic effect), atrioventricular ganglion affects conduction

1.2 Blood Vessels

Deoxygenated blood from the head and upper body and from the lower limbs and lower torso is brought to the right atrium by the superior (SVC) and inferior (IVC) venae cavae. When the pulmonary valve is open, the right ventricle ejects blood into the pulmonary artery. The pulmonary veins carry oxygen-rich blood from the lungs to the left atrium. The aorta receives blood ejected from the left ventricle. The right and left hearts, with their serial chambers, play the role of a lock between low-pressure circulation and high-pressure circuit. The atrioventricular coupling sets the ventricle for filling and pressure adaptation, and ventriculoarterial coupling for ejection.

The heart works as a pump to distribute blood to body organs and perfuse the organ tissues. Blood is propelled under high pressure, especially in sys-

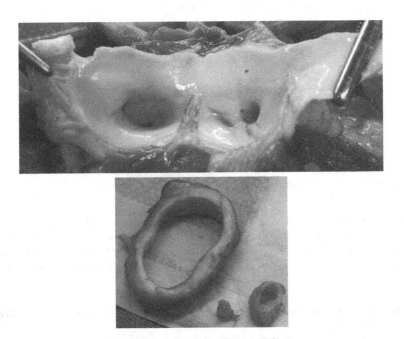

Figure 1.10. Coronary artery calibers. (**Top**) Ostia in the Valsalva sinuses between the valve insertion lines and sinotubular ridge are separated by commissures of the right and left coronary cusps of the aortic valve. In this 3-month-old calf, the bore of the entrance segment of the right coronary artery is smaller than the left. (**Bottom**) Differences in bore of the entrance segment slices of the aorta and right and left coronary arteries (from left to right) by artery slice sizes. Axial cut of vessel segments, like those here displayed, induces a circumferential lengthening of 10% to 20% (azimutal prestress in unstressed configuration).

(negative dromotropic effect), and cranioventricular ganglion affects left ventricule contractility (negative inotropic effect) [15].

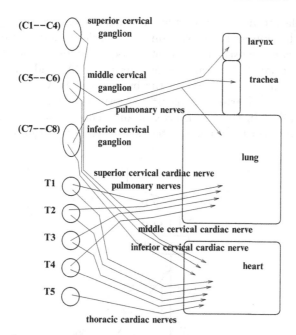

Figure 1.11. Sympathetic innervation of heart and respiratory tract (C: cervical nervous root; T: thoracic (dorsal) nervous root/ganglion).

temic circulation, which is driven by the aorta through a network of branching arteries of decreasing size, arterioles, and capillaries to the tissues, where it delivers nutrients and removes catabolites. Blood is collected through merging venules and returns to the heart through veins under low pressures (Fig. 1.12). Large blood vessels of the upper part of the mediastinum, efferent from and afferent to the heart, are displayed in Figs. 1.13 and 1.14.

Each blood circuit in systemic and pulmonary circulation is composed of three main compartments, arterial, capillary, and venous. The arterial tree can be split into a distribution compartment, which includes arteries irrigating main body regions, and a perfusion compartment, which is composed of smaller arteries irrigating tissues. The main architectural features of the arterial tree are: (1) taper, (2) branchings with a between-branch length of a few vessel radii, and (3) bifurcation after giving birth to five to ten branches. The venous collector can also be subdivided into serial compartments. The vascular networks are a closed tortuous multigeneration system of branching or bifurcation (with given flow division ratios) or merging deformable vessels, more or less short. Blood flows depend on vascular architecture and local geometry, especially curvature[19] and branching angles, area ratios, vessel lengths with possible taper and prints of neighboring organs. Order of magnitude in vessel sizes is given in Table 1.8.

---

[19] Both hip and apex curvatures influence flow behavior in branching segments.

The vasculature has some peculiar features. *Arteriovenous anastomoses* are found in the skin and gut; wall smooth muscles usually close such a bypass. *Arterial anastomoses* are frequently observed between the head arteries, in particular between the branches of internal (ICA) and external (ECA) carotid arteries, as well as between the branches of ICA and of the basilar trunk (an example of artery merger)[20] to form the Willis circle under the brain (Fig. 1.15).[21] In the brain, arterial and venous distributions are separate, whereas in other body parts, arteries and veins run together with nerves

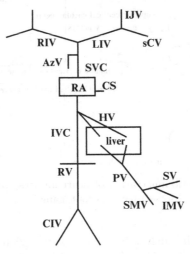

Figure 1.12. Schematic drawing of the large vein system. The left (LIV) and right (RIV) innominate veins, formed by the junction of the internal jugular (IJV) and subclavian (sCV) veins, unite to form the superior vena cava (SVC). The azygos vein (AzV) enters and ascends in the thorax to finally arch and end in the superior vena cava. Most of the cardiac veins end into the coronary sinus, which goes to the right atrium (RA). The common iliac veins (CIV), formed by the union of the external iliac and hypogastric veins, unite to form the inferior vena cava (IVC). The inferior vena cava receives the renal (RV), suprarenal, inferior phrenic, and hepatic veins. The portal vein (PV) is formed by the junction of the superior mesenteric (SMV) and splenic (SV) veins, which receives the inferior mesenteric vein (IMV).

Table 1.8. Vessel geometry. Approximate magnitude (mm) of caliber d and wall thickness h for different vessel compartments.

Vessel	Aorta	Artery	Arteriole	Capillary	Venule	Vein	Vena cava
d	10–25	3–5	0.30–0.01	0.008	0.01–0.30	5	30
h	2	1	0.02	0.001	0.002	0.5	1.5

---

[20] The vertebral arteries merge to give birth to the larger basilar trunk.

[21] The large arteries supplying the brain are the carotid and vertebral arteries, leading to the network of pial arteries on the surface of the brain. On the cerebral

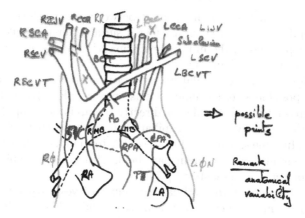

Figure 1.13. Scheme of large blood vessels of the upper thorax with the airways (coronal view, usual configuration). RA, LA: right and left atria; Ao: aorta (ascending aorta, arch of aorta and descending aorta); BCT: arterial brachiocephalic trunk; RSCA: right subclavian artery; RCCA, LCCA: right and left common carotid arteries; PT: pulmonary trunk; RPA, LPA: right and left pulmonary arteries; SVC: superior vena cava; RBCVT, LBCVT: right and left brachiocephalic trunks; RSCV, LSCV: right and left subclavian veins; RIJV, LIJV: right and left internal jugular veins; T: trachea; RMB, LMB: right and left main bronchi; RR, L Rec: right and left recurrent nerves; RΦ, LΦN: right and left phrenic nerves; X: vagus nerves.

in a connective sheath as a neurovascular bundle. *Portal systems* deliver venous blood to the liver and pituitary gland drained from other organs (i.e., Fig. 1.25). Sinusoids acts on sensors such as the *glomus caroticum*. *Venous sinuses* are endothelium-lined spaces in connective tissue where blood is collected (e.g., coronary and dural sinuses, and erectile tissue).

The vasculature is characterized by three properties: diversity, complexity and variability. Diversity is the consequence of large between-subject variability in vessel origin, shape, path and branching. Since the flow dynamics strongly depend on vessel configuration, subject-specific models are required for improved diagnosis and treatment. The variability is due to environmental actions and regulation.

Different bore (d) values of blood vessels are used in the literature to define the limit beween macrocirculation and microcirculation: 500, 300 or 250 μm. *Microcirculation* starts with the arterioles ($10 < d < 250$ μm) and ends with the venules. Local microvascular networks have been observed using suitable microscopy[22] in thin tissues, such as the mesentery, the cremaster, etc.

---

cortex, the arteries branch into smaller penetrating arteries, which enter the brain.

[22] A fibered confocal fluorescence microscopy system dedicated to imaging at the microscopic level, with non- or minimally invasive access, has been developed by Mauna Kea Technologies (www.maunakeatech.com), in particular for microvascular research.

The capillary network is interposed between small arterioles and small venules. Capillaries have the thinnest wall suitable for molecule transfer. In particular, capillaries in the lung parenchyma are very closed to the alveolar wall for efficient gas exchanges.[23]

A sketch of the microvasculature shows the afferent arteriole and efferent venule with possible arterioloveinular anastomosis[24] (Fig. 1.16). The arteriole give birth either to metarterioles or directly to capillaries [16]. The entrance segment of the capillary has a precapillary sphincter, which regulate local flow distribution. In some territories, 40–50% of the total blood volume is contained in the microcirculation, the main part being either in the capillaries (heart and lungs) or venules (mesentery) [17].

Taking into account the estimated number of vessels per vascular compartment, the cumulated cross-sectional area is the highest and blood velocity

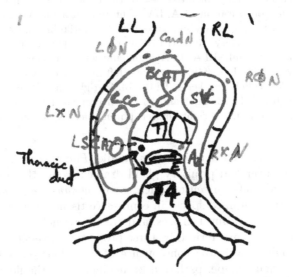

Figure 1.14. Scheme of large blood vessels of the upper mediastinum (transverse plane). Thoracic (dorsal) vertebra 4 (T4) level. LL, RL, left and right lung parenchymae; Card N: cardiac nerve; RΦ, LΦN: right and left phrenic nerves; LXN, RXN: left and right vagus nerves; E: esophagus; T: trachea; BCAT: brachiocephalic arterial trunk; LCC: left common carotid artery; LSCA: left subclavian artery; SVC: superior vena cava; Az: azygos vein.

---

[23] The double wall comprises the capillary endothelium, a basement membrane, and the alveolar wall lined by a liquid film with the surfactant. The latter reduces the surface tension and adapt the latter to the lung volume. A given edge of the lung capillary in the interalveolar septum is related to a given alveolus, the opposite edge to the adjoining alveolus. In connecting parts of the interalveolar septa, the capillary can even have gas exchanges with more than two neighboring alveoli. Flow in the alveolar septum has then be modeled by sheet flow.

[24] Arteriovenous shunts connect small arterioles and venules.

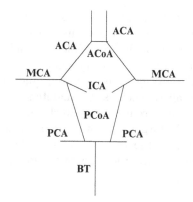

Figure 1.15. (**Left**) Magnetic resonance image of Willis circle in a healthy female volunteer. The arteries are characterized by huge curvature and branching. This anastomotic network has various anatomic variations among individuals (frequent absence in communicating arteries, branching variability). (**Right**) Schematic drawing of the cerebral artery network. The internal carotid artery (ICA) gives birth to the middle cerebral (MCA or sylvian artery) and anterior cerebral (ACA) arteries. The anterior communicating artery (ACoA) connects the two ACAs. The posterior communicating artery (PCoA) anastomoses ICA with the posterior cerebral arteries (PCA), the terminal branch of the basilar trunk (TB). The basillaris is formed by the union of the right and left vertebralis, largest branches of the subclavian arteries (the arteries merge rather than branch off).

the lowest in the capillaries, the exchange zone. The ratio between the wall thickness and lumen caliber, as well as the resistance are the greatest in the arterioles.

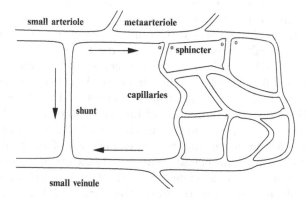

Figure 1.16. Capillary network interposed between a small irrigation arteriole and a small drainage venule, with an arterioloveinular anastomosis (shunt) and precapillary sphincters.

The vasculature adapts to physiological and pathological situations. The uterine arteries increase in bore during pregnancy and return to their original size after birth (reversible remodeling). Cardiac output and stroke volume also increase during pregnancy. Hypertrophic and hypotrophic remodeling of the uterine arteries during and after pregnancy is controlled by different substances, such as the angiotensin converting enzyme [18] (Chap. 2).

Some circulatory networks are used as illustrations because they are frequently targeted by wall diseases (atheroma, aneurisms, and varicose veins; Sect. 7.5): the aorta and its main branches, the carotid arteries and their main branches, and the veins of the lower limbs.

1.3 Circulatory Networks

1.3.1 Aorta

The aorta is the main trunk of the systemic circulation. It starts at the upper right corner of the left ventricle (bore ~ 25 mm). After ascending for a short distance, it arches backward and to the left, over the main left bronchus (with a possible print). The aortic arch has a variable configuration according to the subject, especially children (Fig. 1.17). It then descends within the thorax on the left side of the vertebral column. It crosses the abdominal cavity and ends (bore ~ 10 mm) by dividing into the right and left common iliac arteries at the lower border of the fourth lumbar vertebra (L4).

The ascending aorta (length ~ 5 cm) goes obliquely upward, forward, and to the right. Its entrance segment presents three mild dilations, the aortic sinuses (Valsalva sinuses), which are bounded by the sinotubular junction. The ascending aorta is contained within the pericardium, together with the pulmonary artery. Its branches are the two coronary arteries.

The aorta arch runs at first upward (height 25–40 mm), backward, and to the left in front of the trachea, then directs backward on the left side of the trachea, and finally passes downward at the lower border of the fourth thoracic vertebra (T4). It is in contact with nerves, the left phrenic, the lower of the superior cardiac branches of the left vagus, the superior cardiac branch of the left sympathetic, the trunk of the left vagus, and its recurrent branch (Fig. 1.13). Below are the bifurcation of the pulmonary artery, the ligamentum arteriosum[25] (Fig. 1.18), which connects the left pulmonary artery to the aortic arch, the left bronchus, and the superficial part of the cardiac plexus. Due to its strong non-planar curvature and the branching set leading to large intrathoracic arteries, time-dependent flow structure is complicated. The distribution of wall shear stress at the top of the aortic arch is characterized by great spatial and temporal variations [19].

---

[25] The ligamentum arteriosum corresponds to the remains of the fetal ductus arteriosus.

The arch gives birth to the innominate (IA) or brachiocephalic trunk, and the left common carotid (LCCA) and left subclavian (LSCA) arteries (Fig. 1.19). These branches can start from the ascending aorta. LCCA can arise from the innominate artery. The left carotid and subclavian arteries can arise from a left innominate trunk. Conversely, the right carotid and subclavian arteries can branch off directly from the aorta (huge between-subject anatomy variability). In a few cases, the vertebral arteries originate from the arch. The bronchial, thyroid, and internal mammary can also come from the aorta arch. The innominate trunk, the arch's widest branch (length 4–5 cm; Fig. 1.20) ascends obliquely upward, backward, and to the right, and divides into the right common carotid (RCCA) and right subclavian (RSCA) arteries.

The descending thoracic aorta is located in the posterior mediastinum. It ends in front of the lower border of the twelfth thoracic vertebra (T12). It approaches the median line as it descends to be directly in front of the column at its termination. The curved descending aorta has a small tapered shape, as the branches given off are small (Fig. 1.21). The descending aorta is in relation with the esophagus and its plexus of nerves, with the hemiazygos veins, azygos vein, thoracic duct, heart, and lungs. The descending aorta gives birth to the costocervical, internal mammary, intercostal (IcA), bronchial

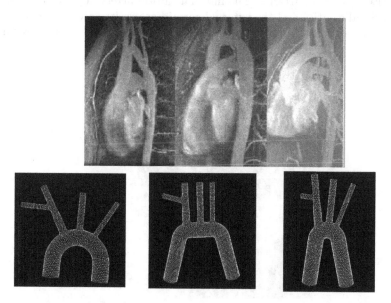

Figure 1.17. Possible configurations of the aorta arch with its three major branches (in most subjects). (**Top right**) Set of radiographies. (**Bottom left**) Mesh of usual shape modeled by a regular circular curvature. (**Bottom mid-panel**) Mesh of aortic arch with a straight segment. (**Bottom right**) Mesh of an angular aortic arch (from M.S. Miette and J. Pichon).

(BrA), pericardiaco-phrenic, mediastinal, esophageal, and epigastric arteries (Fig. 1.19).

The abdominal aorta begins at the aortic hiatus of the diaphragm, diminishes rapidly in size, and gives off many branches (Fig. 1.19). (1) There are unpaired branches, such as the celiac artery (CA), which quickly gives birth to three branches, the left gastric (LGA), hepatic (HA) and splenic (SA) arteries, the superior (SMA) and inferior (IMA) mesenteric, and the middle sacral arteries. (2) The paired branches include visceral arteries such as the suprarenal, or adrenal (AdA), renal (RA), internal spermatic/ovarian (or genital or gonadal; GA) on the one hand, and parietal, such as the inferior phrenic and lumbar (LA) arteries on the other hand. The aorta finally divides into the two common iliac arteries (L/RCCA), which give birth to the internal (IIA) and external (EIA) branches.

1.3.2 Carotid artery

Blood is supplied to the brain via: (1) four main arteries, two internal carotid arteries and two vertebral arteries; and (2) more ore less numerous anastomosis according to the subject between the deep and superficial perfusion networks.

The common carotid artery (CCA) bifurcates into the internal (ICA) and external (ECA) carotid arteries. The internal carotid artery supplies the cranial and orbital cavities and the external carotid artery irrigates the exterior of the head, face and neck. At the carotid bifurcation, the walls of the carotid

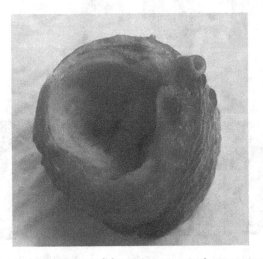

Figure 1.18. Aorta branching and ligamentum arteriosum: a small side branch (upper right edge) and the closed entry of the large fetal ductus arteriosus (opposite edge). The wall thickness is not uniform, due to wall remodeling in this three-month old calf by flow stresses, according to the non-planar strong curvature of the aortic arch, using the locus of the ligamentum arteriosum as a landmark.

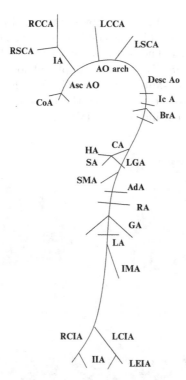

Figure 1.19. Schematic drawing of the aorta and its main branches. The aorta ascends (Asc Ao) and forms a huge bend, the aorta arch, directed to the left side of the trachea to step over the left main bronchus and then downward, becoming the descending aorta (Desc Ao).

sinus have receptors innervated by the glossopharyngeal nerve. The carotid body lies behind the common carotid bifurcation. ECA branches can be divided into: (1) anterior (superior thyroid SThA, lingual LiA, and facial FaA); (2) posterior (occipital OcA and posterior auricular PAuA); (3) ascending (ascending pharyngeal APhA); and (4) terminal (superficial temporal STeA and maxillary MaA) arteries (Fig. 1.22).

1.3.3 Lower Limb Veins

The limb venous network is composed of a superficial and a deep compartment (Fig. 1.24). The main superficial veins of the lower limbs are the short (SSV or saphena interna) and the long saphenous vein (LSV or saphena externa), which runs from foot to knee (saphenopopliteal junction) and from foot to groin (saphenofemoral junction), respectively. Venous blood currently also moves from the superficial to the deep venous network via *perforating* or *communicating veins* that cross the deep fascia.

The popliteal vein (PoV) is formed by the union of the anterior tibial vein (ATiV) and the trunk formed by the confluence of the posterior tibial (PTiV) and peroneal (PeV) veins. The popliteal vein becomes the femoral vein. The small saphenous vein (SSaV) cross the popliteal fossa and drains into the popliteal vein. The great saphenous vein (GSaV), the longest vein, ascends from the foot to the groin to enter into the femoral vein. The femoral vein (FV) becomes the external iliac vein (EIV).

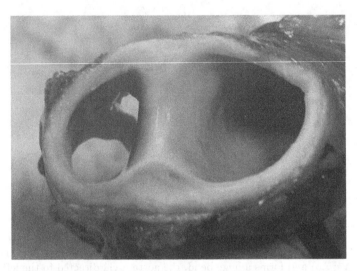

Figure 1.20. Aorta branchings. The brachiocephalic trunk entry with its curved smooth carina. The slice shape of isolated arteries does not depict the in vivo configuration due to different loadings.

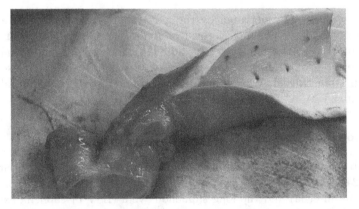

Figure 1.21. Aorta branchings. The aortic arch downstream from the branching of the brachiocephalic trunk in the calf. A longitudinal incision of the aorta wall, from the dowstream end up to the arterial ligament, exhibits ostia of small branches, such as intercostal arteries.

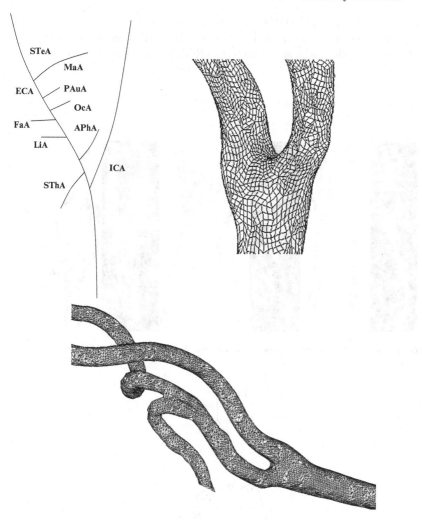

Figure 1.22. (**Top left**) Carotid arteries and its main branches. (**Top right**) zoom of a quadrangle surface mesh of the carotid bifurcation suitable for fluid-structure interaction computations using MIT shell element. (**Bottom**) 3DR reconstruction and facetization of the regional artery network.

1.3.4 Portal System

The portal system includes the veins that drain the blood from most of the digestive system. The portal vein (PV) is formed by the junction of the superior mesenteric (SMV) and splenic (SV) veins. The tributaries of the splenic vein are: the gastric veins (GV), left gastroepiploic vein (LGEV), pancreatic veins (PV), and inferior mesenteric vein (IMV). The inferior mesenteric vein receives the hemorrhoidal (HV), sigmoid (SiV), and left colic (LCV) veins. The superior mesenteric vein receives the intestinal (IV), ileocolic, right colic (RCV), middle colic, right gastroepiploic (RGEV), and pancreaticoduodenal veins. The gastric coronary (GCV), pyloric (PyV), and cystic (CyV) veins end in the portal vein.

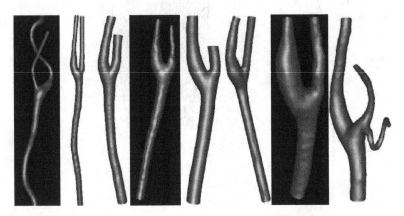

Figure 1.23. Anatomic variations of the carotid bifurcation [20] (with permission).

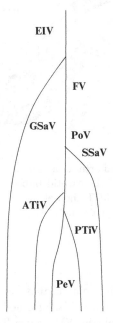

Figure 1.24. Veins of the lower limbs. The venous network is composed of a superficial (great and small saphenous veins) and a deep circuit (femoral and popliteal veins).

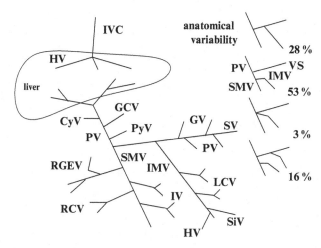

Figure 1.25. Schematic drawing of the portal system and its anatomical variability.

Cardiovascular Physiology

The cardiovascular system provides adequate blood input to the different body organs, responding to sudden changes in demand for nutrient supplies. For a stroke volume of 80 ml and a cardiac frequency of 70 beats per minute, a blood volume of 5.6 l is propelled per minute. The travel time for oxygen delivery between the right atrium and peripheral tissues has a magnitude of $\mathcal{O}(s)$.

Cardiac functioning depends on several factors, including: (1) ions carriers, which determine ion fluxes and intracellular concentrations; (2) sarcomere activity, particularly the crossbridge cycling rate; (3) extracellular matrix; (4) wall perfusion, which is responsible for nutrient inputs; and (5) cardiac loads. Ventricle filling and emptying indeed depend on the state of the vascular bed upstream and downstream from the heart. For example, decrease in intrathoracic pressure at inspiration increases venous return and the converse during expiration. Ventricle filling and emptying are strongly coupled. Ventricle re-expansion helps to refill blood from the atrium. Recoiling force due to the downward motion of the beating heart associated with the stretch imposed on upper vessels favors the atrium filling.

The autonomous nervous system controls the cardiovascular system and adapts its output to the body's needs according to its activity and the environmental stimuli. The autonomous nervous system, with its two subsets, the sympathetic and parasympathetic systems, is made of afferent and efferent neurons that associate the body organs with the central nervous system, and the converse.

2.1 Heart

Two heart pumps, propelling blood into the pulmonary and systemic circulation, are combined into a single muscular organ to synchronously beat. Due to pressure differences between the vasculature entry and exit, which drive the blood flow through it, atria are auxiliary pumps that allow rapid ventricle filling, especially at rest when the cardiac frequency is low. The atrial chambers generate a blood pressure increment upstream from the ventricle inlet.

The heart has an average oxygen requirement of 6 to 8 ml/min per 100 g at rest. Approximately 80% of oxygen consumption is related to its mechanical work (20% for basal metabolism). Myocardial blood flow must provide this energy demand. The myocardium also uses different substrates for its energy production, mostly fatty acid metabolism, which gives nearly 70% of energy requirements, and glucids.

The *cardiac output* (CO) is the amount of blood that crosses any point in the circulatory system and is pumped by each ventricle per unit of time. In a healthy person at rest, cardiac output ranges from 5 to 6 l/mn. Cardiac output is determined by multiplying the *stroke volume* (SV; blood volume pumped by the ventricle during one beat) by the heart rate (f_c). The stroke volume is related to ventricular contraction force and blood volume. The stroke volume is the difference between the end-diastolic volume (EDV) and end-systolic volume[1] (ESV; Table 2.1). The ratios of blood volume to SV in the serial compartments of both systemic and pulmonary circulations are shown in Table 2.2.

Various factors determine cardiac output. *Preload* corresponds to a stretching force exerted on the myocardium at the end of diastole, imposed by the ventricular blood volume; and *afterload* to the resistance force to ejection. Moreover, heart contractility is affected by different molecules. As f_c increases, cardiac output rises until a critical f_c is reached; then it decreases. These factors can be combined. The *cardiac index* (CI) is calculated as the ratio between blood flow rate (q) and body surface area ($2.8 < CI < 4.2 \, l/mn/m^2$). *Cardiac reserve* refers to the heart's ability to quickly adjust to immediate demands. It is defined by the maximum percentage of cardiac output, which in a healthy young adult is 300% to 400%.

The heart has a chaotic behavior. Its non-periodic behavior characterizes a pump able to quickly react to any changes of the body's environment. The normal heartbeat indeed exhibits complex non-linear dynamics. At the

---

Table 2.1. Physiological quantities at rest in healthy subjects: f_c decreases and then increases with aging; SV decreases with aging ($q \sim 6.5 \, l/mn$ at 30 years old and $q \sim 4 \, l/mn$ at 70 years old).

EDV	70–150 ml
ESV	20–50 ml
SEV	50–100 ml
f_c	60–80 beats/mn, 1–1.3 Hz
q	4–7 l/mn (70–120 ml/s)
ejection fraction	60–80%

---

[1] EDV is the maximal volume achieved at the end of ventricular filling. ESV is the residual volume of the ventricle at the end of systolic ejection.

opposite, stable, periodic cardiac dynamics give a bad prognosis. A decay in random variability over time, which is associated with a weaker form of chaos, is indicative of congestive heart failure [21]. This feature, positive with respect to heart function, is a handicap in signal and image processing, and ensemble averaging is used to improve the signal-to-noise ratio.

Most variability is due to diastole duration changes. The relative difference in mean systolic and diastolic durations reaches values of about 5% and 35% (with a ratio between the standard deviation and the mean of 0.1–0.2), respectively.

2.1.1 Cardiac Cycle

The heart beat is a two-stage pumping action over a period of about 1 second or less: a longer first *diastole* and a *systole*. More precisely, the heart rhythm focuses on left ventricle activity, which consists of four main phases: (1) *isovolumetric relaxation* (IR), with closed atrioventricular and ventriculoarterial valves; (2) ventricular filling (VF), with open atrioventricular valves and closed ventriculoarterial valves; (3) *isovolumetric contraction* (IC), with closed atrioventricular and ventriculoarterial valves; and (4) systolic ejection (SE), with closed atrioventricular valves and open ventriculoarterial valves. Durations of these four phases of the cardiac cycle are given in Table 2.3. Both VF and SE can be subdivided into rapid and reduced subperiods (rapid [RVF] and slow [SVF] ventricular filling, and rapid [RSE] and slow [SSE] ejection). With atrial contraction (AC), mechanical events are divided into seven phases.

Phase 1 (IC) is the onset of ventricular systole and coincides with the R wave peak. When the left ventricle begins to contract, the mitral valve is tightly closed to prevent a back flow of blood. The phonocardiogram shows the first heart sound (S1). CMC tension is developed proportionally to EDV. The intraventricular pressure (p_V) quickly increases (Fig. 2.1), while both the

Table 2.2. Approximative blood compartment volume relative to SEV (%) with SEV of 80 ml and total volume 4.4 l.

Pulmonary circulation	16.3
Arteries	5
Capillaries	0.8
Veins	10.5
Systemic circulation	38.7
Aorta	1.3
Arteries	5.6
Capillaries	3.7
Veins	28.1

mitral and aortic valves are closed. The closed ventricle volume is kept constant (incompressible blood) but ventricle shape varies. IC creates a rotational motion of left ventricle wall. As the left ventricle contracts, blood impacts the closed mitral valve, which bulges into the atria, in its filling process, with a further intra-atrial pressure p_A increment: the c wave. Blood does not regurgitate back into the atria because the parachute-like shape of the closed mitral valve is maintained by papillary muscles and chordae tendineae.[2]

Phase 2 (RSE) occurs when p_V overcomes the end-diastolic aortic pressure p_a. The aortic leaflets are pushed open. Traces of p_a and p_V follow one another closely. They reaches a maximum. The blood flow in the ascending aorta increases up to a peak, with a delay with respect to p_V maximum. Aortic blood inflow is greater than the outflow due to windkessel effect (blood storage in elastic arteries). From the p_V rounded summit, the intraventricular volume reduces quickly. After c wave, x descent is observed on p_A plot, which corresponds to atrium diastole; x event is imposed by the systolic downward displacement of the right ventricle base.

During **phase 3** (SSE), p_a and p_V decrease. The aortic blood flow reduces. Meanwhile, the venous return gradually fills the right atrium and v wave is engendered on the atrial pulse. ECG shows the beginning of T wave.

Phase 4 (IR) is characterized by a strong decline in p_V. The second heart sound (S2) and dicrotic notch of p_a trace are due to aortic valve closure. Flow rate in the aorta root exhibits complete back flow, but blood regurgitation into the ventricles is prevented by VAV closure (both aortic and mitral valves are closed). The constant ventricular volume is the residual volume (ESV). The p_V decline continues until $p_V < p_A$.

Phase 5 (RVF) starts with mitral valve opening, p_V becoming lower than p_A. The aortic valve remains closed. Atrial pulse tracing shows y descent, which is generated by AVV opening and ventricular filling. The third sound (S3) is recorded. Atrial blood quickly fills the ventricle. Pressures p_A and p_V are very similar, with p_A slightly greater than p_V. RVF accounts for 80% of total ventricular filling.

Table 2.3. Duration (ms) of the four main phases of the cardiac (left ventricle) cycle ($f_c = 1.25\,\mathrm{Hz}$, i.e., 75 beats/mn).

Phase	Cycle timing	Duration	Starting event
IC	0–50	50	Mitral valve closure ECG R wave peak
SE	50–300	250	Aortic valve opening
IR	300–400	100	Aortic valve closure
VF	400–800	400	Mitral valve opening

---

[2] Valve cusps do not evert into the atrium during the ventricular systole by contraction of the papillary muscles, which are related to ventricular myocardium.

Phase 6 (SVF) is also called *diastasis*. Pressures p_A and p_V raise gradually. The ventricular volume curve increases slowly and slightly. Additional blood is forced from the pulmonary veins into the left ventricle through the opened mitral valve. As blood is collected in the left ventricle, the sinoatrial node sends out the action potential, which leads to atrial contraction. Diastasis sends a small blood volume into the left ventricle ($\sim$15% of ventricular filling) with a slight increase in p_V. Diastasis (duration $\sim$180 ms, is shortened by f_c increase.

Phase 7 (AC) is the final stage of the cardiac cycle. A large amount of blood ($\sim$70% of the filling capacity) has already filled the ventricles prior to atrial contraction. The atrial systole contributes slightly to ventricular filling at resting f_c, but maintains cardiac output during exercise. The fourth heart sound (S4) recorded by the phonocardiogram and a wave of the atrial pulse result from atrial contraction. The action potential travels in the ventricular myocardium. The following left ventricle systole begins when the ventricles are full of blood. The cycle begins again.

The heart wall is composed of a matrix with cells, fibers, and blood vessels. Microcirculation represents an important compartment for blood volume storage. Microcirculation filling, enhanced during diastole due to myocardium relaxation, could help the ventricle expansion by a straightening effect.

2.1.2 Stroke Volume

Stroke volume can be modified by changes in ventricular contractility (rate of tension development), i.e., force generation associated with sarcomere length prior to contraction (*Frank-Starling effect*)[3] and velocity of myofiber shortening. Increased inotropy augments the ventricular pressure time gradient and therefore the ejection velocity. Inotropy increase causes ESV reduction and SV increase, as displayed by pressure-volume loops. Increased stroke volume

---

[3] In 1895, Frank found that under isovolumetric conditions, the larger the EDV, the greater the developed tension and pressure. Starling's later experiments demonstrated that the heart intrinsically responds to venous return (to EDV) increases by increasing the stroke volume (heart autoregulation). The relationship between EDV and stroke volume is associated with the relationship between sarcomere length and calcium ion influx and sensitivity. Myofilament length-dependent activation is explained by the separation distance between actin and myosin along the sarcomeric filament axis. The intrinsic ability of the heart to develop greater tension at longer myocardial fiber lengths over a finite range of fiber lengths is due to sliding filament arrangement in cardiomyocytes, with increase of cross-bridge number between actin and myosin filaments. Frank-Starling mechanism refers to the heart's intrinsic capability of increasing inotropy and stroke volume in response to venous-return increase. The Frank-Starling effect describes static filling mechanisms in an isolated motionless heart. It works for high filling pressure and low flow rate (cardiac failure); however, cardiac functioning is an unsteady phenomenon.

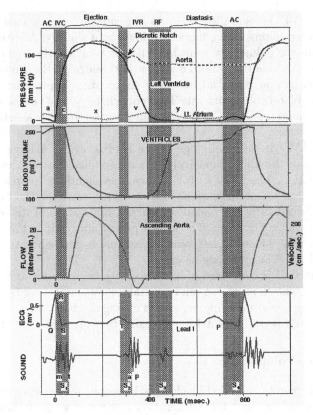

Figure 2.1. Cardiac cycle. Evolution of the pressure in the left cardiac cavities and aorta (top), of left ventricle volume (second row), aortic flow (third row), and ECG trace and phonocardiogram (bottom; from [22]).

causes EDV reduction. Increases in inotropic state maintain stroke volume at high heart rates, which alone decreases the stroke volume because of reduced time for diastolic filling and reduced EDV. When the afterload is increased, ESV initially rises and SV decreases. Afterward, increased ESV raises EDV, if venous return remains constant, and stroke volume can be restored. The left ventricle responds to an increase in arterial pressure by rising contractility, hence SV, whereas EDV may return to its original value (*Anrep effect*). An increase in heart rate also stimulates inotropy.[4] When the heart rate is high, ion carriers are not efficient enough to remove all the calcium which creates a positive inotropic state (*Bowditch effect* or *Treppe effect* or frequency-dependent inotropy).[5] Most of the signals that stimulate inotropy involve Ca^{++}, either

---

[4] This frequency-dependent enhanced contractility helps to offset the decreased ventricular filling time at higher heart rates by shortening the systole time duration, thereby increasing the time available for diastole.

[5] Positive chronotropy C+ induces positive inotropy I+.

by increasing Ca^{++} release by the sarcoplasmic reticulum and influx via Ca^{++} channels during the action potential or sensitizing TN-C to Ca^{++}. The heart is also able to manage excessive blood volume due to increased venous return through sympathetic stimulation with its C+ and I+ effects.

The *stroke work* (SW) is depicted by the area inside the pressure–volume curve. It is approximated by the product of mean arterial pressure and stroke volume. Such estimations explain the SW differences between left and right ventricles. *Cardiac efficiency* is defined as the ratio of stroke work to myocardial oxygen consumption (SW/q_{O_2}). Various factors influence q_{O_2} especially wall tension. Increased inotropy and increased heart rate augment q_{O_2}.

Systole and diastole are dynamically related. The systolic contraction provides heart recoil and energy that is stored for active diastolic dilation and aspiration [23]. Moreover, heart motion during systole[6] pulls the large blood vessels and surrounding mediastinal tissues that react by elastic recoil. Heart diastolic rebound can participate in ventricular filling. The aspiration function of the heart during the diastole can be exhibited by a negative pressure zone on the pressure-volume loop (Fig. 2.2).

2.1.3 Pressure–Volume Curve

At the nanoscale, heart contractility is associated with varying concentrations of Ca^{++} (systolic calcium entry and diastolic calcium reuptake) and ATP in the myocyte cytosol in response to the depolarization wave, associated with great and quick sodium ion fluxes (Na^+ do not interfere with intracellular involved processes) through Na^+ channels. ATP, produced by mitochondrial oxidative phosphorylation, is used to generate the mechanical energy with high efficiency, whatever the loading conditions.

At the macroscale, the physiological concept of heart contractility, the capacity of the myocardium to develp a contraction force whatever the preload or afterload, is depicted by the pressure-volume loops for given cardiac frequencies (Fig. 2.2). Measurements of the cardiac hemodynamics still require invasive techniques, at least to get good estimates of intracavital pressures (Sect. 3.2.1). The systolic, diastolic, and mean values of the pressure in the four heart chambers are given in Table 2.4.

Ventricular performance is displayed in clinical practice by the pressure -volume diagram (Fig. 2.2). LV pressure-volume loops illustrate three features: myocardium work, myocardial characteristics (inotropy and lusitropy), and blood circulation influences (pre- and afterload, Table 2.5). The volume range corresponds to the stroke volume. The end-diastolic pressure–volume relationship (EDPVR) depicts both venous return and lusitropy. The end-systolic pressure–volume relationship (ESPVR) represents both afterload and ventricle inotropy.

---

[6] During systole, the heart moves downward.

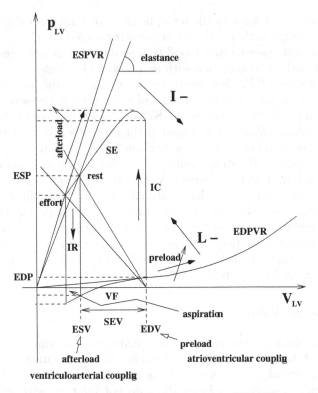

Figure 2.2. Pressure–volume curve of the left ventricle. EDPVR slope increase is associated with decreased lusitropy (higher pressure with same stroke volume). When preload rises, EDV increases. EDPVR slope is the reciprocal of ventricular compliance. ESPVR gives the maximal pressure generated at a given inotropic state. EDV and EDP are indices of heart preload. ESPVR slope lowering and elevation represent decreased inotropy and increased afterload, respectively (adapted from [23]).

Different types of tests have been carried out: (1) when EDV is changed and the afterload is kept constant (constant ESP), ESV remains constant

Table 2.4. Pressures (kPa) in the cardiac cavities and large vessels (datum variability).

Site	Systolic	Diastolic	Mean
RA	0.2–1.3	0–0.5	0–1
RV	0.3–4.0	0–0.9	1.1
PA	1.3–4.0	0.7–2.1	1.1–2.7
LA	0.4–2.2	0.1–1.3	0.3–1.6
LV	10.4–18.2	0.3–1.6	5–5.3
Ao	10.4–18.2	7.8–11.5	11.3–13.3

Table 2.5. Influence factors on pre- and afterload.

Heart load	Factors	Indices
Preload	Venous return	EDV, EDP
	Blood volume	
	Venous tone	
	LV ejection	
	RA/LA aspiration	
	Muscle contraction	
	(inferior limbs)	
	Respiration	
	Tissue activity	
Afterload	Downstream resistance	$p_{LV_{max}}$, $p_{Ao_{max}}$
	Circulating hormones	
	Neural activity	
	Humoral factors	
	(pH, p_{O_2}, p_{CO_2})	

whatever EDV; (2) when afterload varies (varying ESP) at constant EDV (constant EDP), the points {ESP, ESV} run along the ESPVR curve.

Left ventricle power can be estimated by the product of the ventricular pressure ($p_V = 13.3\,\text{kPa}$) by the flow rate ($dV/dt \sim VES/T = 8 \times 10^{-5}/8 \times 10^{-1} = 10^{-4}\,\text{m}^3/\text{s}$), which gives a power of $1.33\,\text{W}$. The RV power is about one-sixth of the LV power.

2.1.4 Myocardial Oxygen Consumption

Total oxygen consumption (2–$10\,\text{ml/mn/100\,g}$) is subject- and age-dependent. The heart has the highest arteriovenous O_2 difference. A large amount of ATP in the CMC is yielded from aerobic metabolism. Contraction accounts for at least about 75% of the myocardial oxygen consumption (MVO2). MVO2 is related to the stroke work.[7] The blood is supplied by the coronary arteries. These artery entrances are located in the upper part of the Valsava sinuses of the aortic root, immediately above the open valve cusps.

The coronary blood flow is mainly produced by two successive fluid–structure interaction phenomena, systolic blood ejection produced by pump contraction, and diastolic recoil of the elastic arteries. The latter produces back flow toward the closed aortic valve with additional input in the coronary arteries. The coronary arteries run on the heart surface. The main arteries branch off in several large vessels that remain on the epicardium. The main effect undergone by these arteries is the deformation of their paths. The large coronary branches give birth to a set of small arteries that penetrate into the

---

[7] MVO2 can be assumed to be an affine function of the heart's mechanical work W: MVO2 = κW + MV02b (MV02b: basal consumption).

cardiac wall. Both the downstream compartment of coronary macrocirculation and microcirculation bear strong variations of the transmural pressure
induced by the working myocardium. The blood vessels are compressed during
myocardium contraction. Therefore, perfusion is hampered by systolic compression. Drainage is more or less transiently enhanced by the compression
of the upstream part of the venous network, like inferior limb venous return
(improved by the contraction of surrounding muscles, which compresses the
valved veins). Conversely, intramural artery expansion during diastole produces a beneficial aspiration effect associated with stretchening of the superficial coronary bed.

O_2 extraction in the capillary bed is more effective during diastole because
capillaries, which cross the relaxed myocardium, are not collapsed. The coronary blood flow is equal to about 5% of the cardiac output. Whenever O_2
demand increases, various substances promote coronary vasodilation: adenosine, K^+, lactate, nitric oxide, and prostaglandins. Coronary perfusion pressure depends on the aorta root pressure, which varies strongly during the
cardiac cycle. The activation of sympathetic nerves innervating the coronary
arteries causes transient vasoconstriction mediated by α1-adrenoceptors. The
brief vasoconstrictor response is followed by vasodilation due to augmented vasodilator production and β1-adrenoceptor activation. Parasympathetic stimulation of the heart induces slight coronary vasodilation.

Oxygen consumption of the myocardium (MVO2) can be assumed to be
an affine function of the mechanical work (W) done during contraction and
basal oxygen consumption (MVO2b). Energy consumption by the sarcomere is
determined by the kinetics of cross-bridging. The number of cross-bridges and
the rate of cross-bridge recruitment is determined by the kinetics of calcium–
troponin bonds and sarcomere length (number of available myosin binding
sites on actin). The Fenn effect is related to the adaptation of energy liberation
by the myocardium contraction when the load undergone by the myocardium
varies during contraction.

Oxidative phosphorylation (Part I), which depends on oxygen transfer, is
regulated to adapt to fluctuations of either oxygen supply or needs. A normal
heart can rise about 20-fold its oxygen use. Oxygen consumption increases particularly with the cardiac frequency, but electron transfer in the respiratory
chain can be reduced due to insufficient oxygen supply. Oxygen availability in
the mitochondria depends on its transport from the adjoining coronary capillaries across the extracellular medium and cardiomyocyte. However, both
the extra- ($<2\,\mu m$) and intra-cellular ($<10\,\mu m$) spaces represent a very short
transport distance. Furthermore, a mitochondrium population is located near
the sarcolemma and myoglobin carry oxygen in the cytosol down to the mitochondria in the cell core. Yet, the intracellular transport is not fast enough to
match the rate of consumption. Consequently, gradients in oxygen concentration can be observed in the cytosol of cardiomyocytes, although the myoglobin
near the sarcolemma is almost fully saturated with oxygen [24]. Nevertheless, intrinsic respiratory regulation in the mitochondria of the cell center can

compensate for relatively slow oxygen transport within cardiomyocytes, electron transfer being sustained despite reduced oxygen supply. The electrochemical gradient across the mitochondrial membrane is thereby maintained for suitable mitochondrial functioning.

2.2 Large Blood Vessels

The blood pressure maintains a suitable blood flow, which is distributed among the different parts of the body and physiological systems (Table 2.6). Blood irrigates the heart pump, brain (the control center), endocrine organ (remote regulation), bone marrow (source of blood cells), lungs of the strongly energy-consuming human body, muscles, ligaments and bones (life gestures), digestive tract (nutrient input), kidneys, liver, etc. The main organs regulating blood volume are the kidneys. The kidneys receive about 20% of the cardiac output for blood processing by filtration and reabsorption. The liver acts as a filter (detoxifying various substances), a storage (for glucids, vitamins, etc.), and an excretory gland (bile).

Beyond the large arteries, blood pressure abruptly drops and the systolo-diastolic pressure difference decays. The pressure drop occurs mostly in arterioles and moderately in capillaries. In the venous bed, most of the pressure decrease is observed in the venules with little further drop in large veins. Therefore, a small pressure difference is sufficient to fill the atrium. Blood velocity decreases from the arteries, with a scale (peak value) of $\mathcal{O}(10 \text{ cm/s})$, to the capillaries with a magnitude of $\mathcal{O}(0.1 \text{ mm/s})$. The pressure in pulmonary circulation is much lower than that in systemic circulation. Flow rate, although the right pump is weaker, is identical because the pulmonary resistances are smaller.

Pressure variations follow external constraints. The main factors that affect the pressure values are blood volume, ejection volume, and vascular resistance. Blood volume mainly depends on $[\text{Na}^+]$ and cardiac function, and the vessel tone on $[\text{K}^+]$ (Part I).

Table 2.6. Estimated blood flow distribution (%) among body parts (datum variability).

Compartment	[25]	[26]	[27]
Heart	3	4	4
Brain	14	13	13
Skeletal muscle	15	21	21
Kidneys	22	19	20
Other abdominal organs	27	24	24
Skin	13	10	
Lungs, pelvis, etc.	6	9	

2.2.1 Arterial Circulation

Left ventricle pressure soars during isovolumetric contraction to reach a value close to the minimum of the aortic pressure. During systolic ejection, the pressure of the left ventricle is slightly higher than that of the aorta. Both pressures reach their maxima and decrease.

In certain subjects who have thick-walled hearts, the pressure wave displays two peaks. Several indices can then be calculated. The *augmentation index* (AIx), derived from the ascending aortic pressure waveform, is the difference between the first and second systolic pressure peaks (the first peak being lower than the second), expressed as a percentage of the pulse pressure. (Pulse pressure is defined as systolic pressure minus diastolic pressure.) This index has been used as a measure of systemic arterial stiffness and additional load imposed on the left ventricle. Increased AIx is related to the risk of coronary heart disease. The so-called *time to reflection* (TR) is the time from the foot of the pressure wave to the first systolic peak. The *diastolic pressure time interval* (DPTI) is the time from the foot of the pressure wave to the dicrotic notch. The *systolic pressure time interval* (SPTI) is the time from the dicrotic notch to the end of the waveform. The *subendocardial viability ratio* (SEVR) is the ratio of the diastolic pressure time interval to the systolic pressure time interval. Applanation tonometry assess the augmentation index and pulse wave velocity; but the use of this technique to derive the central waveform from non-invasively acquired peripheral data needs to be validated [28].

The arterial pressure p_a evolves between its systolic p_s and diastolic p_d values during the cardiac cycle (Table 2.7). The systolic pressure reflects the cardiac output and distensibility of elastic arteries, whereas p_d is an index of peripheral vessel state. The pulse pressure is the difference between p_s and p_d. It mainly depends proportionally on stroke volume and is inversely proportional to arterial compliance. The mean arterial pressure (mAP) is currently estimated by $p_d + (p_s - p_d)/3$.[8] mAP decreases from the aorta (∼13 kPa) to the arterioles (∼5 kPa). The arterioles are the main site of blood flow resistance and control blood input into the capillaries. mAP can be considered in a first approximation as the product CO × SVR (SVR: systemic vascular resistance). Arterial pressure has a circadian pattern. Pressure values increase progressively with aging up to 120% after 60 years old. Vessel pressure variations along the circulatory network are given in Table 2.8.

Pressure variations drive the unsteady blood flows, the flow time variations being complex (multi-harmonic signal; Chap. 5). Moreover, the waveform varies with the vessel station. Ranges of cyclic variations in blood velocity in several arteries is given in Table 2.7. Values of the flow rate extrema in the femoral artery and duration of the acceleration and deceleration phases are provided in Table 2.9.

---

[8] mAP is underestimated using 0.333 as a multiplier rather than 0.412: mAP = $p_d + 0.412(p_s - p_d)$ [29].

Table 2.7. Blood flow quantities in proximal arteries (Sources: [30, 31]).

Artery	Velocity (cm/s)	Pressure (kPa)
Aortic arch	−20–60	10.7–16.0
Descending thoracic aorta	−10–60	9.3–15.3
Abdominal aorta	−10–60	9.3–15.3
Common iliac artery	−7.5–60	8.7–13.3

2.2.2 Venous Return

The veins are distensible (the venous compartment is used as blood storage) and collapsible, especially thin-walled superficial veins. Gravity causes blood pooling in the legs in the upright position. The main vein goal is to provide the heart filling flow (venous return), with a lower velocity than in arteries, the right heart pumping blood into the pulmonary circulation. The venous return is the blood volume reaching the right atrium which is equal to the cardiac output. The venous return is inversely proportional to the central venous pressure (CVP~0.25–0.80 kPa for CO of 5 l/mn). The central venous pressure decreases with inspiration, thereby increasing venous return, due to

Table 2.8. Indicative pressure (kPa) evolution in the vasculature.

Compartment	Mean	Systolic	Diastolic
Systemic circulation			
Aorta	12.7	16	11.2
Artery	12	16.8	10.6
Arteriole	9.1	12.5	7.7
Capillary	3.5		
Vein	1.5		
Right heart			
RA	0.9	1.2	0.4
RV	1.1	3.9	0.8
Pulmonary circulation			
Artery	1.7	2.9	1.2
Arteriole		2	0.7
Capillary		1.2	0.5
Vein		1	0.2
Left heart			
LA	1	1.9	0.3
LV	5	16	0.5

Table 2.9. Flow features in the femoral artery in a healthy volunteer at rest (Source: [32]).

Accelerating flow duration	90–100 ms
Latency between ECG R wave and systolic ejection	190 ms
Decelerating flow duration	140–190 ms
q_{smax}	40 ml/s
q_{min}	−12 ml/s
q_{dmax}	6 ml/s
$\bar{q}$	4 ml/s

negative intrathoracic pressure at inspiration and increased intra-abdominal pressure.

The flow in large veins undergoes cardiac activity. The jugular vein pulse is characterized by an a wave generated by atrial contraction, which is followed by a x valley associated with the peak jugular velocity, due to atrial relaxation and descent of the atrioventricular floor during ventricular ejection, the down slope being interrupted by a c wave induced by the carotid artery pulse (acceleration phase of the arterial flow) [33]. The following v wave occurs during ventricular relaxation, and the y valley, shallower than that of the x, during rapid ventricular filling.

Several agents act on venous return (Tables 2.10 and 2.11). The main factors affecting venous flow include: (1) the pressure difference between venules and the atrium, (2) muscular tone of the venous wall, (3) venous compliance and venous blood volume, (4) external pressure, in particular pressures in the abdominal and thoracic cavities, which vary during the respiratory cycle, and (5) massage by skeletal muscle contraction. Muscle contractions help venous return by compressing the surrounding veins equipped with valves. Both intra-abdominal and intra-thoracic pressures decrease during the inspiration (lung inflation) and increase during the expiration [34]. Respiration-induced flow rise and decay in the inferior vena cava are observed with a phase lag in the thoracic segment with respect to the abdominal one. Moreover, the hepatoportal contribution explains the quantitative difference between these two IVC segments. The respiratory cycle affects venous return via the systemic and hepatoportal venous supplies in the opposite way [34]. The systemic venous flow rises, whereas the hepatoportal venous contribution decays during inspiration. Like in the arteries, venous tone is governed by the autonomous nervous system. Venous flow is thus a time-dependent flow, especially in veins close to the heart due to rhythmic activity of right cardiac pump and respiration, as well as in leg veins due to muscle contraction-assisted flow. In the leg veins, the peak velocity $\widehat{V}_q = 20\text{--}30$ cm/s and the mean velocity $\overline{V}_q = 2\text{--}4$ cm/s.

Table 2.10. Influence agents on venous tone.

Constriction	Deep inspiration
	Hypoxemia
	Hypopressure in carotid sinus
	Central hypercapnia
	5HT
	Cold
Dilation	Rest
	Hyperpressure in carotid sinus
	ACh

2.2.3 Coronary Circulation

The left and right coronary arteries and their main epicardial branches function as distribution vessels. The epicardial arteries branch into smaller arteries, which dive into the myocardium and give birth to the microvasculature with its resistance vessels and dense capillary network along the cardiomyocytes. The huge capillary density with respect to cardiomyocyte population minimizes transport distance. A close contact between capillaries and cardiomyocytes is, indeed, necessary to provide oxygen for suitable ATP synthesis, a major factor of myocardium contraction, whereas calcium ion is a major determinant of excitation-contraction coupling (Chap. 6). Tissue perfusion for nutrient transport and drainage for catabolite removal are effected by a complex vasculature. The microcirculation with arterioles, capillary bed, and small and large draining venules, are wholly intramural. The large venules merge to form the venous collector. Large veins are epicardial like large arteries.

The volume of blood in epicardial coronary arteries has been estimated to be equal to about 1.6 ml/100 g of left ventricle tissue for a perfusion pressure of 13.3 kPa, the right coronary bed ususally contributing about 15 % of the volume in the common structure of the coronary tree [35]. The left coronary artery irrigates both ventricles; the coronary flow that perfuses the

Table 2.11. Influence factors on the venous pressure.

Vis a tergo	Capillary pressure
Vis a fronte	Cardiac function
Vis a latere	External pressure
	Muscle contraction (limbs)
	Breathing (abdomen, thorax)
	Vein tone
Vis a parte interiore	Vein volume
	Energy dissipation

right ventricle is about 65% of that of the left ventricle. Blood volume in the intramyocardial bed is assessed to be 4–5 ml/100 g of left ventricle tissue. The capillary density ranges from 1.3 to 5 per square millimeter in mammal hearts.

Blood flow at the entrance of the coronary arteries deeply varies during the cardiac cycle. The entry segment of the coronary arteries is perfused during ventricular ejection. They receive a portion of the flow which crosses the open aortic orifice during the systole, the valve reaching at most about the level of the artery axis (Fig. 2.3).

The coronary circulation has two special strongly time-dependent features. Epicardial vessels undergo strong path deformation over the beating heart during the cardiac cycle (Chap. 5). Time-dependent vessel curvature is associated with a variable motion amplitude due to changes in heart cycle period.

Intramural vessels can be compressed during systole by contracted cardiomyocytes and expand during myocyte relaxation. Capillary bore varies from 4.5 to 9 µm, especially due to myocardium activity. Systolic perfusion is then hampered, whereas drainage is transiently stimulated by the collapse of the upstream venous network. The arterial inflow reaches its greatest values during diastole, at least in the larger left coronary artery (Fig. 2.4), and venous outflow during systole [36]. During systole the blood flow rate has been found to be greater in deformable epicardial arteries than in intramural

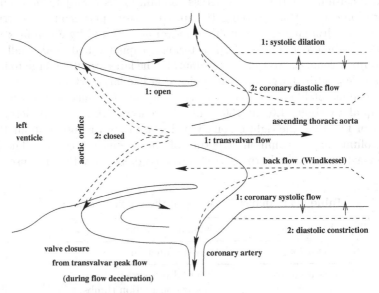

Figure 2.3. Coronary arteries receive blood during systolic ejection from the left ventricle and during the diastole from restitution of stored blood by the elastic arteries (strong windkessel effect). Aortic valve leaflets begin to close from the peak transvalvar flow, then reducing the coronary input. The first inflow is followed by a second distolic inflow (Fig. 2.4), characterized by a second peak flow.

arteries, and conversely during diastole. The amplitude of the pressure cyclic variations is much higher in intramural arteries than in epicardial arteries.

Myocardial contraction reduced the caliber of penetrating arteries (diastolic bore range: 140–650 μm) to about 17% of end-diastolic values, whereas the bore of epicardial coronary vessels increased during systole to about 8% of the end-diastolic value [37]. The compression degree is stronger in the deep wall than in the superficial wall. The longitudinal dimension increases up to about 5% of the end-diastolic value.

Conversely, luminal expansion of intramural arteries associated with a curvature increase of epicardial arteries during ventricular myocardium relaxation enhances coronary filling. After the peak flow through the aortic valve, the pressure difference between the aorta and any capillary bed reaches its highest values and favors tissue perfusion. When the aortic valve closes, the elastic arteries shrink and the left ventricle enters its isovolumetric relaxation phase. The pressure decays abruptly in the left ventricle and more slowly in the arteries due to the windkessel effect, whereas it decreases much more in the coronary circulation owing to myocardium relaxation. Coronary perfusion is particularly enhanced during this time interval of the cardiac cycle.

Phasic vasomotor activities might explain heterogeneity in transit times of blood particles between myocardium regions. Coronary flow is mainly characterized by a heterogeneous regional distribution of blood among the myocardium layers from the inner part (subendocardium) to the outer one (subepicardium). The pressure within the heart wall, indeed, decrease from

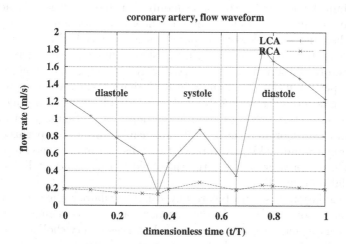

Figure 2.4. Flow rates in the left and right coronary artery (adapted from [38]). The phasic flow pattern of the coronary flow is related to the compression of intramural vessels by the contracting myocardium. During diastole, the coronary vessels can be irrigated, although the pump is not propelling blood, because of the windkessel effect associated with the large elastic arteries, mainly the aorta.

the endocardium to the epicardium, at least during the systole [35]. A triple layer model has been developed that includes the deep subendocardium, mid-myocardium, and superficial subepicardium.

A retrograde flow can occur in the arteries of the deep and middle compressed myocardial layers toward the arteries of the subepicardial layer [35]. However, the retrograde flow is mostly concealed by the distensible epicardial arteries. The distensible epicardial arteries can indeed store about 75% of stroke volume during the systole. During the following diastole, the blood stored in the epicardial arteries is discharged with the flow incoming from the aorta into the intramural arteries. The flow remains anterograde in arteries of the subepicardial layer as well as in veins of all layers.

Certain small coronary arteries of the mid-myocardium might convey retrograde flow during systole. Cardiomyocyte sets form branched bundles, along which are located the capillaries. The cardiomyocyte shortening direction corresponds to the main axis of the cardiac myofiber. The distance of the between-cardiomyocyte space decays. The capillary between two myofibers can be more or less compressed. Nevertheless, the collagen struts between the cardiomyocytes and capillaries could adapt the capillary path and cross-section, avoiding strong narrowing of the capillary lumen and reduction in needed oxygen supply.

The perfusion pressure that drives the blood from the entrance of the coronary arteries to the right atrium varies from a minimum of 10 to 10.5 kPa to a maximum of 14 to 15 kPa. However, coronary flow responds not only to upstream coronary perfusion pressure but also to the energetic needs of the myocardium. Myocardial perfusion is mainly adjusted to oxygen extraction by changes in coronary resistance. Coronary flow is proportional to the oxygen supply when the arterial oxygen content is normal [35]. Oxygen demand is more relevant to heart perfusion than the blood driving pressure (Sect. 2.6.1).

The autoregulation of the coronary flow refers to the maintenance of myocardium perfusion despite changes in perfusion pressure when myocardium metabolism is kept constant. Autoregulation extends for a given range of perfusion pressure. The autoregulation range depends particularly on mammal species and age. Coronary autoregulation usually works for perfusion pressures between 8 and 20 kPa. Autoregulation fails first in the deep myocardium when perfusion pressure reaches the threshold (either upper or lower) of the autoregulation range [35].

Autoregulation is set up before any usual regulatory effect occurs. Otherwise, the vasomotor tone of the coronary arteries is regulated by sympathetic nerves and several vasoactive substances (acetylcholine, adenosine, prostaglandins, leukotrienes, histamine, serotonin, atrial natriuretic peptide, angiotensin, endothelial-derived vasoactive molecules, etc.).

Coronary reserve at any perfusion pressure is expressed by the difference between autoregulated and maximally vasodilated flow. Coronary reserve hence is the ability to increase flow for maximal extraction of oxygen from the coronary blood. Normal hearts have a four- to fivefold reserve.

Flow resistance is hence composed of three varying components: (1) friction; (2) the resistance associated with the autoregulated vasomotor tone for flow adaptation to the myocardium demand and coronary flow reserve; and (3) transient additional systolic resistance induced by the contracting myocardium.

Blood flow distribution can be centrally controlled at the local level, particularly targeting arterioles, independently of neural and hormonal signals. Increase in cardiac activity augments coronary blood flow. Activation of sympathetic nerves of the coronary vasculature induces transient and slight vasoconstriction mediated by α1-adrenoceptors. Vasoconstriction is followed by vasodilation caused by vasodilators associated with β1-adrenoceptor activation of the myocardium to increase the coronary flow for the raised metabolic activity. Parasympathetic stimulation generates a small coronary vasodilation. Additionally, endothelial production of nitric oxide mediates coronary vasodilatation.

2.2.4 Pulmonary Circulation

Pulmonary blood volume is shared between the arterial, capillary and venous compartments with volume of about 150, 100, and 200 ml, respectively. Blood flow rates are nearly equal in serial pulmonary and systemic circulation due to permanent adaptation of blood ejections between both hearts. The circulation time between the pulmonary valve and the left atrium is less than 4 s. The mean capillary circulation time is about 700 ms, whereas the time required for a balance between alveolar gas and blood is lower than 300 ms. At rest in the lying position, $\sim0.7 < p_{pa} < \sim3.3\,\text{kPa}$ and $\sim0.4 < p_{pv} < \sim1.3\,\text{kPa}$. The pulmonary vessel resistance (PVR) is smaller than 8 kPa/1s. The pulmonary blood flow is subjected to cyclic variations of intrathoracic pressure and nondirect upstream (venous return and preload) and downstream effects of the systemic circulation. During inspiration, the intrathoracic pressure decreases and the vessel transmural pressure increases; and the reverse during expiration. Pulmonary vasoconstriction is induced by alveolar hypoxia, pH increase, α-stimulating substances, and angiotensin-2, whereas α-blocking molecules, acetylcholine, and bradykinin generate vasodilation.

The pressure range in the pulmonary microcirculation is lower than in the systemic one. Because the pressure drop in the alveolar capillaries is small, RBC transit time rises for efficient gas exchanges. Due to slow travel through the alveolar capillaries, leukocyte concentration in the lung parenchyma is greater than in the organs perfused by the systemic circulation.

Nonuniform distribution of the perfusion in the different lung regions is due to the hydrostatic pressure difference between the lung apex and basis.[9] Ventilation distribution is aimed at partially matching the perfusion variations. The lung is generally subdivided into three zones [39]. Zone 1, where

---

[9] Hydrostatic pressure decrease with height at a rate of $\sim100\,\text{Pa/cm}$ for arterial (p_a) and venous (p_v) pressure and $\sim0.1\,\text{Pa/cm}$ for alveolar pressure (p_A).

$p_v < p_a < p_A$ is characterized by collapsed pulmonary capillaries. Zone 2, where $p_v < p_A < p_a$ is a transitional region where capillaries may undergo distal partial closure. In zone 3, where $p_A < p_v < p_a$, the capillaries are dilated.

2.3 Microcirculation

The microcirculation, with its four main duct components: arterioles, capillaries, venules and terminal lymphatic vessels, regulates blood flow distribution within the organs, the transcapillary exchanges, and the removal of cell wastes. Arterioles are small precapillary resistance vessels. They are richly innervated by sympathetic adrenergic fibers and highly responsive to sympathetic vasoconstriction via both $\alpha 1$ and $\alpha 2$ post-junctional receptors. It is thus a major site for SVR regulation. The primary function is flow regulation, thereby determining nutrient delivery and catabolite washout. They partially regulate capillary hydrostatic pressure and fluid exchanges. In some organs, precapillary sphincters can regulate the number of perfused capillaries. Venules are collecting vessels. Sympathetic innervation of larger venules can alter venular tone which plays a role in regulating capillary hydrostatic pressure.

Large lymphatic vessels have muscular walls. Spontaneous and stretch-activated vasomotion in terminal lymphatic vessels helps to convey lymph. Lymph circulation is mainly under local control. Smooth muscle cells of the stretched lymph vessels, indeed, rhythmically contract at low frequency due to lymph accumulation. Sympathetic nerves cause contraction. Valves direct lymph into the systemic circulation via the thoracic duct and subclavian veins. Lymph flow is very slow. Lymph has a composition similar to plasma but with a small protein concentration. The protein concentration in the lymph is about half that of plasma.

The capillary circulation, characterized by: (1) a low flow velocity, and (2) a short distance[10] between the capillary lumen and tissue cells, is adapted to molecular exchanges (Table 2.12). Capillaries indeed are the primary site of exchange for fluid, electrolytes, gases, and macromolecules, mainly by filtration, absorption, and diffusion. Fluid can move from the intravascular compartment to the extravascular spaces, composed of cellular, interstitial, and lymphatic subcompartments. The transport of fluid and solutes (electrolytes and small molecules) is determined by hydrostatic and osmotic transendothelial pressures,[11] as well as endothelium permeability (Fig. 2.5). Fenestrated capillaries obviously have a higher permeability than continuous capillaries.

---

[10] The capillary wall is constituted by: (1) the glycocalyx, (2) the endothelium with possible pores or tight junctions according to the perfused territories, and (3) a basal membrane.

[11] Capillary hydrostatic pressure is normally much greater than tissue hydrostatic pressure. The net hydrostatic pressure gradient across the capillary is positive from the lumen to the tissue in the upstream capillary segment. It drives fluid

In most capillaries, there is a net filtration of fluid by the capillary endothelium (filtration exceeds reabsorption). Excess fluid within the interstitium is removed by the lymphatic system.

Table 2.12. Main features of the capillary circulation.

Length	0.2–0.4 mm
Radius	4 μm
Between-capillary distance	10–30 μm
Capillary density	300–5700 (heart)
Total surface	60–1100 mm^2/mm^3 of tissue (heart)
Volume	10^6 μm^3 (∼500 RBCs in a capillary)
Compartmental blood volume	∼300 ml (∼5% of total blood volume)
Transit time	0.6–3 s
Pressure	3.5–4 kPa (arteriolar side)
	1.5–2.9 kPa (venular side)
Capillary set flow	100 ml/s
Mean velocity	≤1 mm/s
Hematocrit	<Ht in large vessels (Fahraeus effect)
Viscosity	μ=μ(R$_h$,Ht) (Fahraeus-Lindqvist effect)

Diphasic flow of deformed cells
Plasma skimming & cell screening
Flow regulation by recruitment (precapillary sphincter) and possible shunt

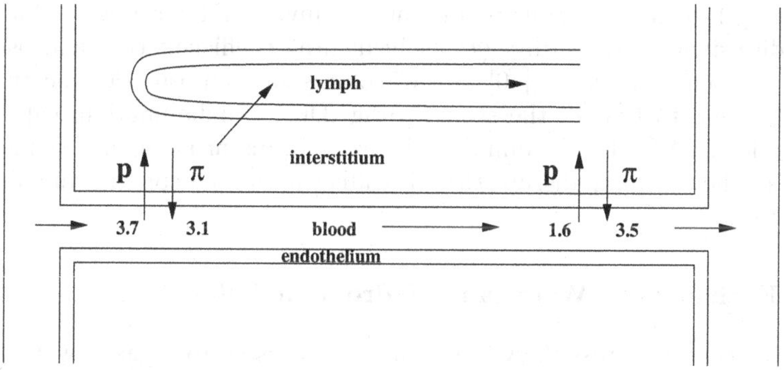

Figure 2.5. Capillary exchanges associated with hydrostatic and osmotic pressures value (kPa) distribution.

out of the capillary into the interstitium. The plasma osmotic pressure is usually much greater than the interstitial osmotic pressure. The osmotic pressure gradient across the capillary favors fluid reabsorption from the interstitium into the capillary. The osmotic pressure difference must be multiplied by a reflection coefficient associated with capillary permeability to the proteins responsible for the osmotic pressure. The net driving pressure for fluid motion is determined by the sum of the hydrostatic and osmotic contributions.

As tissue fluid volume is three to four times larger than plasma volume, tissue fluid serves as a reservoir that can supply additional fluid to the circulatory system or draw off excess. In pathological circumstances,[12] the interstitium can become edematous. Excess accumulation of water in the intersticium is due to lack in reabsorption into the capillaries and/or filtration into the lymphatics.

The capillary flow locally depends on the upstream resistance in the arterioles and the downstream resistance in the venules. Local control of substance transport to the tissue is done by: (1) recruitment of terminal arterioles (the higher the number of open capillaries, the greater the solute delivery and waste removal); (2) possible autoregulation (maintainance of a constant flow despite changing vascular pressure) associated with the cellular activity; and (3) vascular permeability.

2.3.1 Microvascular Permeability

With the exception of the liver, adrenal, and bone marrow sinusoids, in which the endothelium has large pores, the endothelium constitutes a selective barrier between blood and tissue. A filter at the luminal entrance to the endothelial clefts provides low permeability to macromolecules. In most microvessels, macromolecule transport is done by transcytosis and not porous clefts. Microvascular exchange is mainly passive. The permeability coefficients relate the net fluxes of fluid (J_w) and solute (J_s) driven by concentration (c) and pressure (p) differences. Four coefficients are involved, hydraulic conductivity (G_h), diffusional permeability ($\mathcal{P}$), solvent drag coefficient (κ_d) and osmotic reflection coefficient (κ_o)[13] [40]. "Ideal" solutes have an osmotic reflection coefficient equal to that of the solvent drag. During inflammation, capillaries become leaky. VEGF, histamine, and thrombin disturb the endothelial barrier [41]. The thrombin disrupt the VE-cadherin–catenin complex in adherens junctions.

2.3.2 Exchanges of Water and Hydrophilic Solutes

Materials that can cross the vessel wall include respiratory gases, water, ions, amino acids, carbohydrates, proteins, lipid particles, and cells. The capillary permeability is high for water and moderate for ions, lipids, and proteins. Fluid motion between blood and tissue fluid is mainly determined by two opposing

---

[12] Increase in fluid volume within the interstitium leads to tissue swelling, or edema. Such an increase can be due to: (1) an increased capillary hydrostatic pressure and venous pressures associated with heart failure or venous obstruction; (2) decreased plasma osmotic pressure; (3) increased capillary permeability caused by pro-inflammatory mediators and damaged leaky capillaries; and (4) lymphatic obstruction.

[13] The osmotic reflection coefficient of a porous membrane is a measure of the selectivity of the membrane to the solute of interest.

forces, hydrostatic and osmotic pressure differences between blood and tissue fluid. Water flux results from the imbalance between hydrostatic and osmotic pressures, for given features of the vessel wall layer and exchange surface area. A net motion of water out of (positive water flux) or into (negative water flux) the vessel leads to filtration and absorption, respectively. Plasma is hence filtered out of the capillary entry segment and reabsorbed back into capillary exit segment. *Hydraulic conductivity* measures the porosity of the capillary wall for the water flux. The *filtration coefficient* is the hydraulic conductance per unit exchange surface area. The ultrafiltrate has the same ion concentration as plasma and does not contain large molecules.

Nutrients and metabolic wastes are transported between blood and tissue cells by convection associated with fluid motions and diffusion according to the concentration gradient, facilitated by the very slow blood flow in the capillary and thereby by a high residence time. The diffusion from a compartment of higher concentration to a region of lower concentration across a membrane of infinitesimal thickness is governed by the Fick law. Brownian motion states that substance flux is proportional to the solute permeability coefficient, which depends on diffusion coefficient and solute solubility, transport surface area and concentration gradient. The reflection coefficient, or selectivity coefficient, measures the probability of solute penetration across the vessel wall. A reflection coefficient equal to 1 means that the vessel wall is impermeable to the molecule (most of the proteins are responsible for the osmotic pressure); a coefficient equal to 0 means that there is no transport restriction (such as small solutes like salts or glucose), and in the range bounded by these two values, that the wall layer is semipermeable to solute with some amount that does not penetrate (reflected off) the barrier.

In microvessels with continuous endothelium, the main route for water and solutes is the endothelium cleft, except when tight junctions exist. Transcapillary water flows and microvasculature transfer of solutes, from electrolytes to proteins, in both continuous and fenestrated endothelium, can be described in terms of three porous in-parallel routes: (1) a water pathway across the endothelial cells, (2) a set of small pores (caliber 4–5 nm), and (3) a population of larger pores (bore 20–30 nm) [40]. The estimated between-cell exchange area is on the order of 0.4% of the total capillary surface area. The pore theory simulates the cleft passages between adjacent endothelial cells in continuous endothelium. The array of junctional strands between endothelial cells is indeed interrupted at intervals, allowing water and solute fluxes. The associated fiber matrix model of capillary permeability associated with the endothelium glycocalyx (thickness ~100 nm; fiber spacing of 7 nm) provides a basis for the molecular size selectivity, especially in a fenestrated endothelium. A thin matrix at the cleft entrance can indeed be a major determinant of the permeability properties of the capillary wall. Such a fiber matrix model can also be applied for the cleft itself, because large portions of the cleft contain matrix components. Models have predicted that the fiber layer (typical thickness 100 nm), which extends from the endothelium surface into the cleft entrance region, sieves solutes [42]. Pore density is defined by the effective fraction of the cleft width used for molecular exchange. Pore size is determined by the interfiber spacing in a fiber matrix at the cleft entrance. The whole membrane permeability coefficient is the sum of the different route coefficients. The membrane reflection coefficient is the sum of the individual coefficients of the in-parallel paths weighted by the fractional contribution of each path to membrane hydraulic conductivity. Despite its limitations,[14] the pore theory is a useful pedagogical tool. Moreover, any fiber matrix-based modeling must not only take into account fiber size and volume, but also matrix organization [40]. The effective pore radius, a selectivity measure, is determined by the interstices in the fiber matrix, which depend on matrix composition and arrangement. The effective pore number is determined by the size and frequency of endothelium passages, in the intercellular spaces as well as through the cells. Water and small solutes can indeed cross the endothelial cells, using specific channels.[15] However, the contribution of the transcellular transfer to the net flux is supposed to be either small or negligible.

Elevated transmural pressures on endothelial cell culture on porous, rigid supports increase the endothelial hydraulic conductivity [44]. It is postulated that elevated endothelial cleft shear stress induced by increased transmural flow causes the hydraulic conductivity increase via a NO–cAMP-dependent mechanism.

2.3.3 Macromolecule Permeability

Macromolecules can cross the endothelium between the cells (paracellular transport) or through endothelial cells (transcytosis), using receptors, specific or not, and vesicles. The macromolecule flux from the blood to the interstitium can occur in post-capillary venules because their endothelial cells have simple intercellular junctions. Microvascular wall models have been proposed with pores for small and intermediate-sized molecules and transendothelial channels for macromolecules. However, the macromolecule transport, which is convection-independent, needs endothelial vesicles. Vacuole-like structures have been observed, isolated inside the endothelial cell or from luminal membrane invaginations, or connected to the abluminal compartment [45]. The luminal and abluminal surfaces of the capillary endothelium indeed are dynamic, with invaginations and protrusions associated with environmental stimuli. Moreover, vesicle translocation between luminal and abluminal membranes is accelerated as the transendothelial pressure is raised. Endothelium permeability can increase by vesiculo–vacuolar organelle without formation of trans- or intercellular gaps.

---

[14] Macromolecular transport is not always coupled to water flows. Furthermore, there are capillaries with no large pores.

[15] Aquaporins (Aqp) are membrane water-transport proteins [43]. Aqp-1 is found in endothelia.

2.4 Rhythmicity

The cardiovascular system is subjected to circadian rhythms due to molecular clock gene activities, as well as to shorter rhythms via, in particular, the functions of the autonomous nervous system and the adrenal glands, characterized by time variations of hormone release (catecholamines and corticosteroids). Besides, exposure to environmental stress leading to adrenal responses also varies during the day.

Molecular clocks exist not only in the suprachiasmatic nucleus of the hypothalamus (central pacemaker) but also in most tissues, especially aorta, liver, kidney, and heart for cardiovascular system functioning. The clock gene mPER1 is found in the heart, veins, and arteries. Peak PER1 activity occurs during the late night in cultured heart tissues and all arterial samples, whereas the phases of the rhythms in veins vary according to the anatomical location [46]. The molecular clock includes positive and negative feedback loops governing circadian gene expression [47]. In particular, genes of the molecular clock regulate enzymes of the synthesis of catecholamines.

Hormones control the expression of vascular clock genes, such as CLOCK and MOP4, via nuclear receptors RARα and RXRα [48]. A suprachiasmatic nucleus clock can entrain the phase of peripheral clocks via chemical cues, such as rhythmically secreted hormones. However, circadian gene expression in peripheral cells can be uncoupled from cyclic gene expression in the suprachiasmatic nucleus [49]. For example, the glucocorticoid hormone analogue dexamethasone transiently changes the phase of circadian gene expression in liver, kidney, and heart, but does not affect cyclic gene expression in neurons of the suprachiasmatic nucleus [50].

Time variations in cardiac frequency and arterial blood pressure are the most well known cardiovascular rhythms.[16] Stroke volume and cardiac output undergo circadian rhythms even at rest but they can be less easily monitored. Blood flow peaks are observed in early afternoon [51]. Vascular reactivity to adrenaline is greater in the early morning (3–4 h). Sympathetic tone and cathecolamine concentration have higher values during early day-time hours than during the night. Blood volume increases during evening and decreases at midnight, in association with circadian renal activity. Plasma levels of hormones (renin, angiotensin, aldosterone, and atrial natriuretic peptide), as well as of plasma proteins (clotting and fibrinolytic factors)[17] and hemoglobin content bear daily rhythms. Consequently, the circadian clock can explain the fluctuations in hemodynamic values.

Seasonal variations in lipid concentrations have been observed. Total cholesterol level, as well as LDL and HDL cholesterol levels, are higher in the winter in most studies (spring in some investigations) and lower in the

---

[16] Nocturnal heart rate and blood pressure can decay down to 0.5 Hz and 4 to 7 kPa.

[17] Fibrinolytic activity falls during early morning hours.

summer (autumn in certain works), with substantial monthly fluctuations. The change magnitude and month for high and low levels can depend on the subject, especially on the subject's age, gender, and ethnic group. Other biomarkers of cardiovascular risk have less been explored.

The circadian clock influences cardiovascular diseases. The clinical onset of both myocardial infarction and stroke occurs more frequently in the early morning than at other times of day. Endogenous rhythms of numerous physiological parameters (cardiovascular, pulmonary, hematologic, and endocrine) determine timing of medications. β-Blockers and calcium channel blockers lower blood pressure and heart rate much more during the day than during the night owing to the time variation in sympathetic tone, metabolism, and pharmocokinetics [51]. Other cardiovascular drugs also exhibit daily variations in activity. Treatment timing coordinated with body clock hence increases its efficiency.

2.5 Convective Heat Transfer

Convective heat transfer by the blood acts in homeothermy of the human body. Circulatory heat exchangers and conservers allow the heat or cooling of the body, depending on environment and body[18] conditions, heat being convected by blood circulation. The main exchange surfaces (with a given thermal conductivity) are the lungs and skin. Blood heat is transferred to the alveolus air to be exhaled or to the environmental air across the exchange tissues, the thin alveolo-capillary membrane, or the skin with its numerous cell layers. Heat exchangers can be bypassed, especially in the limbs with two venous networks, the superficial under the skin and the deep accompanying the arteries, with anastomoses between them.

Cancer treatments can use thermal ablation,[19] which are less invasive than surgery. Therapy efficiency depends on the local blood flow. Heat transfer models use heat source(s) and sink(s) and effective conductivity to describe the thermal influence of blood flow. A continuum model cannot account for the

---

[18] During exercise, body temperature rises.

[19] Image-guided radiofrequency ablation treats cancers particularly localized to the liver, kidney, and adrenal glands by heating. One or more radiofrequency needles are inserted into the tumor. Cryotherapy uses gas refrigerated cryoprobes, which are inserted inside the tumor, initiating the formation of ice balls to destroy cancerous cells by freezing and thawing processes. One of the main difficulties is the determination of the optimal position of the probes and treatment duration for complete destruction of cancerous cells without damaging too much surrounding normal cells. Another kind of tumor therapy consists of thermal and mechanical exposure to high-frequency focused ultrasound (HIFU). Tumor antigens and other compounds released from destroyed cells can stimulate the anti-tumor immunity. The optimal exposure time is an important parameter to avoid damage of normal cells, especially walls of neighboring blood vessels.

local thermal impact of the vasculature, with countercurrent vessel segment pairs. The vasculature must also be modeled, down to a certain diameter, and be combined with a continuum model for heat transfer in the irrigated tissues.

The bioheat transfer equation proposed by Pennes in 1948 is used in physiological heat transfer modeling [52]. The Pennes model provides the rate of heat change in a given body tissue from the sum of the net heat conduction into the tissue, metabolic heat generation, and heating (or cooling) effects of the arterial supply.[20] The blood flow-associated convective heat transfer term is given by: $q_b C_b (T - T_a)$ (q_b: organ perfusion rate per unit volume of tissue; c_b: specific heat of blood; T_a: arterial temperature; and T: local tissue temperature). The energy field in the perfused tissue domain is given by the bioheat transfer equation:

$$\rho_{tis} c_{tis} \frac{\partial T}{\partial t} = k_{tis} \nabla^2 T + q_b C_b (T - T_a) + q_{met} + q_s,$$

where c_{tis} the specific heat of the explored biological tissue, ρ_{tis} its density, k_{tis} its thermal conductivity, q_{met} the metabolic heat source (rate of energy deposition per unit volume assumed to be homogeneously distributed throughout the tissue of interest, but usually neglected), and q_s a possible heat source (e.g., in the case of thermal ablation). This equation is coupled to the equation of energy field in the flowing blood domain, which includes: (1) a directional convective term due to the net flux of equilibrated blood $\rho_b c_b (\mathbf{u} \cdot \nabla T)$ ($\mathbf{u}$: blood velocity, which can be composed of two terms, a hemodynamic component and acoustic streaming component generated by mechanical effects of high-frequency focused ultrasound when this therapeutic procedure is used to destroy the tumor); (2) the contribution of the nearly equilibrated blood in a tissue temperature gradient $k_b \nabla^2 T$ (k_b: perfusion conductivity); and (3) heat deposition q_h due to an imposed source:

$$\rho_b c_b \frac{\partial T}{\partial t} = k_b \nabla^2 T - \rho_b c_b (\mathbf{u} \cdot \nabla T) + q_h.$$

Values of the thermal conductivity and diffusivity of cardiac and arterial walls are given in Tables 2.13 and 2.14.

The heat transfer coefficient for the blood can be evaluated from the Sieder-Tate equation, when $q_m c_p / (\lambda_T L) > 6$:

$$h_h d / \lambda_T = 1.75 (q_m c_p / (\lambda_T L))^{1/3} (\mu_w / \mu)^{0.14},$$

where d is the vessel bore, L its length, h_h the heat transfer coefficient, λ_T the thermal conductivity, q_m the mass flow rate, and μ_w the near-wall blood/plasma viscosity at wall temperature.

---

[20] Pennes underestimated the magnitudes of the conduction and convection terms in the energy balance, using inappropriate values of tissue thermal conductivity and tissue perfusion rate.

Spatial variations in temperature distribution has been computed in a 3D muscle vascular model [54]. The tissue domain is composed of a twin artery and vein, which give birth to dichotomic trees of arterioles and venules, with eight generations of paired, closely spaced vessels, assuming property constancy at each generation. The arteriovenous spacing, vessel bore and density, and flow rate depend on tissue depth. An efficiency function, which depends on the volumic tissue blood perfusion rate and radial coordinate, is proposed to improve the bioheat equation. Using an appropriate procedure to analyze Pennes data, the data support the theory [55]; but the Pennes model, like new bioheat transfer models, lacks experimental validation and reliable evaluation of tissue properties [56].

2.6 Regulation of the Circulation

The heart adjusts the body requirements by increasing the ejection volume and its beating frequency. Stroke volume depends on heart inotropy and on pre- and afterload. The afterload is determined by the arterial resistance, which is mainly controlled by the sympathetic innervation (the higher the resistance and arterial pressure, the smaller the ejected volume). The preload affects diastolic filling, and consequently, the end-diastolic values of the ventricular volume and pressure. Stroke volume can rise from about 70 ml to 100 ml, i.e., an increase of almost 50%, and the cardiac frequency from about 60 to 180 or more, i.e., a threefold augmentation. The cardiac output is thereby adjusted using mainly the beating rate rather than stroke volume.

Table 2.13. Mean and standard deviation (SD) of thermal conductivity (mW/cm/C) and diffusivity ($\times 10^3$ cm2/s) of aortas and atherosclerotic plaques at 35 C (Source: [53]).

Tissue	Thermal conductivity Mean	SD	Thermal diffusivity Mean	SD
Normal aorta	4.76	0.41	1.27	0.07
Fatty plaque	4.84	0.44	1.28	0.05
Fibrous plaque	4.85	0.22	1.29	0.03
Calcified plaque	5.02	0.59	1.32	0.07

Table 2.14. Mean and standard deviation (SD) of thermal conductivity (mW/cm/K) and diffusivity ($\times 10^3$ cm2/s) of myocardia at 37 C (Source: [53]).

Thermal conductivity Mean	SD	Thermal diffusivity Mean	SD
5.31	0.37	1.61	0.20

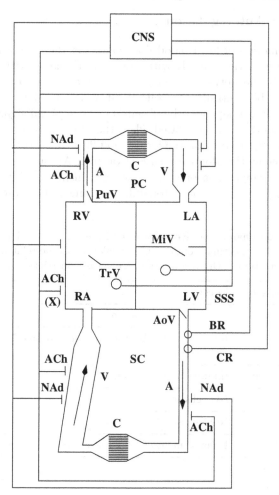

Figure 2.6. Neural control of the blood circulation (SSS: stretch-sensitive sensor; BR: baroreceptor, which responds to p_a; CR: chemoreceptor, which reacts to pH, p_{O_2}, and p_{CO_2}).

The blood circulation is controlled by a set of regulation mechanisms, which involve the central command (the nervous system; Fig. 2.6), the endocrine organs via hormone release (Table 2.15) and local phenomena (mechanotransduction; part I). Acting like very fast wired and wireless communications, neural and endocrine signals allow the cardiovascular system to adapt to environmental changes, regulate, and synchronize the functions of more or less autonomous cell sets.

2.6.1 Vascular Autoregulation

The primary function of local blood flow control in the circulation is to provide an adequate nutrient supply with respect to tissue activity, especially to maintain equilibrium between oxygen delivery and consumption. Blood flow distribution can be centrally controlled and at the local level, particularly targeting arterioles, independently of neural and hormonal signals.

Organ perfusion is autoregulated when the blood driving pressure variations has only a slight effect on the local blood flow. A nearly constant flow is maintained whereas the blood pressure is varying. Autoregulation occurs for a given pressure range. The pressure–flow relationship exhibits a very small slope (nearly a plateau) in the autoregulatory range, and more or less steep

Table 2.15. Additional vasoactive substances from remote, regional or local origin (VIP: vasoactive intestinal peptid; CGRP: calcitonin gene-related peptid; NMJ: neuromuscular junction; VDt: vasodilation; VCt: vasoconstriction; Source: [57]).

Molecule	Effect
	Vasodilation
Histamine	Inhibition of NAd release from arteriole NMJ NO release
Dopamine	Inhibition of NAd release
Bdk	NO release
Ach	NO release
VIP	Enhancement of ACh release and effects
Enkephalin	
CGRP	
5HT	NO release Arteriole vasodilation Enhancement of action of NAd and ATn2 Artery vasoconstriction
NAd	Major $\alpha 1$ and moderate $\beta 1$ effects NO release
Ad	α (VCt) and β (VDt) effects
	Vasoconstriction
ATn2	PLC activation
ADH	PLC activation

slopes outside the autoregulatory range[21] (Fig. 2.7). Autoregulation, indeed, fails whether the perfusion pressure is too low, dropping below the lower limit of autoregulation (the local blood flow decreases with decaying blood pressure), or too high, rising above the upper limit (the local blood flow increases with the soaring blood pressure). Autoregulation is done by modifying the vascular resistance (ratio between the perfusion pressure and blood flow: $R = \Delta p/q$). When the blood pressure increases, the blood flow augments and the stretched arteriole walls contract. The vasoconstriction elevates the vascular resistance. The local blood flow subsequently decreases. Conversely, when blood pressure decays, the resulting vasodilation decreases the vascular resistance and the local blood flow increases.

The heart, kidneys, and brain are organs that exhibit a strong autoregulation. Skeletal muscle and viscera circulation shows a moderate autoregulation. Skin circulation displays slight autoregulation. When hypotension occurs, baroreceptor reflexes induce vasoconstriction in the systemic vasculature. The blood flows to the brain and myocardium do not significantly decay because of autoregulation, at least if the arterial pressure does not decline below the autoregulatory range. Partial occlusion (stenosis) of large distribution arteries causes reduced pressure in irrigating arteries. The downstream

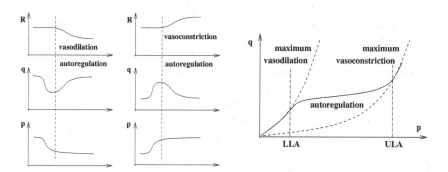

Figure 2.7. Autoregulation of local blood flow. (**Left**) Perfusion of certain organs is autoregulated when the blood driving pressure either decays or rises, the local blood flow being maintained almost constant by either vasodilation or vasoconstriction of the organ vasculature. (**Right**) Autoregulation occurs for a pressure range, between the lower (LLA) and upper limit of autoregulation (ULA).

---

[21] The blood vessel is maximally vasodilated below the lower limit and maximally vasoconstricted above the upper limit of autoregulation. Below the autoregulatory range, the pressure–flow relationship of the selected vessel indeed follows the non-linear pressure–flow curve of the vessel which is maximally dilated after vasodilator infusion into the explored organ, that disables autoregulation. The blood vessels passively dilate with increasing pressures, hence reducing flow resistance and augmenting flow rate. Above the autoregulatory range, the pressure–flow relationship of the vessel follows the pressure–flow curve of the vessel that is maximally constricted.

resistance vessels dilate in response to reduced pressure and blood flow to avoid local hypoxemia. Therefore, autoregulation ensures adequate blood flow and oxygen delivery to the essential organs.

Smooth muscle tone is controlled by the blood stresses applied to the vessel wall and by metabolic and myogenic mechanisms. However, autoregulation mechanisms vary among organs. The pressure-dependent activation of vascular smooth muscle cells of the cerebral arterioles is the main regulator of blood flow in the brain. The greater the wall stretch, the stronger the vasoconstriction. Increased vessel wall stretch caused by rising perfusion pressure activates phospholipase-C which leads, via cytochrome-P450-4A, to the production of 20-hydroxy eicosatetraenoic acid, which in turn, activates protein kinase-C [58]. Protein kinase-C inhibits potassium channels, especially the calcium-activated potassium channel, modifying the membrane potential[22] of the smooth muscle cells and inducing vasoconstriction. The metabolic activity of the nerve cells also acts. Glutamic acid is released from metabolically active neurones, stimulating astrocytes for arachidonic acid production. Arachidonic acid is converted by cytochrome-P450-2C11 to epoxyeicosatrienoic acid, which stimulates calcium-activated potassium channels, antagonizing 20-hydroxy eicosatetraenoic acid. Cerebral autoregulation thereby results from a balance between vasoconstriction generated by 20-hydroxy eicosatetraenoic acid and vasodilation induced by epoxyeicosatrienoic acid. The astrocytes also release other vasoactive molecules, such as thromboxane-A2, prostacyclin, and prostaglandin-E and -F. The calcium-activated potassium channels can be affected by various mediators (nitric oxide, adenosine, and prostacyclin).

Renal autoregulation is mainly located in glomerular arterioles. Stretch-activated vasoconstriction and tubulo–glomerular feedback maintain nearly constant blood flow. The tubulo–glomerular negative feedback deals with increased delivery of water and salts to the distal tubule, which causes vasoconstriction. This feedback can be affected by angiotensin-2, nitric oxide, thromboxane, 20-hydroxy eicosatetraenoic acid, and ATP.

Coronary blood flow is tightly coupled to oxygen demand, the myocardium having a very high basal oxygen consumption. Increase in cardiac activity augments coronary blood flow. Coronary autoregulation works for perfusion pressures between 8 and 26 kPa. Coronary circulation produces vasoactive substances. Endothelial production of nitric oxide mediates coronary vasodilatation. Coronary endothelial cells can also produce epoxyeicosastrienoic acids, which cause coronary vasodilatation. ATP-gated potassium channels and adenosine are implicated. Ischemic myocardium releases increased amounts of adenosine, which regulates the coronary flow. Oxygen is a major mediator. Myocardial oxygen consumption increases when coronary blood flow augments, antagonizing autoregulation. In abnormal states, coronary autoregulation can change. Hypoperfused human epicardial coronary arteries change

---

[22] The membrane is maintained in a relatively depolarized state, partially because of inhibition of K^+ channels.

their autoregulatory responsiveness. Vasoconstriction distal to the site of coronary angioplasty results from altered autoregulation [59].

2.6.2 Neural Effects on Heart and Blood Vessels

The nervous control of the circulation operates via: (1) afferents (from the cardiovascular apparatus and other endocrine glands), (2) interconnecting neurons, and (3) efferents (from the nervous centers).[23] These neurons are either peripheral or central, with excitatory and inhibitory synapses. The afferent neurons are constituted of three main types: barosensitive (afferents from the arterial baroreceptors), thermosensitive (cutaneous vasoconstrictors activated by hypothermia, emotions, and hyperventilation),[24] and glucosensitive (adrenaline release from the adrenal medulla stimulated by hypoglycemia and physical exercise) afferents from the blood vessels, heart, kidneys and adrenal medulla. The neurons include two usual kinds: (1) parasympathetic cholinergic neurons, which mainly innervate the heart, and (2) sympathetic noradrenergic neurons which innervate the heart and the vessel walls.[25]

Sympathetic and parasympathetic neurons modulate cardiovascular dynamics. Parasympathetic efferent pre-ganglionic neurons in the medulla oblongata project axons via the vagi to intrinsic cardiac parasympathetic postganglionic neurons. Sympathetic pre-ganglionic neurons in the spinal cord send axons to post-ganglionic neurons in paravertebral ganglia. The interdependent sympathovagal command concept states that the activated sympathetic inhibits the parasympathetic and vice versa. Intrathoracic ganglia process centripetal and centrifugal informations using short loops.

The main targets of the nervous control are: (1) nodal tissue, (2) cardiomyocytes, and (3) vascular smooth muscle cells. The brain gets signals from sensors and adjusts the circulatory parameters to match its needs. Activation of sympathetic efferent nerves to the heart increases heart rate (positive chronotropy C+), contractility (positive inotropy I+), and conduction velocity (positive dromotropy D+) (Table 2.16). Parasympathetic nerves have mostly negative effects but quicker than those of the sympathetic nerves (Table 2.17).[26] Both sympathetic and parasympathetic nerves act synergetically.

---

[23] The head nervous system is here considered as the center of the regulation loop. It receives signals from peripheral organs via afferent nerves and sends cues to visceral effectors through efferent nerves. The nervous regulation of the blood circulation is composed of cholinergic pre-ganglionic neurons of the central nervous system, which lead either to peripheral control nodes, such as para- or prevertebral sympathetic ganglia, endocrine glands, especially adrenal glands (catecholamine secretion), or visceral ganglionic networks.

[24] Skin circulation is mostly regulated via the rostral ventromedial medulla and medullary raphe [60].

[25] When it is not caused by vascular or renal disorders, hypertension can be due to a strong sympathetic tone.

[26] Parasympathetic effects on inotropy are weak in the ventricle and significant in the atria.

Table 2.16. Neural activation of circulatory organs.

Effect	Sympathetic	Parasympathetic
Heart		
Bathmotropy (CMC excitability)	B−	B+
Chronotropy (emission frequency of action potential)	C+ (major)	C− (major)
Dromotropy (conductibility)	D+ (moderate)	D− (major)
Inotropy (CMC contraction force)	I+ (major)	I− (minor)
Tonotropy (distensibility)	T+	T−
Blood vessels		
Resistance	+ (major)	Non-significant
Capacitance	− (minor)	Non-significant

Whenever the sympathetic system is activated, the parasympathetic activity is downregulated and reciprocally. In blood vessels, sympathetic activation constricts arteries and arterioles. The vasoconstriction causes an increase in resistance and pressure, and a decrease in distal blood flow. Sympathetic-induced constriction of capacitance veins decreases venous blood volume and increases venous pressure. Most blood vessels in the body do not have parasympathetic innervation. The overall effect of sympathetic activation is to increase the: (1) cardiac output, (2) systemic vascular resistance, and (3) arterial blood pressure.

Perivascular adrenergic and cholinergic nerves release many types of neurotransmitters, including peptides, purines, and nitric oxide (usual cotransmission),[27] for blood flow regulation at the regional or general scale. The endothelium deals with blood flow control at a local scale;[28] however, both regulators interact.

Sympathetic nerves express both nerve growth factor receptor TrkA and Sema3a receptor neuropilin-1. The semaphorin Sema3a promotes the ag-

Table 2.17. Response speeds of various commands of the cardiovascular system.

Command	Response
Parasympathetic	Very quick (f_c)
Sympathetic	Quick (f_c, SV)
Biochemical	Slower (f_c, SV)

---

[27] ATP is released as a co-transmitter with noradrenaline for sympathetic vasoconstriction in small arteries and arterioles.

[28] Strong, local blood flow variations, as well as hypoxia lead to changes in vascular tone mediated by the endothelium.

gregation of neurons into sympathetic ganglia during early embryogenesis. Cardiomyocyte-derived chemoattractant nerve growth factor is required for sympathetic axon growth and innervation in the heart. The neural chemore-pellent Sema3a is abundantly expressed in the trabecular layer in early-stage embryos, then restricted to Purkinje fibers after birth. Sema3a builds a trans-mural sympathetic innervation patterning, characterized by an epicardial-to-endocardial innervation gradient [61]. Alterations in Sema3a expression triggers various kinds of arrhythmias.

Nervous signals are integrated in the cardiovascular center, which is lo-cated in the brainstem. Various subgroups of nerve cells determine: (1) the *cardio-inhibitory*, (2) *cardio-excitatory*, and (3) *vasomotor* areas. The cardio-inhibitory center sends an inhibitory efferent pathway to the SAN via vagal parasympathetic fibers. The other two centers project sympathetic fibers to the SAN and the myocardium on the one hand, and the smooth muscles of blood vessel walls on the other hand.[29] The cardiac centers maintain a bal-ance between the inhibitory effects of the parasympathetic nerves and the stimulatory effects of the sympathetic nerves.

Limbic (App. A.3), cortical, and midbrain structures function for the short-term regulation of blood pressure by the sympathetic tone. The background activity of the sympathetic tone for long-term control of the blood pressure is driven by neurons of the rostral ventrolateral medulla, the spinal cord, hypothalamus and nucleus of the solitary tract [60] (Fig. 2.8). The nucleus of the solitary tract is an integrative center for blood circulation control. It directly receives cues from baroreceptors, voloreceptors, and chemoreceptors, as well as many synaptic inputs. The spinal cord receives chemical (tissue oxygen content) and physical (tissue stretch) outputs. The hypothalamus, with its paraventricular and dorsomedial nuclei, is another integrative center for the regulation of blood circulation. The dorsomedial nucleus is implicated in environmental stresses. Neurons of the paraventricular nucleus are affected by blood volume, pressure, and osmolality.

The regulation of renal sympathetic activity by arterial baroreceptors uses the rostral ventrolateral medulla. The renal nerve response is also induced by hepatoportal osmoreceptors, arterial baroreceptor, and voloreceptors. The re-sponse to atrial voloreceptors involves the nucleus of the solitary tract and the paraventricular nucleus of the hypothalamus to regulate sodium reabsorption by the kidney, and subsequently blood volume. Peripheral and brain osmore-ceptors and hypothalamic sodium receptors, particularly in the median pre-optic nucleus, affect renal sympathetic activity. Other central osmoreceptors and sodium receptors are located in circumventricular organs.

Circulating hormones, such as aldosterone, control blood circulation via the circumventricular organs (subfornical organ, organum vasculosum lamina terminalis, and area postrema). Circulating hormones, such as angiotensin-2,

---

[29] Both the heart and the blood vessels are effectors of excitatory sympathetic fibers.

also directly affect the sympathetic ganglia. Angiotensin-2 also acts on the median preoptic nucleus, the nucleus of the solitary tract, the rostral ventrolateral medulla, and the paraventricular nucleus of the hypothalamus. The central nervous system has its own receptors, responding to changes in blood gas levels (O_2 and CO_2) via brainstem chemoreceptors, and in sodium and osmolality via hypothalamic receptors. Endothelial nitric oxide synthase in the rostral ventrolateral medulla reduces blood pressure.

Post-ganglionic sympathetic vasoconstrictor fibers use noradrenaline as a neurotransmitter. The resting frequency of 1 to 4 Hz can increase to 10 Hz. Inactive units may be recruited. Vasodilator fibers are not involved in the bulk regulation of the peripheral resistance, but locally increase the blood flow. Sympathetic vasodilator fibers arise from the cerebral cortex, and, after a synapse in the hypothalamus, go through the medulla to alter the activity of the pre-ganglionic vasodilator fibers. The post-ganglionic fibers of the sympathetic vasodilator system are cholinergic. The parasympathetic vasodilator fibers participate in autonomic reflexes, such as digestive secretion.

Exercise activates the sympathetic system, thus increasing blood pressure and blood flow to skeletal muscles, whereas it reduces the blood flow to other body organs, such as the kidneys. Two main mechanisms increase sympathetic nerve activity: (1) the exercise pressor reflex, and (2) the central command. The exercise pressor reflex arises from chemo- (muscle metaboreflex) and mechano-receptors (muscle mechanoreflex) of skeletal muscles, which stimulates the nervous centers via afferent fibers. The central command simultaneously activates the locomotor and cardiovascular systems.

Muscle metaboreflex desensitization and mechanoreflex sensitization are observed after myocardial infarction. During exercise, exaggerated sympathetic activation, with augmented muscle sympathetic signaling and excessive renal vasoconstriction,[30] occurs in heart failure. Moreover, the rise in muscle blood flow decays with respect to normal vasculature. The renal and lumbar sympathetic responses associated with the central command increase in heart failure, causing excessive peripheral vasoconstriction [62].

2.6.2.1 Heart Rate Regulation

The heart is able to beat independently. The cardiac frequency is higher than the activity rhythms of the main processes involved in the regulation of blood flow (Table 2.18). Adaptation needs a slight delay.

The nervous system regulates the cardiac frequency (f_c), in superimposition of heart automatism, to adapt f_c to the changing needs of the body. However, the frequency increase is bounded by the necessary *diastolic filling* associated with venous return and the *diastolic perfusion* of the coronary arteries. Various sensors of the circulatory system send messages to the cardiac centers, which respond by sending messages to the heart.

---

[30] Renal vasoconstriction reduces the blood supply, causing excessive renin secretion and inappropriate salt and water retention.

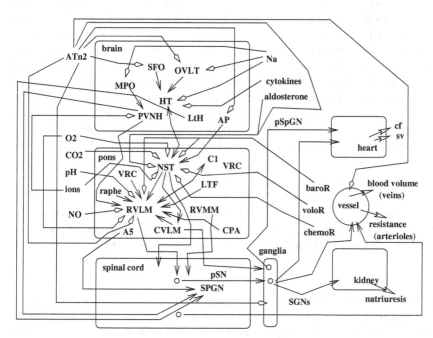

Figure 2.8. Nervous regulation of the blood circulation, especially sympathetic activity and its influence agents (adapted from [60]). An increase in blood pressure activates baroreceptors and inhibits cardiac, renal, and vascular sympathetic efferents. The baroreceptor reflex dampens short-term fluctuations of blood pressure. However, it can be reset via both neural and humoral mechanisms such that the operating range is shifted without reduction in reflex sensitivity. The nucleus of the solitary tract (NST), hypothalamus (HT), with its paraventricular nucleus (PVNH) and median preoptic nucleus (MPO), are integrative centers for the regulation of blood circulation. The interneurons (releasing γ-aminobutyric acid) of the caudal ventrolateral medulla (CVLM) inhibit barosensitive neurons of the rostral ventrolateral medulla (RVLM). RVLM neurons coexist in a pons region with the ventral respiratory column (VRC) neurons for coordination of respiration and circulation, and a cluster of adrenaline-synthesizing neurons (C1). Circumventricular organs are implicated, such as subfornical organ (SFO), organum vasculosum lamina terminalis (OVLT), and area postrema (AP). Efferents neurons include sympathetic pre-ganglionic (SPGN), sympathetic ganglionic (SGN), parasympathetic (pSN), and parasympathetic post-ganglionic neurons (pSpGN). Others involved nervous structures comprise caudal pressor area (CPA), lateral hypothalamus (LtH), lateral tegmental field (LTF), a noradrenergic cluster located at the pontomedullary junction (A5). The cutaneous circulation is mainly regulated via the rostral ventromedial medulla (RVMM) and raphe. Stimuli of neuronal activity include aldosterone, angiotensin-2 (ATn2, which acts on median preoptic nucleus, PVNH, NTS, RVLM, and SPGN), ions (especially sodium and pH), and blood gas (cf: cardiac frequency, sv: stroke volume).

Additional factors such as hormones and body temperature also influence f_c. Under stresses, *cathecolamines* are released from the adrenal medulla into the circulation to produce an increase in heart rate. *Thyroid hormones*, thyroxin (T4) and triiodothyronine (T3), accelerate f_c and modulate heart contraction. Elevation of the *body temperature* is associated with f_c increase. Conversely, hypothermia is accompanied by f_c reduction. The *ion concentrations* in the extracellular environment may have a significant influence on cardiac function. Potassium excess in the extracellular environment ($[K^+]_e$) reduces f_c and contractibility, as well as calcium level $[Ca^{++}]$ reduction. Excessive $[Na^+]$ depresses cardiac function, whereas Na^+ deficiency in the extracellular space leads to cardiac fibrillation.

2.6.2.2 Arterial Pressure Regulation

Blood pressure is regulated by short-, mid-, and long-term mechanisms (Fig. 2.9). Both blood pressure and the sympathetic activity undergo a circadian rhythm. Arterial pressure depends on both cardiac output[31] and vascular resistance:

$$p_a = f_c \times SV \times SVR.$$

In the short-term control, the arterial pressure is monitored by suitable receptors, mainly baroreceptors;[32] adjustments are made via neural mechanisms, which change the cardiac output and peripheral resistances. Long-term control of the blood pressure involves indirect monitoring of blood volume. Hormonal mechanisms restore the blood volume and indirectly the blood pressure. The long-term regulation involves: (1) the renin–angiotensin system, (2) natriuretic peptides and (3) antidiuretic hormone (Sect. 2.6.6).

Table 2.18. Rhythmic processes associated with blood flow.

	f (Hz)
Heart cycle	1
Respiratory cycle	0.3
Vessel myogenic activity	0.1
Neurogenic control activity	0.04–0.4
Endothelium metabolic activity	≤ 0.01

---

[31] $q = f_c \times SV$. The heart rate is mainly regulated by the nervous system (both parasympathetic and sympathetic components), as well as by hormones. Stroke volume is controlled by myocardial contractility and its regulating factors (sympathetic command), pre- and afterload, with their sympathetic command, and circulating regulators.

[32] Baroreceptors have a rest activity associated with the sympathetic tone of the vascular smooth muscle cells. They react with bursts synchronized with the arterial pressure pulse and respiration.

Blood volume and, subsequently, blood pressure are controlled by fluid and electrolyte (particularly sodium and potassium) excretion by the kidney (Sect. 2.6.6). Defective salt reabsorption in a portion of the distal nephron

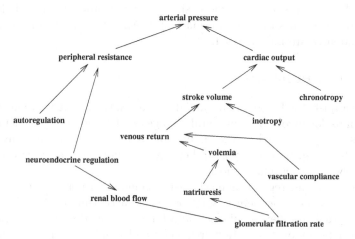

Figure 2.9. Influence factors and regulation of arterial pressure. The arterial pressure is determined by the cardiac output and total systemic peripheral resistance. Cardiac output depends on cardiac functioning (chronotropy, inotropy, etc.) and venous return, itself particularly affected by volemia and vascular compliance. Blood pressure thereby is controlled by the autonomic nervous system, each factor being regulated by either the sympathetic system or both the sympathetic and parasympathetic systems. The sympathetic baroreflex is a feedback loop, with afferents from baroreceptors. The distention of the arterial wall following an increase in blood pressure activates the baroreceptors, and subsequently inhibits cardiac, renal, and vasomotor sympathetic efferents to restore the blood pressure. With the parasympathetic counterpart, the sympathetic baroreflex is first aimed at dampening short-term fluctuations of blood pressure with appropriate resetting when the blood pressure adapts to the imposed conditions, such as during exercise. The new operating range then does not disturb reflex sensitivity. The sympathetic activity fluctuates because of the time-dependent nature of arterial baroreceptor activity, polysynaptic transmission through the baroreflex loop, and breathing. These factors are also regulated by endocrine (renin–angiotensin–aldosterone system, cathecolamines, vasopressin, and natriuretic peptides) and paracrine factors. The total peripheral resistance is mainly determined by the activity of the sympathetic system and by local autoregulations. The kidney is the primary regulator of extracellular fluid volume and electrolyte balance. The kidney participates in blood pressure regulation via the relationship between the renal blood pressure and natriuresis. Any increase in sodium retention produces an initial blood volume expansion, then an increase in blood pressure associated with a rise in cardiac output. Subsequent tissue overperfusion leads to an increase in peripheral resistance, and cardiac output returns to its resting values. Renal blood flow and glomerular filtration are controlled by the renal sympathetic nerves, concentrations of circulating hormones, and renal paracrine and autocrine factors (ATP, nitric oxide, etc.).

leads to hypertension. Most high blood pressure disorders are, indeed, due to either disturbed regulation of peripheral vascular resistance or defective salt and water reabsorption by the distal nephron. Dysregulation in the renin-angiotensin system is often implicated in arterial hypertension. Angiotensin-2 increases blood pressure via angiotensin type 1 receptors. Renal AT1Rs are primary determinants of hypertension [63].[33] Angiotensin-2-mediated aldosterone responses are not involved.

2.6.2.3 Other Regulations

The CVS adjusts the blood volume and distribution to provide nutrients to the working muscles involved in postural changes and locomotion. Anticipatory responses are obvious when the body motion is associated with any emotional context [64]. However, postural changes are not necessarily preceded by preparatory cardiovascular actions.

Other autonomic regulation mechanisms include thermoregulation and respiration frequency-dependent respiratory sinus arrhythmia (RSA)[34] which, together with baroreflex regulation, affect heart rate. Respiratory sinus arrhythmia depends on respiration frequency and amplitude. The respiratory sinus arrhythmia can require several mechanisms. Pressure variations experienced by thoracic blood vessels during respiration, tereby changing afterload, can influence the mechanoreceptors of the vasculature walls and mediate a baroreflex. Coupling between the respiratory and circulatory autonomic nervous centers can also be involved. The respiratory sinus arrhythmia can also contribute to respiratory arterial pressure fluctuations [65]. Thermoregulation operates at very low frequencies (below 40 mHz), whereas baroreflex regulation and RSA are low- ([40–150 mHz]) and high-frequency ([150–400 mHz]) components, respectively.

2.6.3 Adrenergic and Cholinergic Receptors

The actions of autonomic nerves are mediated by the release of neurotransmitters that bind to specific receptors in the heart and blood vessels (Table 2.19). These receptors are coupled to signal transduction pathways. In the heart, catecholamine such as adrenaline (Ad; or epinephrine) or noradrenaline (NAd;

---

[33] AT1Rs are expressed by epithelial cells throughout the nephron, in the glomerulus and renal blood vessels. Once activated, they promote sodium reabsorption by stimulating both sodium-proton antiporter and sodium-potassium ATPase on the apical (luminal) and basolateral plasmalemma, respectively, in the proximal tubule of the nephron. AT1Rs stimulate epithelial sodium channels in the collecting ducts. Furthermore, activated vascular AT1Rs induce vasoconstriction, which subsequently reduces renal blood flow and sodium excretory capacity.

[34] The respiratory sinus arrhythmia is the variation in heart rate occurring simultaneously with respiration. On the ECG traces, it induces fluctuations of the RR interval series.

or norepinephrine) released by sympathetic nerves preferentially binds to β1 receptors, causing I+, C+ and D+ effects (Table 2.20). G-proteins are activated, which in turn activate adenylyl cyclase. β2 Receptor stimulation has similar cardiac effects and becomes increasingly important in heart failure as β1 receptors become downregulated. NAd can also bind to α1 receptors on cardiomyocytes causing increased contractility. In blood vessels, NAd preferentially binds to α1 receptors inducing vasoconstriction. α receptors are linked to intracellular calcium stores. Similar responses occur when NAd binds to postjunctional α2 receptors located on some blood vessels. NAd can also bind to postjunctional β2 receptors which causes vasodilation. β Receptors are linked to adenylyl cyclase. Relaxation can be mediated by cAMP-dependent phosphorylation and inactivation of myosin light chain kinase. NAd can also regulate its own release by acting on prejunctional α2 (inhibition) and β2 (stimulation) receptors. Circulating Ad binds to β2 receptors, causing vasodilation in some organs.

Table 2.19. Vessel innervation.

	adrenergic receptor	muscarinic receptor
Artery	α1	M3
Vein	α1, α2, β	M3

Table 2.20. Types and responses of adrenergic receptors. Adrenaline is released by the adrenal medulla, and noradrenaline is secreted by the nerves and adrenal medulla (Source: [27]).

	α1	α2	β1	β2
Ligand	Ad			
	NAd			
Second messenger	IP3		cAMP	
Vessel	Vasoconstriction		Vasodilation (β1 < β2)	
Heart			C+, D+, I+, B+	
Lipolysis	Decreased		Increased (β1 < β2)	
Glycogenolysis	Increased (β1 < β2)			
Insulin release	Inhibition		Stimulation (β1 < β2)	
Synapse	Inhibition of ACh and NAd release			
Micellaneous	TC aggregation		stimulation of renin release	

The stimulation of β-adrenergic receptors increases cytosolic $[Ca^{++}]$ and cardiac contraction, whereas the excessive activation of β receptors induces myocardial hypertrophy and dysfunction in the case of infarction. The Ca^{++}–calmodulin-dependent protein kinase-2 (CamK2) belongs to the β receptor signaling cascade associated with maladaptive myocardial remodeling [66]. CamK2 inhibition might hamper such a pathological remodeling.

The myocardium also contains muscarinic receptors associated with adenylyl cyclase and a K^+ channel in the sarcolemma. Acetylcholine (ACh) released by parasympathetic nerves bind to muscarinic receptors. ACh reduces [cAMP] and increases K^+ currents. This produces negative inotropy (I−), C− and D− effects. In blood vessels, muscarinic receptors are coupled to the formation of nitric oxide, which causes vasodilation. Prejunctional muscarinic receptor activation inhibits NAd release. This is one mechanism by which vagal stimulation overrides sympathetic stimulation in the heart. Arteries in skeletal muscle are innervated by sympathetic nerves, which release ACh to induce hyperemia, particularly at the onset of exercise.

Vasomotor tone can be locally regulated (Part I). It is also controlled by the autonomic nervous system (ANS), which acts by nervous[35] and humoral path.[36] Adrenoceptors, stimulated by NAd, are located on arterial, arteriolar and venous smooth muscle. They induce slow depolarizations that last for several seconds. ATP and NAd are co-stored in synaptic vesicles in sympathetic nerves. When they are co-released, they act post-junctionally for contraction of the vascular smooth muscle. Purinoceptors, activated by ATP, generate fast depolarizations. Conversely, muscarinic receptors, present in arteries and veins have inhibitory effects. Five distinct but related muscarinic receptors have been identified [67]. G-protein muscarinic (M3) receptor is located on the surface of the endothelial cell. In summary, NAd and ACh induce contraction and relaxation of the vascular smooth muscle, respectively.

Heart failure is characterized by cardiac overstimulation by the sympathetic nerves for compensation of decreased cardiac function, associated with increased blood concentrations of catecholamines. During heart failure, α2-adrenoceptors in chromaffin cells of the adrenal medulla are disturbed by increased activity of G-protein-coupled receptor kinase GRK2, contributing to elevated blood levels of catecholamines [68]. Normally, α2-adrenoceptors generate via Gi/o-protein an autocrine feedback inhibition of catecholamine secretion induced by activation of nicotinic cholinergic receptors. During heart failure, α2-adrenoceptors of the adrenal gland loose their inhibitory function on the sympathetic system. GRK2 inhibition in adrenal glands during heart failure restores the inhibition of catecholamine release by activating

---

[35] Nerve fibers in adventitia act by electrochemical stimulation at neuromuscular junctions and biochemical processes (release of neurotransmitters) preferentially at external layers of the media.

[36] Flowing vasoactive hormones act after transmural migration up to internal layers of the media.

α2-adrenoceptors. In the heart, GRK2 is also upregulated and reduces the ventricular function. In the cardiomyocyte, GRK2 phosphorylates and desensitizes β-adrenoceptors, thus reducing catecholamine-induced signaling via Gs-subunit and ACase-PKA pathway, causing reduced contractility.

2.6.4 Circulation Sensors

Mechano-[37] and chemosensors[38] in walls of the cardiovascular system, especially in diverse cardiac regions, coronary and large intrathoracic (particularly along the inner aortic arch and at the bases of both vena cavae) and cervical vessels, continuously record the hemodynamics regime, transduce the signals and fed the information to the corresponding efferent neurons.

Chemosensors, or *chemoreceptors*, transduce a chemical signal into an action potential. Impulses are transmitted via the vagus into the vasomotor centers, as well as the respiratory centers. The vascular chemoreceptors are located in the carotid sinuses[39] and aortic arch, i.e., the same sites as the baroreceptors. Reduced O_2 concentration and increased CO_2 and H^+ concentrations stimulate the chemoreceptors (Table 2.21).

Sensing is also done by the *baroceptors* in the high pressure system (aorta and carotid sinuses) and the *stretch receptors* in the low pressure system (pulmonary artery, ventricles, and venae cavae). Sympathetic nerve activity is inhibited by activated lung stretch receptors and carotid and aortic baroreceptors. The baroceptors allow quick control by the central nervous system to adjust the arterial blood pressure and maintain it at physiological values (baroreflex negative feedback).[40] The magnitude of the baroreceptor responses depends on the targeted organ. The stretch receptors in the low pressure system are more involved in the regulation of blood volume. These receptors can,

Table 2.21. Baroreceptors and chemoreceptors.

Receptor	Signal	Cardiac effect	Vessel effect
Baroreceptor (threshold $\sim 8\,kPa$)	p Wall deformation	$p \searrow \Rightarrow \Sigma c \oplus, p\Sigma c \ominus$ (I+, C+)	$p \searrow \Rightarrow \Sigma c \oplus$ (vasoconstriction)
Chemoreceptor	p_{O_2}	$p_{O_2} \searrow \Rightarrow p\Sigma c \oplus$	

---

[37] Fast-responding mechanosensors transduce the stretch undergone by the wall and papillary muscles under increasing luminal pressure and tension exerted by cordae tendinea. They ensure a beat-to-beat coordination of the heart rate and contractility.

[38] Multiple chemicals can be followed up.

[39] The carotid sinus is located at the carotid artery bifurcation.

[40] The feedback loop is a regulatory loop that feeds the system, either negatively (an increasing output has a suppressing effect on the triggering signal) or positively (an increasing output produces a further rise in the output), modulating the input by the output.

in the longer term, change blood circulation pressure. Signal acquisition depends on receptor sensitivity. The biomechanical signals are then transduced into electrochemical events with given neural firing rates.

Baroreceptors are sensitive to the rate of pressure changes, as well as mean pressure. The combination of reduced mean pressure and reduced pulse pressure reinforces the baroreceptor reflex [69]. The background activity can either decay or rise for blood pressure stabilization. The firing rate of the baroceptor nerves increases with the blood pressure, and hence the wall deformation, from a threshold up to a maximum (saturation). Increased firing not only inhibits sympathetic activity to the blood vessels, heart, and kidneys, but also increases vagal tone to the heart. Conversely, a fall in arterial pressure reduces afferent signals, which relieves inhibition of sympathetic tone, increases the peripheral resistances, and restores the cardiac output and subsequently the arterial pressure.

The baroreceptors are mainly located in the heart, in the aortic arch and in the carotid sinuses. The receptors of the carotid sinus respond to pressures ranging from 8 to 24 kPa. Receptors of the aortic arch, less sensitive than those of the carotid sinus, have a higher threshold pressure. Aortic baroreceptor neurons exhibit mechanosensitive ion channels that are gadolinium-sensitive and have non-specific cationic conductances [70]. The fibers of the aortic nerve enter the adventitia, between the left common carotid and left subclavian arteries, and separate into bundles generally containing one myelinated fiber and several unmyelinated fibers [71]. When they are close to the aortic media, the myelinated fiber loses its myelin sheath. Both unmyelinated and premyelinated axons branch off. Sensory nerve endings of aortic baroreceptor neurons are located in the adventitia of the aortic arch, between the left common carotid and left subclavian arteries. The basal lamina exist around the sensory terminals. The central axon terminals are located in the nucleus of the solitary tract in the central nervous system.

Once the baroreceptors are stretched, the Hering or carotid sinus nerve (a branch of the glossopharyngeal nerve, IX cranial nerve pair) stimulates inhibitory areas of the vasomotor center (the nuclei tractus solitarius and para-median in the brain stem) [25]. The aortic arch baroreceptors are innervated by the aortic nerve, which then merges with the vagus nerve (X cranial nerve), traveling to the brainstem. Efferent limbs are carried through sympathetic and vagus nerves to the heart and blood vessels, controlling heart rate and vasomotor tone. The cardiovascular nervous center responds by increasing sympathetic and decreasing parasympathetic outflux. Barosensitive sympathetic efferents control the activity of the heart and kidneys, as well as the release of noradrenaline from adrenal chromaffin cells, and constrict the arterioles, except those of the skin.

Multiple transmitters regulate the barosensitive neurons. Glutamate, γ-aminobutyric acid, acetylcholine, vasopressin, serotonin, corticotropin-releasing factor, substance-P, oxytocin, and orexin have been found in nerve ends with synapses on pressure-regulating neurons, such as C1 cells[41] and the hypothalamus [60].[42]

2.6.5 Short-Term Control of the Circulation

The control of the circulation here deals with overall circulation (rather than local controls of blood flow in the skeletal muscles and head). The time scale of the short-term regulation of the circulation is $\mathcal{O}(s)$ to $\mathcal{O}(mn)$, whereas for the long term, it is $\mathcal{O}(h)$ to $\mathcal{O}(day)$. The short-term control includes several reflexes, which involve the following inputs and outputs: arterial pressure, heart rate, stroke volume, and peripheral resistance and compliance. So the autonomic nervous system can receive complementary information from the circulation and has several processing routes. The importance of a given feedback loop with respect to the different other reflexes can become primary in certain circumstances or secondary in others. Nervous control of the circulation must therefore take into account the whole set of involved factors. The control of the peripheral resistance and compliance is slower than the command of the heart period and the stroke volume. Because the heart period can be non-invasively measured, using the RR interval of the ECG records, most investigations consider the relation between arterial pressure and heart period. This short-term control of circulation requires: (1) receptors, (2) nervous signaling, and (3) corresponding feedback loops. There are several types of mechanosensitive receptors in the circulation. The interaction between the receptor varieties is not clearly defined. Studies commonly are focused on baroreceptors, which act as starting elements in the reflex regulation of arterial pressure.

Both arterial pressure and RR interval vary from one heart beat to another. The power spectrum of the RR interval serves as a measure of its neural modulation. The relationship between arterial pressure and RR interval via the baroreflex[43] is exhibited by low- (LF) and high- (HF) frequency spectral components. Normalized indices of LF and HF components of the spectral analysis of RR interval variability are used, dividing the quantities by the difference between power variance and power of very low-frequency (VLF) component. Low-frequency oscillations in RR interval not only depend on the baroreflex control loop but also on a central rhythmic modulation of neural autonomic activity.

---

[41] C1 cells belong to a cluster of adrenaline-synthesizing neurons in the pons region where the rostral ventrolateral medulla is located. Many of C1 cells regulate the kidneys.

[42] γ-aminobutyric acid signaling from the caudal ventrolateral medulla is important for baroreflex.

[43] Arterial pressure and RR interval are considered baroreflex input and output, respectively. The effect of arterial pressure on RR intervals and reciprocally RR interval on arterial pressure (closed loop interaction between the two physiological signals) is described by two transfer functions [72].

Mayer waves, low-frequency arterial blood pressure oscillations, are observed in response to decreased central blood volume, seen in upright posture [73]. The stability of the arterial baroreflex feedback is then reduced.

During exercise, the mesencephalic locomotor region inhibits the baroreceptor reflex by activating interneurons of the nucleus tractus solitarius (NTS), which inhibit NTS cells receiving baroreceptor input [74].

2.6.6 Delayed Control of the Circulation

Delayed mechanisms involve circulating hormones as catecholamines, endothelins, prostaglandins, nitric oxide, angiotensin, and others.

2.6.6.1 Nephron

Urination removes water, certain electrolytes, and certain wastes from the body.[44] Urine is produced in the nephrons (Fig. 2.10) by three regulated processes: filtration, reabsorption, and secretion. Renal blood is first filtered from the glomerulus, a capillary ball formed from an afferent arteriole and leading to a narrower efferent arteriole, by the Bowman capsule (glomerular filtration). The glomerular filter consists of three layers, the: (1) fenestrated endothelium, (2) glomerular basement membrane, and (3) interdigitated podocyte extensions, which completely enwrap the glomerular capillaries. The filtration barrier restricts the passage of molecules according to their size, shape, and charge. Water, electrolytes, glucose, amino acids, wastes (urea), and other filtered chemical species form the glomerular filtrate. The juxtaglomerular apparatus of the arteriole walls contains granular cells that secrete renin.

Molecules (water, glucose, amino acids, ions, and other nutrients) are reabsorbed from the renal tubules back into the peritubular capillaries, which drain into a venule. The tubule is composed of several segments, the: (1) proximal convoluted tubule; (2) loop of Henle, with its descending limb, with its thick (in the outer medulla) and thin (in the inner medulla) segments, and ascending limb, with its thin (in the inner medulla and the inner stripe of the outer medulla) and thick (in the outer stripe of the outer medulla) segments; (3) distal convoluted tubule, (4) connecting tubule, and (5) cortical and medullary segments of the collecting duct.

About 60% to 70% of salts and water are reabsorbed at the proximal convoluted tubule. The descending limb of the loop of Henle is permeable to water, but impermeable to salts. The asscending limb of the loop of Henle is impermeable to water, the active pumping of sodium to concentrate salts in the hypertonic interstitium, with Na^+-Cl^- co-transporters in particular. The distal convoluted tubule secretes hydrogen and ammonium. Molecules (hydrogen ions, potassium ions, urea, and ammonia) are secreted from peritubular capillaries into the distal and collecting tubules via either active transport

---

[44] Nitrogenous wastes are excreted as ammonia, urea, or uric acid.

$(Na^+-K^+$ pumps) or diffusion. The collecting duct is permeable to water owing to antidiuretic hormone and urea in its downstream segment. The distal nephron is the site of endocrine regulation. Mineralocorticoid receptors are located in the Henle loop, distal convoluted tubules, connecting tubules, and collecting tubules and ducts [77, 78].

The macula densa is the specialized area of the downstream segment of the thick ascending limb and the upstream part of the distal tubule in the neighborhood of the afferent and efferent arterioles of its own glomerulus. The macula densa regulates arteriolar resistance.

Solute and water transport between ascending and descending urine and blood pipes are not only governed by osmotic forces and interstitial hydro-

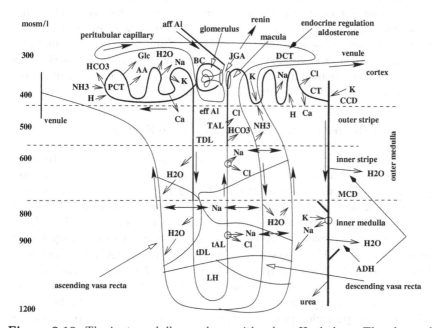

Figure 2.10. The juxtamedullar nephron with a long Henle loop. The glomerulus with its afferent arteriole (aff Al) and efferent arteriole (eff Al), and with the Bowman capsule (BC). Water, electrolytes (Na, K, Cl, etc.), glucose (Glc), amino acids (AA), urea, and other filtered chemicals form the glomerular filtrate. The tubule is composed of several segments: (1) the proximal convoluted tubule (PCT); (2) loop of Henle (LH), with its descending (thick [TDL] in the outer medulla and thin [tDL] in the inner medulla) and ascending (thin [tAL] in the inner medulla and the inner stripe of the outer medulla, thick [TAL] in the outer stripe of the outer medulla) limbs; (3) distal convoluted tubule (DCT); (4) connecting tubule (CT); and (5) cortical (CCD) and medullary (MCD) segments of the collecting duct. The juxtaglomerular apparatus (JGA) secretes renin. The macula densa regulates arteriolar resistance. Interstitial osmolarity of 300 mosm/l is observed at the level of cortical nephrons with short loops and ordinary peritubular capillaries. At the hairpin curve, the interstitium osmolarity is equal to 1200 mosm/l or more (Source: [79]).

static pressure, but also by active processes and membrane permeability. An osmotic gradient is caused by Na^+ and Cl^- transport out from the ascending tubule to the interstitium. Interstitial osmolarity of 300 mosm/l is observed at the level of cortical nephrons with short loops and ordinary peritubular capillaries. Interstitial hyperosmolarity pulls water from descending vasa recta and descending tubules. In the medulla, the interstitial sodium level balances sodium concentrations in both vasa recta. Water from the descending tubule and descending vasa recta enters in the interstitium, and then the ascending vasa recta. At the hairpin curve, interstitium osmolarity reaches 1200 mosm/l or more (urea leaving the collecting duct intensifies the osmotic gradient) [79].

In the nephron, sodium uptake is associated either by chloride reabsorption or potassium secretion. Simultaneous sodium and chloride ion reabsorption by sodium–chloride co-transporters is an electroneutral process. Electrogenic sodium reabsorption via the sodium channel generates a negative charge in the urinary lumen and subsequent secretion of potassium via potassium channels, such as the renal outer medullary potassium channel. Both sodium carriers are found in the distal convoluted tubule. When sodium–chloride co-transporter activity is augmented, the sodium channel function decays and vice versa.

2.6.6.2 The Renin–Angiotensin–Aldosterone System

Late-adaptive mechanisms are provided by the kidneys, which control the volemia through Na^+ and water reabsorption under action of the renin–angiotensin–aldosterone system (RAAS; Fig. 2.11 and Table 2.22). Sympathetic stimulation via $\beta 1$-receptors, renal artery hypotension, and decreased Na^+ delivery to the distal tubules stimulate the release of renin by the kidney. Renin cleaves *angiotensinogen* (ATng) into *angiotensin*-1 (ATn1). Angiotensin converting enzyme (ACE) acts to produce angiotensin-2 (ATn2). ATn2 constricts the arterioles, thereby rising SVR and p_a. ATn2 acts on the adrenal cortex to release *aldosterone*, which increases Na^+ and water retention by the kidneys. ATn2 stimulates the release of *vasopressin* (or antidiuretic hormone, ADH) from the posterior pituitary, which also increase water retention by the kidneys. ATn2 favors NAd release from sympathetic nerve endings and inhibits NAd re-uptake by nerve endings, hence enhancing the sympathetic function.

The *angiotensin-converting enzyme* (ACE) regulates blood pressure. It cleaves small peptides, such as angiotensin-1 and bradykinin. ACE also shed various glycosylphosphatidylinositol-anchored proteins from the plasmalemma. This activity is enhanced by the membrane raft disruptor *filipin* [75].

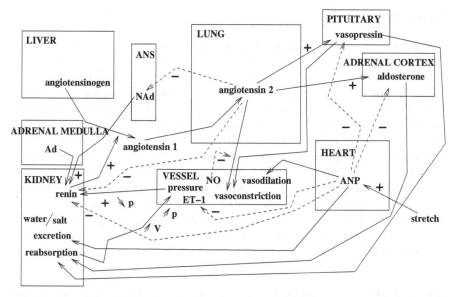

Figure 2.11. The atrial natriuretic peptide (ANP) and the renin–angiotensin system. Angiotensin-2 induces vasoconstriction associated with changes in renal hemodynamics, vasculature remodeling, and release of aldosterone by the adrenal cortex and vasopressin by the pituitary gland. Vasopressin increases water retention by the kidneys. Its effect on vasomotor tone at physiological concentrations is negligible. ANP generates vasoconstriction. In the kidneys, it inhibits renin release and decreases sodium reabsorption. In the adrenal cortex, it inhibits aldosterone release.

Kinase WNK4[45] regulates the sodium–chloride cotransporter of the distal convoluted tubule, hence switches the balance from electroneutral to electrogenic sodium reabsorption in the distal convoluted tubule (Fig. 2.12) [80]. Kinase WNK4 inhibits the sodium–chloride co-transporter and renal outer medullary potassium channel ROMK1. Its predominant effects target the sodium–chloride co-transporter. Kinase WNK1 might be an inhibitor of kinase WNK4.

Aldosterone is released by the adrenal gland either by volume loss, owing to angiotensin-2,[46] or hyperkalemia. Aldosterone favors sodium reabsorption via the sodium channel in the mineralocorticoid-sensitive segments of distal nephron. Kinase WNK4 can mediate the renal response to aldosterone.

---

[45] Serine/threonine kinases WNK (with no K) are characterized by the absence of lysine usually found in the catalytic domain of all other serine/threonine kinases. WNK1 and WNK4 are expressed in the distal convoluted tubule, connecting tubule, and collecting duct of the nephron.

[46] Angiotensin-2 also directly stimulates renal sodium reabsorption, independently of aldosterone.

2.6.6.3 Vasopressin

Neurohypophyseal vasopressin exerts its regulation via three receptors, vascular V1, renal V2, and pituitary V3.[47] These receptors interact with specific kinases PKC and GRK5. Vasopressin, via V1-receptor, stimulates steroid secretion in adrenal glands. Cardiovascular effects of vasopressin are low, even at high concentrations. Vasopressin regulates blood pressure, especially in pathophysiological conditions, such as severe hypovolemia. V1a-receptor-mediated vasoconstriction increases blood pressure (Fig. 2.13). Conversely, V2-receptor-mediated release of nitric oxide from the vascular endothelium decreases blood pressure via vasodilation. Therefore, the magnitude of blood pressure changes due to vasopressin results from a summation of vasoconstriction mediated by V1-receptors and vasorelaxation mediated by V2-receptors and NO. Moreover, vasopressin acts on the brain. Its indirect vasodilator effect is caused by inhibiting sympathetic efferents and enhancing the baroreflex [83].[48] V1-receptors maintain the blood pressure at physiological levels, not via direct

Table 2.22. Renin–angiotensin–aldosterone system and cardiac natriuretic peptides (Source: [76]).

	Angiotensin-2	ANP
Artery	Vasoconstriction	Vasodilation
Myocardium	I+, C+	
Adrenal medulla	Aldosterone release	Aldosterone release inhibition
Adrenal cortex	Cathecolamine release	
Kidney	Glomerulus filtration reduction Water and salts reabsorption Renin release	Glomerulus filtration increase Water and salt excretion Renin release inhibition
Pituitary (hypophysis)	ADH release ACTH release Prolactin release	ADH release inhibition
Sympathetic	NAd release	

---

[47] V1- and V3-receptors are also called V1a- and V1b-receptors.

[48] Reflex control of the cardiovascular system mainly involves baroreceptors, afferents to the central nervous system, cardiovascular centers, and sympathetic and parasympathetic efferents to the heart and vasculature. The potentialization of baroreflexes is done via central action, activating V1-receptors in the area postrema, and sensitization of the arterial baroreceptors, as well as cardiac afferents.

vasoconstriction, but by regulating the neural and hormonal actions of vasopressin [84].[49]

Vasopressin acts jointly with noradrenaline to produce vasoconstriction. ADH potentiates NAd action on vascular smooth muscle cells. Therefore, significant ADH effects can be observed. Besides, electrolytes in the blood and extracellular fluid modifies the surface polarity of smooth muscle cells and force. Blood concentrations of O_2 and CO_2 also affect the force developed by smooth muscle cells.

2.6.6.4 Natriuretic Peptides

The endocrine heart acts as a modulator of the activity of the sympathetic nervous system and the renin-angiotensin-aldosterone system in particular [85]

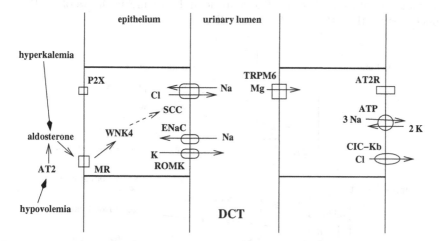

Figure 2.12. The distal convoluted tubule (DCT) with certain ion carriers and receptors, as well as aquaporin and kinase WNK4, a possible effector of aldosterone (Sources: [80, 81]). Ion carriers in epithelial cells of the distal nephron include inward rectifier renal outer medullary potassium channel (ROMK) and sodium–chloride co-transporter (SCC), epithelial sodium channel (ENaC). (Others such as Na^+–K^+ AT-Pase, Na^+–Ca^{++} exchanger, Na^+–H^+ exchanger, cation-chloride (Na^+–K^+–$2Cl^-$) co-transporter, amiloride-sensitive Na^+ channel, Ca^{++}-sensitive K^+ channel, pH-sensitive K^+ channel, epithelial Ca^{++} channel, plasma membrane Ca^{++}-ATPase, calbindin-D28k, Ca^{++}-dependent Cl^- channel, ATP-sensitive Cl^- channel, and H^+ ATPase, are not represented here.) Mg^{++} is transported transcellularly by TRPM6 (it is reabsorbed paracellularly in the thick ascending limb of Henle loop [82]). Plasmalemmal receptors include the mineralocorticoid receptor (MR), activated by aldosterone, angiotensin-2 receptor (AT2R), and nucleotide receptors P2X4, P2X5, and P2X6, among others.

---

[49] V1a receptors are strongly expressed in the nucleus of the solitary tract. V1a receptors are thereby involved in the regulation of the baroreflex control.

(Fig. 2.11). Natriuretic peptides[50] control the body fluid homeostasis and blood volume and pressure. Atrial (ANP) and brain natriuretic peptides (BNP) are synthesized by cardiomyocytes as preprohormones, which are processed to yield prohormones and ultimately hormones.[51] They are then released into the circulation at a basal rate. Augmented secretion follows hemodynamical or neuroendocrine stimuli. They relax vascular smooth muscle cells. They also regulate SMC proliferation. They decrease baroreflex activity. They have direct and indirect renal actions. ANP increases renal blood flow. It inhibits renin release by the kidneys, raises the glomerular filtration rate, and decreases the tubular sodium reabsorption. In the adrenal cortex, natriuretic peptides inhibit aldosterone synthesis and release (functional RAAS antagonist; Table 2.22). The endothelial production of C-type natriuretic peptide (CNP) is stimulated by TGFβ and TNFα [87]. Endothelial CNP can regulate the local vascular tone (relaxation) and growth via cGMP production by vascular smooth muscle cells. The diuretic and natriuretic effects of CNP are much weaker than those of ANP and BNP.

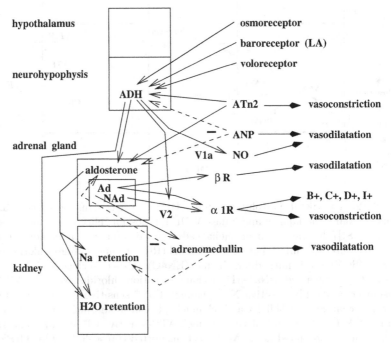

Figure 2.13. Interactions of hormones (angiotensin-2 [ATn2], atrial natriuretic peptide [ANP], vasopressin [ADH], aldosterone, and cathecolamines) and effects on blood circulation.

---

[50] In the literature, natriuretic peptides are also known as natriuretic factors.
[51] In cardiomyocytes, activated PKB increases the expression of the atrial natriuretic peptide using the phosphatidylinositol-3 kinase signaling pathway [86].

Images, Signals, and Measurements

Si l'activité scientifique expérimente, il faut raisonner ; si elle raisonne, il faut expérimenter.[If the scientific activity experiments, one must reason; if it reasons, one must experiment.] (Bachelard) [1]

3.1 Cardiovascular Imaging and Geometry Modeling

Computerized medical imaging provides subject-dependent 3D geometry of any body organ, in particular the heart and large blood vessels. Medical images taken of the human body are mainly displayed in three main planes: (1) coronal planes which divide the body into front and back regions; (2) sagittal planes which separate the body into left and right parts; and (3) transverse planes, perpendicular to the body axis, which split the body into upper and lower domains. Nowadays, numerical simulations are performed in computational domains based on imaging data after 3D reconstruction. Input data for the surface reconstruction of the target CVS compartment usually come either from X-ray computed tomography [88–90], magnetic resonance images [91–94], or 3D ultrasound images [95].

3.1.1 Imaging Techniques

3.1.1.1 Computed Tomography

Computed tomography (CT) uses special X-ray equipment to obtain cross-sectional pictures of regions inside the body with gray-level scaling.[1] To enhance vascular anatomy, an intravenous injection of a radio-opaque solution can be made prior to or during the scan (computed angiography [CA]).

In *spiral CT* (SCT, or helical CT), X-rays have helical path because data acquisition is combined with continuous motion of the acquisition system.[2] SCT image entire anatomical regions in a 20- to 30-s breath hold. The patient can hold his/her breath for the entire study, hence reducing motion artifacts.

---

[1] The Hounsfield unit scale in each pixel expresses the attenuation coefficient coded in 4096 gray levels. It is calibrated upon the attenuation coefficient for water and air, with water reading $0\,\mathrm{HU}$ and air $-1000\,\mathrm{HU}$.

[2] The scanner rotates continuously as the patient's couch glides.

Data can be reconstructed to get 3D organ displays and virtual endoscopy provided.

Multi-slice spiral CT (MSSCT) scanners can acquire many slices in a single rotation. Sixteen-slice MSSCT has high spatial resolution and reduces examination time. Pre-contrast scanning (low-dose acquisition) is advisable when bleeding is suspected; it can evaluate wall calcifications and intramural hematomas and dissections. Contrast scanning is triggered when the bolus reaches the region of interest (e.g., ROI intensity measurement > 180 HU), the injection duration matching the acquisition time. Special protocols can: (1) evaluate the vessel caliber; (2) exhibit mural thrombus on aneurism wall and demonstrate arterial supply of aneurism as well as initial flap, false and true channels of arterial dissection; and (3) help to determine the therapy choice between surgery and endovascular procedures (Chap. 7).

In *electron beam CT* (EBCT), an electron beam is generated and focused on a circular array of tungsten X-ray anodes. Emited X-rays, which can be triggered by ECG, are collimated and detected as in conventional CT. EBCT, which has no mechanical motion of the X-ray source and a stationary detector array, allows very quick scanning in 50 to 100 ms. The high temporal resolution is associated with a limited spatial resolution. The scan speed allows to get a complete image set of the heart during a single breath-hold, or any moving and deformable ducts [96]. EBCT is useful to evaluate right and left ventricular muscle mass, chamber volumes, and systolic and diastolic ejection fractions. EBCT can also measure calcium deposits in the coronary arteries. Features of these main CTs are summarized in Tables 3.1, 3.2, and 3.3.

3.1.1.2 Magnetic Resonance Imaging

> "Bref l'art poétique de la physique se fait avec des nombres, avec des groupes, avec des spins, etc."
> (Bachelard) [97] *[Brief, the poetic art of physics is done with numbers, sets, and spins, etc.]*

Table 3.1. Main features (size [mm] and time) of computed tomography.

	SCT	MSSCT	EBCT
Slice thickness	10	$1.5 \rightarrow 0.6$	8
Between-slice thickness		30–50% Overlapping	Pair gap of 4, contiguous dual slicing
Scanning acquisition time (s)	≥ 160	20	8–12
Slice acquisition time (ms)			100

Nuclear magnetic resonance imaging (MRI) uses magnetic fields and radiofrequency waves to stimulate hydrogen nuclei in selected regions of the body that emit radio waves [98]. The emerging signals have frequency, amplitude, and phase components that are processed to construct images of the human body. Magnetic field gradients are introduced to determine the spatial location of re-emitted microwaves. Frequency and timing characteristics of the excitation pulse are modified to image particular types of molecules and motions. Each slice usually represents a thickness of 2 to 10 mm. Each pixel represents from 1 to 5 mm. *MR angiography* (MRA) focuses on blood vessels without using any contrast material, although contrast agents may be given to provide better MR images [99].

Functional MR (FMR) can display the myocardium function [100]. High-performance magnetic resonance imaging now allows high resolution and quick image acquisition, which can provide particularly good delineation of organs and thin-walled vessels. *Phase-locking MRI* can then be used to record the vasculature deformation by imaging the selected region of the cardiovascular system at different phases of the cardiac cycle. Imaging of soft tissue dynamics uses *MR tagging* which allows for encoding of a grid of signal voids on cardiac MR images produced with various techniques [101–105].

Diffusion tensor magnetic resonance imaging (DTMRI) is particularly used to estimate the myofiber orientation in the heart wall. Effective diffusion tensor ($\mathbf{D}_{\text{eff}}$) MRI non-invasively evaluates the structure of biological tissues by measuring water diffusion[3] using magnetic field gradient and

Table 3.2. Spatial resolution of X-ray sanners.

CT	$\Delta x, \Delta y$ (mm)	Δz (mm)	Matrix	Slice number
MSSCT function mode	0.5	0.6	512×512	4
MSSCT volume mode	0.5	0.6	512×512	$\mathcal{O}(100)$
SS-EBCT	0.7	1.5–3	512×512	~ 100
MS-EBCT	1	8	360×360	2–8

---

[3] Diffusion of water within a tissue excited by a magnetic field gradient causes MRI signal attenuation. The eigenvectors and eigenvalues of the voxel-averaged diffusion tensor specify the principal directions and rates of water diffusion in each voxel of the tissue image. The eigenvector corresponding to the maximum eigenvalue of the diffusion tensor points in the direction of maximum rate of diffusion assumed to be the direction of the axis of a cylindrical fiber. The orientation of the eigenvectors can be defined by inclination and transverse angles. The inclination angle of the myofiber is the angle between: (1) the intersection line of the image plane and the plane parallel to the epicardial tangent plane at the corresponding azimuthal position (tangent plane); and (2) the projection of the eigenvector onto the tangent plane. The transverse angle is the angle between: (1) the intersection line, and (2) the projection of the eigenvector onto

Table 3.3. Temporal resolution of X-ray sanners.

CT	slice scan time (ms)	temporal image number
MSSCT function mode	300	60
MSSCT volume mode	300	4
SS-EBCT	50	1
MS-EBCT	30	6–40

diffusion- and non-diffusion weighted images [106]. The water Brownian motion in a medium characterized by ordered rod-like elements has a preferential path. Consequently, the probable molecule location is in an ellipsoid displacement domain rather than a sphere profile when the medium is isotropic. The effective diffusion tensor is computed from the measured apparent diffusion tensor once the eigenvalues[4] have been determined using six different directions.

The regional cardiac deformation can be estimated from 3D images to derive the wall displacement field, and in association with a biomechanical model, compute the strain field. The viability of the myocardium when any coronary artery becomes occluded can then be assessed. The wall strain field provides a better analysis than the endocardial motion and allows updating of the remodeling [111]. Using MRI, a shape-tracking approach has been proposed [112]. After segmentation of the left ventricle inner and outer walls in each slice of the initialization cycle phase, contours are propagated in the corresponding slice of the different acquisition instants, and after checking, assembled into endo- and epicardial surfaces. Each small patch of the original surface is mapped to a plausible window of the ventricle deformed surface at a given time, and the patch of the deformed surface having the most simi-

---

the image plane. The correlation has been checked by experiments performed in an excised portion of the right ventricle by comparison of DTMRI and histology myofiber angles [107, 108], but DTMRI time and space resolutions, especially in vivo, were too large to get an accurate map of myofiber angles. More recently, space resolution of 310 to 390 μm in the slice plane with a slice thickness of 0.8 to 1 mm has been obtained in isolated dog hearts [109]. These authors extract two local angles, the myofiber main axis angle and the cross-sheet (from endocardium to epicardium) angle. The myofiber angle is defined by the angle between the local circumferential tangent vector of the reconstructed mesh of the heart wall and projection of the primary eigenvector of the voxel-related water diffusion tensor onto the epicardial tangent plane. The cross sheet angle is determined by the radial vector and the projection of the tertiary eigenvector of the diffusion tensor, which is parallel to the cardiac sheet normal, onto the plane defined by the radial and circumferential vectors. Images can be obtained using a slice-selection fast spin-echo diffusion-weighted technique coupled to gradient recalled acquisition in the steady-state (GRASS) imaging mode to define epicardial and endocardial surfaces [110].

[4] The diffusivities along the three principal axes of the ellipsoid.

lar shape is selected. This method is questionable. The motion of the heart wall can be tracked by phase-constrast cineMRI, like blood movement. Heart cyclic motion can be also measured at grid points from tagged MRI [113]. MR tagging modifies the magnetization of selected targets (tags) within the heart wall. The wall motion between the tagging and image acquisition is captured by tag displacement. Spatial modulation of magnetization is used to create tagged plane.

With a relatively large acquisition window, MRI is not perfectly suited to image the thin mobile heart valves located in rapidly moving regions during the cardiac cycle.[5] Valve position tracking and one-dimensional motion-compensated transvalvar velocimetry have been developed using Comb[6] tagging [114]. Coronary arteries undergo complex displacement due to both respiratory and cardiac motion. Variability in heartbeat frequency is not only associated with chaotic heart behavior, but also with the subject's psychological response to unusual environment. The variability in respiratory and cardiac cycle durations within and between subjects hinders movement prediction. However, real-time low-resolution tracking of coronary grooves and surrounding fats has been proposed in specific orientations, assessing minimal motion acquisition windows and vessel locations for high-resolution imaging (so-called image-based navigators). Proposed Cartesian frames are defined by the long axis of the left ventricle and the atrioventricular groove.

Another new technique is magnetic particle imaging (MPI) [115]. Using magnetic resonance imaging, contrast agents that incorporate strongly magnetic particles can be introduced into the body to highlight specific anatomical structures (blood vessels) or serve as markers for nanoscale processes. An external time-constant magnetic field (selection field) is applied with a space-dependent strength. It vanishes in the center of the field of view, the field-free point (FFP), and increases in magnitude toward the edges, where the magnetization reaches saturation. When the particles in a magnetic field are further excited by an additional oscillating radiofrequency field (modulation field), saturated magnetic particles remain in the same state, whereas unsaturated ones, in the FFP area, reply to the modulation field with oscillating magnetization. The latter thereby induce a signal in the detector, which can be assigned to the poorly magnetized regions. The FFP position can be changed through the sample, for example, by moving the object within the coil assembly, to generate a tomographic image. The resulting map can give the spatial distribution of the magnetic particles.

---

[5] The displacement of the aortic valve has been estimated to be equal to 15–20 mm.

[6] Comb excitation enables simultaneous tagging in multiple parallel planes during breathhold.

3.1.1.3 Ultrasound Imaging

Ultrasound imaging (USI) involves US propagation through biological tissues where US are partially reflected at each acoustical interface on its path. The echoing waves are then interpreted to create anatomical images. The quality of echographic images depends on the: (1) axial (in the US propagation direction) and transverse resolution, (2) ultrasonic attenuation, and (3) echo dynamics. Because the resulting images are associated with the interaction between US waves and tissues, the collected information can reveal the mechanical properties of the tissues through which US travel (compressibility and density). The US attenuation, by diffusion and absorption, can also provide additional data on tissue heterogeneity level at the wavelength scale and between-cell cohesion, respectively. The ultrasound transducer functions as both a stereo loudspeaker, to generate streams of high-frequency sound waves, and a microphone, to receive the echoing waves back from the internal structures and contours of the organs. Because of the freely maneuverable probe, spatial sampling of the produced data is both inhomogeneous and unpredictable. Mechanical arms can be fixed to the probe and its location and orientation can be measured. Landmarks can be attached to the probe and cameras can track it. However, once the position and orientation of the probe are known, the data are still noisy.

Three-dimensional transthoracic and transesophageal echocardiographies are used for cardiac valve explorations. *Intravascular ultrasound* (IVUS) imaging is a catheter-based technique that provides real-time high resolution images of both the lumen and arterial wall of a vascular segment. Axial and transverse resolutions are 80 to 100 µm and 200 to 250 µm, respectively. Thirty images per second can be obtained. IVUS is used to detect atherosclerotic plaques (Chap. 7), which give a much better estimate of stenosis degree than angiography. The angiographic evaluation indeed depends on the imaging incidence angle with respect to the stenosed artery. IVUS is also used in interventional cardiology to control stent placement and assess stent restenosis.

Tissue-Doppler US is applied for tissue motion estimation using appropriate signal processing. The inherent limitation of the measurements is an inability to estimate more than one velocity component. However, combined with conventional US echography the tissue-Doppler US technique may estimate cardiac wall motion. Intravascular ultrasound palpography assesses the mechanical properties of the vessel wall using the deformation caused by intraluminal pressure. Regions of higher strains are found in fatty than in fibrous plaques [116]. *Contrast echocardiography* can be useful for the check ups of acute myocardial infarction in patients suffering from chest pain without obvious ECG signs. This echocardiography technique requires the injection of a contrast agent bolus.

Ultrasound ECG-triggered *elastography* of the beating heart provides real-time strain data[7] at selected phases of the cardiac cycle. Periodic myocardial thickening associated with normal heart function as well as tissue ischemia or infarction can be detected [117].

Ultrasound guidance in interventional medicine is the cheapest and easiest procedure. However, two main drawbacks, border and mirror-image shadows, limit its use.

3.1.1.4 Nuclear Medicine Imaging

In nuclear medicine imaging (NMI), radioactive tracers, which have a short life time, attached to selected substances, are administred into the patient. Tagging molecules seek specific sites. The distribution within the body of the radioactive isotope provides information on irrigation and the chemicophysical functions of the explored organ. The patient is placed in a detector array and the radiation emitted from the body is measured. NMI produces images with low resolution and a high amount of noise, due to necessary low-radiation doses.

The two most common types of NMI are *single photon emission computed tomography* (SPECT) and *positron emission tomography* (PET). SPECT uses photon-emitting radiotracers, whereas PET utilizes radiotracers that produce positron-electron pairs. Cardiac PET combines tomographic imaging with radionuclide tracers of blood flow metabolism and tracer kinetics for quantifying regional myocardial blood flow, substrate fluxes, and biochemical reaction rates. PET assesses the regional blood volume and flow on the one hand, and local O_2 extraction rate and consumption on the other hand. Cardiac SPECT studies myocardial perfusion with agents such as thallium-201 and technetium tracers, at rest and during testing. Data fusion with CT or MRI images allows us to couple physiologic activity to the underlying anatomy.

3.1.2 3D Reconstruction

Imaging devices provide non-invasively accurate and very large datasets of discrete information on explored organs. However, output data are usually not suitable for archiving and data processing, as well as for representing 3D geometry; polygonal models are largely preferred. This requirement for piecewise approximations of the domain boundaries is reinforced by numerical applications. Most of the current reconstruction algorithms convert the initial sampled data into surface triangulations having the same degree of complexity as the original data (i.e., a number of triangles on the order 10^5 to 10^6). Hence,

---

[7] Displacement of the tissue between two images can be used to assess the bulk rheology of a region of interest of the explored tissue. Elastographic scanning map strain magnitude (image brightness) and sign (color hue associated with compression or distention, for instance).

to be easily manageable, the complexity of such polygonal models needs to be simplified drastically. Surface simplification algorithms are aimed at finding a compromise between the minimal number of triangles and the preservation of the geometric accuracy of the surface model. In addition, specific requirements can be imposed on resulting meshes, for instance, element shape and size for numerical simulations. Moreover, smoothing is required to limit computational flow errors [118].

The common technique to create a mesh from imaging data is segmenting[8] and faceting. This two-step method consists first in segmenting the selected organ in the images, and then using the segmentation surface to create the facetization. The automatic mesh generator must be able to cope with such surface, which is frequently full of gaps, overlaps, and other imperfections. Various algorithms have been proposed to reconstruct a polygonal model (a piecewise linear approximation) depending on the nature of the sampled data (series of slices, range images, point clouds, etc.). Three classes can be defined. *Slice connection algorithms* work for a series of planar parallel sections of the target vessel. At first, a closed contour is extracted in each slice, then contours are connected to each other between each pair of adjacent slices [119, 120]. *Marching-cube approaches* attempt to extract an implicit surface from a 3D range image based on a "voxelhood" analysis. *Delaunay tetrahedralization algorithms*, which first generate 3D triangulation over a point cloud and then extract a bounded surface triangulation from this set of tetrahedra using suitable topological and geometrical criteria [121].

3.1.2.1 Level Set Methods

Level set methods are numerical techniques designed to track the evolution of fronts [122]. Level set methods exploit a strong link between moving fronts and equations from computational fluid equations [123]. This technique, based on high-order upwind formulations, is stable and accurate, and preserves monotonicity. Furthermore, it handles problems in which separating fronts develop, the existence of sharp corners and cusps, and topology changes. Level set methods are designed for problems in which the front can move forward in some places and backward in others. The solution starts at an initial position and evolves in time (initial value formulation). The level set method tracks the motion of the front by embedding the front as the zero level set of the signed distance function. The motion of the front is matched with the zero level set of the level set function, and the resulting initial value partial differential equation for the evolution of the level set function resembles a Hamilton-Jacobi equation. In this setting, curvatures and normals may be easily evaluated; topological changes occur in a natural manner. This equation is solved using entropy-satisfying schemes borrowed from the numerical solution of hyperbolic conservation laws. The interface between a vessel and the surrounding tissues can be detected, the border being defined by intensity gradient.

---

[8] Vessel bore and wall smoothness depend on the threshold.

3.1.2.2 Marching Cubes

The marching cube algorithm is used in volume rendering to reconstruct an isosurface from a 3D field of values [124]. The basic principle behind the marching cube algorithm is to subdivide the space into a series of cubes. In the framework of medical image processing, the matrix of cubes or cells is defined by the set of voxel barycenters. The imaged region is represented as a field of values through which the surface to be determined is defined by a threshold, which is provided by a previous step of the image processing. The first step is to calculate the corner values. The mean intensity values of the voxels are assigned to the corresponding barycenters. The algorithm then instructs to "march" through each of the cubes, testing the corner points and replacing the cube with an appropriate set of polygons, most often triangles. This step is done by inserting vertices at the cell edges using linear interpolation; each vertex is positioned according to the ratio between the selected threshold and the values of the neighboring corners. The result formed by joining the vertices with facets is a piecewise surface that approximates the isosurface. The isosurface can be defined by the set of intersection points between the voxel mesh and a 3D implicit function, the value of which is given from image thresholding. This operation is easy in case of high-quality acquired images.

3.1.2.3 Slice Connections

In some circumstances, the surface is reconstructed from a point set, which defines planar parallel contours of vessels. A triangulated surface is drawn between each pair of consecutive sections, ensuring that each point in a contour is connected to its closest point in the next contour. To fit up a surface on a set of contours amounts to constructing a volume enclosed by these contours. The global volume is considered as the union of independent pieces resulting from pair treatment. Consider a set of input points that define the vessel contours obtained by the level set method. After a cubic-spline fit of each contour associated with smoothing, a new node set can be defined by equally spaced points along the vessel contour. Then two successive slices are projected orthogonally to the local axis in a same plane and a 2D constrained Delaunay triangulation is built. The surface triangles are finally extracted by elevation of the two planes. The projection direction based on the set of intrinsic axes of the vessels gives a better slice-pair treatment than using a projection direction normal to the slice planes, as is done for general purpose. Furthermore, entry and exit sections are rotated to be normal to the local axis.

3.1.2.4 Deformable Models

Contour shape can be a priori known, as closed contour with a regular surface of a given region of the image. It can then be considered as an elastic contour in equilibrium under a set of forces. Among the strategies used to create a computational mesh from imaging data, there is a direct generation procedure, which starts from an average template mesh for the organ of interest and performs an elastic deformation of the mesh onto the image set [125].

3.1.2.5 Finite Octree Method

The vascular segment of interest is placed in a cube that is subdivided into octants of length scale determined by the mesh size to suit the possible caliber changes of the explored vascular segment [126]. The octants containing the vessel walls are trimmed to match the wall surface with a given tolerance level. Smoothing can then be performed. The octants within the vessel are subdivided into tetrahedra.

3.1.2.6 Implicit Surfaces

Smooth implicit functions for 3DR have been associated with spectral/hp high order elements for blood flow computations [127]. The vessel edges are first detected, segmented, and smoothed using B-spline interpolation in each image of the slice set. The resulting contours are fitted by an implicit function defined as a linear combination of radial basis functions associated with the contour nodes (~30 points) and a set of interior constraint points along the normal direction to the spline at the corresponding nodes. An isosurface extraction leads to vessel surface triangulation using an implicit surface polygonizer. The vessel surface is then smoothed using a boundary representation of its edges and surfaces by spline curves and surfaces, preserving the main curvatures. Bicubic spline patches, which are interactively defined and projected onto the implicit surface, serve as the initial element for meshing [128]. Mesh size and shape optimization is determined by the eigenvalues of the Hessian matrix of the implicit function.

3.1.2.7 Snakes

Snakes, or active closed contours defined within an image domain, can be used in image processing, particularly for image segmentation [129]. Snakes move under the influence of internal forces coming from the curve and external forces computed from the image data. The functional to be minimized is analog to the deformation energy of an elastic material subjected to a loading. The properties of the deformable snakes are specified by a potential associated with a contraction–expansion term by analogy to a mechanical thin heterogeneous membrane. The internal and external forces are defined so that the snake will conform to an object boundary (image intensity gradient) within an image. The initial contour can be a small circle centered on a selected point. The iterative deformation of the initial curve in the force field can be done by convoluing gradient images to Gaussian-type functions and modeling the deformable curve by splines. The coefficient number of these splines rises during the iterations to progressively decrease the curve energy. An additional force has been proposed to deform the snakes avoiding to track spurious isolated edge points [130]. This method has been successfully applied to heart ventricule extraction.[9] External force, like gradient vector flow (GVF), can be computed as a diffusion of the gradient vectors of a gray-level or binary edge map derived from the image [131]. Superposition of a simplicial grid over the image domain and using this grid to iteratively reparameterize the deforming snakes model, the model is able to flow into complex shapes, even shapes with significant protrusions or branches, and dynamically change topology [132].

3.1.2.8 Axis-Based Method

Bioconduit modeling can be based on vessel-axis determination. Once the set of axes is known, a "response function" is computed for each vessel slice from a vector rotating around the axis point and the intensity gradient. The voxel positions for which the response function is maximum give the vessel contour in the investigated slices when the axis is correctly determined. Precisely approximated contours a posteriori confirm that the vessel axis was reasonably well estimated. Several techniques may be used for axis determination. The vessel contours can be detected by derivative operators.[10] The eigenvalues of the Hessian matrix $\nabla^2 i$ of the intensity i can be computed; the eigenvalue that is the nearest to zero estimates the axis location [133].

Reconstruction of complex vessel tree in any organ can use a marking procedure to detect all the connected components of the lumen of the vessel network and set up a mark for each of them serving as starting subset for high-order reconstruction. A suitable morphological filter is then needed. A morphological filter based on selective marking and depth constrained connection cost, which labels the vessel by binarization of the difference between original image and connection cost image, can be proposed [134]. An energy-based agregation model is applied to the marking set for tree 3DR with respect to voxel values. The marking set progressively grows by state change of boundary voxels, according to local minimization of an energy functional (Markov process-like method). The energy functional is composed of three propagation potentials $\mathcal{E} = \mathcal{U}_r + \mathcal{U}_l + \mathcal{U}_c$ ($\mathcal{U}_r$: potential associated with pipe topography, $\mathcal{U}_l$: potential based on similar density for all ducts, $\mathcal{U}_c$: limiting growth potential for bounded growth within lumen limits) [134]. The state is determined with respect to the states of 26-connected adjoining voxels $y \in \mathcal{V}_{26}(x)$, and their gray levels $F(y)$. The smoothing is adaptive according to vessel caliber

---

[9] The snakes is dilated by external and internal forces using a finite element method to solve the minimization problem, only taking into account the suitable edge points that have been extracted by an edge detector.

[10] The contour is then defined as the location of the maxima of the gradient of the image intensity in the gradient direction.

based on adaptation of Gaussian kernel to size of labeled vessels. Vessel axis computation is based on geodesic front propagation (GFP) with respect to a source point, the axis being defined by the set of centroids of successive fronts. The determination of the axis tree uses GFP combined to 3D distance map associated with vessel wall geometry, which allows space partitioning. This method provides robust branching point detection with axis hierarchy preservation [135].

3.1.2.9 Limitations

Despite being efficient and robust, 3D reconstruction generally suffers several drawbacks. Discrete data may be very noisy, i.e., points that are off the surface. The accuracy of scanning and sensing devices leads to unnecessary dense datasets, with a density that is not related to the local geometric complexity, and consequently to very large polygonal models. Reconstruction algorithms often introduce artefacts in the polygonal approximation, such as "staircases" effects. The element shape quality do not always fit the requirement in numerical simulations. To overcome these problems, a two-step integrated approach consists of first generating a simplified geometric surface mesh, possibly preceeded by a surface smoothing stage, and then constructing a computational surface mesh by taking into account shape and size requirements for the mesh elements.

3.1.3 Meshing

Two main issues in mesh studies applied to numerical analysis are the surface mesh quality, which must not significantly affect the problem solution, and the algorithms associated with volumic meshes which are automatically generated [136–139]. The second work class focuses on mesh adaptation and adaptativity, including mesh refinement/coarsening, edge refinement and swapping, and node displacement, based either on metrics and error estimations independently of equation types [140, 141] or minimization of the hierarchical estimator norm [142–145].

The objective is to construct surface meshes that match strong requirements related to the accuracy of the surface approximation (geometry) and boundary conditioning as well as the element shape and size quality (computation). Patches such as B-splines and NURBS, which fit the facetization can be meshed. However, facetization can be directly processed. Determined corners and ridges define patches that are triangulated by an advancing front technique [146]. In any case, the first stage consists of simplifying the initial dense surface mesh to produce a geometric surface mesh. First, the initial reference mesh must be simplified to remove redundant elements while preserving the accuracy of the geometric approximation of the underlying surface. A simplification procedure based on the *Hausdorff distance* can thus be used. This algorithm involves vertex deletion, edge flipping, and node smoothing local mesh modifications. Often, especially for surface triangulations supplied by marching-cube algorithms, a smoothing stage based on a bi-Laplacian operator is required to remove the "staircase" artifacts [147]. This stage yields to a geometric surface mesh that is a good approximation of the surface geometry and contains far fewer nodes than the initial reference mesh [148].

Surface discretization obtained from the 3D reconstruction needs further treatment to be suitable for numerical simulations. Boundary conditions must be set sufficiently far from the exploration volume, otherwise they affect the flow within the fluid domain. Moreover, any geometry change along a vessel (bends, branching segment, lumen narrowing or enlargment, wall cavity, taper, etc.) induces flow disturbances over a given length both upstream and downstream from the causal segment. Consequently, short straight ducts in the direction of the local axis can be connected to every vessel end. Furthermore, vessel-end sections must be cross-sections because of stress-free boundary conditions usually applied at outlets. The blood vessels are continuously curved; curvature induces transverse pressure gradient in any bend cross-section as well as in upstream and downstream cross-sections of possible straight pipe over a given length, which depends on the values of the flow governing parameters [149]. Besides, axial pressure gradient is exhibited in vessel sections that are non-perpendicular to the vessel axis.[11]

Numerical simulations are the final objective, so element shapes and sizes must be controlled as they usually impact the accuracy of the numerical results.[12] Therefore, an *anisotropic geometric metric map* based on the local principal directions and radii of curvatures is contructed in tangent planes related to mesh vertices. This metric map prescribes element sizes proportional to the local curvature of the surface [150]. The metric map can also be combined with a computational metric map, e.g., supplied by an a posteriori error estimate, and eventually modified to account for a desired mesh gradation. Then, a surface mesh generation algorithm is governed by the metric map and based on local topological and geometrical mesh modifications. This approach can be easily extended to *mesh adaptation* as it already involves mixing geometric and computational metrics to govern the mesh generation stage.

Dynamic meshing is aimed at meshing an organ during its displacement, like the cardiac pump, without computing the whole mesh at each iteration. Three main steps are required: (1) node displacement, (2) mesh coarsening with removal of collapsed element, determined from a triangle degradation criterion, and (3) mesh enrichment after a mesh smoothing, especially in high-error regions.

---

[11] In straight pipes, the axial pressure difference, which varies either non-linearly in the entry length or linearly when the flow is fully developed, is exhibited in any duct section that is not normal to the centerline.

[12] The surface element size depends on the local surface curvature. The stronger the curvature, the smaller the size.

Heat and wave propagation can be efficiently computed using mesh adaptivity and anisotropic mesh adaption to the propagation front, avoiding inaccurate predictions due to numerical diffusion, especially for free-surface problems [151, 152]. Accurate predictions of the interface require a refined mesh in the vicinity of the interface (Fig. 3.1).

3.2 Hemodynamics Signals

3.2.1 Volume and Pressure

In back decubitus, the blood vessels are assumed to be approximately located at the same height, which is supposed to be given by that of the right atrium. The reference point of the intravascular pressure is located at the level of the tricuspid valve [25]. The mean pressure at this point does not significantly vary with body position (variations lower than 0.15 kPa). In the lying position, the gravity can be neglected, whereas in the upright position the hydrostatic component varies linearly with height H ($-\rho g H$, $H < 0$ and $H > 0$ under and over the reference point, respectively). Hydrostatic pressure then rises or decays with the vertical distance from the reference point, whether the vascular site is located under or over the reference point. The generating pressure takes blood pump action into account.

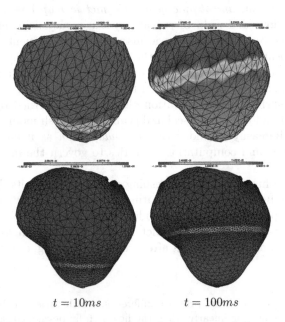

$t = 10ms$ $t = 100ms$

Figure 3.1. Mesh adaptivity for numerical simulation of the propagation of the action potential in the heart wall, based on Fitzhugh-Nagumo equations (from Y. Belhamadia).

Time variations of pressure and of volume of heart cavities are estimated from catheter-based measurements.[13] Pressure–volume loops throughout the complete cardiac cycle are then plotted. Alternatively, chamber volume changes can be assessed using echocardiography, radionuclide imaging, or tagged MRI.

Blood volume in the heart, and pulmonary and systemic circulation is nearly equal to 0.05, 0.25, and 0.70 with respect to total volume, respectively. Blood volume with respect to total volume in the arterial, capillary, and venous compartments is equal to about 0.30, 0.05, and 0.65 in the pulmonary circulation and about 0.17, 0.09, and 0.74 in the systemic circulation [153]. An important between-author variability in the distribution of blood volume between the main compartments of circulation is found in the literature (Table 3.4). Estimates of the circulation times between two compartments of the circulation also varies among the literature data (Table 3.5).

The arterial pressure is routinely measured using sphygmomanometer. Korotkoff sounds indicate systolic blood pressure; they might be generated by instabilities of coupling between blood and collapsed artery [154] and disappear at diastolic pressure. *Impedance plethysmography* is another non-invasive technique to measure the blood pressure via vessel volume changes, and

Table 3.4. Estimated blood volume distribution (%) in the circulation compartments (between-author variability).

Compartment	[26]	[27]
Heart	18	7
Pulmonary circulation	12	9
Systemic circulation	70	84
Arteries	15	15
Capillaries	5	5
Veins	50	64

Table 3.5. Circulation time (s) estimated from indicator bolus measurements (Source: [153]).

Thoracic aorta to femoral artery	2–4
Cubital vein to carotid sinus	12–33
Right heart to aorta	6–9

---

[13] Quickly varying pressures must be measured by transducers, which convert pressures into electrical signals, able to accurately sense high frequencies. Sensitivity, linear output for the whole pressure range, suitable frequency response, and spatial resolution are the main features of transducers. In particular, the transducer size must be smaller than the distance over which exist spatial pressure variations.

using several measuring sites, to estimate the wave speed [155]. Artery bore changes are small and produce low impedance variation. The recorded signal is then very sensitive to noise and must be adequatly filtered. Moreover, the wave speed depends on the chosen reference point, electrode distribution, and artery compartment (proximal/distal). It is also affected by the subject's age and position. In normal subjects, the wave speed is higher in a standing than supine position. Other pressure measurement techniques include *tonometry* [156] and arterial *photoplethysmography*, which is based on the optical determination of blood volume changes and pulsations in superficial arteries (in fingers) [157]. Photoplethysmography uses the reflection or transmission of infrared light. Artery volume variations modulate light intensity recorded by the photodetector. To be valid, the method requires several measurements made over long periods. The reliability of the Finapress technique has been studied [158–160]. Vessel caliber can be simultaneously measured by US echography. Blood pressure and wall radial velocity simultaneous measurements can evaluate the elastic modulus of the vessel wall.

The *Korteweg-de Vries equations* (KdV equation: $g_t + \kappa_1 g_x + \kappa_2 g_{xxx} = 0$) have traveling solutions,[14] the solitons,[15] the shape and speed of which are preserved after interactions. Two- and three-solitons[16] are peculiar solutions of the KdV equation. Their analytical expression can be found in [162]. Soliton-based signal processing, coupled to a windkessel model, can be used to compute the blood pressure wave in the large proximal arteries, indeed even at the left ventricle outlet, the pressure being distally measured (at the finger), using 2- and 3-solitons [163].

Heart pressures are invasively measured using catheter-mounted microtransducers with suitable frequency response. Catheters are introduced in peripheral veins or arteries for pressure measurements in right or left heart, respectively. Pressure time variations and gradients (dp/dt) have been assessed

---

[14] The non-linear soliton equation was developed by D.J. Korteweg and G. de Vries at the end of the nineteenth century. The non-linear term balances the dispersion. The existence and uniqueness of solution to the Cauchy problem for the non-linear Korteweg-de Vries equation and local controllability around the origin given by the non-linear term has been proved [161]. Signal processing using progressive wave speed analysis is more appropriate than employing frequency analysis suitable for stationary waves.

[15] The soliton is a wave that propagates without dispersion. Solitons interact without losing their identity, keeping shape and amplitude. An n-soliton solution refers to n components of different amplitude that interact. Propagation speed is proportional to wave amplitude. The higher the amplitude, the faster the propagation. Soliton solutions are used to model fast dynamics of pressure wave propagation and an associated windkessel model to take slow dynamics into account.

[16] The second wave of the 2-soliton model is associated with the dicrotic wave due to the aortic valve closure. The 3-soliton model is used to fit bifid pressure waves. A bifid curve exhibit an incisure in the ascending systolic part near the peak value rather than an usual monotonic soaring aspect.

using MR measurements of velocity and acceleration within the ascending aorta [164].

3.2.2 Flow Rate and Velocity

Flow rates have been previously estimated according to the Fick principle or dilution methods, using injected tracers. Calibration and positioning of catheter flowmeters are difficult. Remote flowmetry is now used. Vascular ultrasound imaging (US angiography) is used to monitor the blood flow and evaluate possible flow blockages. The quality of velocity measurements, whatever the technique, depends on the resolution: (1) spatial resolution (size of the sample volume), (2) temporal resolution (quickly varying flow), and (3) amplitude resolution (signal-to-noise ratio).

3.2.2.1 Doppler Ultrasound Velocimetry

The Doppler ultrasound (DUS) technique detects the Doppler effect[17] on circulating blood cells. Several depictions of blood flow are used in medical Doppler imaging: (1) color Doppler (CDUS), for a global description of blood flow, for an estimation of the mean velocity in the investigated vessel region and display of bulk motion using graphical color map; (2) pulsed Doppler (PDUS), for detailed analysis at a selected site (velocity distribution within the measurement volume; Fig. 3.2); (3) and power Doppler, which determines the amplitude of Doppler signals rather than frequency shift. Duplex scanning pulsed US velocimetry is used for real-time imaging guidance. Multichannel pulsed US Doppler velocimetry with phase look loop frequency tracking of the Doppler signal gives good agreement with LDV [165]. Phase-shift averaging reduces physiological variability. Phase errors are minimized by a cross-correlation between the velocity and its ensemble average $\langle v(t) \rangle = (1/N) \sum_{k=0}^{N-1} v_k(t \pm \Delta t_w/2 + kt)$ (Δt_w: time window, Fig. 3.2).

Duplex or color flow Doppler ultrasound is used for blood velocimetry. Every ultrasound transmission from the probe is directed forward and contains

---

[17] The Doppler effect is a change in the frequency of a wave, resulting in the case of a reflected wave, from the motion of the reflector. The Doppler shift frequency f_D (the difference between transmitted and received frequencies) depends upon the transmitted frequency f, blood velocity v, and Doppler angle θ, which is the angle of incidence between the US beam and the estimated flow direction (the local vessel axis): $f_D = 2fv\cos\theta/c$ (Doppler equation), where c is the speed of sound. Local blood velocity is calculated using the Doppler equation: $v = f_D c/(2f\cos\theta)$. The angle θ is evaluated by the sonographer by aligning an indicator on the duplex image along the longitudinal axis of the vessel. If the US beam is perpendicular to the blood-stream direction, there is no Doppler shift. The angle also should be less than 60 degrees, since the cosine function has a steeper curve above this angle and errors in angle correction are therefore magnified.

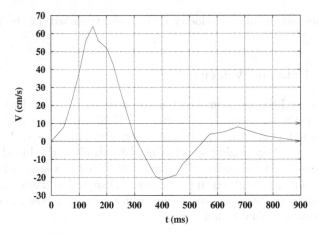

Figure 3.2. Ensemble-averaged (25 cycles) cross-sectional averaged velocity $\langle V_q \rangle$ in the femoral artery of a healthy volunteer measured by US Doppler technique. The ensemble averaging assumes a constant cycle period (here 900 ms, $f_c \sim 1.1$ Hz), neglecting chaotic heart behavior. In pulsatile flows, the boundary layer moves forward and backward, especially in certain proximal arteries, such as the femoral artery. Positive velocities during the systolic ejection are followed by negative velocities during the diastole before the final recovery stage with a very small velocity amplitude. The time-mean cross-sectional averaged velocity is then small ($\overline{V}_q \sim 10$ cm/s).

several pulses to measure Doppler shift. *Intravascular Doppler US* (IDUS) is used to assess flow pertubations induced by arterial lesions, to provide continuous monitoring of flow velocity throughout angioplasties, etc. However, IVUS probes operate in a high-resolution B mode, emitting at right angles from the transducer tip only one or two pulses per transmission. Real-time imaging can be done with up to 30 conventional IVUS frames. Differences in the position of blood cells between sequential images are computed to assess the local magnitude of blood flow, without any quantification.

3.2.2.2 Nuclear Magnetic Resonance Velocimetry

Magnetic resonance velocimetry (MRV) is able of 3D blood[18] flow velocity measurements across whole selected vessel section [166]. MRV allows quantification in deep vessels that cannot be explored by DUS techniques. Among several operation mode, phase-contrast magnetic resonance velocimetry (PCMRV) is used to analyze and quantify blood flow [168, 169]. The PVMRI employs the signal from blood-conveyed protons stimulated by specially designed magnetic field gradients. The output phase shift is proportional

---

[18] Inhaling hyperpolarized helium, MRV can be performed on respiratory flows [167].

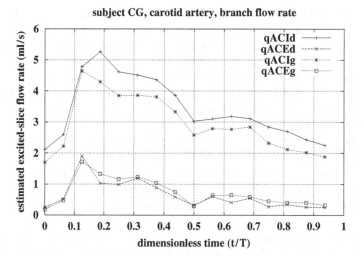

Figure 3.3. Variations of the flow rates in the external (E) and internal (I) left (g) and right (d) carotid arteries measured by MRI in a healthy volunteer.

to flow velocity. In contrast, the net phase shift of the nuclei in stationary tissue is equal to zero. At multiple instants during the cardiac cycle, a phase image and corresponding magnitude image are constructed. The magnitude images depict the anatomy at the specified location. On phase images, the intensity of each voxel within the vessel lumen corresponds to the velocity of blood flow at that location (Fig. 3.3, Table 3.6). Dynamic MRA analyzes transient image features, such as retrograde flow [170]. Bipolar flow-encoding gradients can be used in a three-dimensional MRI procedure to visualize small vessels having relatively slow flow [171]. Moreover, multiple station acquisition with possible wave velocity calculation can be done using comb excitation and Fourier velocity encoding [172]. Multiple components of velocity and acceleration by Fourier phase encoding can be simultaneously measured with a few encoding steps and efficient velocity-to-noise ratio [173]. Among the influence parameters, the number of flow-encoding steps, signal samples for averaging, and dimensions of the flow field, maximizing the number of steps is the most efficient way of improving the precision of measurements, keeping a reasonable acquisition time [174].

3.3 Measurement of Heart Electric and Magnetic Properties

Basic bioelectric source models include the single dipole with variable orientation and magnitude, either with a fixed location or moving, and multiple dipoles, fixed in space, each representing an anatomical region of the heart.

The cardiac generator has been modeled in first works at the end of the nineteenth century as a dipole. The human body is considered as a resistive, piecewise homogeneous (isotropic or not) conductor. The thoracic conductor can incorporate the following components: heart wall, with pericardium and fat, high-conductivity intracardiac blood,[19] low-conductivity lung parenchyma,[20], intermediate-conductivity thoracic muscle layer, and non-conducting bones,[21] as well as other organs such as large vessels. The electric generator corresponds to the heart wall.

The propagating depolarization of cardiomyocytes can be modeled as a double layer. A double layer at the activation surface can be approximated by a single resultant dipole. At the wavefront, a lumped negative- and positive point source constitute a dipole in the direction of propagation. A double layer, with a positive side pointing to the recording electrode, produces a positive signal.

Unlike depolarization, repolarization is not a propagating phenomenon but rather a propagating-like process. Any cardiomyocyte repolarizes at a certain time after its depolarization, independently of the repolarization of the adjoining cells.

Table 3.6. Coefficients of the Fourier series of cross sectional velocity (cm/s) in femoral and internal carotid arteries:

$$U_q(t) = A_0 + \sum_1^{12} \Big(A_k \cos(\omega k t) + B_k \sin(\omega k t) \Big).$$

	femoral	carotid		femoral	carotid
A_0	2.95	33.29	B_1	−7.29	9.52
A_1	15.41	−5.27	B_2	5.72	0.72
A_2	29.40	−5.10	B_3	29.30	0.29
A_3	22.87	−2.81	B_4	42.25	−3.25
A_4	0.10	−2.15	B_5	35.66	−0.95
A_5	−36.97	−0.12	B_6	−13.21	−1.17
A_6	−45.89	0.79	B_7	−32.07	−0.50
A_7	−8.91	0.27	B_8	−16.09	−0.41
A_8	10.67	0.29	B_9	−4.71	−0.18
A_9	11.10	0.06	B_{10}	2.18	−0.17
A_{10}	10.31	0.12	B_{11}	8.11	−0.20
A_{11}	4.97	0.07	B_{12}	7.30	−0.16
A_{12}	−1.66	0.09			

---

[19] Blood resistivity depends on the blood flow [175] and on the hematocrit [176].
[20] The resistivity of the lungs is approximately twenty times that of the blood.
[21] The resistivity of bones increases about a hundred-fold with respect to blood.

3.3.1 Electrocardiogram

Body surface electrodes record the electrical activity generated by traveling action potentials (combination of all electrical signals) to provide an electro-cardiogram (ECG). ECG thereby depends on the electric behavior of tissues between the thoracic skin and heart wall. Owing to very narrow between-cell space and the existence of many gap junctions, the cardiomyocytes can be represented as clusters of myofibers receiving at different times the action potential according to their situation with respect to the electrochemical wave propagation, with preferential propagation along the myofiber axis.

The ECG signal relies on several observations and assumptions. Nodal cells and cardiomyocytes are characterized by two electrophysiological states: polarization (rest) and depolarization followed by repolarization (variable duration according to the site, up to several hundred ms) during the genesis of the action potential in the pacemaker (normally the sinoatrial node) and its propagation in the cardiac wall. The propagating activation front can be defined by its resultant vector. The propagation speed of the activation front is assessed in the main regions of the conduction tissue and in the myocardium. In the standard 12-lead ECG system the source is a dipole in a fixed location and the volume conductor is either infinite homogeneous or spherical homogeneous. The thorax is supposed to be homogeneous. The atria and ventricles are activated during separate phases of the cardiac cycle, the physiological diastole and systole, respectively. Their sequential activation hence produces different signals, at least for the depolarization. The size, the location, and the orientation of the heart have a limited variability between human subjects.

However, many quantities affect the measurements. The resistivities of the intracardiac blood, of the cardiac wall, and of the lungs are about 1.6, 5.6, and 10 to 20 Wm, respectively. The stronger conducting intracardiac blood mass leads to an increased sensitivity to radial dipole elements (decreased sensitivity

Table 3.7. Signals of ECG leads (standard (L1, L2, L3) and augmented (aVR, aVL, aVF) limb leads, precordial leads V_i, $i = 1, \ldots, 6$) (u: voltage, subscript F: standard left leg lead, subscript L: standard left arm lead, subscript R: standard right arm lead, $u_s = (u_F + u_L + u_R)/3$.

Lead	Voltage
L1	$u_R - u_L$
L2	$u_R - u_F$
L3	$u_L - u_F$
aVR	$3(u_R - u_s)/2$
aVL	$3(u_L - u_s)/2$
aVF	$3(u_F - u_s)/2$
V_i	$u_{V_i} - u_s$

to tangential dipoles) with respect to the homogeneous model (Brody effect) [177]. However, the Brody effect is reduced when other heterogeneities are included. Besides blood mass, lung resistivities, and position of the lungs and heart change during cardiac and respiratory cycles. These changes affect electrocardiology [178].

3.3.1.1 ECG leads

The *Einthoven standard limb leads* explore cardiac electrical activity in the coronal plane. The standard bipolar leads at the right arm (R), left arm (L), and left leg (F) are supposed to constitute the vertices of an equilateral triangle, the so-called Einthoven triangle [179–181]. L1 is positive in the direction R to L, L2 in the direction R to F and L3 in the direction L to F (Table 3.7). A simple model assumes that the cardiac dipole is located at the center of a homogeneous sphere representing the thorax, hence at the center of the equilateral triangle. Hence, the voltages measured by the three limb leads are proportional to the projections of the electric heart vector on the sides of the lead triangle.

After the sinoatrial node depolarization, the action potential spreads in the atrial walls. The projections of the resultant vector of the atrial electric activity on each of the three Einthoven limb leads is positive. Depolarization reaches the atrioventricular node. Propagation through the atrioventricular junction slows down, allowing complete ventricular filling. A delay in activation progression is thus observed. Once activation has reached the ventricles, propagation proceeds along the Purkinje fibers to the inner walls of the ventricles. Activation wavefronts proceed from endocardium to epicardium and mainly from apex to base. Ventricular depolarization starts from the left side of the interventricular septum. Therefore, the resultant dipole from this septal activation points to the right, causing a negative signal in left-to-right arm and left foot-to-right arm leads. As depolarization on both sides of the septum, then apex occurs, the resultant vector points to the apex. Depolarization propagates through the wall of the right ventricle to reach the epicardial surface of the right ventricle free wall, whereas it continues to move in the thicker wall of the left ventricle. The resultant vector points leftward with a maximal magnitude. The amplitude then decreases until the whole ventricular myocardium is depolarized. Once the ventricle basis reached, there is a second delay.

The action potentials of the epicardial cells have shorter duration than those of the endocardial cells. The repolarization appears to proceed from epicardium to endocardium. Recovery and activation dipoles are thus in the same direction. Because repolarization is more disperse, the signal has a much smaller amplitude and longer duration than those of depolarization.

Three *Goldberger augmented limb leads* - aVR, aVL, and aVF - are currently added to the standard limb leads [182]. Goldberger unipolar leads use the same electrodes; each one is a positive pole and the two others negative. Leads aVR is oriented north-west, aVL north-east and aVF south.

Precordial leads - V1 to V6 - are located on the chest wall. V1 and V2 are located in the fourth intercostal space on the right and left side of the sternum. V4 is located in the left fifth intercostal space at the midclavicular line. V3 is located halfway between V2 and V4. V5 and V6 are at the same horizontal level as V4 but on the left anterior axillary line and at the mid-axillary line, respectively.

Commonly, ECGs are carried out with a 12-lead technique. Leads R, L, F, aVR, aVL, aVF are derived from the same three measurement points. Limb leads R, L, F more or less reflect the frontal components. Precordial leads target the transverse components. In the standard 12-lead ECG, the source is supposed to be a dipole in a fixed location and the volume conductor homogeneous and either infinite or spherical homogeneous. However, the thorax is heterogeneous. Tissue resistivity change between the skin and the cavital blood is, at least partially, responsible for the Brody effect, with an increased and decreased sensitivity to radial and tangential components, respectively. Lung resistivity, position, and shape change during the respiratory cycle, and hence affect the electric heart vector. Heart motion during the cardiac cycle also influences ECG [183].

Endocavital exploration used catheter-based sensors. The travel timing of the action potential is thus much more precise. In particular, conduction in the atrioventricular node can be differentiated from propagation in the trunk of the His bundle.

3.3.1.2 ECG Trace

Normal ECG consists of three basic features, a P wave, QRS complex and T wave, corresponding to the atrial, ventricular depolarisation and ventricular repolarization, respectively. The shape of the different waves, which depends on electrode location, and the time intervals between them (PQ, or PR in absence of Q wave, QT, ST, as well as QRS duration) are analyzed for medical check up (Table 3.8). Any alteration in action potential transmission and cardiac frequency is revealed by ECG.

Normal P wave amplitude and duration are about 100µV (<250 µV) and 85 to 100 ms. PR interval ranges 120 to 200 ms. Normal QRS complex amplitude and duration are about 1 mV and 800 to 900 ms. Q wave normally has an amplitude lower than 10 to 20 µV and a duration shorter than 50 ms. All complexes are normally quasi-evenly spaced with a rate of 60 to 100 per minute. T waves follow QRS-complex after about 200 ms.

3.3.1.3 Mathematical Electrocardiography

Mathematical electrocardiography is aimed at simulating the electrical activity of the heart using the so-called bidomain model (Sect. 6.1.1). The bidomain

Table 3.8. Depolarization and repolarization of the atria and the ventricles produce the usual ECG sequence P–QRS–T. Amplitude of ECG waves (μV). Duration of ECG waves and intervals (ms).

P wave	<110
PR spacing	120–200
Q wave	<40
QR wave	<30 (V1–V2), <50 (V5–V6)
S wave	<40
QRS wave	<100
QT interval	350–450

model involves transmembrane and extracellular electrical potentials, intracellular and extracellular conductivities, as well as a simplified model of ion tranport across the cardiomyocyte membrane and in the extracellular medium and gating variables. An additional equation governs the electrical potential in the torso. The thoracic cage is commonly assumed to contain the heart, lungs, ribs, and remaining tissues [184]. Adequate boundary conditions are defined at each interface.

One goal of mathematical electrocardiography is to compute, at least roughly, the location and size of myocardium infarctions. A three-dimensional reconstruction of the thorax and its main components is used with possible adaptive mesh to assess electrical activity in several selected points of the torso. A model of the cardiomyofiber structure is incorporated because conductivity in the heart wall is anisotropic. Another objective of numerical simulations is to optimize the position of the probes of an implanted pacemaker, which detects and corrects conduction defect of the action potential.

3.3.2 Vectorelectrocardiography

The measurement and display of the electric heart vector is called vectorcardiography (VCG), or vectorelectrocardiography [185]. The three components of the electric heart vector are measured with respect to a selected Cartesian frame, which is defined by coordinate axes corresponding either to body axes (usually) or cardiac axes. Although the VCG information is similar to that of the ECG, VCG allows better analysis of the activation front. Seven electrodes are used, one serving as a reference.

3.3.3 Magnetocardiography

Sarcolemmal depolarization–repolarization cycles produce temporal changes in magnetic field around the heart. The magnetocardiography (MCG) non-invasively records the magnetic field (a vector field), whereas ECG estimates the electric potential field (a scalar field).[22] Both methods can be combined to give a better measurement of the electric activity of the myocardium, hence a better diagnosis in patients [186, 187].

The magnetocardiogram corresponds to the first detected biomagnetic signal [188]. The local magnetic fields created by the small electrical currents[23] in the heart wall can be detected using superconducting quantum interference device sensors (SQUID) without any contact with the skin. An array of multiple SQUID sensors is placed at selected positions over the torso. However, MCG technology is more complicated than that of the ECG and requires an expensive equipment.

3.3.4 Impedance Plethysmography

Impedance plethysmography is aimed at determining changes in body tissue volumes by measuring tissular frequency-dependent electric impedance at the body surface. Impedance plethysmography measures tissue impedance. Impedance cardiography has been proposed for the non-invasive estimation of stroke volume, using a constant current fed to the thorax by an electrode pair with a frequency range of 20 to 100 kHz and measuring the resulting thoracic voltage via separate electrode pairs [190]. Impedance plethysmography has also been used for the detection of thromboses in leg veins. However, modeling addressed for impedance cardiography relies on crude assumptions, more or less neglecting changes in blood volume in the thoracic vessels and organs, and above all flow-dependent changes in blood conductivity.

3.3.4.1 Magnetic Susceptibility Plethysmography

Magnetic susceptibility plethysmography is aimed at measuring blood volume changes in the thorax. The motions of the heart and other thoracic tissues and organs during the cardiac cycle induce variations in magnetic flux when a strong magnetic field is applied to the thorax. Magnetic susceptibility plethysmography does not currently have any clinical applications.

---

[22] Electric and magnetic leads are different. Signal-to-noise ratio for the electrical and magnetic recordings are affected by different factors. Although the electric resistivity of lung parenchymae are relatively high, the magnetic permeability of the biological tissues ressembles to the one of a free space, allowing easy recordings from the posterior face of the thorax.

[23] Magnetocardiographic signals have been computed using a model with a source represented by an uniform double layer and with a heterogeneous, multicompartmental model of the thorax, the geometry of which is derived from magnetic resonance imaging [189]. Computed and measured magnetic signals were in good agreement. The magnetocardiogram and electrocardiogram have a common basis.

3.4 Phantom Experiments

"savoir, c'est connaître par le moyen de la démonstration." (Aristote) [191] [To know is cognize by means of demonstration.]

A physical model is designed in an explicit context with an objective set for a better understanding of the physiological phenomena, and to make a testable prediction about it. The realism level depends on the objective set. Once set up, any model must be evaluated and refined if necessary. Physical tests are performed not only to validate numerical predictions, but also to provide realistic estimates of the uncertainties. Experimental models work with real fluid when the fluid has suitable physical properties and the dimensionless parameters have similar values (similarity principle). Physical model are now made by stereolithography or other rapid prototyping techniques.[24]

Pressures and flow rates are measured by transducers and flowmeters with suitable range and frequency response. Laser Doppler velocimetry (LDV) and particle image velocimetry (PIV) are used to measure velocity fields.

3.4.1 Photochromic Tracer Method

The photochromic tracer method allows simultaneaous flow visualization and velocity measurements.[25] Once illuminated by a pulsed laser, a seeded tracor undergoes reversible photochemical reaction with color changes. The traces are recorded at selected time intervals (≤ 5 ms) with sufficiently small dye trace thickness (0.2 mm) [192]. This is thus a technique similar to the *hydrogen bubble method* used to visualize complex flow patterns. A stream of small hydrogen bubbles is produced by a cathode wire located within the fluid, normal to the direction of flow. Such a method is invasive, inducing local perturbations, although they are minimized by an appropriate design of the immersed system. This technique, which is older than the photochromic tracer method, has been used in biomechanics [193].

---

[24] Stereolithography and rapid prototyping are techniques to manufacture physical models from surface mesh files (in current standard format STL). The file data are sorted out in a slice stack to be loaded into the machine that drives the motions of a laser. The laser beam strikes the surface of a bath with liquid photopolymers, quickly solidifying the photopolymers. Stereolithography then builds a layer at a time, slice after slice. The self-adhesive property of the material allows between-layer binding to form a complete three-dimensional object. Other low-cost rapid prototyping techniques exist, also building models layer by layer using wax, plastic, starch, etc. From starch-based models, transparent polymer models can be obtained.

[25] The error in velocity v measurement is, at least, of $(0.15^2 + (0.015v)^2)^{1/2}$ [192].

3.4.2 Laser Doppler Velocimetry

Laser Doppler velocimetry (LDV) was the reference method in the past decades to non-invasively measure velocities with a good time and space resolution [194]. Laser beams interfere at the focal point to produce dark and bright fringes. Created constructive or destructive interferences along a given coordinate allow measurements of the particle speed across the exploration volume. This technique is suitable for flow modulations induced by time-dependent pressure inputs and velocity fluctuations observed when the flow loses its stability. Laser Doppler velocimetry is still used to measure the velocity field in certain flow regions characterized by large spatial velocity gradients using a setup with appropriate space resolution.

However, measurements require tedious work over a long time, during which flow conditions and fluid properties can vary. Velocity components are indeed measured in various stations, and at each stations in different positions within the cross-section. To avoid non-obvious calculation of the position in complex tube geometry, the refractive indices of both the fluid and duct wall are matched.

A laser source produce a laser beam that is splitted into two beams of equal intensity. One beam undergoes a frequency shift in a Bragg cell. The pedestal and high-frequency noise are removed. The two beams are brought to intersection inside the pipe lumen in a selected point by a lens. The optical probe, of given size, operates in the fringe mode. The lens features are selected according to the velocity range, size of the test section, and frequency response of the processing unit, knowing that the size of the sampling volume increases with the focal length.

The flow is seeded with small (~ 1 μm) neutrally buoyant particles, which follow the fluid streams very closely. Incident light is diffused in every direction (but with variable intensity) by flowing particles, which cross the exploring volume. Velocity is measured using the dual-beam either forward or backward scattering mode. The main features of an example of laser Doppler velocimetry is given in Table 3.9. The frequency of the diffused light is shifted with respect to the incident beam by the Doppler frequency f_D ($f_{dif} = f_{inc} + f_D$. The Doppler frequency is given by the ratio of the particle velocity to the gap between the interference fringes:[26] $f_D = v/d_f$. Forward or backward scattered light is collected by a lens and focused on and transmitted through a small pinhole to a photomultiplier. The output of the photodetector is then processed by a frequency tracker.

3.4.3 Particle Image Velocimetry

Particle image velocimetry (PIV) is a non-intrusive technique that measures the velocities of seeded micron-sized particles. The particle-seeded flow is

---

[26] Let θ be the intersection angle between the two beams. The gap between the interference fringes is $\lambda/(2\sin(\theta/2))$ (λ: beam wavelength).

Table 3.9. Main features of the optical system of a laser Doppler velocimeter (i: light intensity, n: refraction index, Source: [195]).

He–Ne laser	$5\,\text{mW}$, $\lambda = 632,8\,\text{nm}$
Between-beam initial gap (mm)	$s \sim 20\text{–}50$
Focal length (mm)	$f \sim 10\text{–}50$
Beam diameter (mm) $D_{e-1} \leftrightarrow i_{\max}/e$, $D_{e-2} \leftrightarrow i_{\max}/e^2$ Focus diameter	$\sim 1,1\,\text{mm}$ $d_{e-2} = (4f\lambda)/(\pi D_{e-2})$
Beam intersection angle	$\theta(n)$
Between-fringe gap	$d_f = \lambda/(2\sin(\theta/2))$
Fringe number	$N_f = c/d_f$
Doppler frequency	$f_D = u/d_f = 2u\sin(\theta/2)/\lambda$
Measurment volume (μm) Cross length Axial length Height	$a = d_{e-2}/\sin(\theta/2) = 920$ $b = d_{e-2} = 91$ $c = d_{e-2}/\cos(\theta/2) = 90$
Pinhole size (mm)	$\sim 0,2\,\text{mm}$

illuminated with a light sheet. When neutrally buoyant, the particles follow the fluid paths; PIV thus provides instantaneous velocity vector measurements in explored sheets of finite thickness through which the laser beam is supposed to have a light intensity with a Gaussian distribution. Values of measurement parameters are given in Table 3.10.

Stereoscopic arrangement with two cameras is needed in 3D flow, characterized by out-of-sheet motions, to measure the three velocity components.

Table 3.10. Particle image velocimetry. Magnitude order of measurement parameters. The homogeneous suspension of neutrally buoyant, uniform, small particles, of speed v, is illuminated by a laser sheet of thickness equal or less than the light beam diameter d_l (M: magnification of the recording camera). To get a correlation between particle images during displacement, the same particles must be imaged, avoiding out-of-plane displacement, and particle images must have the same shape and intensity (Source: [196]).

Minimal concentration	$c > 5 \times 10^3/\text{cm}^3$
Exposure time	$0,5d_l/(Mv)$ (few tens of ms)
Sweeping frequency	$\Delta t = 10\,\text{ms} \implies \nu = 100\,\text{Hz}$
Image size	$d_p/\Delta x = 1/10$
Displacement	$\Delta x = Mv\Delta t \approx 150\,\mu\text{m}$

A charge-coupled device (CCD) camera records separate images that show the positions of the illuminated particles at two different times, t and $t + dt$. The images are then processed to extract velocities from apparent displacements of the particles during the time intervals between exposures. The velocity vectors are computed from a chosen number of voxels (or analysis windows AW) of the light sheet by measuring the movement of particles between two light pulses with a known time interval Δt. A given number of illuminated particles travel through each AW. At least ten particle images should be seen in each AW. An average particle displacement is computed in each AW. In the resulting image, a set of imaged particles is recorded with variable intensity, depending in the particle position in the explored flow plane related to light-intensity distribution. The between-pulse time is set such that particle displacement is lower than $L/4$ (L: the pixel size in the streamwise direction) to satisfy the Nyquist criterion. Processing can involve various types of data-reduction techniques: autocorrelation and cross-correlation among others, which are performed in each AW. Autocorrelation is used to process double-exposure images (pairs of particle images are recorded on the same flow image), whereas cross-correlation is applied to pairs of single-exposure images (a sequence of two light-pulse images, I1 (t) and I2 ($t + \Delta t$) is recorded). Auto- or cross-correlation operation is performed in each AW. The autocorrelation function shows a central peak and two symmetric secondary peaks. The position of these secondary peaks with respect to the center of the AW gives the displacement of the particles. The symmetry of the autocorrelation function do not provide displacement direction, which has to be determined from the flow. Cross-correlation produces a single intense peak on the correlation plane, the position of which with respect to the AW center gives access to both the pixel-averaged motion direction and the particle displacement in the AW, once length calibration has been made (i.e., from the known pipe diameter) due to image magnification. The velocity in the AW during the time Δt between laser pulses is then computed from the location of the displacement peak on the correlation plane. A velocity-vector map over the whole target area, which is discretized into a regular grid of AWs, is generated by repeating the image processing for each AW over the image pair (Fig. 3.4).

3.4.4 Electrochemical Sensors

Time-dependent wall shear stress (WSS) affects the transendothelial mass transport. Blood velocity can be faithfully measured except at positions close to the vessel wall. A suitable device is then required to measure time and space changes in WSS due to the 3D quasiperiodic nature of blood flow. Moreover, flow reversal (bidirectional flow in the vessel lumen) can occur in certain arteries during the diastole near the wall, as well as flow separation. Additionally, the signal frequency content must be accurately sensored, especially to reproduce peak values.

A method to measure WSS in model experiments is related to heat trans-fer, a constant-temperature hot-film probe being put at the wall. The wall velocity gradient can also be measured by electrochemical probes. The elec-trochemical technique is based on mass transfer of a flowing electrolyte. The measuring system is composed of a wall working electrode, a downstream counter electrode, and a reference electrode. The measurement requires a very quick-kinetic oxidation–reduction (redox) reaction of an electrolyte, which is chemically inactive with respect to the solvent. The solution commonly used is an equimolar concentration of potassium ferro/ferri-cyanides ($25\,\mathrm{mmol/m^3}$). Charge transfer between the flowing fluid and polarized electrode is measured, the current being almost entirely due to concentration diffusion of the carrier ions. The circular sensor geometry (diameter $300\,\mu m$, thickness $50\,\mu m$) can be partitioned into three regions (gap $10\,\mu m$) to sense both the magnitude and direction of the local fluid velocity [197–199].

Mass tranfer is performed by three mechanisms: convection, diffusion, (re-sulting from the gradient of a chemical potential, generally the concentration gradient), and ion migration due to the electrical potential gradient. The cur-rent of migration transport depends on electrolyte concentration. Using an inert electrolyte reduce the migration flux. Hypotheses associated with the measurement principle are given in Table 3.11.

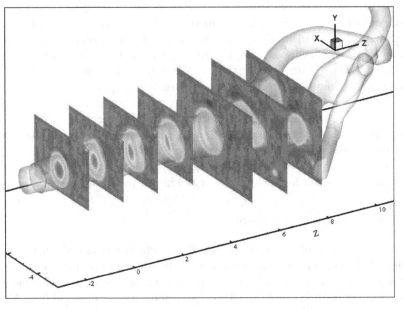

Figure 3.4. Isovelocity contours in a subset of measurement planes using PIV in an elastomere model of the carotid artery network conveying a steady flow (from J. Vetel).

Table 3.11. Electrochemical sensor. Hypotheses associated with the wall shear rate measurement by transport of a chemical species i.

$$c_{i,t} + \nabla \cdot (\mathbf{J}_{i_d} + \mathbf{J}_{i_c} + \mathbf{J}_{i_m}) = \mathcal{R}_i$$

Spliting between mass transfer and flow	
Neither loss nor production	$\mathcal{R}_i \sim 0$
No migration flux	$\mathbf{J}_{i_m} \sim 0$
Constant physical properties	
Incompressible fluid	$\nabla \cdot \mathbf{v} = 0$

$$c_{i,t} + \mathbf{v} \cdot \nabla c = D\nabla^2 c$$
$$\mathbf{v} \cdot \nabla c \sim u_x(c/l) + u_y(c/\delta_D) + u_z(c/L)$$

Single influence of WSS	
Linear velocity profile	$u_z(t) = \dot{\gamma}_w(t)n$
$Sc = \nu/D \ll 1$	
No local curvature	$\delta_D \ll \delta_\nu \ll R$
Axial $\gg$ transverse convection	$u_x \sim 0$
Local homogeneous flow	
Sensor size $d_s \gg \delta_D$	$c/l \ll c/\delta_D \rightarrow c_{,x}, \, c_{,xx} \sim 0$
	$c_{,zz} \sim c/L^2 \ll c_{,yy} \sim c/dg_D^2, \, c_{,zz} \sim 0$
No mass transfer outside the probe	

[1a]	$u_y/\delta_D \ll u_z/L \implies$	$c_{,t} + \dot{\gamma}_w y c_{,z} = D c_{,yy}$
[1b]	$u_y/\delta_D \ll u_z/L \implies$	$c_{,t} + \dot{\gamma}_w y c_{,z} = D(c_{,yy} + c_{,zz})$
[2a]	$u_{y,y} + u_{z,z} = 0 \implies$	$c_{,t} + \dot{\gamma}_w y c_{,z} - \dot{\gamma}_{w,z}(y^2/2)c_{,y} = D(c_{,yy} + c_{,zz})$
[2b]	$u_{y,y} + u_{z,z} = 0 \implies$	$c_{,t} + \dot{\gamma}_w y c_{,z} - \dot{\gamma}_{w,z}(y^2/2)c_{,y} = D c_{,yy}$

Consider a simple case. Under steady conditions, the mass balance is given by $\dot{\gamma}_w n \partial c/\partial z = D\partial^2 c/\partial n^2$ ($\dot{\gamma}_w$: local wall shear rate). The mean diffusion mass transfer rate $\mathbf{J}$ per unit surface area is $J = -D\partial c/\partial n|_{n=0}$ (n: local normal direction to the wall). The flux at the probe surface controls the electrochemical process and thus the current i is given by $J = i/(n\mathcal{F})$ (n: charge number, $\mathcal{F}$: Faraday constant). The concentration c is supposed to vary linearly inside the concentration boundary layer. The summation of the mass transport equation in the diffusion boundary layer shows that the mass flux, and consequently the current, are proportional to the velocity gradient $j = K_m/\Delta c \propto D^2 \dot{\gamma}_w ML^{-1}$ (L: sensor length between the leading and trailing probe edges, K_m: mass transfer coefficient).

In unsteady flows, phase lag errors arising from the use of a steady solution must be corrected to calculate the time-variation of WSS [200]. In pulsatile flows, mass transfer depends on the velocity modulation rate and flow frequency.

3.4.5 Tracking Methods of Wall Deformations

White light speckle method is exploited for displacement measurements of the heart wall [201]. Once the speckle pattern is created on the specimen surface, speckle displacements associated with the body deformation are recorded. Subsequently, the pair of speckle patterns is processed. Digital reconstruction of dynamic deformations of a biological conduit can also be based on the optical stereoscopic method and projection of a grid with a binary code on the conduit surface [202].

Two shear wave sources located at opposite sides of a sample of a biomaterial create interference patterns, which can be visualized by the vibration sonoelastography technique [203, 204]. When the two vibration sources have slightly different frequencies, the interference patterns move in proportion to the shear velocity.

> *In order to accommodate new experimental proofs, one must ... deform the primitive concepts. One must not only study the conditions of application of these concepts, but one must incorporate the conditions of application of a concept into the very meaning of the concept itself.* (G. Bachelard) [205]

4

Rheology

Rheology ($\overset{\mathsf{C}}{\rho}\epsilon\omega$: to flow, water/blood flow, $\overset{\mathsf{C}}{\rho}\epsilon\hat{\upsilon}\mu\alpha$: flow, flux, λόγοσ: dissertation) is the study of material deformation and flow under stresses. The rheology of both the flowing blood and loading walls of the cardiovascular system has been investigated. Soft biological tissues are characterized not only by time-dependent non-linear rheological behavior exhibited by their composite anisotropic poroviscoelastic (commonly involving flowing fluid) materials, which are irrigated (living tissues) and innervated (short-range history of such controlled materials), but also by tissue growth, remodeling, and aging (mid- and long-range history).

4.1 Blood Rheology

Different types of apparatus have been used for blood viscosimetry, subjecting the blood sample to laminar shear. Capillary viscosimeter is based on a precision bore capillary tube of given diameter and length that conveys a laminar blood flow at constant temperature. However, due to the shear rate range experienced by the fluid between the wall and tube axis, this technique is not suitable. The rotational coaxial concentric cylinder viscometer allows a blood sample placed in the annular space between an immotile (stator) and rotating (rotor) cylinder, of known sustained steady rotation speed, to be subjected to a quasi-linearly varying velocity throughout the gap. The shear rate is then nearly constant across the between-cylinder gap. Shear stress is obtained from the measured torque acting on the stationary cylinder. The cone-and-plate viscosimeter is designed to subject the blood in a narrow space between a rotating flat circular plate and a cone (usual angle of less than three degrees) to laminar shear. The rotation speed and torque are measured, in isothermal conditions, currently for low and moderate shears. The viscosity of a Newtonian fluid in a laminar steady flow (Re < 250) in an axisymmetric stenosis in a straight tube can be non-invasively evaluated using a viscosity vs.

centerline velocity abacus made from UD Doppler velocimetry and numerical simulations [206].

Erythrocytes play a major role on the rheological behavior of the blood due to both their size and number. The blood indeed behaves like a concentrated dispersed RBC suspension in a solution (plasma), which is composed of ions and macromolecules interacting between them and bridging the erythrocytes. RBC aggregation can be evaluated by optical techniques based on reflected and transmitted light energy by blood samples subjected to shear [207].

Several factors affect the blood rheology (Table 4.1). The rheological properties of non-Newtonian blood is dictated by the flow-dependent evolution of the blood internal microstructure, i.e., the state of the flowing cells (possible aggregation and deformation) rather than the conformation of the plasma macromolecules. The interactions between the conveyed plasma molecules and RBCs indirectly govern blood rheological behavior. In large blood vessels (macroscale), the ratio between vessel bore and cell size ($\kappa_{vp} > \sim 50$) is such that blood is considered as a continuous homogeneous medium. In capillaries (microscale), $\kappa_{vp} < 1$ and the blood is heterogeneous, transporting deformed cells in a Newtonian plasma. In the mesoscale ($1 < \kappa_{vp} < \sim 50$), flow is annular diphasic (rigid particle flow) with a core containing cells and a marginal plasma layer.

Shear-step experiments (steady state after a short transient regime) show that the blood has a *shear-thinning* behavior [208]. Furthermore, it is *viscoelastic*[1] and *thixotropic*.[2] Time dependency, a feature associated with viscoelasticy and thixotropy, is also present in physiological conditions (flow pulsatility, vasomotricity, etc.). Blood exhibits *creep* and *stress relaxation* during stress formation and relaxation [209]. Blood exposures to sinusoidal oscillations of constant amplitude at various frequencies reveal a strain-independent *loss*

Table 4.1. Factors affecting the blood viscosity.

Plasma	Cell	Vessel	Flow
Temperature	Cell concentration	Vessel bore	Shear field
Protein concentration	Cell aggregability		
	Cell deformability		

[1] A loaded purely elastic body instantaneously deforms up to a given strain proportional to the applied force (it behaves as a spring). A loaded viscoelastic body progressively deforms up to an equilibrium state (as if a dashpot exists as a representative material component). Whereas elasticity is expressed in real parameters, viscoelasticity is described by frequency-dependent complex numbers. Its viscoelastic character can be estimated by the Weissenberg number, which is the product of the relaxation time scale and characteristic deformation rate.

[2] A thixotropic fluid is a time-dependent viscous material. Fluid behavior is affected by fluid structural changes, which are reversible. Shear stress decreases with the shear duration in steady states.

modulus and strain-dependent *storage modulus*. The existence of both moduli characterizes a viscoelastic material. Hence, the concentrated suspension of deformable RBCs in a macromolecule environment is a viscoelastic material that experiences a *loading history*. Thixotropic behavior is explained by changes in blood internal structure, which is experimentally sheared, thereby by the kinetics of both reversible RBC aggregation and deformation, with their time scales.

The velocity profile in a pulsatile flow of a non-Newtonian fluid in a straight pipe is flat in the core. The shear rate is then low in the core. If the fluid has shear-thinning behavior, the fluid viscosity is then higher in the core than in the unsteady boundary layer where the shear rate is large.

The shear-thinning behavior is exhibited by a sigmoid relationship between shear rate $\dot{\gamma}$ and fluid viscosity μ (Fig. 4.1). It is explained by RBC aggregation at low shear rates associated with high viscosity values, whereas RBC deformation and orientation lead to a low viscosity plateau μ_∞ at high shear rates $\dot{\gamma} > 10^2/s$ (Newtonian behavior). There is huge between-subject variability in the μ vs. $\dot{\gamma}$ relationship. This variability points out the need for standardized procedure (temperature, pH, Ht, protein concentration, etc.). The μ–$\dot{\gamma}$ relationship depends on Ht [210]. The viscosity maximum μ_0, which is difficult to measure with available experimental devices, is about ten times greater than μ_∞ in steady states. When both RBC aggregation and deformation are inhibited, μ_0 is close to μ_∞ [211].

Input rheological data, which are provided by experimental results obtained in steady state conditions,[3] are far from the physiological ones in the arteries [213]. Shear-step experiments do not mimic the flow. The flowing blood is characterized by a smaller convection time scale than the characteristic time of cell bridging. Moreover, in large arteries, RBCs, expelled from the left ventricle where they have been shaken, are expected neither to deform nor aggregate. A blood volume that enter a low-shear region often has previously traveled across high-shear zones in which possible rouleaux have been disrupted. Therefore, non-Newtonian effects can be neglected owing to the RBC transit time in the absence of stagnant blood regions. The blood, in large vessels, can then be considered to have a constant viscosity, greater than the dynamic viscosity of water and plasma (Table 4.2). However, in the low flow region of or dowstream from arterial wall diseases, where the residence time of conveying particles and cells is much greater, RBC aggregation characteristic time might have the same order of magnitude than the local flow

---

[3] Blood behavior depends on the structure of the RBC set. At rest, the blood is composed of a rouleau network. Low shear rates deform and partially break the rouleau network into smaller rouleau networks and isolated rouleaux. At mid shear rates, only isolated rouleaux of various size are observed. At high shear rates, RBCs are scattered and oriented in the shear direction. A shear rate step induces a transient and stable strain regime, with a viscoelastic, elastothixotropic, and Newtonian response at low ($\dot{\gamma} = 0.05/s$), mid ($\dot{\gamma} = 1/s$), and high shear level ($\dot{\gamma} = 20/s$), respectively [212].

time scale. Non-Newtonian fluid flow must then be simulated, but dynamical data that suit blood in motion are not yet available.

In the framework of continuum mechanics, shear-thinning behavior and loading history of blood are incorporated via an appropriate formulation of constitutive law. The Carreau model, which belongs to the family of *generalized Newtonian models*, has been proposed for numerical simulations. Power values are obtained by fitting the power-law region of the shear-thinning behavior. Generalized Newtonian model used to mimic the supposed shear-

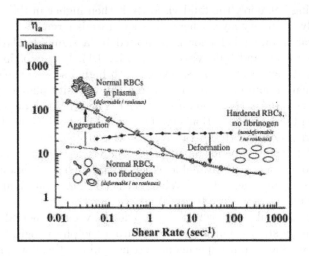

Figure 4.1. Blood shear rate–relative viscosity relationships with its shear-thinning behavior in static conditions (from [208]). The relation depends upon the kinetic of formation and rupture of RBC aggregates at low shear rates and kinetics of RBC deformation and RBC orientation at high shear rates. Two ratios are involved: (1) the ratio between RBC size and flow length scale (vessel radius), and (2) the ratio between aggregation time constant and flow time scale (convection characteristic time for macrocirculation, transit time for microcirculation) in the explored vessel segment. Kinetics are governed by relaxation phenomena. Blood rheology is thus governed by variation in RBC suspension structure (thixotropic medium with viscoelastic behavior). The value at low shear rates is questionable because of the resolution limit of measurements. A between-subject variability exists, in particular in slope and curve inflection point values ($\dot{\gamma}_{(1/2)}$)

Table 4.2. Blood (Ht = 45%, $T = 37°$C) and water ($T = 37°$C) physical properties.

	μ ($\times 10^{-3}$ Pl)	ρ ($\times 10^3$ kg/m^3)	ν ($\times 10^{-6}$ m^2/s)
Blood	3–4	1.055	2.8–3.8
Plasma	1.2	~1.03	~1.2
Water	0.692	0.993	0.696

thinning blood behavior does not take into account the rheological history of the fluid. However, this model is used to handle a simplified equation set:

$$\frac{\mu(\dot{\gamma}) - \mu_\infty}{\mu_0 - \mu_\infty} = \frac{1}{\left(1 + (\gamma_{(1/2)}\dot{\gamma})^2\right)^p} , \tag{4.1}$$

where $\gamma_{(1/2)}$ is the shear rate at mid-slope and p the slope of the shear-thinning regime. A value set has been proposed [213]: $\mu_0 \sim 40\,\text{mPa s}$, $\mu_\infty \sim 4\,\text{mPa s}$, $\gamma_{(1/2)} \sim 0.25/\text{s}$, $p \sim 1/3$.

Shear-induced platelet activation has been modeled using a complex viscoelastic model previously developed [214] and threshold-dependent-triggered activation function [215].

4.2 Heart Wall

Biosolid mechanics investigate loadings and deformations of the heart and vessel walls. The relevant variables are stress ($\mathbf{c}$: stress vector; $\mathbf{C}$: stress tensor), strain (ϵ; $\mathbf{E}$: strain tensor), elastic modulus (E), the bulk modulus (B) and the Poisson ratio (ν_P). Stress is a measure of forces applied to a biotissue region per unit area induced mainly by surface forces induced by adjoining regions and body forces.

4.2.1 Heart Valves

Planar biaxial stress–strain relationships of the mitral valve anterior leaflet exhibit a slight hysteresis, a strain rate-independent response, negligible creep, but significant stress relaxation [216]. Stress-relaxation is slightly greater in the radial direction than the circumferential direction. Valvular tissues can thus be modeled by an anisotropic quasi-elastic material. A dynamic three-dimensional finite element analysis of a pericardial bioprosthetic heart valve based on a generalized non-linear anisotropic elastic constitutive model shows that the leaflet free edge is subjected to significant flexural deformation and high stress magnitude [217].

In vitro uniaxial traction tests sort the valve strips in increasing order of stiffness: (1) axial strips of pulmonary valves, then (2) of aortic valves and (3) circumferential strips of aortic valves, then (4) of pulmonary valves [7]. However, axial and azimuthal strips of porcine aortic valve leaflets are stiffer than the corresponding strips of pulmonary cusps, the circumferential strips being stiffer than the axial ones [218]. Porcine valves do not model human ones because their stress–strain relationships for the same loading experiments are different.

4.2.2 Myocardium

Rheological tests performed on small samples of myocardium allow 1D analysis in the passive state. However, such samples do not represent the bulk myocardium in its environment. Sample dissection removes interaction with the myofiber and collagen networks, as well as perfusion. Like every biological soft tissues, these simple tests exhibit non-linear viscoelastic behavior. Myocardium viscoelasticity has been studied using relaxation tests after step traction [219]. The relaxation function depends not only on the strain, but also on the experimental environment (pH, temperature, and composition of the surrounding fluid).

4.2.3 Small Deformation

The material constitutive equation is described by a mathematical model. Models of the behavior of the left ventricule subjected to small deformations are based on linear elesticity and are associated with a hypothesis set: (1) the reference geometry is commonly cylindrical with a free rigid end without mass which undergoes rotation and translation; (2) in a given wall layer, muscular fibers are parallel; (3) the orientation angle θ of the muscular fiber varies linearly across the wall: $\theta(r) = \theta_i(1 - (2(r - R_i)/(R_e - R_i))$ ($\theta_i = 70$ degrees:

Table 4.3. Human aortic and pulmonary valve elements. The elastic modulus (MPa) has been evaluated from in vitro uniaxial traction tests, using stress of 1 MPa (Source: [7]).

Valve element	Structure	Function	Elastic modulus
Cusp	Endothelium CnF bundles Several EnFs	Tightness	AoV circumferential strips: 15.3 ± 4.0 AoV axial strips: 2.0 ± 0.2 PuV circumferential strips: 16.1 ± 2.0 PuV axial strips: 1.3 ± 0.9
Fibrous ring	EnFs CnFs	Support	AoV: 12.5 ± 3.0 PuV: 10.1 ± 2.6
Commissures	CnFs	Support	AoV: 13.8 ± 3.2 PuV: 10.0 ± 2.8
Sinotubular junction	EnFs		AoV: 7.4 ± 2.3 PuV: 5.9 ± 1.6
Sinuses	EnFs		AoV: 10.5 ± 3.2 PuV: 14.3 ± 3.5

near-endocardium orientation angle, $R_e = 26\,\text{mm}$, $R_i = 14\,\text{mm}$, $R_e/R_i = 1.9$); (4) the muscular fibers are synchronized; and (5) the inertia effects are neglected. The cardiac cycle is then split into a set of successive equilibrium states.

The myocardium, subjected to small deformations, can be modeled as a set of muscular fibers embedded in a perfect fluid [220]. The stress tensor $\mathbf{C}$ is given by: $\mathbf{C} = -p\mathbf{I} + T_f\hat{\mathbf{f}} \otimes \hat{\mathbf{f}}$, where T_f is the muscular fiber tension per unit area and $\hat{\mathbf{f}}$ the unit direction vector of the myofiber. The term p incorporates the fluid pressure and solid contribution associated with a Lagrange multiplier due to the incompressible muscular fiber matrix. Such a constitutive law does not take into account the extracellular matrix fibers. The myocardium indeed consists of myocytes interconnected by a dense collagen weave that goes in every direction. A third term $2G\epsilon$ (G: shear modulus, ϵ: small deformation tensor) has then been added to the equation right-hand side (RHS) to consider a fluid–fiber–muscle incompressible medium [221, 222]. During the cardiac cycle, the constitutive law of the muscular fiber can be expressed as a linear combination of two passive (end-diastolic) and active (end-systolic) states [223]: $T_f = [(1-\beta)E^p + \beta E^a]\hat{\mathbf{f}} \cdot \epsilon \cdot \hat{\mathbf{f}} + \beta T_0$ ($E^p = 20\,\text{kPa}$, $E^a = 200\,\text{kPa}$ are the passive and active state elastic moduli, $T_0 = 20\,\text{kPa}$: maximum developed tension). $\beta(t)$ is the activation function that models $[\text{Ca}^{++}]$, which is equal to 0 and 1 at the end of the diastole and systole, respectively, and has a piecewise linear evolution.

Ventricular geometry affects performance [223]. The greater the relative wall thickness (R_e/R_i), the larger the ejection fraction. The maximal orientation angle of 70 degrees corresponds to an optimum of the ejection fraction. The greater the passive elastic modulus of the muscular fiber E^p and the matrix shear modulus G, the lower the cardiac performance. The higher the active elastic modulus of the muscular fiber E^a and the muscular tone T_0, the larger the ejection fraction. The limitation of such models is due to small ejection fraction associated with deformation lower than 10%.

4.2.4 Large Deformation

Models of the behavior of the myocardium subjected to large deformations have also been developed [223]. The Cauchy stress tensor $\mathbf{C}$ is given by $\mathbf{C} = \mathbf{F}(\partial W/\partial \mathbf{G})\mathbf{F}^T - p\mathbf{I}$, where $\mathbf{F}$ is the transformation gradient tensor and $\mathbf{G}$ the Green-Lagrange strain tensor, W the deformation energy function, and p is a Lagrange multiplier. When the material is incompressible, as are most biological materials that are rich in water, $\det(\mathbf{F}) = 1$. The first Piola-Kirchhoff stress tensor $\mathbf{P}$ is expressed with respect to the deformed configuration: $\mathbf{P} = \det(\mathbf{F})\mathbf{F}^{-1}\dot{\mathbf{C}}$. The *passive state* is defined by the "unstressed"[4] configuration (absence of internal and external stresses). The *active*

---

[4] Residual stresses exist. Their quantification needs to cut the myocardium in cross slices and then cut the slice wall and measure the opening angle [224].

state means myocardium contraction associated with a new configuration but without applied stress (free contraction without environmental constraint). The *loaded state* corresponds to an active state that undergoes internal and external stresses. They are several expressions in the literature of the deformation energy function for the myocardium. The resulting deformation energy function during the cardiac cycle is the sum of three terms, which simulate passive ground matrix, and passive elastic and active components of the myofibers [225].

Most often, the left ventricle is assumed to be cylindrical, and the cylindrical shape is supposed to be kept during deformation. Displacement and deformation in the virtual active state are given with respect to the passive reference state. The transformation Φ from the passive to the active state is, by definition (Sect. 4.2), done without stress. A single activation-dependent reduction in cylinder height is assumed to result from this transformation [223]. Good agreement is found with experimental data [226]. The radial stress E_{rr} is the strain component that has the greatest magnitude during the cardiac cycle. Cauchy stress reaches its highest value at the beginninig of systolic ejection, between the mid-wall and the endothelium.

A cursory review of theoretical constitutive laws of the myocardium is given in Sachse's textbook [227]. The heart wall has been modeled by a homogeneous, incompressible, transversally isotropic material to derive a constitutive law for the myocardium during the whole cardiac cycle [228]. The wall behavior during the cardiac cycle is continuously described using three states: (1) a passive unstressed state, (2) a virtual state defined by a constant geometry but a rheology change, and (3) an active state of contraction without rheology change. The time-dependent strain energy function is composed of two terms, a passive and an active strain energy function (SEF) associated with passive fibers and cardiomyocytes, respectively: $W(\mathbf{G}, t) = W_{pas}(\mathbf{G}) + \beta(t)W_{act}(\mathbf{G})$ ($\beta(t)$: activation function).

A fibrous, anisotropic, diphasic model of the left ventricle shows that cavity twisting is the dominant motion [229]. Idealized models show that cardiac performance parameters (compliance, wall strains and stresses, ejection fraction, etc.) depend on cavity geometry and wall rheology [230].

Mathematical homogenization theory for heterogeneous media based on N-scale asymptotic expansion considers objects of length scale L_0, which have a relative periodicity, and thus are made from repeatible basic units of length scale L_u. A small parameter $\epsilon = L_u/L_0$ is used to expand the equation in powers of ϵ. In the myocardium, the basic unit contains a limited number of cardiomyocytes, inside which the electric field is computed. A constitutive law for the myocardium has been derived from discrete homogenization. A CMC set has been modeled by a quasi-periodic discrete lattice of elastic bars [231].

The glycosaminoglycan content of the pericardium is mainly composed of hyaluronan, chondroitine sulfate, and dermatan sulfate. Dynamic tensile tests in the frequency range 0.1 to 20 Hz evaluate an initial storage modulus of 6.5 MPa and an initial modulus of the high linear stress-strain phase of 14 MPa [232].

4.2.5 Poroelasticity

The multilayered composite wall with its three main constituents (cells, elastin, and collagen) and its aqueous ground matrix can be modeled by a poroviscoelastic material. The viscoelasticity theory for the biological tissues introduced by Fung does not take into account the microstructure, polyphasic porous nature, and fluid phase effects on the fluid-infiltrated porous tissue behavior. The short-time relaxation tissue behavior after quick loading is associated with the viscoelastic properties of the biological tissues, but, the interstitial fluid affects long-time relaxation behavior. In particular, stress relaxation experiments are characterized by a characteric time that depends on dynamic fluid viscosity, tissue permeability, and appropriate elastic modulus $E(1 - \nu_P)/(1 + \nu_P)(1 - 2\nu_P)$ [233]. The poro(visco)elasticity theory is then used because it takes into account the fluid in viscoelastic tissues. The heart and vessel walls indeed contain a large amount of free liquids, the interstitial fluid and blood conveyed in the small arteries and veins, which penetrate into the wall, and microcirculation. Moreover, the relaxation myocardium properties are more pronounced in cold-blooded animals (batracians), the myocardium of which have a smaller connective tissue content and a larger porosity, than in warm-blooded mammals [234]. Vasculature wall layers have different permeabilities due to different composition. Large length-scale structural heterogeneities (strata) can be differentiated from the local, small length-scale, random heterogeneities. Furthermore, these permeabilities are strain-dependent.

Any deformable porous medium, more or less saturated with an incompressible fluid, gradually deforms when it undergoes loading. The medium reaction depends on the rate at which the fluid is squeezed out of the voids of the porous structure. The flow model in deformable, porous materials can be applied to the fluid transport through the blood vessel wall. The flow is assumed to obey the Darcy law. The theory of deformation (*poroelasticity theory*) of a purely elastic, isotropic, porous material containing an incompressible fluid in isothermal conditions was pioneered by Terzaghi [235] and further developed by Biot [236–238], first considering material isotropy and incompressible materials, and then anisotropic body and compressible fluid. The constitutive equation relates strain and fluid content to stress and pore pressure. The linear biphasic poroelastic theory requires a set of material parameters (elastic and shear moduli, Poisson ratio, and Biot constant) and the material permeability to determine the adaptation of the poroelastic medium to the loading (consolidation) and interstitial fluid flow patterns. Body stiffness and permeability are supposed to depend on local strain gradient. Constituent compressibility was later taken into account [239–241].

The strain distribution in a deformable porous material subjected to a steady compression is uniform in the absence of flow (undrained deformation,[5]) whereas it is heterogeneous when the deformation results from flow (drained boundary-unloaded deformation [242, Fig. 1]). The transport between a porous material and its environment depends not only on diffusion but also on possible transport induced by convection outside the porous medium, in particular when the pressure at the body boundary fluctuates [243]. Consequently, in blood flow applications, unsteady conditions must be taken into account. Non-zero mean sinusoidal gas motion experiments through a porous medium have shown that the flow depends on imposed amplitude and frequency [244]. A constitutive law for porohyperelastic materials based on strain energy density function has been proposed, the shear stress including the pore fluid stress [245]. Poroelasticity theory has been applied to the arterial wall [246] and heart [234, 247]. The osmotic pressure is generally not incorporated in the deformation model.

4.3 Vessel Wall Rheology

In vitro rheology measurements on excised or isolated vessels need suitable vessel conservation and preconditionning.[6] Imaging velocimetry (US, MRV) can be used as an indirect method to estimate compliance of the vessel walls (Table 4.4). Two exploring stations are not sufficient because a single value of the wave speed does not take into account the non-linear pressure-dependency of the wave speed. It has been proposed to use three stations, two giving the boundary conditions and one the pressure-dependency, assuming a 1D flow. It is also possible to process the pressure signals from two stations in three identifiable wave points (foot, peak, and notch) with different time delay between the two waves, the absence of reflexion being assumed [248].

Lagrangian (f/A_0) or Cauchy (f/A) stresses are used whether rest or deformed configuration is the reference state. The strain is a dimensionless measure of deformation, usually normalized in 1D experiments. Various strain measures have been introduced (App. C.5) [249]. The elastic modulus E, or Young modulus, is a measure of the wall stiffness given by the slope of the stress–strain relationships in the elastic range. The compressibility is the ability of a material to undergo change in volume (simultaneous change in size in all directions) when it is subjected to loading. Poisson ratio ν_P quantify the compressibility. It is defined as the ratio of the resultant strain in lateral direction to strain in primary direction.

---

[5] Undrained deformation refers to either a short loading time scale or a sample confinement such that the microscopic fluid exchanges between neighboring body elements and macroscopic fluid leakage outside the body do not occur.

[6] Experiments must be quickly performed after vessel excision. During the period between sample removal and testing, deterioration of the tissues must be avoided ($T < 4°C$, $t < 48\,h$). Rheology tests must be standardized. In particular, because of the stress-history dependency, a period of adaptation to loading must be kept. Only accommodated materials provide suitable values for rheological quantities.

The wall rheology affects: (1) the blood circulation via the coupling between the wall mechanics and blood dynamics; and (2) mass transport within the wall in both normal and pathological conditions. The within-wall stress field is involved in development of arterial diseases and complication occurrence. The vessel wall is a composite[7] soft body that exhibits non-linear,[8] anisotropic,[9] nearly incompressible, viscoelastic[10] behavior over finite strains. Moreover, it adapts to environmental changes. The wall is, indeed, a living active material.[11] Vessel walls exhibit longitudinal and circumferential prestresses.[12]. With aging, the wall becomes stiffer and the wave speed c_p increases. Wall rheology has been described using several tests.

Table 4.4. Rheology tests on vessel wall.

Isolated vessels	
1D static tension	Incremental elastic modulus
Harmonic loading	Dynamic elastic modulus, damping coefficient
2D traction	Directional elastic modulus
Dilation test	Bulk modulus ($C_V = dV/dp$)
non-invasive methods	
Photoplethysmography and US echography	Compliance ($C_R = dR/dp$), Incremental elastic modulus
Imaging (MRI)	Compliance ($C_A = dA/dp = A/(\rho c^2)$, $c = \Delta L/\Delta t$)

---

[7] Walls are characterized by material heterogeneity, being a layered multicomponent structure (cells, fibers, and matrix).

[8] Wall stiffness increases with rising loading. For small pressures, elasic fibers strenghten, whereas collagen fibers line up. At high pressures, the reaction of the collagen fibers is dominant.

[9] The wall behavior is direction-dependent. Three elastic moduli ($E_r(p)$, $E_\theta(p)$, $E_z(p)$ using the cylindrical coordinates) can be involved in straight vessels. The vessel wall is often stiffer in the circumferential than in the longitudinal direction.

[10] The response to loading is time dependent, progressive, and delayed. Damping and phase lag occur between stress and strain. The wall exhibits stress relaxation under constant deformation, creep under constant load, and hysteresis during cyclic loading around a given deformation state.

[11] The wall is irrigated and innervated. It experiences remodeling as well as local and central control.

[12] Under zero transmural pressure, residual tensions exist in an uncut vessel both in the axial and azimuthal directions (T_L, T_c). After cross cuts of the vessel, it lengthens. The ratio between in vivo length and excised length is ~ 0.6–0.7 (after removal of neighborhood connections, $L/L_{\text{in vivo}} \sim 0.9$) After an axial cut of the isolated vessel, it springs out.

Pressure–diameter and pressure–volume relationships provide, at given pressure values, parameters such as the bulk modulus $B = \Delta p/(\Delta V/V)$ and vascular specific compliances based on the cross-sectional lumen area $C = (\Delta A/A)/\Delta p$.[13] The pulse wave speed (PWS) $c^2 = (A/\rho)(\Delta p/\Delta A)$ can be estimated from C or B, assuming an uniform, circular cross-section. Various experimental methods have been developed; they are summarized in a literature review [250]. Additionally, MRI shows that wall deformation is not uniform as well as circumferential variation in wall strains during the cardiac cycle [251].

Muscular tone affects vessel compliance, especially in muscular arteries and arterioles. Phenomenological models of smooth muscle cell behavior are based on the Kelvin model (combination of a spring in parallel with a Maxwell model composed of a spring and a dashpot in series). It is composed of a parallel elastic element (PEE), a serial elastic element (SEE) and a contractile element (CE). PEE describes the non-linear wall smooth muscle response in the absence of muscular tone. SEE couples smooth muscle cells to other wall constituents, to model additional elasticity associated with the contraction. SEE is usually characterized by an exponential constitutive law.

Near the unstressed state, when the stresses are small (but not the deformations), stress–strain relationships lead to linear constitutive equations. Vessel rheology depends not only on its intrinsic properties, but also on the surrounding medium in which the vessel deforms more or less freely. Combination of pressure and axial tension do not induce significant torsion of the vessel wall embedded in a sheath and connected to the surrounding tissues, but radial and longitudinal strains. Consequently, the vessel wall is commonly assumed to be orthotropic. The vessel wall is also supposed to be incompressible due to its composition. Moreover, the physiological magnitude of the applied stresses in situ is not expected to remove free water. Rheological experiments have shown that volume changes are less than 1.5%, mainly due to water removal [252].

Uniaxial loadings exhibit non-linear force–deformation relationships. The relationships between stresses and strains have been mainly explained by the wall microstructure. The ability to bear a load is mostly done by elastin and collagen fibers. Non-linearity is commonly understood as an initial response of elastin fibers and a progressive recruitment of collagen fibers. At low strains, collagen fibers are not fully stretched and elastin fibers play a dominant role. At high strains, the higher stiffness of the stretched collagen fibers affects the elastic properties [249]. When the elastin is higher than the collagen content, the elastic modulus decreases, and distensibility increases, and vice versa.

Incremental stress-strain experiments determine constant piecewize elastic moduli, the so-called incremental moduli E_{inc}. E_{inc} has been calculated from a formula that is derived from the linear theory for a long thick-walled uniform

---

[13] In the literature, vascular specific compliances are also expressed using either the vessel caliber $C = (\Delta d/d)/\Delta p$ or volume $C = (\Delta V/V)/\Delta p$.

tube of incompressible isotropic material $E_{\text{inc}} = 1.5d_i^2 d_e(\Delta p/\Delta d_e)/(d_e^2 - d_i^2)$ [253]. Imaging determination of the wall thickness can only be done by US echography. The formulation was later modified for an orthotropic cylindrical pipe with a non-linear stress–strain relationship:

$$E_{\text{inc}} = 2d_i^2 d_e(\Delta p/\Delta d_e)/(d_e^2 - d_i^2) + 2pd_e^2/(d_e^2 - d_i^2)$$

[254]. Linearized relations between the incremental stresses and strains can only be used when the wall is subjected to small loadings around a given state. Rheological parameters are indeed only valid for the selected loading range centered on a mean value. Furthermore, they depend on the loading history.

Incremental modulus of vessel wall is usually determined when the wall material with its entire set of layers or in selected layers subjected to small strain increments in axial, circumferential, or radial directions leading to small variations in stresses such that the behavior is supposed to be linear. Under homeostatic conditions, incremental moduli are layer- and direction-dependent[14] [255]. Multi-layer models are then necessary to describe wall rheology. Unfortunately, wall dissection into layers affects the results.

Deformation of wall layers can be differentiated by bending experiments of immersed strips of excised artery cut transversally, using the beam theory on heterogeneous vessel wall, once the neutral axis of the vessel wall is determined [256]. Bending experiments are performed, assuming a two-layered wall, the intima and the media being merged to form the internal layer, due to the situation of the neutral axis, and the adventitia corresponding to the external layer. The elastic modulus of the inner layer of the pig aorta has been found to be 10 times that of the outer layer [257]. In the rat aorta, the ratio was found to be smaller, being equal to three to four [258]. The difference in animal species, the aorta wall having similar structures, does not wholly explain the variation in elastic moduli of the adventitia relative to the intima–media layer between the two mammals obtained in the same laboratory (but not the same co-workers), using the same technique of bending of biological samples, and possibly the same setup. In addition, 50% errors in residual stresses are observed if the wall aorta is considered homogeneous. Circumferential prestresses are higher in the internal mid-media layer than in the external media [259]. Moreover, the internal layer is stiffer. Arterial caliber decreases after adventitia removal [260]. Adventitia removal also affects the rheological properties of the wall of brachiocephalic arteries.

Uniaxial loading is widely used because carefully controlled 2D/3D experiments are difficult to carry out on biological tissues. Tissue gripping can damage the soft tissues and induce strong end effects. Nevertheless, biaxial testing has been carried out on thin tissue slices. The biotissue is immersed in a physiological solution at body temperature and attached to the measuring

---

[14] The incremental modulus is highest in the circumferential direction. The media is stiffer than the whole wall, which is stiffer than the adventitia in the circumferential direction and conversely in the axial direction.

device in a trampoline-like fashion. However, the mechanics underlying biaxial loadings remain to be better understood.

In any case, rheological properties differ whether tests are performed on isolated segments of the cardiovascular system or in vivo. The vasculature wall is a living tissue. In in vitro experiments, connections and interactions between regions of the anatomical system and the wall surface and its neighborhood are removed, although they affect wall rheology. Furthermore, excised tissues are more or less dry and not perfused. Any loading squeezes the free water out of the voids of the porous structure and out of the sample if the sample is not confined. Conversely, in vivo measurements are carried out in a noisy environment due to blood circulation and respiration on targeted regions of limited surface areas, most often without preconditionning and without control of influence factors.

Analytical description of time-dependent strain-stress relations are usually derived from phenomenological models. Experimental curves are fitted with analytical functions such as third degree polynomials and exponentials. The most practical expression for the artery subjected to an internal pressure and axial stretching has been found to be an exponential formulation [261], but the associated parameters do not necessarily have a physical meaning. The vessel wall, like most biological tissues, has viscoelastic behavior, characterized by creep, stress relaxation, hysteresis, and loading-rate dependency. Another approach takes into account the microstructure of the wall to describe static rheological behavior. Non-linear elastic model and quasi-linear viscoelasticity have then been developed, considering a ground matrix with a more or less continuous, stress-bearing network of incompressible fibers (rods) that have identical or various initial lengths, cross sections, and orientations in the unstressed configuration and that are uniformly or randomly distributed [262]. Uniaxial stretching induces change in rod size, shape, and orientation.

Linear viscoelastic models have been used, the vessel wall viscoelasticity being represented by dynamic elastic (E_{dyn}) and viscous (loss or damping) moduli[15] ($E(\omega) = E_{\mathrm{dyn}} + \imath\omega\eta$). Early works are based on Voigt model defined by a spring and a dashpot in parallel. This model is characterized by two parameters, the dynamic elastic modulus E_{dyn} and the damping coefficient, which depends on the oscillation frequency. E_{dyn} does not change strongly at frequency above $2\,\mathrm{Hz}$ and the dynamic-to-static modulus ratio is quasi-independent of the mean stress.

Non-linear viscoelastic material theory was used for a 1D tensile loading with step increments of deformation [249]. The history of stress response is described by a relaxation function $\mathcal{R}(\lambda, t) = G(t)T^{(\mathrm{el})}(\lambda)$, $G(0) = 1$, where

---

[15] For sinusoidal loading $\mathbf{c} = \hat{c}\exp\{\imath\omega t\}$ of amplitude $\hat{c}$, the resulting deformation $\epsilon = \hat{\epsilon}\exp\{\imath\omega t + \varphi\}$, has an amplitude of $\hat{\epsilon} = \hat{c}/(E_{\mathrm{dyn}}^2 + (\omega\eta)^2)^{1/2}$ and a phase lag of $\omega\eta/E_{\mathrm{dyn}}$ (rad). The dynamic elastic modulus and the loading on the one hand, the loss modulus and the deformation rate on the other hand, have the same phase. $E_{\mathrm{dyn}} = (\hat{c}/\hat{\epsilon})\cos\varphi$.

$T^{(el)}(\lambda)$ is a pure elastic instantaneous response determined by experimental results and $G(t)$ is the reduced relaxation function identified by experimental data fitting [249].

1D cyclic loading and unloading experiments at a given strain rate and for a given strain range, after preconditioning, exhibits hysteresis. The loading and unloading processes of the strain–stress relationship can be split, considering different elastic material for each process (*pseudoeleasticity*) [249]. Using the pseudoelasticity concept which separately treats the loading and unloading, two different elastic materials are considered for each process instead of using viscoelastic laws. The loading rate is not taken into account. The hysteresis loop can be almost insensitive to the strain rate, a 10^3-fold change in strain rate inducing a stress variation at any strain that does not exceed a factor of 2 for the papillary muscle of the rabbit [263]. However, the literature data exhibit discrepancies due to the type of biological tissue, modulation rate, and magnitude.

3D constitutive equations are more realistic, although they are derived under a hypothesis set. They are based on *strain energy function* (SEF) W, which links stresses to strains via a differentiation. When the cylindrical vessel is assumed to be symmetrically loaded, $W = W(\widehat{e}_r, \widehat{e}_\theta, \widehat{e}_z)$ and the constitutive equation $c_i = \lambda_i(\partial W/\partial e_i) + p$, $i = r, \theta, z$, where c_i is the principal Cauchy stress component, e_i the principal Green strain component and p a scalar function determined at equilibrium from BCs. Logarithmic and exponential formulation have been proposed in the literature, but they are not fully appropriate for numerical simulations. Polynomial expressions are preferred. Blood vessel walls are commonly described by *pseudo-strain energy function* (PSEF) to derive the within-wall stress distribution. Dynamics of smooth muscle cells change rheological properties of the vessel wall according to its tone level, which depends on biomechanical, neural, and chemical stimuli. A constitutive equation of the vessel wall must account for its structure and composition. The vessel wall can be considered to be made of three elements: elastin, collagen (the response of which depends on the fiber stretch level), and smooth muscle cells (the reaction of which is affected by deformation-dependent tone level) [264]. Usually, collagen fibers, embedded in the ground matrix and undulated in the rest configuration, are supposed to be gradually recruited. The vessel wall has been modeled as an isotropic elastic material containing an anisotropic helical network of stiff collagen fibers with a given orientation with respect to the circumferential direction [265]. The structural type of SEF, a sum of an elastic and anisotropic SEFs, has also been developed to take into account the elastic moduli of collagen and elastin as well as collagen waviness and orientation [266].

Constrained mixture models, which meld classical mixture and homogenization theories, consider the specific turnover rates and configurations of the main constituents to study stress-dependent wall growth and remodeling [267, 268], but appropriate knowledge of constituent material properties is still lacking.

Mechanical non-linear model is currently based on hyperelastic incompressible materials. The Mooney-Rivlin material has an elastic stored energy density $\mathcal{W}$ given by:

$$\mathcal{W}(\mathbf{F}) = (|\mathbf{F}|^2 - 3)C_1/2 + (|\operatorname{cof}\mathbf{F}|^2 - 3)C_2/2 \qquad (4.2)$$

where $\mathbf{F}(x) = \mathbf{I} + \nabla\mathbf{u}$ denotes the transformation gradient, and C_1, C_2 are material constants, which can be determined by matching experimental stress–strain relationships [269].

4.4 Microrheology

Cell deformability is demonstrated by cell migration across endothelium clefts. Cell deformability is assessed by filtration through filters of few μm pore size for given filtration pressures. The cell is a complex body that is commonly decomposed into two major rheological components, the plasma membrane and the cytosol. Cell membranes and cytosols can be assumed to be poroviscoelastic and poroplastoviscoelastic materials, respectively [270]. With such a decomposition, macroscopic laws, in particular constitutive laws, are supposed to be valid because the cell size is much greater than the size of its microscopic components. However, the nucleus has a greater size than the various other kinds of cell organelles, which are neglected. When dealing with cell rheology, at least the viscoelastic nucleus must be taken into account. Rheology at the microscopic scale deals with bond forces (Table 4.5).

Rheological sensors must have a suitable size, greater than the size of the cell organelles, as demonstrated by micro- and macrorheometric measurements of the storage and loss moduli in a frequency range $0.01 < f < 4\,\mathrm{Hz}$ of entangled solutions of the bacteriophage-fd in the concentration range $5 < c < 15\,\mathrm{mg/ml}$ [271]. Several rheological techniques have been developed recently to explore cell rheology, which include among the usual methods, a micropipette technique, [272], twisting magnetocytometry [273], and optical tweezer [274] (Table 4.6).

Table 4.5. Interactions at the microscopic scale involved in cohesion, attraction, or repulsion (Source: [270]).

Type	Feature
Covalent bond	10^3–10^4 pN
van der Wals attraction	Short distance 10–10^2 pN
Ionic bond	
Hydrogen bond	
Steric repulsion force	Short distance
Hydrophobic interaction	Long distance

Micropipette aspiration allows the study of continuous deformation and penetration of a cell into a calibrated micropipette (bore<10 μm) at various suction pressures ($[10^{-1}$–10^4 Pa]) to determine an apparent cell viscosity by measuring the rate of cell deformation and pressure. The leading edge of the cell is tracked in a microscope to an accuracy of ±25 nm. Associated basic continuum models, which assume that the cell is a viscous fluid contained in a cortical shell, yield apparent viscosity, shear modulus, and surface tension [275]. The results depend on the ratio of the cell size to the micropipette caliber. Soft cells, such as neutrophils and red blood cells, develops about 16 times smaller surface tension than more rigid cells, such as endothelial cells [276]. Some cell rheology data are given in Table 4.7. Micropipette modeling is not suitable when incompressibility is assumed and viscoelasticity is neglected.

Atomic force microscope (AFM), a combination of the principles of the scanning tunneling microscope and stylus profilometer, provides a force range of [10 pN–100 nN] [277]. Various operating mode can be used. Force dynamic spectroscopy measures time-dependent forces under stretching to provide static and dynamic elastic modulus and adhesion forces [278].

Twisting magnetocytometry (TMC) uses ligand-coated ferromagnetic beads to apply controlled mechanical stress to cells via specific surface receptors. The sampled cell is subjected to a magnetic field and the bead position is recorded using videomicroscopy. The torque resulting from the shear is measured to determine the viscosity and the elastic modulus using a Kelvin model. Mechanical linkage between the receptor and cytoskeleton can be tested as well as cell rheology (shear modulus, viscosity, and motility) over a wide range of frequencies. The bead size affects the results.

Laser beams can be used to move cells and biological molecules. Optical tweezers trap micro- and nanometer-sized dielectric bodies by a focused laser beam through the microscope objective. Light induces fluctuating dipoles that then interact with the electromagnetic field gradient. Tweezing is due to the

Table 4.6. Cell rheology techniques.

Technique	Mechanical properties
Micropipette aspiration	Material constants, elastic modulus, flexural rigidity
Microindentation	Bulk modulus, compliance
Magnetocytometry	Shear modulus
Optical tweezer	Shear modulus

Table 4.7. Indicative physical properties of blood component (Source: [282]).

Cellules	$\rho \sim 1,09 \times 10^3 \, \text{kg/m}^3$
Neutrophils	$\nu = 153 \pm 55 \, \text{Pa.s}$
Lymphocytes	$\nu = 117 \pm 62 \, \text{Pa.ms}$

dipole force of light incident on the dielectric object. The object is pulled toward the beam center (of higher light intensity) when it has a greater index of refraction than the surrounding medium. Optical tweezers are used to apply calibrated forces to cells, with beads bound to the membrane. A target part is attached to a dielectric microsphere held in the trap, whereas another region interacts with an immobilized attractors. The position and stiffness of the trap can be used to measure movements and forces in the investigated object, the trap acting as a spring. The object length is inferred from the force applied by the optical trap and the position of the microsphere relative to the fixed end. The optical trap can be used to make quantitative measurements of displacements ($\mathcal{O}(1)$ nm) and forces ($\mathcal{O}(1)$ pN) with time resolution ($\mathcal{O}(1)$ ms). When an external force is applied to a micron-sized bead in an optical trap, bead displacement from the trap center is proportional to the applied force. When force is quickly applied, the trapped bead can be used as a force transducer. The restoring force in the trap depends on the size, shape, and optical index of the trapped object and is proportional to the incident light power. For instance, the shear modulus of RBC membrane was found to be $2.5 \pm 0.4 \, \mu N/m$ [279].

Both room temperature stability and instrument mechanical stability must be controlled. Brownian motions of the microsphere add noise. Single-trap optical tweezers isolation from environmental and instrumental sources of noise to reach the Brownian noise limit is quite difficult. Dual-trap optical tweezers, in which the studied object is held at both ends by microspheres in two separate optical traps, reduce environmental noise [280]. The dual-trap method is less sensitive to Brownian fluctuations than a single trap.

Measurements have been carried out on round and spread endothelial cells, as well on isolated nuclei [281]. Non-linear force–deformation curves have been found to be affected by cell morphology, the nucleus influence being much greater in spread cells, the most common in vivo shape. Moreover, the living cell is tightly linked to the extracellular matrix (Part I), which can be supposed to be a gel-like medium. Cell adhesion (Part I) affects cell rheological properties. Owing to the cell adaptation to its environment associated with cytoskeleton structural changes, material parameters depend in particular on the cytoskeleton polymerization state (thixotropy). Last but not least, the results of the rheological tests depend not only on the cell state, but also on the techniques, and for a given technique, the experimental procedure (cell environment, loading conditions, impacted region size, etc.).

5

Hemodynamics

Hemodynamics: (1) explores the flow features in the heart and blood vessels, in normal conditions and in pathologies; (2) studies the pressure–flow relationships and the transport of substances by blood; and (3) is a major factor in therapy optimization. Hemodynamics differ at the different length scales of the circulatory circuit. The blood circulation can indeed be decomposed into two major compartments: (1) microcirculation, at the cellular level, where the suspension of blood cell (strongly deformed or not) flows at low Reynolds number (Re); and (2) macrocirculation, in which blood can be supposed to be Newtonian in normal conditions and unsteadily flows at high Re.

5.1 Flow Modeling

Due to its complexity, the cardiovascular system, after splitting, is generally modeled by parts. In large vessels, the flow length scale is such that the fluid is assumed to be a homogeneous continuum. The motion of the vascular wall and blood is commonly described by classical mechanics laws. In particular, blood flows in large vessels are computed using Navier-Stokes equations, which provide a good approximation when the local flow length scale remains greater than the flowing cell size.

5.1.1 Physical Quantities

The main blood variables, the velocity vector $\mathbf{v}$ and the stress tensor $\mathbf{C}$, use the Eulerian formulation. The main wall quantities are the stress $\mathbf{C}$ and displacement $\mathbf{u}$. At interfaces, blood, compliant vessel walls, and possible flowing particle domains are coupled by constraint continuity. The set of conservation equations is closed by the relationships between the transmural pressure p^1

---

[1] $p = p_i - p_e$, where the external pressure p_e, the distribution of which is currently supposed to be uniform, is assumed to be equal to zero. This assumption, which

and the cross-sectional area A (state law). The connected segments of the blood circuit are decoupled with accurate boundary conditions (BC) most often unknown. Constitutive laws of involved materials depend on the microstructure (Chap. 4).

5.1.2 Flow Equations

Blood flows are governed by mass, momentum, and energy balance principles, which are expressed by partial differential equations (PDE). The governing equations of a vessel unsteady flow of an incompressible fluid (with mass density ρ, dynamic viscosity μ, and kinematic viscosity $\nu = \mu/\rho$), which is conveyed with a velocity $\mathbf{v}(\mathbf{x}, t)$ ($\mathbf{x}$: Eulerian position, t: time), are derived from the mass and momentum conservation:[2]

$$\boldsymbol{\nabla} \cdot \mathbf{v} = 0,$$
$$\rho(\mathbf{v}_t + \mathbf{v} \cdot \boldsymbol{\nabla})\mathbf{v} = \mathbf{f} + \boldsymbol{\nabla} \cdot \mathbf{C}, \tag{5.1}$$

where[3] $\mathbf{v}_t \equiv \partial\mathbf{v}/\partial t$, $\mathbf{f} = -\boldsymbol{\nabla}\Phi$ is the body force density (Φ: potential from which body force per unit volume are derived, which is most often neglected) and $\mathbf{C}$ the stress tensor. The constitutive equation for an incompressible fluid is: $\mathbf{C} = -p_i'\mathbf{I} + \mathbf{T}$ where $p_i' = p_i + \Phi$ (when Φ is neglected, $p_i' = p_i$), $\mathbf{I}$ is the identity tensor and $\mathbf{T}$ the extra-stress tensor. When the fluid is Newtonian, the stress tensor is a linear expression of the velocity gradient and pressure; $\mathbf{T} = 2\mu\mathbf{D}$ where $\mu = \mu(T)$ (T: temperature) and $\mathbf{D} = (\boldsymbol{\nabla}\mathbf{v} + \boldsymbol{\nabla}\mathbf{v}^T)/2$ is the deformation rate tensor. The equation set (5.1) leads to the Navier-Stokes equation:

$$\rho(\mathbf{v}_t + (\mathbf{v} \cdot \boldsymbol{\nabla})\mathbf{v}) = -\boldsymbol{\nabla}p_i' + \mu\boldsymbol{\nabla}^2\mathbf{v}. \tag{5.2}$$

5.1.3 Governing Parameters

The formulation of the dimensionless equations depends on the choice of the variable scales ($\bullet^\star$). The dimensionless equations exhibit a set of dimensionless parameters (App. C.1).

---

is a good approximation for superficial vessels, becomes questionable when the vessel is embedded in an environment that constraints the vessel or that, such as the thorax, undergoes cyclic changes. Inflation and deflation of respiratory alveoli are, indeed, induced by nearly oscillatory variations in intrathoracic pressures generated by respiratory muscles, with a more or less gradient in the direction of the body height.

[2] Mass and momentum conservation equations are written from the analysis through infinitesimal control volume on the one hand, and fluid particle loading on the other hand.

[3] $\boldsymbol{\nabla} = (\partial/\partial x_1, \partial/\partial x_2, \partial/\partial x_3)$, $\boldsymbol{\nabla}\cdot$ and $\boldsymbol{\nabla}^2 = \sum_{i=1}^{3} \partial^2/\partial x_i^2$ are the gradient, divergence, and Laplace operators, respectively.

The *Reynolds number* Re $= V^\star L^\star / \nu$ ($V^\star \equiv V_q$: cross-sectional average velocity, $L^\star \equiv R$: vessel radius) is the ratio between convective inertia and viscous effects. When the flow depends on the time, both mean $\overline{\text{Re}} = \text{Re}(\overline{V}_q)$ and peak Reynolds numbers $\widehat{\text{Re}} = \text{Re}(\widehat{V}_q)$, proportional to mean and peak V_q respectively, can be calculated. $\text{Re}_{\delta_S} = \text{Re}/\text{Sto}$ is used for flow stability study (Sto $= R/\delta_S$, δ_S: Stokes *boundary layer thickness*).[4]

The *Stokes number* Sto $= L^\star (\omega/\nu)^{1/2}$ is the square root of the ratio between time inertia and viscous effects. The *Strouhal number* St $= \omega L^\star / V^\star$ is the ratio between time inertia and convective inertia (St $= \text{Sto}^2/\text{Re}$).

The *Dean number* De $= (R_h/R_c)^{1/2} \text{Re}$, for laminar flow in curved vessels, is the product of the square root of the vessel curvature ratio by the Reynolds number. The Dean number De is calculated in phantom tests but not in image-based flow models because of the complex curvature of the vessel axis which varies continually in every direction.

The *modulation rate* (or amplitude ratio), easily determined when the blood flow is approximated by a sinusoidal component, of amplitude $V_\sim$, superimposed on a steady one, $\gamma_v = V_\sim / \overline{V}$ plays a role in flow behavior.

An unsteady Reynolds number has been identified for a nonzero-mean sinusoidal flow:[5] $\text{Re}_\omega = \text{Re}_\sim^2(\delta)/\gamma_v = \overline{V}V_\sim/(\omega\nu)$ [283]. A flow waveform dimensionless parameter has been proposed $\kappa_{(-)}\text{Sto}\big(\overline{\text{Re}}/(\text{Re}_{\max} - \text{Re}_{\min})\big)$, where $\kappa_{(-)}$ is the number of negative flow portions during the flow cycle [284].

5.1.4 Boundary Conditions

The boundary of the fluid domain Ω is partioned into three surfaces: the entry cross-section Γ_1, the exit cross-sections Γ_2, and the vessel wall Γ_3.[6] The classical no-slip condition is applied to the rigid vessel wall. A time-dependent uniform injection velocity $\mathbf{v}_\Gamma(t)$ can be prescribed, at least, at the inlet, that is obtained from the Fourier transform of in vivo US or MR

---

[4] Both Stokes and Rayleigh boundary layers thickness are $\propto (\nu t)$. The Rayleigh boundary layer deals with a flow over a flat plate which suddenly moves in its own plane, with a constant speed (transient regime). The Stokes boundary layer deals with a harmonic motion of a flat plate in its own plane, with an angular frequency ω (periodic flow).

[5] This dimensionless parameter is also the ratio of the stroke length ($V_\sim/\omega$) to the steady diffusion length ($\nu/\overline{V}$).

[6] Γ_3 is the fluid-structure interface, i.e., the moving boundary when the deformable wall is taken into account.

flow signals.[7] At the outlet cross-sections,[8] either the pressure is set to zero or, better, a stress-free condition is commonly applied. Blood is conveyed through a vessel network. The standard BCs at the intlet(s) and outlet(s) of a small explored CVS region do not take into account input and output impedances. Bulk interactions between the region of interest and network parts upstream and dowstream from it are neglected. When BCs are computed from DUS or MRV data, they display transient informations associated to the measurement period. Moreover, such measurements are performed more or less far upstream from the computational domain entry and not necessarily in the same subject. Flowing blood interacts with the vessel wall via mass transport and transmitted stresses. The size of the artery lumen can vary up to about 10% from its diastolic to its systolic shape. Such a deformation enters in the field of large displacements. Appropriate modeling is thus based on a set of equations that describes blood motion and wall displacement. A multiphysics multiscale approach is then necessary to provide suitable BCs both at the moving interface and domain inlets and outlets (Sects. 6.2.4 and 6.3).

Womersley Solution

The Womersley solution corresponds to the Poiseuille flow with a harmonic flow (App. C.3). It has been used to provide entry velocity distribution in computational models of blood flows:

$$\tilde{v}_\sim = \imath \widehat{\widetilde{G}_{p\sim}} \left(\frac{J_0(\imath^{3/2} \tilde{r}(\omega/\nu)^{1/2})}{J_0(\imath^{3/2}\mathrm{Sto})} - 1 \right). \tag{5.3}$$

The pressure signals recorded at two vessel stations can be transformed into a flow signal. The harmonic analysis of the pressure waveforms is followed by the computation of the flow generated by each harmonic, using a simple flow model. The pulsatile flow is then reconstructed by the synthesis of the flow harmonics (Womersley-type technique). Other simple methods are based on lumped parameter models (Sect. 6.2.4.1).

---

[7] Most often the spatial resolution of in vivo velocity measurements is not high enough to provide velocity distribution at vessel ends, especially at the domain entry of greater cross-section. In absence of measurements, the inlet velocity profile is provided by the Womersley solution, which implies a fully developed flow without body forces in a long smooth straight pipe of circular cross-section, a set of properties never encountered in blood circulation and vessel geometry. Such inlet BC is thus not more appropriate than time-dependent uniform injection velocity. However, the uniform velocity condition is associated with high wall shear at the entrance susceptible to induce larger flow separation if an adverse pressure gradient is set up, in particular in the transition zone between trunk and branches. Furthermore, high entry wall vorticity diffusion and convection in the vessel may rise the swirling amount found in the explored volume. In addition, the transverse mesh must limit velocity discontinuity between zero Dirichlet wall and inlet BCs.

[8] The outlet sections must be perpendicular to the local vessel axis and be short straight pipe exits to avoid pressure cross-gradient.

Table 5.1. Flow response to the imposed pressure difference ($p(t)$).

Near wall	Viscosity-dominant pattern
	Important phase lag
Boundary layer	Inertia balances viscosity
	Moderate phase lag
Core	Inertia-dominant pattern
	Quick response
	Small phase lag

5.2 Flow Features

Due to wall friction, the near-wall fluid particles slow down, whereas the fluid particles in the core accelerate. Viscous forces are dominant in the boundary layer, whereas inertia forces are greater in the core. In the *boundary layer*, the fluid particles then respond to the pressure time changes with a phase lag with respect to those in the *core* flow. The phase lag of the motion of the fluid particles to the time-dependent imposed pressure at the vasculature inlet by the heart thus depend on their location within the vessel lumen (Table 5.1).

The vasculature is made of successive bends and branchings. The embranchment can be, at a first approximation, supposed to be constituted of two juxtaposed bends, with a slip condition on the common wall within the stem. Bends present either gentle or strong curvature, and various curvature angles up to 180 degrees (aortic arch, intracranial segment of the internal carotid artery). The bend then represents the most simple basic unit of the circulatory system.

> "*Cet ordre est déterminé par le degré de simplicité, ou ce qui revient au même, par le degré de généralité des phénomènes, d'où résulte leur dépendance successive et, en conséquence, la facilité plus ou moins grande de leur étude*" (A. Comte) [285]

Energy dissipation is generated by fluid shearing within the vessel lumen; geometrical changes (bends, tapers, branchings, etc.) with possible flow separation add head losses. The pressure loss in bends, embranchments, and confluences depends on: (1) vessel caliber, (2) respective flow rates, (3) curvature angle and branching/merging angles, (4) wall roughness, and (5) fluid physical properties. The influence agents can be combined into dimensionless ratios (curvature ratio, area ratio, flow distribution, head loss/friction coefficient, Reynolds number, Dean number, etc.). Values of the head loss coefficient ζ[9] in various types of singular vessel geometries in steady flow can be found in Idel'cik's textbook [286].

---

[9] $\Delta p = \zeta \rho V_q^2 / 2$.

Table 5.2. Values of the moments of simple cases of the velocity distribution.

	Uniform profile	Parabolic profile
M_1/Rq_v	0	0
$M_2/(R^2 q_v)^{1/2}$	0.707	0.577
$M_3/(R^3 q_v)^{1/3}$	0	0

The bend skews off the velocity distribution in any cross-section. The velocity distribution can be described and quantified by the moments of the velocity distribution (Table 5.2):[10]

$$M_0 = \int_0^{2\pi} \int_0^R v(r,\theta) r \, dr \, d\theta = q_v \,, \quad M_1 = \int_0^{2\pi} \int_0^R rv(r,\theta) r \, dr \, d\theta \,.$$

M_1/Rq_v provides the location of the flow mass center from which are computed M_2, and M_3. $\left(M_2/(R^2 q_v)\right)^{1/2}$ gives the standard deviation of the cross velocity distribution and $\left(M_3/(R^3 q_v)\right)^{1/3}$ the skewness around the flow mass center.

Bends, embranchments, and junctions induce 3D flows. Change in cross-section along the vessel length, as displayed in straight transition zones of planar symmetrical vessel bifurcation and straight collapsed segments, also generates 3D flows. However, the types of the local velocity fields projected in the cross-section plane, the so-called *secondary motion*,[11] are different whether they are produced by vessel curvature or gradual change in transverse shape.

5.2.1 Bend Flows

The first investigations of bend flows dealt with steady laminar flow in rigid curved tubes. In a bend with a curvature radius of its axis R_c, any fluid particle experiences a lateral acceleration $\propto \rho V_q^2/R_c$. In any cross section of the curved vessel, the centrifugal forces are balanced by a reacting transverse pressure difference, which decays from the outer to the inner wall. The vessel curvature induces a helicoidal motion of the fluid particles [287]. At any location of the fluid domain, the velocity vector can be decomposed into a component parallel to the local vessel axis (streamwise component normal to

---

[10] The higher the fluid velocity in an area of infinitesimal surface dA of the explored cross-section of the fluid domain labeled by its radial and azimuthal positions, the larger the probability of the fluid particle associated with a bolus of infinitesimal volume dV injected usptream from the explored cross-section to cross this pipe section.

[11] The secondary flow does not depict any actual flow, but corresponds to a representation mode of the flow field, using projections of the local velocity field on planes of selected cross-sections.

the cross-section) and a component in the plane of the cross-section. The latter component constitutes the secondary motion. In steady bend flows, the secondary motion is defined by a vortex pair. Each vortex, of axis parallel to the streamwise direction, occupies a half cross-section. The vortex direction near the centerplane is outward and returns to the inner wall in a small near-wall layer.

Most often, the vessel is constituted by serial bends, more or less gentle, mostly non-planar. Due to the velocity distribution induced by the centrifugal forces in the upstream bend, impingement is produced in the following curved vessel segment.

The steady laminar motion of an incompressible Newtonian fluid through a rigid smooth curved pipe of circular cross-section, with a small uniform curvature, has been studied by Dean [288–290], who pointed out that a parameter, the so-called Dean number (De $=$ Re $\times \kappa_c^{1/2}$) governs the laminar motion[12] in a bend. The higher De, the stronger the secondary motion [292]. The parameter De is the combination of two dimensionless factors, Re and the curvature ratio $\kappa_c = R/R_c$. Similar values of De with different values of Re and κ_c lead to different flow behavior [195]. The effects of curvature are expressed even at small κ_c ($\kappa_c < 0.02$), but significant bend-induced modifications in velocity fields occur at a distance $(RR_c)^{1/2}$ from the bend entry in steady flows [293]. The distortion of the velocity distribution in the bend cross-section is associated with a greater pressure drop [294]. The helical fluid motion in a bend induces more uniform distribution of an injected tracer [295]. In a steady laminar flow in a plane curved pipe, the curvature increases and decreases the axial component of the wall shear stress (WSS) on the outer and inner edge, respectively. Furthermore, the duct curvature generates a cross-component. Flow variations produced by the bend are felt upstream [296]. The pressure drop is influenced by a 90 degree bend several diameters upstream from the bend inlet at Re $= 2 \times 10^5$ [297]. Significant influences are observed in straight ducts connected to a curved tube, both upstream and downstream [195].

In a plane uniform bend conveying a zero-mean sinusoidal (purely oscillatory) fully developed flow, the *secondary motion* is governed by the Reynolds number Re_2 and the Strouhal number St_2 of the secondary flow[13] [298]. Twin

---

[12] In fully developed turbulent flow, the friction factor depends on the Ito number Ito $=$ Re $\times \kappa_c^2$ [291].

[13] When the secondary velocity scale (App. C.1.2) $V_2 = V^2/\omega R_c$, $Re_2 = V^2 R/\nu \omega R_c$ and $St_2 = \omega^2 RR_c/V^2$. St_2 is also the ratio of the product of the curvature and pipe radii to the square of the particle displacement amplitude (stroke length).

The secondary motion Reynolds number $Re_2 = \widehat{V}/(\omega\nu).\kappa_c$ can be also expressed as:

$$Re_2 = \frac{\widehat{V} R^2}{\nu^2 R^2 (\omega/\nu)}\kappa_c = \left(\frac{\widehat{Re}}{Sto}\right)^2 \kappa_c.$$

vortices are found in each half cross-section. In the transverse boundary layer, the fluid is driven by centrifugal forces toward the outer edge and returns near the wall to the inner edge. Viscous effects lead to a second transverse vortex in the core with an inward secondary motion near the centerplane. The secondary motion described in bend fully developed purely oscillatory flow has also been theoretically found in fully developed pulsatile flow (nonzero-mean sinusoidal flow), when the pressure modulation rate γ_p and De are small, and Sto is great [299]. This theoretical analysis leads to a classification of bend fully developed pulsatile flow according to the values of three governing dimensionless parameters, De, Sto and Re_2, knowing γ_p.[14] This classification has been adopted for entry bend flow, from experimental observations in a 180-degree bend [300]. Two types of pulsatile flows were used: test 1[15] and 2.[16] Test 2 is defined by greater curvature effects (κ_c) as well as larger unsteadiness (Sto), associated with a stronger tendency of flow reversal during part of the flow cycle (γ_u). Secondary motions are more complex. Complex secondary flows have also been observed in a computational model of a bend fully developed flow,[17] either nonzero-mean sinusoidal flow or using a physiological-type input [301]. The lower Sto, the earlier the occurrence of the Lyne-type secondary motion[18] [302]. The lower Sto, the higher $\overline{Re}$ necessary to get this secondary motion.

The fully developed laminar pulsatile flow (nonzero-mean sinusoidal flow) can be classified into four flow types (App. C.1.2): (1) viscosity-dominated motion, in which unsteady, convective, and centrifugal inertias are small; (2) unsteady inertia-dominated movement; (3) convective inertia-dominated flow; and (4) combination of influences of the involved forces without strongly prevailing term [303]. The boundaries between the regions of the De–Sto diagrams defining the fully developed motion types 1, 2, and 3 depend on γ_u.

The impedance (pressure drop/flow rate ratio) of 180-degree curved tubes conveying nonzero-mean sinusoidal flows is greater than in bends through which the fluid flows steadily, which is greater than the impedance measured in straight pipes [304]. The maximum secondary velocity has been found to be much smaller than the maximum axial velocity $V_{2,\max}/V_{z,\max} < 0.04$, whatever the flow cycle, in a computational model of the flow in the left coronary

---

[14] Let $G = \partial p/\partial z$ and $\gamma_p = G_\sim/\overline{G}$, with $G_\sim \propto \rho\omega V$ (time inertia reaction) and $\overline{G} \propto \mu V/R^2$ (fully developed steady component). Then, $\gamma_p = Sto^2$. γ_p can also be expressed by:

$$\gamma_p \propto \frac{\omega}{\nu^2}\frac{R^3}{R}\nu\frac{V}{V}\frac{\kappa_c^{1/2}}{\kappa_c^{1/2}} = \frac{Sto^3 Re_2^{1/2}}{De}.$$

[15] $\gamma_u = 0.34$, $\kappa_c = 1/20$, Sto $= 8$ and $\overline{De} = 60$.
[16] $\gamma_u = 1.02$, $\kappa_c = 1/7$, Sto $= 12.5$ and $\overline{De} = 190$.
[17] $\kappa_c \geq 3.8$, Sto ≤ 21.
[18] The authors developed a finite difference-based computational model of a fully developed flow in a bend with $\kappa_c = 1/7$, $\gamma_u = 0.98$, $9.5 \leq \overline{De} \leq 85$, $7.5 \leq$ Sto ≤ 25.

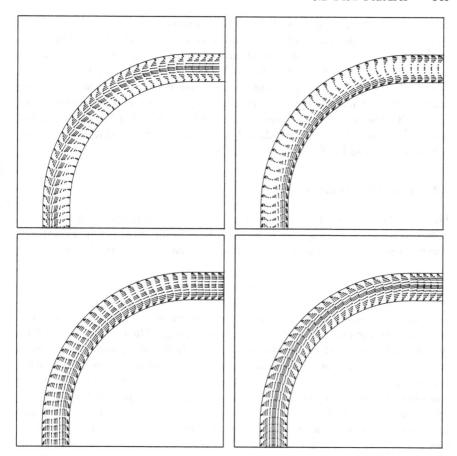

Figure 5.1. Oscillatory axial velocity profiles in a bend (90 degree angle, $\kappa_c = 1/10$). Finite element method. From top to bottom and left to right: End of decelerating and accelerating phase in the reverse direction. The maximum velocity moves from the inner bend to the outer bend. The flow reversal region is larger at the inner bend than at the outer edge.

artery, using a physiological-type input associated with uniform injection velocity[19] [305]. The velocity maximum moves from the outer wall to the inner edge in oscillatory flow, each bend edge being transiently subjected to strong shears during the flow cycle (Fig 5.1).

Time-Varying Curvature

The epicardial coronary vessels undergo large axis deformation during the myocardium contraction-relaxation cycle. The steady flow in a bend with time-

---

[19] The bend is tapered ($R_{exit}/R_{entry} = 0.96$). $\kappa_c = 1/45.5$.

varying curvature (relative variation in curvature radius $\Delta R_c / R_c \leq 0.15$), with or without displacement of the curvature center, is affected by the vessel motion [306]. The zero-mean sinusoidal flow of an incompressible Newtonian fluid in a bend characterized by time-dependent curvature ($\kappa_c(t) = \overline{\kappa_c}(1 + \gamma_R \sin \omega t)$), modeling the motion of the coronary vessels on the heart surface, depends on four main governing parameters, $\overline{\kappa_c}$, Sto and two parameters associated with the curvature, the secondary motion-associated Reynolds number Re_2 and a time-dependent curvature Dean number $De_t = \widehat{\kappa_c} Re$, or a combination of Re_2 and De_t[20], $Rct = De_t / (Re_2 \times \overline{\kappa_c})^{1/2}$ [307].

5.2.2 Embranchment Flows

The vasculature does not present any symmetrical bifurcation or junction. However, some branching sites have an almost symmetrical geometry, at least in certain subjects, such as aortic bifurcation (quasi-symmetrical bifurcation).

5.2.2.1 Symmetrical Bifurcation

The symmetrical bifurcation requires the specification of a smaller number of geometry parameters than asymmetrical branchings. The geometry quantities that must be specified are: branching angle,[21] flow divider (apex) sharpness, curvature radius of the outer bifurcation edge, gradual change in transition zone shape, area ratio,[22] curvature of the stem and branches and planarity or non-planarity of the vessel axes. In addition to geometry parameters, dynamical factors are input flow rate, entry velocity profiles (inlet vorticity, inlet wall shear rate), values of the flow governing parameters, and flow distribution or flow ratio ($FR_i = q_{bi}/q_t$).

In steady flows, the flow splitting into two streams is associated with the generation of a new boundary layer at the branching inner wall and a maximum velocity near it even when the parent and daughter tubes are straight. The branching curvature induces a bend-like motion of the fluid particles. The secondary flow then corresponds to the display mode of the helical motion. Flow separation can occur in the entrance segment of the branches when Re is high enough [308]. Furthermore, the larger the branch angle, the greater the wall vorticity at the entrance section of the trunk, the wider the flow separation region. The transition zone generates 3D flows, with a source-sink type

---

[20] Because $Re_2 = (\widehat{Re}/Sto)^2 \kappa_c$,

$$Rct = \frac{\widehat{\kappa_c} \widehat{Re} Sto}{\widehat{Re \kappa_c}} = \gamma_R Sto.$$

[21] The branching angle is the angle between the branch and trunk axis.

[22] The area ratio (AR) is the ratio of the sum of the branch cross-sectional area A_b to the trunk cross-sectional area A_t: $AR = \sum_i A_{bi}/A_t$.

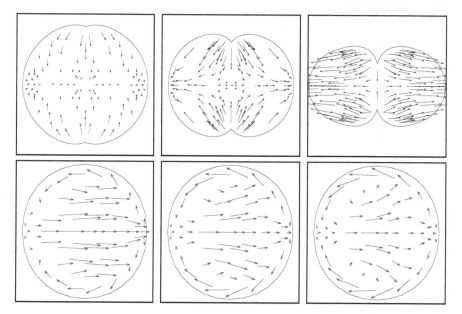

Figure 5.2. Types of secondary motions in a plane symmetrical bifurcation (between branch-axis angle of 70 degrees, area ratio of 0.8) with a transition zone $(L = 1\,d_t)$ with converging top and bottom walls and diverging edges. Steady flow (Re = 1200). (**Top**) Source-sink type of secondary motions in the transition zone. (**Bottom**) Bend-type secondary motions in the entrance segment of the branch. Stations at 0.25, 0.5, and 0.75 d from the transition zone inlet and branch entry.

of secondary motion (Fig. 5.2). When the flow is blocked in one branch, an induced "dynamical curvature" at the bifurcation, superimposed to the branch curvature, generates twin vortices in a half cross-section of the permeable branch (Fig. 5.3).

Asymmetrical Branching

Several kinds of geometric asymmetry exist, which can be combined: (1) bifurcation with branches of different cross-sectional areas, and thus given area ratios $(AR_i = A_{bi}/A_t)$; (2) bifurcation with branches of similar bores but with different branching angles (e.g., carotid bifurcation [309, 310]); (3) side branches with a smaller caliber $(AR \ll 1)$ and a given branching angle, whereas the trunk undergoes a moderate change in direction (e.g., superior mesenteric unpaired branching, renal paired branching).

The flow field[23] is affected by the flow distribution, which can vary during the cardiac cycle, as demonstrated in carotid bifurcation [311]. The

---

[23] $\widehat{\mathrm{Re}} = 800$ and Sto $= 4$ in the trunk. Flow part in the external carotid artery varies from 45% at peak flow to 10% during the diastole.

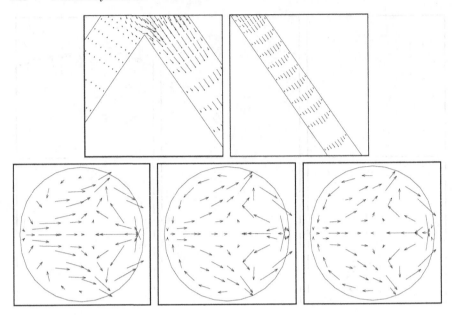

Figure 5.3. Steady flow (Re = 1200) in a symmetrical bifurcation (between branch-axis angle of 70 degrees) with one (left) blocked branch. (**Top**) axial component (**Bottom**) secondary motion in the permeable branch at stations 0.25, 0.5, and 0.75 d from the branch inlet. One vortex in each half cross-section can be assumed to be generated by dynamic curvature corresponding to the fluid particle paths in front of the obstructed branch. Absence of flow separation.

importance of the branch-to-trunk flow ratio has also been educed by flow visualization in a mold of the abdominal aorta with its renal branches[24] [312]. In a model of carotid bifurcation, computed secondary motions of a Newtonian fluid flow are greater than those of a shear-thinning fluid flow, using the Casson law[25] [313]. The velocity fields have also been measured in: (1) a deformable model of the aorta with the three main branches of the aortic arch and the terminal branches, associated with a ball aortic valve,[26] with a physiological input, using USV [314]; (2) a model of the abdominal aorta with the coplanar coeliac and mesenteric arteries, which receive 20% and 10% of the nonzero-mean sinusoidal flow[27] respectively, using LDV and electrochemical sensors [315]; (3) a model of the coronary artery with an embranchment, using

---

[24] $\widehat{\mathrm{Re}} = 1700$ and Sto = 6.7 or 7.9.
[25] $\overline{\mathrm{Re}} = 300$ and Sto = 4.8.
[26] The values of the governing parameters are not given.
[27] Sto = 16 and $\gamma_u = 1$.

LDV[28] [316]; and (4) a model of the pulmonary artery with its two coplanar branches and an inlet trileaflet valve[29] using LDV [317].

In a 2D symmetrical bifurcation conveying a nonzero-mean sinusoidal[30] flow, with a flow division ratio of 2:1, the flow separation are swept downstream at each flow cycle [318]. Flow separation can occur in the branches and stem when the branching corner is sharp or the branching angle is high such as in T-shaped branching. In T-junctions,[31] LDV shows that the pulsatile flow separation is small and located at the wall opposite to the branch when a viscoelastic fluid[32] flows in a glass model and is larger and located at the branching wall, immediately downstream the embranchment when it is conveyed in a silicone model [319]. When a Newtonian aqueous glycerine solution is convected through the same model, flow separation remains at the same location, at the wall opposite to the branch, with a separation point nearly facing the branch axis, but it is greater in the rigid model than the silicone model.

5.2.3 Wavy Wall

The flow near the rigid wavy wall, which is defined by a sinusoidal undulation of wavelength λ and of amplitude smaller than the Stokes boundary layer thickness, depends on two dimensionless parameters, Sto $= \lambda(\omega/\nu)^{1/2}$ and Re $= V_\infty^2/(\nu\omega) = L_s V_\infty/\nu$ (V_∞: free-stream velocity, L_s: stroke length scale) [320]. Flow separation occurs in hollows of significant size (one fifth of the channel width) during the acceleration phase, and grows to occupy most of the hollow at peak flow [321]. During the deceleration phase, it continues to grow and an ejected vortex appears that is more or less rapidly eroded, depending on the value of the Strouhal number. Such a wavy wall model represents a wall bulging rather than the endothelium coating with between-cell-nucleus hollows, such small-sized hollows being neglected in macrocirculation, but not in the smaller vessels of microcirculation.

5.2.4 Entry Flow

In straight pipes conveying a steady flow of a Newtonian fluid, two flow regions can be defined: (1) a developing flow region in a tube length equal to the entry length (Le), and (2) a fully developed flow region characterized by an invariant

---

[28] But the vessel bore is too small for suitable measurements ($\overline{Re} = 180$ and Sto $= 2.7$).

[29] Branching angle of 120 degrees, $A_{br}/A_t = 0.48$, $A_{bl}/A_t = 0.39$, flow ratio of 50/50, $\widehat{Re} = 1660$

[30] $\widehat{Re} = 1000$, Sto $= 7.9$, and $\gamma_u = 1$ in the trunk.

[31] Bifurcation angle of 90 degrees, $A_b/A_t = 0.25$, $q_{b1} = q_{b2}$.

[32] A mixture was made of 0.0375% separan AP30, 0.01% separan AP45, 0.01% MgCl$_2$, 4% isopropanol, and 43% glycerine.

velocity distribution in the cross-section and a constant pressure drop. In entry steady flows, the flow domain can be subdivided into two subdomains, a *boundary layer*[33] of thickness $\delta \propto (\nu L/V)^{1/2}$ (L: distance form the tube entry) and a core where the fluid is assumed to be inviscid. In fully developed flows, the boundary layer spreads across the whole pipe lumen.

Pulsatile Flow

In pulsatile flow, two boundary layer thicknesses δ_S[34] and δ are estimated either from the balance between the inertia forces associated with the local acceleration in the core flow $\propto \rho \omega V_\infty$ and the viscous forces in the boundary layer $\propto \mu V_\infty / \delta_1^2$ (μ: fluid dynamic viscosity, V_∞: free stream velocity) or from the balance between the inertia forces due to the convective acceleration in the boundary layer and the viscous forces respectively: $\delta_S \propto (\nu/\omega)^{1/2}$ and $\delta \propto (\nu L/V_\infty)^{1/2}$. $\delta_1 = \delta_2$ when $z \equiv L_+ = 2.64 V_\infty/\omega$. When $z < L_+$, the temporal inertia forces are dominant; when $z > L_+$, the convective inertia forces are greater. The entry length in pulsatile flow in a straight rigid cylindrical pipe of circular cross-section has been found to have similar expression as for the steady flow, using Le$/\delta$Re(δ) rather than Le$/R$Re(R) when Sto ≤ 14 [324]. The entry length in a straight tube conveying a steady flow can be defined by $(v_{ax}(\infty) - v_{ax}(\text{Le}))/v_{ax}(\infty) \leq 0.01$. Similarly, the entry length in any periodic flow of amplitude $\widehat{V}_q$ can be defined by the centerline velocity, whatever the time t, using the proposed ratio $(v_{ax}(\infty, t) - v_{ax}(\text{Le}, t))/\widehat{V}_q$ [325]. Expressions of Le$/(R$Re$_R)$ with respect to Sto have been proposed, knowing that, for a given Sto, the value of Le$/(R$Re$_R)$ can vary according to the cycle phase [326].

Bend

The ratio of the entry length of a bend of uniform plane curvature and the entrance length of a straight pipe, both conveying a steady flow, Le$_c$/Le$_{st} \propto$ De$^{-1/2}$ for sufficiently large De [294] (Le$_c \propto R \times$ Re$^{1/2} \times \delta^{-1/4}$). The flow development of a zero-mean sinusoidal motion in a 180-degree bend de-

---

[33] The boundary layer is a layer adjacent to the wall beyond which the fluid is considered inviscid, of uniform velocity V_∞ (plug core flow). It produces a vorticity that diffuses by action of the viscosity (momentum diffusion). The boundary layer is then a vorticity source and momentum sink. Boundary layer equations have been proposed as a simplified version of Navier-Stokes equations by L. Prandtl [322]. Blasius used them for the flow over a flat plate [323].

[34] The Stokes boundary layer thickness $\delta_S = (\nu/\omega)^{1/2}$ measures the radial distance from the wall toward the centerline, which can be traveled by means of diffusion by any fluid particle (any disturbances) within the flow cycle of period T (of circular frequency ω).

pends on the oscillation amplitude and frequency parameter[35] [327]. The entry
length also depends on the amount of inlet vorticity. In a 116-degree bend,
the physiological-type flow is not fully developed [328].

Blood vessels between two branching sites are never long enough for the
complete development of the flow. Blood flow is then never free from entrance
effects. The flow is characterized by a growing boundary layer and a non-
linearly varying pressure with the distance from the vessel inlet. The entrance
conditions can have large influences on fluid flow, especially at high Reynolds
number motions.

5.2.5 Flow Stability

*"Il y a un imbécile en moi, et il faut que je profite
de ses fautes... C'est une éternelle bataille contre
les lacunes, les oublis, les dispersions, les coups de
vent. "* (P. Valéry) [329]

*"La seule chose que nous ayons à faire ici, c'est
d'insister sur un précepte qui prémunira toujours
l'esprit contre les causes innombrables d'erreur...
Ce précepte général, qui est une des bases de
la méthode expérimentale, c'est le doute;..."*
(C. Bernard) [330]

A stable flow is one in which any small disturbance is spontaneously eliminated. Strong local pressure gradients, velocity profile inflections, and shear
reversals affect the flow stability. A major source of instabilities is vortex
production. Vortices arise from concentrated-vorticity regions. The sheared
vortex can then be stretched and rotate. Self-modulated perturbations can
lead to evolving small-scale structures. A disturbed flow is characterized by
transient instabilities, which decay as they propagate downstream, due to the
dissipative action of the viscous forces. The development of velocity/vorticity
perturbations in the flow, with a characteristic magnitude $\varepsilon V^\star$, can lead to a
transitional flow. Transitional flow is defined by preserved perturbations. Turbulent motions are characterizd by a non-linear[36] highly dissipative process
associated with random motions, 3D fluctuation velocities, and high diffusivity, which is accounted for by an eddy viscosity $\nu_T(\mathbf{x}, t)$.

A critical dimensionless parameter is a ratio of destabilizing to stabilizing
forces. When the threshold is exceeded at any point of the fluid domain, the
flow destabilization is not counterbalanced by restoring forces and the disturbances grow. The Reynolds number, the ratio of convective inertia forces
to viscous forces, is adapted to steady flows. The critical Reynolds number

---

[35] Bend flow is characterized by $\kappa_c = 1/7$, $\widehat{De} = 32$, and Sto $= 1$ for test 1,
$\widehat{De} = 5.8$, 38.5 and Sto $= 4.4$ for test 2. Measurements used LDV.

[36] There is no simple combination of signals due to non-linear transfer functions
and energy transport between different scale and frequency components.

Re_{crit}, at which transition from laminar pattern occurs, has a currently recognized value of 2200 in straigth pipes [331, 332]. The critical Reynolds number is the threshold below which the flow is stable to infinitesimal disturbances whatever the frequency and wavelength, but this threshold can be pushed upward to about 10^5 when extreme care is done to reduce disturbances. Because the blood flows with a pulsatile nature in deformable vessels under a pressure difference which has a given harmonic content, the standard Re_{crit} is not a reliable index of transition from laminar flow for the blood circulation. A simple threshold is not suitable because it is strongly related to the flow conditions from which it is determined. However, the transition Reynolds number for a nonzero-mean sinusoidal flow in straight ducts has been defined by the value of $\overline{Re}$ at which the slope of the isoSto St–$\overline{Re}$ relationships changes [333]. The peak Reynolds number[37] for a pulsatile flow is more appropriate. In addition, flow modulations affect the values of Re_{crit} [334]. Re_{crit} is lower for non-harmonic (piston driven by a slider-crank mechanism $L/L' = 4$) than for harmonic (piston driven by a scotch-yoke mechanism) pulsations, the latter providing, for the same mean pressure gradient, a more stable flow than the steady flow.

Like steady flows, time-varying flows are stable if any disturbance decays continuously. Periodic flows are transiently stable if a disturbance grows during a part of the cycle and decays. Periodic flows are unstable when the disturbance grows during each period. Periodic flow patterns have thus been classified into four main types [335].[38] **Type 1** corresponds to a laminar flow, the flow remaining undisturbed throughout the flow cycle. **Type 2** is a disturbed laminar flow with small amplitude perturbations. **Type 3** is an intermittently turbulent flow in which high-frequency velocity fluctuations occur at the beginning of the deceleration phase, increase, and dissipate prior to or during the subsequent acceleration phase. **Type 4** is a fully developed turbulent flow, high-frequency velocity fluctuations existing during the whole flow cycle. Additional types have been proposed [336, 337].[39] **Subpattern 1** is defined by small-amplitude perturbations occurring in the early stage of acceleration phase in the flow core. **Subpattern 2** is characterized by small-amplitude perturbations during the whole acceleration phase.

---

[37] In simple nonzero-mean sinusoidal flow with a modulation amplitude $V_{q\sim}$, $\widehat{Re} = \overline{Re} + Re_\sim$ ($Re_\sim = V_{q\sim}R/\nu$).

[38] This investigation deals with zero-mean sinusoidal flows, explored by LDV, with $380 < \widehat{Re}_\delta < 1420$, and $7 < Sto < 14.2$. Turbulence appears explosively toward the end of the acceleration phase and is sustained throughout the deceleration phase.

[39] Both studies deal with zero-mean sinusoidal flows, using hotwire velocimetry, either with $13 < \widehat{Re}_\delta < 1080$ ($52 < \widehat{Re} < 2920$) and $1.9 < Sto < 8.8$, or with $300 < \widehat{Re} < 32500$ and $2.6 < Sto < 41$. Transiently turbulent flows are characterized by sudden generation of turbulence during the deceleration phase with relaminarization during the acceleration phase.

Reynolds stresses are used to study possible transitional flow in the presence of an implanted medical devices. The Reynolds stress tensor $\overline{\rho \mathbf{v'} \otimes \mathbf{v'}}$ is given in a Cartesian frame by:

$$\rho \begin{pmatrix} v_1 v_1 & v_1 v_2 & v_1 v_3 \\ v_1 v_2 & v_2 v_2 & v_2 v_3 \\ v_1 v_3 & v_2 v_3 & v_3 v_3 \end{pmatrix}$$

In a steady flow, each velocity component v_i, mesured in a Cartesian frame, is decomposed into a mean $\bar{v}_i = (1/N)\sum_{k=1}^{N} v_i^k$ (N: sample number) and a fluctuation v_i'. The Reynolds stress component can then be computed: $\overline{\rho v_i v_j} = (1/N)\sum_{k=1}^{N} \rho v_i^k v_j^k$. Assuming quasi-steadiness in a small time interval, the same decomposition can be made in pulsatile flow, using a very short time window (Δt_w of few ms) and a set of flow cycles ($N > 100$). The quantities of interest are the principal Reynolds stresses $\overline{\rho v_i v_j}|_P$ along the principal axes of the tensor [338].

The turbulent boundary layer is currently decomposed into a three-tiered structure, with: (1) a near-wall viscous sublayer where the laminar shear is greater than the turbulent shear, (2) a buffer sublayer, and (3) an inertial sublayer. The dimensionless viscous sublayer thickness is defined by: $\tilde{\delta}_S = \delta_S(u_*/\nu) = (\dot{\gamma}_w/\omega)^{1/2}$. The dimensionless forcing angular frequency $\tilde{\omega} = \tilde{\delta}_S^{-1}$ determines whether the viscous and buffer sublayers have quasi-steady behavior or not. The vessel lumen crossed by a turbulent oscillatory flow can be decomposed into three zones, which can be identified throughout the flow cycle [335]: (1) a viscous sublayer characterized by a length scale $\delta_{Tv} = \nu/u_*$, the unsteady length scale being defined by $\delta_{Tt} = u_*/\omega$, (2) a logarithmic layer, and (3) a central wake. The flow structure depends on the length scale ratios, $\mathrm{St} = R/\delta_{Tt} = R\omega/u_*$ and $\mathrm{Re} = \delta_{Tt} u_*/\nu = \delta_{Tt}/\delta_{Tv} = u_*^2/\omega\nu$. When $\mathrm{Sto} > 2$, a transitional $\mathrm{Re}_{\delta,\mathrm{crit}}$ of 350 to 400 has been observed.

The simplest quantity to first consider is the available time for disturbance growth and spreading. A dimensionless relaxation time has been proposed, i.e., the product of the ratio of the time available for perturbation growth (the duration of the acceleration phase) to the momentum diffusion time scale (R^2/ν) by the ratio between convective inertia and viscous effects (Re) [339]. The pulsatile flow is indeed unstable if the disturbance experiences a net growth over each flow cycle (otherwise, it is either stable, when every perturbation decays at each instant, or transiently stable when a disturbance grows during a cycle part and decays). The higher Sto, the lower the time available for disturbance growth. The growth rate increases with Re, when the flow unsteadiness acts as a simple modulation factor. The value of the Strouhal number (St) can then be taken into account. When $\mathrm{St} \ll 1$ ($\mathrm{St} \propto T_{\mathrm{conv}}/T$) and $\mathrm{Re} \gg 1$ ($\mathrm{Re} \propto T_{\mathrm{diff}}/T_{\mathrm{conv}}$), the convective time scale for vortex development inside the vessel lumen ($R/\widehat{V}$, $\delta_S/\widehat{V}$ inside the boundary layer) is lower than both the momentum diffusion time scale (R^2/ν, δ_S^2/ν inside the boundary layer) and the cycle period T (even shorter than the duration of the decelerating phase).

Destabilization–stabilization of a pulsatile flow depends thus on the frequency and magnitude of flow modulation. A critical Strouhal number, based on the unsteady boundary layer thickness $((R/\delta_S)^2/(\widehat{V}\delta_S/\nu) = R^2\nu/\delta_S^3\widehat{V})$ has been proposed for time-dependent flows.

A single index being insufficient, diagrams based on the main involved dimensionless parameters are used. The Stokes number, which ressembles Re, the convective inertia being replaced by the unsteady inertia in the force ratio, is a frequency parameter. A combination of Sto and Re, with suitable scales, has been defined to provide critical conditions [340]. These authors observed turbulent bursts in the dog aorta after peak flow when $\widehat{Re} > 250\,$Sto, which disappear before the cycle end. However, such disturbances can have been generated by the measurement sensors. Similar orders of magnitude have been found in pipe flows using the same measurement technique (hotfilm velocimetry) [341]. Demarcation of the flow patterns is commonly made on Sto–Re diagrams.

Pipe curvature damps turbulences in steady flows [342]. The higher κ_c, the greater Re_{crit}. However, when disturbances are small, Re_{crit} in bends can be lower than in straight pipes. Re_{crit} appears thus to depend on the perturbation amplitude in steady flow. Whereas viscous forces have a stabilizing effect, tube curvature can have a destabilizing effect on the Stokes layer due to the centrifugal forces along the inside wall [343]. Analysis of linear and non-linear stability of zero-mean sinusoidal flows in curved pipes have been studied in the case of fully developed flow, based on a WKBJ perturbation solution [344, 345], defining a critical Taylor number $Ta = \widehat{V}^2\delta_S^3/(R_c\nu^2)$. The vessel curvature thus affects the flow stability. Another factor plays a role, the vessel deformability. Perturbations grow more easily in straight deformable tubes than in rigid ducts [346]. Conversely, distensible straight pipes have been found to damp existing turbulence in steady flow, the turbulence intensity being lower than in rigid pipes [347].

The instability limit $Re_\delta = Re/Sto$ has been estimated to be about 100 in a plane boundary layer of thickness $\delta \propto (\nu/\omega)^{1/2}$ as well as in the aorta, from in vivo measurements that strongly disturbs the flow [348]. In pulsatile flows, turbulence bursts could appear when the flow begins to decelerate. Some explanations attribute the turbulence to the stability of the oscillating boundary layer [349]. The phase difference between the response of the boundary layer flow and the core flow to the time-dependent pressure gradient produces a velocity profile characterized by an inflection point, which is considered an instability source.[40] When Sto is high (small flow modulation period), the inflection point of the inflectional velocity profile is ineffective in producing instabilities [350]. When the inflection point is neither very close to the wall nor far away from it, its importance is reduced [351]. The flow in a channel that is suddenly blocked off (at a quicker rate than the aortic flow, the aortic

---

[40] The inviscid linear stability theory states that velocity profiles with an inflection point are candidates for instability.

valve beginning to close after the peak transvalvular flow) gives an estimate of the available time for instability occurrence $(0.023\nu/R^2)$, where R^2/ν is the damping time scale by viscous effects [352].

These inflexion points must be convected at a relatively slow velocity for local perturbation growth. Turbulence burst may quickly die downstream when conditions change. On the contrary, instability has been proposed to be related to the behavior of the inviscid core of the pulsatile flow, i.e., outside the Stokes layer [353]. These investigators provided a flow stability criterion based on the velocity radial gradient and a parameter that depends on the perturbation amplitude, Sto, and velocity modulation rate γ_v. This criterion is thus difficult to handle, especially in actual blood flows. These authors point out that three main governing parameters must be taken into account in flow stability investigations, Re, Sto, and γ_v.

Overestimation of flow instabilities[41] has been shown in the framework of experiments carried out on artificial heart valves compared with the values obtained from a spectral analysis of LDV data and determination of the main frequency modes [354].

In summary, several features affect flow stability: vessel curvature according to disturbance amplitude,[42] wall distensibility, flow period, and the frequency content of the pressure signal. Laminar flow in blood vessels is a weak assumption.

5.2.6 Flow Separation

Flow separation induces abrupt stream divergence from the vessel wall. In steady flows, the area of the flow separation region depends on Re. The *separation point*[43] divides the flow into: (1) a recirculating flow inaccessible to the upstream flow, and (2) a downstream-directed flow with a boundary layer passing over the region of recirculating flow. In vessel flows, the flow reattaches further downstream to the wall at the *reattachment point*.[44] There is a priori no fluid exchange across the separation surface between the separated and unseparated flow regions in 2D flows. The streamlines diverge from the tube wall. The separation surface is an envelope of the limiting streamlines.

The flow separation in time-dependent flows can be defined by the separation of the boundary layer with $du/dn \leq 0$ (n: local wall normal) at the separation point throughout the entire flow cycle [355]. At moderate Re (10^2–10^3), flow separation can occur. At small St (10^{-4}), flow separation is quasi-steady,

---

[41] A shift toward the unstable region of the Sto vs. Re diagram is observed when the scales are cardiac frequency and peak velocity.

[42] Bending may favor or damp disturbances with respect to straight pipe, whether the disturbance amplitude is small or great.

[43] A point of vanishing shear that can appear in unsteady flow in deformable vessels associated with the flow reversal does not mean necessarily separation.

[44] In immersed bodies, flow separation can form a wake.

Table 5.3. Unsteady flow separation. Illustration by a simple case.

dq/dt	dA/dz	$\partial p/\partial z$	Flow separation
Acceleration phase	Divergence	$V_q^2(dA/dz) < dq/dt$	Absence
	(> 0)	$V_q^2(dA/dz) > dq/dt$	Presence
(> 0)	Convergence	< 0	Absence
	(< 0)		
Deceleration phase	Divergence	> 0	Presence
	Convergence	$V_q^2\|dA/dz\| < \|dq/dt\|$	Presence
		$V_q^2\|dA/dz\| > \|dq/dt\|$	Absence

whereas at intermediate St (10^{-3}–10^{-2}), flow separation regions expand during the deceleration phase, and with increasing St, vortex shedding can occur [356]. Flow separations in time-dependent flows have peculiar features. In periodic flows, the flow separation region in the entrance region of the branch of a symmetrical bifurcation with a between-branch flow distribution of 2 : 1 can move [318]. In 3D unsteady flow separation, the separation surface can be incomplete so that the fluid particles are not necessarily tracked down and can flow downstream, escaping in particular during the deceleration phase, having spent more time in the area than those of the main stream.

The unsteady axial pressure gradient ($\partial p/\partial z$), which can induce flow separation, can be estimated to a first appoximation, when viscous forces are neglected, in a rigid ($\partial A/\partial t = 0$) single ($\partial q/\partial z = 0$) vessel conveying a 1D flow, by the following equation:

$$\frac{\partial p}{\partial z} = \frac{\rho}{A}\left(V_q^2\frac{dA}{dz} - \frac{dq}{dt}\right).$$

Flow separation occurrences are given in Table 5.3.

5.2.7 Vessel Deformability

Blood is conveyed in flexible vessels which undergo deformation under varying transmural pressures p, defined as the difference between the internal and external pressures ($p = p_i - p_e$). However, the deformable vessels have controlled lumen size (Part I and Sect. 2.6.3). Arteries dilate when the pressure wave, associated with the systolic ejection from the left ventricle, is traveling through the arterial bed. The cross-sectional luminal area A_i is then inflating because p is evolving in a range of positive values. The cross-sectional shape may be sligthly affected owing to the non-uniform distribution of p over the entire vessel perimeter and action of the neighboring structures. The artery wall undergoes local strains in axial, azimuthal, and radial directions imposed by pressure wave traveling.

Fluid and structure mutually exert reciprocal effects in deformable vessels. A change in velocity induces a variation in internal pressure that generates a change in cross-sectional area, which in turn, produces a variation in fluid velocity.

5.2.7.1 Distensible Vessels

The myocardium activity induces alternate dilation and recoil of the arterial wall, especially of the elastic arteries. The static equilibrium of an elastic vessel and state laws are given in Apps. C.8 and C.9. The blood flow is coupled to the arterial wall motion. The pressure wave, which precedes the blood flow, dilates the blood vessels. Pulse waves are deformation waves coming out from the heart with a speed of order 10 times greater than the blood velocity. When blood enters the aorta, the aorta starts expanding until the blood inflow stops. During the diastole, the expanded aorta retracts and squeezes the blood out into the downstream arteries, where the flow is then uninterrupted.

The information of blood ejection by the left ventricle is received by the peripheral arteries with a more or less great delay according to the distance from the heart due to the wave propagation duration, as the wave transmits the information. Injection pressure remains relatively low (about one hundred times less) with respect to the required value in rigid ducts, although it drives the increase in blood velocity from zero at the end of the cardiac cycle to about $1\,m/s$ (maximal peak value in proximal arteries) in approximately $100\,ms$ in the deformable arteries.[45] The lower the fluid density for a given compressibility, the smaller the fluid inertia, the higher the propagation speed and the faster the motion trigger. Because blood is incompressible, reduction in blood inertia is due to vessel distention, the mass increase for a given pressure occurring in a larger volume.

The artery dilates when the pressure wave travels through it. The Reynolds-type number $(dR/dt)R/\nu$ indicates the relative importance of the viscous forces in a contracting/expanding vessel. In expanding vessels, the near-wall flow is delayed [357]. In a distensible vessel model, backward flows and wall shear stress are reduced to 75% of their magnitude in a rigid model [358]. When the wall is distensible, wall shear stress increases at certain wall points and decreases at others, as shown by LDV with a large exploration volume. It may then be necessary to take into account locally the wall distensibility, especially when the stress field in the wall subjected to the flow stresses must be known to assess the fissuring risk.

However, the pulse wavelength λ in the distensible arteries is several meters per second, which is much greater than the distance traveled by the fastest fluid particles during the cardiac cycle ($\widehat{V}T \ll \lambda$). In the absence of taper, the blood is commonly assumed to be conveyed through vessels of quasi-uniform

---

[45] High injection pressures would be noxious for the vessels and organs on the one hand and would lead to heart fatigue one the other hand.

cross-section over a vessel length on the order of 10 cm. Moreover, the blood volume ($V = 5$ l) is increased by a relative volume change of 0.02 when it receives a stroke volume SV = 100 ml during a duration t_{se} = 200 ms. The additional flow rate generated by an aorta segment of length $L = 10$ cm and radius $R = 10$ mm is $q_{add} = 0.2 V_{Ao}/t_{se} = \pi.10^{-6}$ m^3/s. Assuming a mean $\bar{q}$ and a peak $\hat{q}$ aorta flow rate of $25\pi.10^{-6}$ and $70\pi.10^{-6}$ m^3/s, respectively, $q_{add} = 0.04\bar{q}$ and $q_{add} = 0.014\hat{q}$. The effects of wall compliance on blood flow then appear to be minor with respect to flow periodicity. Yet, fluid–structure interaction must be taken into account to demonstrate whether vessel deformability has significant effects on the velocity and pressure fields.

5.2.7.2 Collapsible Vessels

Veins constitute the compliant compartment of the blood vessel network (vein compliance allows the blood volume to reside mainly - up to 70% - in the venous network). However, veins may experience changes both in cross-sectional area and shape when they are subjected to negative transmural pressures during natural or functional testing maneuvers, although p is uniformly distributed in the entire cross section [359–361]. The easier the collapse, the thinner the vessel wall or the more superficial from the skin the vessel path (deep thin-walled veins can be modeled by thick-walled vein-like tubes).

The dynamics of the fluid are strongly coupled to the mechanics of the flexible vessel wall via the non-linear *tube law*, which relates p to A_i [362–366]. Tube loading can be such that contact occurs between the opposite walls at one point and, afterward with increasing loading, over a line. At slightly negative transmural pressures, any thin-walled flexible vessel is very compliant, small variations in transmural pressures inducing large changes in cross-sectional area.[46]

The tube law depends strongly on both tube geometry (length, shape of the unstressed cross-section - circular or elliptical -, and in the latter case, tube ellipticity, and wall thickness) and rheology in the unstressed state. When the unstressed cross-section is circular and the transmural pressure is slightly negative, the flexible duct keeps its circular cross-section down to the buckling pressure p_b (the compliant tube buckles under a slightly negative transmural pressure).

When the transmural pressure is lower than the contact pressure, the lumen of a straight deformable vessel of elliptical unstressed cross-section is reduced to two parallel narrow tear-drop channels separated by the wall contact region. With further decrement in transmural pressure, the flat region of

---

[46] Huge changes in tube transverse configuration for slightly negative transmural pressure occur in any compliant pipe, whether the unstressed cross section is elliptical [364] or circular [367], whether the deformable vessel has uniform homogeneous walls or is a composite material of non-uniform geometry [269], in vitro as well as in vivo [96], especially when it has thin walls and it is not strongly surrounded by more or less stiff biological tissues.

contact spreads laterally. When the unstressed cross-section is circular, different modes of collapse (bifurcation problem) can occur according to the number N of lobes (i.e., opened regions of the collapsed tube lumen, which are associated with symmetry axes). The buckling pressure is proportional to $N^2 - 1$ in computational shell models of deformation of tube of infinite length and purely elastic wall [362, 368]. The lobe number N depends on the tube geometry and rheology in its unstressed and deformed states. The cross-section shape usually displays three or four lobes either in tubes subjected to longitudinal bending effects [369] or in short and thin-walled pipes [367].

Tube law, one- and three-dimensional data from flow models in collapsible vessels are given in App. C.10. Analysis of details of actual properties of the tube-fluid couple deals with the three-dimensional aspects [370, 371].

5.2.8 Wave Propagation

Wave propagation results from ventricle-aorta coupling, with the ventricular ejection in the distensible arteries. Harmonic analysis has been used to process the assumed periodic waves, which are represented by an infinite Fourier series of sinus and cosinus functions with appropriate coefficients, or more conviniently by the exponential formulation $\sum a_k \exp\{\imath \omega k t + \varphi_k\}$.

The pressure waveform ($p = \hat{p} \exp\{\imath \omega t\} \exp\{\imath \varphi\} \exp\{-\gamma z\}$, φ: phase lag, $\gamma = \alpha + \imath \beta$: propagation coefficient, α: attenuation constant, β: phase constant) in the proximal arteries (i.e., the subclavian artery) presents the following features. During quick ventricular ejection, pressure abruptly rises up to an anacrotic inflection and then more slowly up to a peak. The pressure continues to decay during the diastole and can exhibit an atrial wave at the end of the diastole. Pressure time variations display an isovolumic wave during the isovolumic relaxation. The pressure waveform is smoother in distal arteries (i.e., the radial artery) showing a dicrotic notch (incisura) followed by a dicrotic wave. The notch is produced by aortic valve closure.

In deformable vessels, there are three propagation modes (App. C.7). Due to surrounding tissue tethering, the Young mode associated with the radial displacement is the main propagation mode. Pressure and flow waves associated with blood ejection during the ventricular systole result from the non-linear interaction of blood and deformable artery walls. Flow velocity and pressure waves change with increasing distance from the heart (Figs. 5.4 and 5.5). During propagation, cross-sectional averaged velocity is delayed, the phase lag being due to the wave traveling. The velocity wave is characterized by additional distortions: (1) a reduction in amplitude, due to flow division at branching sites, associated with an admittance decay; (2) increase in width; and (3) disappearance of back flow if any exists in the upstream arteries. These changes are mainly due to wave attenuation. The pressure wave varies with increasing distance from heart with: (1) a *peakening* (amplitude rise) associated with a gradual decay in mean pressure; (2) a wave *steepening*, the slope of the initial accelerating phase increasing with distance from the heart; and

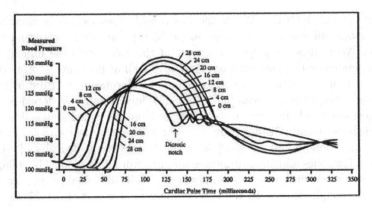

Figure 5.4. Blood pressure waveforms in different sites of the dog descending aorta (Source: [373]).

(3) a smoothing and gradual attenuation of the dicrotic wave, formation of a second rear wave, and possible secondary oscillations. Proportionality varies between the pressure and the flow rate from heart to the periphery.

It is commonly thought that the traveling waves undergo shape deformations due to the dispersion associated with harmonic content, the non-linear pressure–cross-sectional area relationship, the wave reflection at impedance-mismatch sites, a decrease in elasticity of peripheral arteries, and an increase in impedance with distance from the heart. Wave reflections are minimized by changes in elastic properties along the arterial tree [372]. A gradual increase in the elastic modulus in the branch relative to the parent vessel allows maximal admittance, if the geometry variations also favors such a process. Wall inertia introduces a response delay to pressure changes. Wall displacement velocity is small compared with blood velocity. The effects of wall inertia are then much smaller than those of the elastic forces. Viscosity does not significantly influence wave speed but determines wave damping.

Wave distortion during wave propagation is caused by multiple factors: vessel architecture and change in size, structure, and rheology, which induce variations in vessel impedance and admittance. From the heart to the tissues, the wall structure and composition gradually vary from the elastic arteries, to the transition and muscular arteries. Non-linear viscoelasticity and wall stress distribution affect the wave propagation. The waves can be reflected at branching sites. Local admittance coefficients[47] can be introduced for the stem Y_t and its branches Y_b ($Y_b = 0$ for a total reflection). When $\sum Y_b/Y_t < 1$, the wave is reflected with a smaller amplitude. When $Y_b \sim Y_t$, the incident wave is slightly disturbed. In such a case, wave propagation can be modeled by solitons. Pulse harmonics interact non-linearly. The wave peak propagates faster

---

[47] Admittance Y is given by $Y = A/(\rho c) = Cc = ADc$.

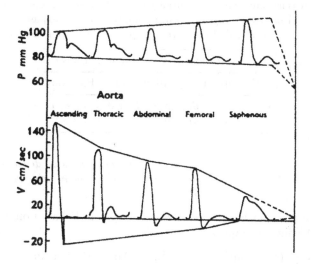

Figure 5.5. Pressure (**top**) and flow rate (**bottom**) in various segments of the canine aorta (ascending and descending thoracic, abdominal) and arteries of the inferior limb (proximal - femoral -, or distal - saphenous -). The pressure and flow rate signals are matched but are acquired in different experiments (Source: [374]).

than the wave foot (steepening). Greater attenuation for higher harmonics smoothes the waveform.

Forward and backward running waves in the arteries have been analyzed using the method of characteristics and a 1D flow model (hyperbolic equation) in an elastic artery, neglecting viscous dissipation [375]. Changes in pressure dp and velocity dv across the incremental wave front are related to $dp = \pm \rho c \, dv$ (Table 5.4). Pressure–velocity curves exhibit loops with a linear part during the early systole, from which the wave speed is computed. Forward and backward traveling components are separated assuming that intersecting wavelets are additive (linearization, $dg = dg_+ + dg_-$) and calculated from measurements. Forward running wavelets are dominant in the aorta. Waves are generally classified by the sign of produced changes (Table 5.4). The energy fluxes carried by the wave $de = dp \, dv$ are positive for all forward running waves and negative for all backward traveling waves. Non-linearities must be taken into account in the analysis and separation of forward and backward running waves [248]. This type of analysis is challenged by soliton modeling.

The wave speed c depends on vessel wall rheology, i.e., on vessel type and size, age, etc. The pressure pulse travels much faster than the blood ($c_p(p(\mathbf{x}, t)) \sim 10\text{–}30\, U_q(\mathbf{x}, t)$). The pulse transit time (PTT) is the time taken by the wave to run between two explored stations. Pulse wave velocity (PWV) is usually measured between the carotid and femoral arteries. PWV is used as an index of arterial distensibility.

Table 5.4. Pressure and velocity variations across the wavelet front, associated with $(dp = \rho c\,dv)$ and backward $(dp = -\rho c\,dv)$ compression and expansion wave fronts (upstream $\{p + dp, v + dv\}$ and downstream $\{p, v\}$ values from the wavelet front).

	Compression wave		Expansion wave	
	Forward	Backward	Forward	Backward
p	$dp > 0$	$dp > 0$	$dp < 0$	$dp < 0$
v	$dv > 0$	$dv < 0$	$dv < 0$	$dv > 0$

> "... c'est-à-dire des points de départ et des trem-
> plins pour s'élever jusqu'au principe universel..."
> (Platon) [376]

The wave propagation in the circulatory system is based on a 1D model, with the following assumptions:

1. The fluid is: (1) incompressible, i.e., the Mach number $\mathrm{Ma} = v/c_{\mathrm{sound}} \ll 1$, the Helmholtz number $\mathrm{He} = \omega L^\star/c \ll 1$, $\rho g L^\star/(\gamma p) \ll 1$ $(\gamma = c_p/c_v),$[48] $\beta_T V_q \nu/(c_p L^\star) \ll 1$ $(\beta_T$: thermal expansivity),[49] $\lambda_T \beta_T \Delta T/(\rho c_p L^\star V_q) \ll 1$[50] [377], (2) homogeneous, and (3) ideal or Newtonian.
2. The vessel is (1) long, straight, cylindrical; (2) has a circular cross-section; (3) its wall is made of a homogeneous structure, and has uniform geometric and rheologic properties; (4) has thin wall $(h/R \ll 1)$; (5) its wall material is purely elastic and isotropic; (6) its axis remains undeformed; (7) it is not prestressed; (8) the axial tension during the radial displacement is negligible; (9) the wall inertia and viscosity associated with the displacement are neglected.[51]
3. (1) The flow is axisymmetry (absence of azimuth dependency), (2) the pressure is constant in every cross-section, (3) there is no secondary motion, (4) most often, velocity distribution is uniform in every cross-section (momentum and kinetic energy coefficients $\alpha_{\mathrm{m}} = \alpha_{\mathrm{k}} = 1$).[52]
4. The wave has (1) a reduced frequency spectrum, (2) an infinitesimal amplitude $(\Delta A/A \ll 1, \partial R/\partial x \sim h/\lambda \ll 1)$, (3) a wavelength $\lambda/R \gg 1$ $(\partial/\partial x^2 \ll \partial/\partial r^2)$, (4) a radial propagation mode.

---

[48] The height scale in the gravity field is much greater than the flow length scale.
[49] Viscous dissipation is small enough.
[50] Heat conduction is not or slightly involved.
[51] Wall inertia is taken into account in the non-linear propagation of solitons. It explains the effective function of virtual valves of the arterial wall distributed along the arterial circuit and the recruitment inducing a soliton peakening.
[52]

$$\alpha_{\mathrm{m}} = \int \frac{v^2}{V_q^2}\frac{dA_i}{A_i}, \quad \alpha_{\mathrm{k}} = \int \frac{v^3}{V_q^3}\frac{dA_i}{A_i}.$$

Table 5.5. Influence factors on wave propagation.

Factor	Effect
Vessel geometry	Impedance and admittance changes reflection (the lower $\phi_{\mathrm{inc}} - \phi_{\mathrm{ref}}$, the greater the wave amplitude) (higher harmonics are more disturbed, $\omega \nearrow$, $\lambda \searrow$)
Viscosity	Damping (wave attenuation)
Inertia	$\rho \omega V \ll \rho V^2/L = \rho V^2 \omega/c$, $V/c \to 1$
Harmonic content	Smoothing ($c = c(\omega)$) (the greater the component frequency, the greater the attenuation)
Fluid–wave interaction	Speed of forward running solitary wave increases ($u + c$)

5. The fluid is weakly coupled to the vessel wall; the effects due to fluid motion on the pressure distribution are much lower than the action of wave propagation ($\rho v^2 \ll \rho v c$, $v \ll c$ - subcritical flow).

Literature studies of wave propagation in blood vessels also deal with the linearized theory. In a cylindrical straight elastic model:

$$\frac{\partial v_r}{\partial t} = -\frac{1}{\rho}\frac{\partial p}{\partial r} + \nu\left(\frac{\partial^2 v_r}{\partial r^2} + \frac{1}{r}\frac{\partial v_r}{\partial r} + \frac{\partial^2 v_r}{\partial z^2} - \frac{v_r}{r^2}\right),$$

$$\frac{\partial v_z}{\partial t} = -\frac{1}{\rho}\frac{\partial p}{\partial z} + \nu\left(\frac{\partial^2 v_z}{\partial z^2} + \frac{\partial^2 v_z}{\partial r^2} + \frac{1}{r}\frac{\partial v_z}{\partial r}\right),$$

$$\frac{\partial v_z}{\partial z} + \frac{v_r}{r} + \frac{\partial v_r}{\partial r} = 0,$$

$$\rho_w h\frac{\partial^2 u_r}{\partial t^2} = c_{rr}(R) - \frac{Eh}{1 - \nu_P}\left(\frac{\nu_P}{r}\frac{\partial u_z}{\partial z} + \frac{u_r}{r^2}\right),$$

$$\rho_w h\frac{\partial^2 u_z}{\partial t^2} = c_{rz}(R) + \frac{Eh}{1 - \nu_P}\left(\frac{\partial^2 u_z}{\partial z^2} + \frac{\nu_P}{r}\frac{\partial u_r}{\partial z}\right),$$

with $v_r = \partial u_r/\partial t$ and $v_z = \partial u_z/\partial t$ ar $r = R(t)$. It takes into account the transverse motion, which is neglected by the 1D model.

5.2.9 Wall Shear Stress

The nine-component stress tensor produces at the vessel wall a two-component shear vector, one component in the plane of the vessel cross-section (cross-component), the second in the local vessel axial direction, which is not necessarily parallel to the tube axis direction (there is no strictly axial component in the general case).[53] The dimensionless extra-stress tensor in a Newtonian

---

[53] In general, the wall normal is not necessarily in the cross-section plane. Only in ducts of uniform cross section, it belongs to the plane of the cross-section.

fluid is equal to $\widetilde{\mathbf{T}} = 2\mathrm{Re}^{-1}\widetilde{\mathbf{D}}$. At each point of the vessel wall, the dimensionless dissipative stress vector $\tilde{\mathbf{c}}_w = 2\mathrm{Re}^{-1}\widetilde{\mathbf{D}}\hat{\mathbf{n}}$ associated with the dimensionless extra-stress tensor $\widetilde{\mathbf{T}}$[54] is defined by:

$$\tilde{\mathbf{c}}_w = \mathrm{Re}^{-1}\left\{ \left[(2\tilde{u}_{x,x}n_x + (\tilde{u}_{x,y} + \tilde{u}_{y,x})n_y + (\tilde{u}_{x,z} + \tilde{u}_{z,x})n_z\right]\hat{\mathbf{e}}_x \right.$$
$$+ \left[(\tilde{u}_{x,y} + \tilde{u}_{y,x})n_x + 2\tilde{u}_{y,y}n_y + (\tilde{u}_{y,z} + \tilde{u}_{z,y})n_z\right]\hat{\mathbf{e}}_y$$
$$\left. + \left[(\tilde{u}_{x,z} + \tilde{u}_{z,x})n_x + (\tilde{u}_{y,z} + \tilde{u}_{z,y})n_y + 2\tilde{u}_{z,z}n_z\right]\hat{\mathbf{e}}_z \right\},$$

where $\hat{\mathbf{n}}$ is the inner unit normal vector of the vessel wall.[55]

The dimensionless stress vector $\tilde{\mathbf{c}}_w$ can be expressed in a local basis. Let $\hat{\mathbf{t}}$ the transverse tangent unit vector, which is perpendicular to $\hat{\mathbf{n}}$ and belongs to the plane of the cross-section; $\hat{\mathbf{t}}$ is oriented in the anticlockwise direction. Let $\hat{\mathbf{b}} = \hat{\mathbf{t}} \times \hat{\mathbf{n}}$. In a non-uniformly deformed vessel (distended or collapsed), in a tapered vessel, etc., $\hat{\mathbf{b}}$ is not parallel to $\hat{\mathbf{e}}_z$. The Frenet trihedron $(P; \hat{\mathbf{n}}, \hat{\mathbf{t}}, \hat{\mathbf{b}})$ is then defined in each point P of Γ_3. $(\tilde{\tau}_t, \tilde{\tau}_n, \tilde{\tau}_b)$ denote the components of the vector $\tilde{\boldsymbol{\tau}}$ in this trihedron. In the local Frenet basis $\tilde{\boldsymbol{\tau}}$ becomes:

$$\tilde{\mathbf{c}}_w = J\left\{ \left[(t_y b_z - t_z b_y)\tilde{\tau}_x + (t_z b_x - t_x b_z)\tilde{\tau}_y + (t_x b_y - t_y b_x)\tilde{\tau}_z\right]\hat{\mathbf{n}} \right.$$
$$+ \left[(b_y n_z - b_z n_y)\tilde{\tau}_x + (b_z n_x - b_x n_z)\tilde{\tau}_y + (b_x n_y - b_y n_x)\tilde{\tau}_z\right]\hat{\mathbf{t}}$$
$$\left. + \left[(n_y t_z - n_z t_y)\tilde{\tau}_x + (n_z t_x - n_x t_z)\tilde{\tau}_y + (n_x t_y - n_y t_x)\tilde{\tau}_z\right]\hat{\mathbf{b}} \right\},$$

where J is the determinant of the Jacobian matrix of the Cartesian-to-Frenet frame change:

$$J = b_x(n_y t_z - n_z t_y) - b_y(n_x t_z - n_z t_x) + b_z(n_x t_y - n_y t_x).$$

The magnitude and orientation of the wall shear stresses (WSS) depend both on the axial and transverse tube configurations and on the flow pattern. In a developing steady flow in a straight pipe of uniform circular cross-section (constant wall curvature), a WSS axial gradient is applied to any point of the wall (entry length). Cross-variation in wall curvature, for instance observed in collapsed tubes due to the transverse bending of the wall, induces transverse gradient of the single WSS axial component [378]. A WSS circumferential component appears when the vessel presents a curvature of its axis. In 3D flows, any point of the wall is then subjected to two shear components. The WSS gradient is a 2D tensor (Fig. 5.6):

$$\nabla\mathbf{c}_w = \begin{pmatrix} c_{wb,b} & c_{wb,t} \\ c_{wt,b} & c_{wt,t} \end{pmatrix}.$$

---

[54] For a 1D flow in a straight pipe at a wall point P (a single component v_z), $\mathbf{c}_w|_P = \mu\nabla v_z \cdot \hat{\mathbf{n}}|_P$.

[55] $\tilde{u}_x$ is the $\tilde{x}$-component of $\tilde{\mathbf{u}}$ in the Cartesian basis $(\hat{\mathbf{e}}_x, \hat{\mathbf{e}}_y, \hat{\mathbf{e}}_z)$. $\tilde{u}_{x,y}$ is the partial derivative of $\tilde{u}_x$ with respect to $\tilde{y}$. The same notation is used for every vector component and partial derivative.

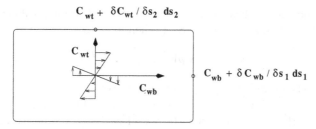

Figure 5.6. Wall shear stress and its spatial gradient.

The diagonal components induces elongation and the off-diagonal components torsion (torque). In unsteady flow, the wall shear stress can strongly vary at any given point with the time and in both main directions (transverse and axial) during the flow cycle (Table 5.6) [149]. Change in direction of the WSS axial component occurs when the pulsatile flow generates a back flow in the whole vessel lumen or a layer near the wall.

Several WSS indices have been proposed. The time-averaged WSS is given by: $|\bar{\mathbf{c}}_w| = (1/T) \int_0^T |\mathbf{c}_w|\, dt$. The oscillatory shear index (OSI) is defined by: $\text{OSI} = \int_{t_{ri}}^{t_{rf}} |\mathbf{c}_w|\, dt / \int_0^T |\mathbf{c}_w|\, dt$, where t_{ri} and t_{rf} are the initial and final instants of the reversed flow period.

In a fully developed steady 1D flow, wall shear stress can be estimated from the pressure drop. The pressure forces are equal to the friction forces. Therefore, in a tube of circular cross-section, $\Delta p/L = (\chi_i/A_i)c_w = (2/R)c_w$. The viscous head loss per unit length is given by the *Darcy-Weisbach formula*:

$$(\chi_i/A_i)c_w = (4/d_h)c_w = \Lambda \rho V_q^2/(2d),$$

Table 5.6. Maxima (M) and minima (m) of variations of wall shear stress components - axial (a) and cross (c) - for a given cycle phase, either for a fixed axial or circumferential position, in a nonzero-mean sinusoidal flow (Sto $= 4$, $40 \leq \text{Re}(t) \leq 360$, $\widehat{\text{De}} = 113$, $0.05 \leq \text{St}(t) \leq 0.45$, $\gamma_u = 0.8$, and $\text{Re}_2 = 181$) in a 90-degree bend ($R/R_c = 1/10$) of a Newtonian incompressible fluid (Source: [149]).

$c_{cM} \sim 0.20\text{--}0.28\, c_{aM}$		
$c_a(z)\big|_{t,\theta}$	max $\sim$ 3	min
$c_c(z)\big|_{t,\theta}$	max $\sim$ 5	min
$c_a(t)\big|_{z,\theta}$	max $\sim$ 17	min
$c_c(t)\big|_{z,\theta}$	max $\sim$ 33	min
$\Delta c_c \sim 2\, \Delta c_a$		

where $\varLambda$ is the friction head loss coefficient.[56] The friction coefficient, or skin friction coefficient, $C_f = \varLambda/4$. WSS is then given by

$$c_w = (\varLambda/8)\rho V_q^2/2 = C_f\,\rho V_q^2/4.$$

Within the 1D boundary layer, friction forces $\propto \rho(V^2/L)$ have the same magnitude order as the inertia forces $\propto \mu(V/\delta^2)$. Therefore, $\delta = L/\mathrm{Re}^{1/2}$ (L: length on the solid wall from the vessel entrance to the exploring site). The boundary layer thickness δ then depends on the Reynolds number Re. The greater Re, the thinner the boundary layer. The longer the distance from the vessel entry, the higher δ. Because $c_w \propto \mu V/\delta = \rho V(\nu V/L)^{1/2}$, $c_w \propto \rho V^2\,\mathrm{Re}^{-1/2}$.

5.2.10 Particle Flow

In microcirculation, blood flow becomes heterogeneous and a fluid-particle mixture must then be considered. The particles are mainly the flowing cells, especially erythrocytes. However, a mixture of a Newtonian fluid, the plasma, and macromolecules can be investigated in tiny capillaries. The particle concentration is such that interparticle distance is greater than particle size. Particle–particle interactions are then lower than fluid–particle interactions. Solid particle flows depend on flow characteristics and particle features, particle shape and size, particle concentration, possible particle deformability, and particle buoyancy with respect to the suspending fluid.

In narrow capillaries, where inertia is negligible, a tiny lubrication film between the endothelium and flowing cell membrane favors motion of the deformed cells. Compared with a homogeneous Newtonian fluid in the same flow conditions, velocity is reduced and resistance is augmented. The velocity reduction factor and resistance rise factor quantify the ratio between particle flow and one-phase flow.

In a steady motion of spherical particles in a Newtonian incompressible fluid, the involved dimensionless parameters are the particle radius size-to-vessel radius $\kappa_s = R_p/R$ and the different Reynolds numbers: (1) the flow $\mathrm{Re} = V_q R/\nu$, (2) the particle $\mathrm{Re}_p = V_q R_p/\nu$, (3) the slip $\mathrm{Re}_{sl} = R_p v_{sl}/\nu$ ($v_{sl} = v - v_p$: slip velocity), (4) the shear $\mathrm{Re}_{sh} = R_p^2\dot{\gamma}/\nu$, and (5) the rotation Reynolds number $\mathrm{Re}_{ro} = R_p^2\omega/\nu$ (ω: particle angular speed). Any flowing particle experiences several forces. The Stokes drag force F_D on a rigid spherical particle, of radius R_p, and the drag coefficient C_D in a steady motion of a Newtonian incompressible fluid are given by $F_D = 6\pi R_p\mu v$ and $C_D = F_D/(\pi R_p^2)(\rho v^2/2)$, respectively. The particles are also subjected to a transverse and a lift force. The lift coefficient $C_L \propto \kappa_s\dot{\gamma}\mathrm{Re}_{sl}^{-1}$. At large Re, when the number of rigid spheres is moderate, the Levitch formula

---

[56] $\varLambda = f_L/\mathrm{Re}$ in laminar flow, $\varLambda = f_T/\mathrm{Re}^p$ in turbulent flow (f_L and f_T are the corresponding shape factors).

gives the viscous force F_v acting on the particle flowing with the velocity v_p:
$F_v = 12\pi\mu R_p v_p$.

Flow-seeded particles steadily conveyed in a vessel migrate transversally due to inertia [379, 380]. The shear rate gradient indeed causes not only rotation of the solid particle but also cross-migration (Magnus effect). The cross-motion can be neglected when Re_{sh} is small. Transverse migration of the seeded particles is produced by a set of factors (not only by inertia): the local gradient in shear rate and the particle volume fraction, inertial lift depending on local velocity distribution, lubrification-associated wall repulsion, motion associated with the deformation of flexible particles, and displacement due to the non-Newtonian properties of the solvent [381, 382]. The particle cross-migration produces a particle-free near-wall layer [383], the thickness of which depends on particle concentration, and blunting of the velocity distribution [384]. Velocity-distribution blunting depends on the particle concentration and relative size, as shown using USV [385]. Steady suspension flows are also characterized by a concentration distribution, with a higher concentration near the channel axis,[57] as displayed by LDV [382]. The cross-migration pattern of non-neutrally buoyant particles is more complex than the motion of neutrally buoyant particles. Complexity depends on the fluid density-to-particle density ratio. For small density differences, the equilibrium position in Poiseuille flow depends whether the particle leads or lags the local fluid [386]. When the density difference is large, the equilibrium position always shifts toward the centerline. In pulsatile flows, neutrally buoyant particles (spheres, rods, and discs) also migrate radially, with several radial equilibrium positions when Sto > 5 [387].

The transport of flexible macromolecules depends on the coupling between different types of motion, translation, rotation, and internal motion of the macromolecules. A general scheme of macrotransport of flexible macromolecules, which are assumed to be a chain of rigid components, has been proposed [388].

The rheological properties of suspensions involved several parameters, the volume fraction of the seeded particles, their shape and size, the presence of electrical charges, and the flow pattern [389].

Erythrocytes can be modeled by capsules.[58] The behavior of a deformed capsule in a shear flow strongly depends on the unstressed shape (spherical, spheroidal, or discoidal) and on the viscosity of its internal fluid relative to the solvent's viscosity $\kappa_v = \mu_i/\mu_e$ [390]. Rheological properties of the membrane control resistance of the capsule to applied stresses. Membrane constitutive laws are based either on Mooney-Rivlin or Skalak [391] laws. These laws lead to quite different membrane behaviors for large deformations [392]. Also, the motion of a capsule freely suspended in a flowing liquid is an example of fluid–structure interaction, the deformation of the membrane being coupled to the

---

[57] The particle volume fraction ranges from 0.1 to 0.3.

[58] A capsule consists of an internal liquid enclosed by an elastic membrane.

motions of the internal and suspending liquids [393]. Deformation depends on the shear rate and membrane rheology. The capsule convection across a pore that generates a capsule deformation has been found to agree with experimental observations [394]. Such process models natural (erythropoiesis) and artificial filtration.

5.3 Cardiovascular Flows

5.3.1 Cardiac Flow

Since the observations of Leonardo da Vinci (1508–1513), vortices are known to appear in the heart. The vortex is associated with the rotation of a fluid region in the flow. Vortices play an important role in mixing and flow stability. Vortex formation when the fluid flows through the valve has been thought to affect the valve dynamics, but the main factor acting on the opening and closing of cardiac valves is the transvalvar pressure acting on this articulated soft membrane. In vivo, a vortex occurs in the left ventricle, early in the diastole, and grows and persists during the whole diastole. The unsteady ventricule Reynolds-like number (Sto^2) can be defined as the ratio between the time for vorticity radial diffusion d_v^2/ν over a distance of the valve diameter d_v and the time for vorticity convection in the flow direction. The unsteady flow in a cylindrical cavity with a closed motile end (piston) with a backward-facing step at the opposite side, modeling the left ventricule cavity, has been numerically and experimentally investigated [395]. A vortex is created and grows during the cavity filling; it disappears during ejection. MRV confirms the existence of vortices [396]. At diastole onset, the transmitral pressure initiated by the relaxation of the ventricular myocardium induces a blood flow from the left atrium to the left ventricle. The blood input in LV decreases the pressure difference between the two cavities. Vortices are associated with ventricular filling. Systolic emptying through the aortic orifice is also characterized by vortices. The time occurrences during the cardiac cycle of the peak pressure and flow rate are given in Table 5.7.

5.3.1.1 Mitral Valve

During isovolumetric relaxation, ventricular pressure p_V becomes greater than that of the aorta p_a. The mitral valve opens and starts to close from the mid-

Table 5.7. Cardiac flow events (t_{SE0}: SE start).

q_{max}	$t_{SE0} + 100\,ms$
p_{max}	$t_{SE0} + 180\,ms$
$(q(p_{max}) \sim q_{max}/2)$	

systole during mitral flow deceleration. Transvalvar pressure decays during diastole. During diastolic filling, mitral orifice widening reduces flow resistance through the orifice. The mitral flow exhibits two maxima when an atrial contraction occurs, the second having a value equal or lower than the first peak. The mitral jet, which is directed to the heart apex (blood velocity of about 80 cm/s), is associated with a large and a small eddy. The eddies disappear during the deceleration phase. Atrial contraction induces a reduction in mitral orifice area.

Transvalvar pressure, as well as chorda tension, act on mitral valve motion. Mitral valve closure has been explained by the fast deceleration of a jet-like motion of a water column through a collapsible tube wholly immersed in a tank, which induces a tube collapse propagating downstream at a speed close to the fluid velocity magnitude [397, 398]. If the time scale is the deceleration phase duration (t_{dp}), the length scale the length (L) of the valve leaflet, and the velocity scale the maximum transvalvar blood velocity (V_{max}), the Strouhal number is defined by St $= L/V_{max}t_{dp}$. St can also be expressed as the ratio of L to the maximum fluid displacement distance during the deceleration phase u_{max} (St $= L/u_{max}$). When St $\ll 1$, the fluid dynamics phenomena occur in the valve vicinity ($u_{max} \ll \mathcal{O}[L]$), whereas if St $\ll 1$, they involve a longer length scale such as the whole left ventricule cavity [398].

5.3.1.2 Aortic Valve

At the beginning of the systole, the systolic ejection flatten the valve cusp toward the aorta wall. During the acceleration phase, the pressure in the Valsalva sinus being lower than the jet pressure, a small portion of the blood ejected volume is conveyed in the sinuses. At the beginning of the deceleration phase of the aortic orifice flow, the transvalvar pressure (luminal pressure minus the sinus pressure) favors the starting of the valve closure. The closure motion occurs during the whole deceleration phase. The space behind the valve enlarges and the coronary flow rate can increase, whereas the intramural coronary vessels are less collapsed by the relaxing myocardium. The aortic valve closure is associated with aorta back flow.

In a model of the aortic valve, the time required for complete valve opening is a small fraction of the acceleration phase, whereas the closure time is about half the duration of the deceleration phase [399]. A vortex is observed in the sinus. Valve motion is related to transvalvar flow. Ventricular volume does not significantly affect valve closure when the flow deceleration remains strong [400]. Open-chest dog observations show that: (1) the higher the ejection volume, the earlier the complete aortic valve opening; (2) the higher the transvalvar flow peak, the more circular the complete opened aortic valve; and (3) the higher the aortic pressure, the earlier the valve closure [401]. In a 2D model of the aortic orifice region, it was shown that the greater the sinus size with respect to the cusp length, the faster the valve closure [402]. However, sinus height can be reduced to half the initial value without noticeable change in valve closure, but a minimum size is required such that the valve closure is not hampered.

Due to heart chaotic behavior [403], the cardiac output is quasi-periodic. The left ventricule ejects blood with $70 <$ SEV $< 100\,$ml. The systolic–diastolic heart flow is a starting–stopping flow type. A given cardiac output, whether the stroke volume is small and the cardiac frequency high or the converse, induces flow changes in the arteries.

5.3.1.3 Unsteady Jet in a Confined Vessel

In circular orifices, jet width and cross-sectional jet area close to the orifice are related to the orifice diameter and orifice area [404], as well as to hemodynamic variables. Let λ be the distance between two points at which the velocity are in phase (longitudinal wavelength): $\lambda = U_{qo}/\omega = d_o/$St ($U_{qo}$: average velocity at the orifice, d_o: orifice bore). St is then the ratio between the jet width at its origin and the longitudinal wavelength. The pulsatile jet expands more than the steady jet, then reaching the growth asymptotic limit quicker, to afterward behave like a steady jet [405]. The pulsation thus influences the jet over a distance of about six times its width at its origin.

5.3.2 Blood Flow in the Large Blood Vessel

The mechanical energy provided by the myocardium is converted into kinetic and potential energy associated with elastic artery distensibility, as well as into viscous dissipation. The aortic flow is a discontinuous flow characterized by a strong windkessel effect, with restitution of systolic-stored blood volume during diastole.[59] The aortic pressure wave is indeed associated with great vessel caliber change. In the arterial tree, the flow is pulsatile with possible bidirectional flow[60] period during the cardiac cycle, and back flow[61] in certain arteries, such as in the femoral artery, whereas the flow rate is always positive in others, such as in the common carotid artery. Due to the succession of bends and branching segments, the flow is developing[62] 3D. The bend is the basic

---

[59] Large elastic arteries have a damping effect that transforms SE into a continuous time-dependent flow. A fraction of the ejection volume is conveyed to the downstream arterial network with a given flow rate $dV_{\mathrm{LV}}/dt = q_{\mathrm{Ao}}(t) = A_{\mathrm{Ao}}(t)U_{q_{\mathrm{Ao}}}(t)$. Another part is stored and reinjected during the following diastole.

[60] In some large arteries, the velocity profile exhibits negative values near the wall during diastole and high shear at the boundary between forward and backward flows.

[61] The flow rate may become negative during diastole (flow reversal throughout the entire artery cross-section).

[62] Entry length has only been investigated in straight pipe and, mostly, for steady flows. Even in this simple case, there are huge variations in literature data [378].

element of the vasculature.[63] Vessel curvature leads to helical blood motion. Wall rheology affects the flow dynamics. The dimensionless flow rate $q/A_0\overline{U}_q$ has been found to be greater in a viscoelastic vessel than a purely elastic tube, and higher in a purely elastic tube than a rigid pipe for a nonzero-mean sinusoidal pressure gradient and straight cylindrical ducts of circular cross-section [406]; however, WSS is not significantly different.

5.3.3 Microcirculation

Microcirculation deals with vessel bores of about 5 to few hundreds μm. The arterioles correspond to the resistive segment of the vascular tree where most of the drop in mean pressure occurs. This peripheral resistance is controlled by the smooth muscle cells, which can widen or narrow the lumen by relaxation or contraction, respectively. Arteriole flow is characterized not only by important pressure loss but also by decrease in inertia forces and an increase in viscous effects. Both the Reynolds number Re and the Stokes number Sto becomes much smaller than 1[64] (App. C.1). Centrifugal forces do not significantly affect the flow in the microcirculation, where the motion is quasi-independent on the vessel geometry. A two-phase flow appears in the arterioles of few tens of μm with a near-wall *lubrification zone* (plasma layer, $h \sim 4$–$5\,\mu$m) and a cell-seeded core. The blood then fills the major part of the vessel lumen. The arteriolar flow is characterized by the *Fahraeus effect* (local Ht < global Ht)[65] and the *Fahraeus-Lindqvist effect* (blood viscosity depends on the vessel radius: $\mu_{\text{blood}}(R_h)$).[66] The Fahraeus-Lindqvist effect can be explained by the interaction of the concentrated suspension of deformable erythrocytes with the vessel wall [407]. The erythrocytes are located in the flow core region, whereas a cell-free layer exists near the wall. The hematocrit within such vessels is reduced. The cell distribution with a vessel-axis concentration can be explained by cell deformation in smaller arterioles. The plasma peripheral layer explains the *plasma skimming*, in particular in the kidneys where blood filtration occurs. In branching sites, RBC distribution is not uniform. Erythrocytes flow in the vessel with a larger core region. Blood flow distribution is regulated by arteriole vasomotor tone and pre- and post-capillary sphincter

---

[63] Any branching segment can be assumed to be a bend juxtaposition, with a slip condition on the common boundary in a first approximation. However, flow separation occurs more easily in branching segments than in curved vessels.

[64] In the microcirculation, $v\colon \mathcal{O}(10^{-2})$–$\mathcal{O}(10^{-3}\,\text{m/s})$ and $R\colon \mathcal{O}(10^{-5}\,\text{m})$. Hence, Re and Sto $\ll 1$.

[65] In the microcirculation, the decrease in local Ht associated with the vessel bore can be explained by a selection between the two phases of the blood, the plasma flowing more quickly than the blood cells. The cellular fraction then decreases with respect to that of the plasma. Such behavior reduces viscous dissipation with respect to a suspension of a higher cell concentration in small vessels.

[66] The relative apparent blood viscosity depends on tube diameter and hematocrit in small pipes.

activity. Peripheral blood flow regulation also uses blood vessel recruitment and possible shunts.

In arterioles and venules, blood effective viscosity depends on Ht, and local hydraulic diameter. Ht decreases due to a higher plasma fraction for a given blood volume. The capillary hematocrit increases when the glycocalyx is removed [408]. Resistance opposed to the flowing blood in the capillaries can then be due to the glycocalyx [409]. The glycocalyx is compressed by the moving erythrocytes; fluid can be expelled from this porous material [410].

In the capillaries, lumen size ($\leq 10\,\mu$m) is smaller than the deformed flowing cell dimension. The blood is non-uniform with a single cell file transported in axial train of deformed cells (RBC elongation and flexion) with between-cell trapped plasma boli. Blood flow is then multiphasic. Interactions between flowing erythrocytes and the capillary wall augment the flow resistance. Blood effective viscosity thus increases in capillaries with respect to its value in arterioles and venules. Venules belong to the intermediary scale of microcirculation. They are merging vessels of similar or different sizes. Junction flow depends on relative vessel size.

The capillaries, of very thin wall, represent the major exchange zone. In this exchange region, where blood velocity is low, the quantity of interest is the *transit time* of conveyed molecules and cells. The mean residence time in a circulatory compartment is the ratio of substance–cell holdup to the flow rate in the compartment.

Solute can be transported by either diffusion or convection through gaps of fenestrated capillaries. A fraction of filtrated plasma is sucked back from the interstitial liquid into capillaries and the remaining part is drained by lymphatic circulation into the large veins. The osmotic pressure Π depends on the wall features.[67] Consider a capillary in which the intraluminal pressure drop Δp_i is linear and the tissue pressure p_w constant. The capillary wall is semipermeable. The solvent transport due to $\Delta p = p_i - p_w$ is decreased by $\Delta\Pi$ due to the presence of macromolecules in the capillary lumen, which do not cross the wall. When the local effective pressure associated with osmotic and hydrostatic pressure differences between the lumen and the wall (subscript lw) $p_{\text{eff}} = \Delta p_{\text{lw}} - \Delta\Pi_{\text{lw}} > 0$, the plasma is filtrated. When $p_{\text{eff}} < 0$, the plasma is reabsorbed from the interstitial fluid. In normal conditions, the

---

[67] Water motion into or out of the capillary, across the endothelium, is partially driven by osmosis. Water moves from regions of low solute concentration to regions of higher solute concentration. Metabolic activity of the endothelial cells is then responsible for this water motion. Osmotic pressure is the pressure required to stop the net flow of water across a membrane (cell layer) separating solutions of different composition. The van't Hoff equation describes the osmotic pressure on one side of the membrane. It is valid for a single chemical species of concentration c: $\Pi = R_g T c$, where R_g is the gas constant and T the temperature. A given solution can lead to different osmotic pressures when the filtration membrane porosity differs.

filtration rate is equal to the sum of the reabsorption rate and lymphatic flow rate.

6

Numerical Simulations

"substituer au visible complexe de l'invisible simple"
[To substitute visible complex with invisible simple]
(Thom) [411]

Numerical simulations yield complete fields of hemodynamical quantities at multiple phases of the flow cycle, and easily test the effects of the involved physical parameters, especially the most important parameters of influence. However, *verification* and *validation* must be performed. Furthermore, the sensitivity of the results to the input parameters must be checked by varying the values of the input data. In some circumstances, the model is calibrated by adjusting the model inputs to the observations. The physics of the problem must be preserved by the chosen modeling and selected simulation technique. The code verification stage uses benchmarks with analytical solutions or "accurate" numerical solutions to check the computational implementation. Computation verification is done by comparing results obtained with various mesh densities. Errors and uncertainties are assessed. Modeling errors are associated with assumptions of the mathematical model derived from the conceptual model of the physical problem (model geometry, flow governing parameter values, material properties, constitutive equations, boundary conditions, etc.). Numerical errors arise from discretization features, convergence degree, computation round-off, modeling hypotheses (such as isotropy), and parameter estimations. Uncertainties are due to the lack of knowledge (unknown material constants of the deformable domain, boundary conditions, etc.). The predictive potential of the computational model is commonly validated by comparison of the numerical results with measurements.

6.1 Heart Pump

Cardiac wall kinematics and electrical activity are measured in vivo with appropriate space and time sampling. However, the stress field in the myocardium and the pressure field within the cavities are not available. Modeling is required to assess these quantities. It is composed of three main elements: (1) The first modeling element refers to the genesis and propagation of the action potential. The epicardial depolarization and deformation of the ventricular wall have been simultaneously measured using electrode brushes and videorecording of optical markers [412] or multielectrode socks and MRI tagging [413]. The nodal tissue remains difficult to locate. (2) The second modeling element is the mechanical behavior of the myocardium, with its constitutive law, which takes into account the electrochemical activation of the myocardium. The deformation of the left ventricle is not uniform [414, 415]. Diastolic deformation, like the systolic one, is heterogeneous, the lateral free wall undergoing greater radial motion and the apical wall more rotation in infants [416]. Myocardial fiber architecture affects cardiac electrochemical and mechanical features. Diffusion tensor imaging has been used to characterize cardiac myofiber orientations, with the reduced encoding imaging (REI) methodology [417]. Cardiac myofiber direction in each mesh element is the mean value of the noisy information contained into the voxels enclosed in the tetrahedron. (3) The third modeling element is the interaction between myocardium activity and blood flow. Blood perfusion of the heart is necessary for the pump activity on the one hand, and the blood ejection is vital for the body on the other hand.

Three-dimensional numerical modeling, which completely describes the myocardium activity must couple: (1) electrochemical wave propagation, (2) myocardium contraction and relaxation, and (3) blood systolic ejection, which is generated by the ventricular contraction tuned by the action potential. Electrochemical wave propagation (command) model provides the arrival time of depolarization in the various parts of the myocardium, the local myofiber orientations affecting the ventricular depolarization timing due to anisotropic myocardium conductivity. Contraction is more synchronous than depolarization. Myocardial fibers are more or less in phase at the end of the isovolumic contraction for a coordinated pumping. However, myofibers are differently stretched whether they are early or lately depolarized, without consequences due to their spatial arrangment. Moreover, even if cross-bridging in different CMCs is simultaneous, the CMCs can differently contract depending on the force applied to each CMC by its own environment. The electromechanical INRIA (ICEMA[1]) model of the electrochemical wave propagation is based on FitzHugh-Nagumo equations and the myocardium functioning model on the Hill-Maxwell rheological model, associated with myofiber direction and dynamics equations [418–420], but most of the biomechanical studies have been focused on the rheological behavior of the heart wall.

6.1.1 Electrochemical Firing and Signaling

Early modeling was carried out during the beginning of the twentieth century. The heart electrical activity has been indeed modeled by three coupled

---

[1] http://www-rocq.inria.fr/sosso/icema2/

oscillators corresponding to the natural pacemaker, the atrium, and the ventricle [421]. The action potential is the command of the myocardium contraction. The basic ionic models included three currents associated with Na^+, K^+ and leakage[2] ion channels, suitable for the nerve cell [422]. The current i is given by $i = C_m \, dV_m/dt + i_{Na} + i_K + i_L$ (C_m: membrane capacitance per unit surface area, V_m: membrane potential resulting from all ion contributions). Each current i_k is given by $i_k = g_k(V_m - V_k)$, where g_k and V_k are the membrane conductance and the constant membrane equilibrium potential for the corresponding ion k. The membrane conductance g_k is expressed by constant conductance and activation/inactivation functions $m(t)$ and $h(t)$ for Na^+, and $n(t)$ for K^+, each depending on the voltage-dependent opening and closing rates. The function $m(t)$ corresponds to a fast scale, whereas $h(t)$ and $n(t)$ are slow-scale functions, which after simplifications lead to a system with slow-fast dynamics. A system of four ordinary differential equations (ODE), one for the current and three for the three gating variables, is solved for the membrane potential as it varies with time.

A model variant for the Purkinje fibers does not represent physiological reality, taking into account an inward Na^+, outward K^+, and leakage currents, and not any Ca^{++} current [423], but it reproduces the main features of the action potential. Such a model has been improved for the heart's natural pacemaker, adding a slow and a delayed inward current [424]. Another model, adapted for the mammalian cardiac cell, includes a Na^+, two K^+ and a Ca^{++} current, associated with six gating variables [425]. It requires the solution of eight ODEs. Further developments based on measurements from cell voltage clamp studies lead to a model of the cardiac cell kinetics that includes nine ODEs, various channel types for each main ion (Na^+, K^+, and Ca^{++}) being incorporated [426]. Ionic models are suitable for at most a CMC set but not to model the normal and pathological rhythm and propagation in the whole myocardium.

The basic tractable phenomenological model of depolarization and repolarization consists of two variables u and v [427, 428], in which fast dynamics are coupled to slow ones (u: "fast" variable, v: "slow" variable):

$$\frac{d\mathbf{u}}{dt} = f(\mathbf{u}, \mathbf{v}), \quad \frac{d\mathbf{v}}{dt} = g(\mathbf{u}, \mathbf{v}).$$

The system behavior in the phase plane (u, v) (stable/unstable singular points, spiral point, stable/unstable limit cycle, bifurcation, etc.) depends on the parameters involved in functions f and g. Such a model has been developed for a general excitable medium, assuming spatial homogeneity. It was improved to more accurately model the electrochemical firing in the nodal pacemaker as well as impulse propagation along the nodal fibers and within the myocardium. The *FitzHugh-Nagumo-type* (FHN) two-variable system corresponds to a monodomain model. The Aliev-Panfilov model gives an example of a dimensionless

---

[2] The leakage current takes into account other ions such as Cl^-.

FHN system [429]:

$$\frac{du}{dt} = \nabla \cdot (D\nabla u) + ku(1 - u)(u - a) - uv,$$

$$\frac{dv}{dt} = -\epsilon(u, v)(ku(u - a - 1) + v),$$

where $\epsilon(u, v) = \epsilon_0 + \kappa_1 v/(u + \kappa_2)$ is a coupling control parameter between the action potential u and the repolarization v (the coupling between the slow and fast phases), D is the diffusion tensor,[3] k a repolarization control variable, and a the reaction parameter.

   Bidomain models have been proposed to simulate electrophysiological waves in the myocardium [430, 431]. The Mitchell-Schaeffer ionic model provides much better ECG signals than the FitzHugh-Nagumo models [432]. The Mitchell-Schaeffer model, with its two variables (u and v) and five parameters (τ_{in}, τ_{out}, τ_{open}, τ_{close}, and v_{gate}), is easy to calibrate and gives a good transmembrane potential. This model of electrical activity of the cardiomyocyte sarcolemma incorporates only an inward and outward current. Mitchell-Schaeffer two-variable model of excitable media is given by:

$$\frac{du}{dt} = \frac{1}{\tau_{in}} vu^2(u - 1) + \frac{u}{\tau_{out}}, \tag{6.1}$$

$$\frac{dv}{dt} = \begin{cases} \frac{1}{\tau_{open}}(v - 1), & u < v_{gate} \\ \frac{v}{\tau_{close}}, & u > v_{gate} \end{cases} \tag{6.2}$$

   Bidomain models take into account intra- and extracellular spaces, separated by the CMC syncytium membrane. Both domains have their own volume-averaged properties, especially the conductivity of the extra- and intracellular spaces. The problem to numerically solve is very complex. Cardiac fibers have anisotropic conduction properties, the impulse propagation being faster in the axial direction than transversally. A conductivity tensor $\mathbf{M}$ is then introduced, assuming that the conductivity values are identical in all directions perpendicular to the myofiber direction [433]. The collection of CMCs, end-to-end or side-to-side connected by specialized junctions, immersed in the extracellular fluid and ground matrix, is modeled as a periodic array that leads to homogenization procedure, with homogenized conductivity tensors $\mathbf{M_i}$ and $\mathbf{M_e}$. The membrane current density J_m is then given by:

$$J_m = -\nabla \cdot \mathbf{i_i} = \nabla \cdot \mathbf{M_i} \nabla u_i = \nabla \cdot \mathbf{i_e} = -\nabla \cdot \mathbf{M_e} \nabla u_e,$$

where u_i and u_e are the electric potentials of the intra- and extracellular spaces supposed to be passive conductors in a quasi-steady state and i_i and i_e the

---

3

$$D = \begin{pmatrix} 1 & 0 & 0 \\ 0 & 0.25 & 0 \\ 0 & 0 & 0.25 \end{pmatrix}.$$

currents $(\nabla \cdot (\mathbf{M}_i \nabla u_i + \mathbf{M}_e \nabla u_e) = 0)$. In its general form, the bidomain model is defined by [431]:

$$\kappa_{av}\big(C_m u_t + 1/R_m f(\mathbf{u}, \mathbf{v})\big) = \nabla \cdot (\mathbf{M}_i \nabla u_i),$$

where $\mathbf{u} = u_i - u_e$ is the action (transmembrane) potential, $\mathbf{v}$ the recovery variable, κ_{av} the averaged surface area-to-volume ratio of the cardiac myofibers, C_m and R_m the plasmalemmal capacitance and resistance (Table 6.1).

Cellular automata

The forward problem of electrocardiography deals with the chest surface potential due to the genesis and propagation of the electrochemical wave throughout the inner part of the heart wall and, afterward, across it. The model can be composed of the image-derived chest with the lungs, cardiac source, and blood masses. Each tetrahedron is characterized by the myofiber orientation assigned to each node, by a given refractory period (or its status, excitable or not), a selected conduction velocity, and its set, either nodal tissue or myocardium. The model of the nodal tissue is subdivided into two subsets, the domain cells that generate the action potential and the conducting cells. Action potential genesis and propagation in the atria and the ventricles, assumed to be either isotropic or anisotropic, can be modeled using cellular automata. Various shapes for atrial and ventricular potentials can be calculated using polynomials, which involve different parameters assigned to

Table 6.1. Values of the bidomain model parameters. R_m corresponds to the ion transport resistance through plasmalemmal ion channels. It is currently obtained by a polynomial formulation $(\kappa_d^2(u - u_r)(u - u_t)(u - u_m))$ of the transmembrane potential using the following set of quantities: depolarization rate κ_d, rest potential u_r, threshold potential u_t, maximum potential u_m ($M_{(i/e)f}$, $M_{(i/e)t}$, $M_{(i/e)n}$: conductivity components of the intra- and extracellular media in the main myofiber axis, normal to the main myofiber axis in the myofiber sheet and perpendicular to the myofiber sheet; Source: [434]).

C_m	$1\,\mu F/cm^2$
κ_{av}	$2000\,cm^{-1}$
M_{if}	$3\,mS/cm$
M_{it}	$1\,mS/cm$
M_{in}	$0.32\,mS/cm$
M_{ef}	$2\,mS/cm$
M_{et}	$1.65\,mS/cm$
M_{en}	$1.35\,mS/cm$
κ_d	0.04
u_r	$-85\,mV$
u_t	$-65\,mV$
u_m	$40\,mV$

mesh nodes, describing the depolarization, the beginning of the repolarization (plateau with or without a repolarization notch), and the final repolarization. Extracellular potentials can be computed using bidomain models.

A cellular automaton is a collection of grid cells with specified simple shape and assigned state that evolves according to a set of simple rules based on the states of neighboring cells during a selected number of discrete time steps. Cellular automata are hence aimed at describing dynamic phenomena, which are discrete in time, space, and states, using rules and structures instead of equations. Computer programs are easier to develop and computing time lower than approximately solving partial differential equations. The main drawback of cellular automata deals with the shape of the simulated spatial structures that depends on the shape of the cells. Two- and three-dimensional cellular automaton models have been developed to simulate conduction disorders, especially in ventricular hypertrophy, the propagation model being composed of the conduction system and bundles of myofibers with anisotropic conductivity [435–437].

6.1.2 Left Ventricle Behavior

Several factors are involved in left ventricle modeling: (1) time-varying geometry (previously simplified, now based on medical imaging); (2) kinematics boundary conditions, taking into account the restrictions imposed by the adjoining cavities (left atrium, right ventricle), the pericardium and the aorta, the extent of the deformation and the pressure distribution on the endocardium; (3) wall material, which is composite and infiltrated by liquids. (The myofibers, with various orientations, are embedded in a matrix with small blood vessels. The myocardial fibers are reinforced (sheath) and connected (struts) by collagen fibers. The cyclic behavior has two main phases, active and passive. The constitutive law must take into account the time-dependent heart behavior.); (4) blood and moving wall interactions during filling and ejection phases; (5) cardiac performance and oxygen consumption, which depend on stress and strain distribution; (6) time and space history of the electrochemical excitation, which triggers the myocardium contraction followed by its relaxation.

Two kinds of problems can be solved. In the direct problem, the stress and strain fields are computed in a given geometry using selected constitutive law and loadings. In the inverse problem, rheological and physiological parameters are deduced from the computed fields of the mechanical variables, input data being the computational domain determined from the medical images and possibly the ECG.

Finite element models of the left ventricule that undergo large displacements have been most often developed to analyze the wall strain and stress fields, neglecting heterogeneity in wall properties, myocardial fiber orientation and wall thickness variation in the axial and azimuthal directions, and assuming uniform transmural pressure. The first simulations were performed in idealized geometries.[4] The heart mechanical behavior was first studied during the diastole, when the myocardial fibers are in a relaxed state, then similarly to any biological soft tissue, the passive rheological properties being used as inputs. Systolic contraction must indeed be associated with an appropriate constitutive law. Stress-dependent stiffness has been approximated by a linear function of the mean-wall stress. Studying normal and pathological hearts, a pressure-dependent stiffness has been implemented using a polynomial function of the pressure [438]. The role of myocardial fiber orthotropy has been investigated, a transverse-to-axial stiffness ratio lower than 0.8 giving better agreement with cineangiography data [439]. With the development of medical imaging, imaging data were used to determine the computational domain, reconstructing ventricle cross-section models, and later heart cavities [440]. Numerical results differ from the findings obtained in idealized geometries. The same team of researchers has developed a three-layered wall model with myocardial fiber orientation of 60, 0, and −60 degrees.

In infarcted and reperfused walls, two methods are used to characterize the myocardial fiber property: (1) an incompressible myocardial bundle-intramural blood composite model, and (2) an inverse determination of the diastolic myocardial elastic modulus using a two-dimensional finite element method and matching the computed and imaged mid-ventricular cross-sectional areas [441].

The finite element method has also been used for stress–strain analysis of the heart valve [442]. The valve is modeled by a thin shell of uniform thickness; however, the values of the material constants are questionable.

6.1.3 Electromechanical Coupling

The INRIA (ICEMA) heart model connects the microscopic and macroscopic description levels via a mesoscopic step [418, 443]. The constitutive equations only require a small number of state variables, so the solution of direct problem can be quickly obtained and inverse problems can be solved.

---

[4] The prolate spheroidal (ζ, η, ξ) coordinate system has been introduced to describe heart anatomy, rather than using the Cartesian (x, y, z) or cylindrical (r, θ, z) frames:

$$x = r \cos \theta = d \sinh \zeta \sin \eta \cos \xi, \qquad y = r \sin \theta = d \sinh \zeta \sin \eta \sin \xi,$$
$$z = z = d \cosh \zeta \cos \eta,$$

where d is the focus location on the z-axis ζ the coordonate directed transmurally, η the coordonate running from the apex ($\eta = 0$) to the heart basis ($\eta = \pi/2$) along an elliptical path and ξ the circumferential coordonate in the cross-section plane (equivalent to θ coordonate of the cylindrical frame).

6.1.3.1 Nanoscale and Microscale Modeling

In the nanoscopic scale, the functioning of the nanomotors, i.e., actin and myosin molecules, is triggered by Ca^{++} and adenosine triphosphate (ATP).[5] $[Ca^{++}]_i$ has been given by a relatively simple function of time [444]. The troponin-C–Ca^{++} interaction can be obtained from chemical kinetics [445]. Sarcomere functioning is modeled by the *sliding filament theory* of sarcomere dynamics [446].

The Huxley sliding filament theory states that the binding f and rupture g frequencies of the actin–myosin bridges are functions of the elongation x. The cross-bridge proportion n with elongation $\tilde{x} = x/\ell$ (ℓ: maximum bridge length) with the kinetics defined by f and g is given by:

$$\dot{n} = \frac{dn}{dt} = (1-n)f - ng. \tag{6.3}$$

Sarcomere contraction generates a shortening $s = s_0(1 + \epsilon_c)$ (ϵ_c: sarcomere deformation), with a velocity $s_0\dot{\epsilon}_c$ assuming a synchronized motion of the set of bridgings (subscript c: contractile component).

The myocardium contraction results from a conformational change of the actin–myosin bridge, coupled with ATP hydrolysis. The ATP hydrolysis rate is regulated by the actin–troponin–tropomyosin complex. The chemical reactions related to the actin (A)–myosin (M) attachment–detachment cycle (cross bridge AM), including ATP, ADP and phosphate (P) generate the following set:

1. M.ATP $\xrightarrow{k_1}$ M.ADP.P   ATP hydrolysis

2. M.ADP.P + A $\xrightarrow{k_2}$ AM.ADP.P Ca^{++}-dependent A-M binding

3. AM.ADP.P $\xrightarrow{k_3}$ AM + ADP + P   sliding (shortening)   (6.4)

4. AM + ATP $\xrightarrow{k_4}$ M.ATP + A   AM unbinding

5. ADP + P $\xrightarrow{k_5}$ ATP   ATP formation.

Reaction 3 of the attachment–detachment cycle is the limiting one. It has been shown that [443]:

$$k_1/(1 + k_1/k_3 + k_1/k_2) \simeq k_1/(2 + k_1/k_3) \simeq k_{ATP},$$

and

$$\frac{d}{dt}\left(\frac{[AM]}{c}\right) = k_{ATP}\left(1 - \frac{[AM]}{c}\right) - k_4[ATP]\frac{[AM]}{c}. \tag{6.5}$$

This equation is similar to (6.3), f being identified to k_{ATP} and g to $k_4[ATP]$.

The attachment–detachment cycle (6.4) occurs when $[Ca](t) > [Ca_\star]$ ($[Ca_\star]$: yield concentration; Fig. 6.1).

---

[5] The number of actin sites available to bind with myosin depends on the total number of actin binding sites, $[Ca^{++}]$, troponin concentration, and the association and dissociation rates of Ca^{++} to troponin.

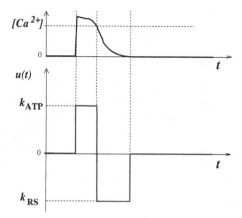

Figure 6.1. Correspondance between intracellular calcium concentration and the single command signal $u(t)$ (from [443]). Step function with a constant atrioventricular delay ($u > 0$: contraction, $u = 0$: passive relaxation, $u < 0$: active relaxation). The heart period, the duration of contraction and active relaxation, the chemical kinetic parameters are assumed to be constant.

1. $[\text{Ca}](t) \geq [\text{Ca}_*]$
   - $\forall \tilde{x} \in [0, 1]$ $\begin{cases} f(\tilde{x}, t) = k_{\text{ATP}} \\ g(\tilde{x}, t) = |\dot{\epsilon}_c(t)| \end{cases}$
   - $\forall \tilde{x} \notin [0, 1]$ $\begin{cases} f(\tilde{x}, t) = 0 \\ g(\tilde{x}, t) = k_{\text{ATP}} + |\dot{\epsilon}_c(t)| \end{cases}$

   Relaxation begins when the sarcoplasmic reticulum pumps begin to extract Ca from troponin at a supposed constant rate k_{SR}.[6] The destruction rate is supposed to be equal to k_{SR}.
2. $0 < [\text{Ca}](t) < [\text{Ca}_*]$

$$\begin{cases} f(\tilde{x}, t) = 0 \\ g(\tilde{x}, t) = k_{\text{SR}} + |\dot{\epsilon}_c(t)| \end{cases}$$

3. $[\text{Ca}](t) = 0$ in the absence of electrochemical stimulation ($f = g = 0$).

A unique command signal $u(t) = |u(t)|_+ - |u(t)|_-$ is then defined:

$$\begin{cases} |u(t)|_+ = k_{\text{ATP}} & \text{if } \text{Ca}(t) \geq \text{Ca}_* \\ |u(t)|_- = k_{\text{SR}} & \text{if } 0 < \text{Ca}(t) < \text{Ca}_* \end{cases}$$

---

[6] The amount of calcium linked to troponin C is assumed to be equal to the cytosolic concentration, i.e., to the extent of the calcium influx driven by the action potential. Fixation and release of calcium from troponin have different time constants.

$$f(\tilde{x}, t) = \begin{cases} |\mathbf{u}(t)|_+ & \text{if } \tilde{x} \in [0, 1] \\ 0 & \text{otherwise} \end{cases}$$

$$g(\tilde{x}, t) = |\mathbf{u}(t)| + |\dot{\epsilon}_\mathbf{c}(t)| - f(\tilde{x}, t).$$

The collective behavior of the sarcomere fibers is governed by the relationship between the stress σ and the strain ϵ in the myofiber direction (viscoelastoplastic behavior). The evolution of the stiffness $k_\mathbf{c}$ and the active stress $\sigma_\mathbf{c}$ of the contractile component, knowing the strain rate $\dot{\epsilon}_\mathbf{c}$ and the command $\mathbf{u}$, are given by the following set of ordinary differential equations [418]:

$$\begin{cases} \dot{k}_\mathbf{c} = -(\alpha|\dot{\epsilon}_\mathbf{c}| + |\mathbf{u}|)k_\mathbf{c} + k_0|\mathbf{u}|_+ \\ \dot{\sigma}_\mathbf{c} = k_\mathbf{c}\dot{\epsilon}_\mathbf{c} - (\alpha|\dot{\epsilon}_\mathbf{c}| + |\mathbf{u}|)\sigma_\mathbf{c} + \sigma_0|\mathbf{u}|_+ \\ \sigma = d(\epsilon_\mathbf{c})(\sigma_\mathbf{c} + k_\mathbf{c}\tilde{x}_0) + \mu_\mathbf{c}\dot{\epsilon}_\mathbf{c} \end{cases} \tag{6.6}$$

The parameters k_0 and σ_0 are related with the maximum available actin-myosin cross-bridges in the sarcomere. $d(\epsilon_\mathbf{c}) \in (0, 1)$ is the modulation function of the active stress that accounts for the length–tension curve[7] and $\mu_\mathbf{c}$ the viscosity ($k_\mathbf{c}(0) = \sigma_\mathbf{c}(0) = 0$). The cross-bridge detachment rate is given by $|\mathbf{u}| + \alpha|\dot{\epsilon}_\mathbf{c}|$.

The action potential $\mathbf{u}$ is modeled by the two-variable Aliev-Panfilov equation system based on the FitzHugh-Nagumo model [429]:

$$\begin{cases} \mathbf{u}_t - \nabla \cdot (\kappa_\mathbf{e}\nabla\mathbf{u}) = f(\mathbf{u}) - \mathbf{v} \\ v_t = \varepsilon(\beta\mathbf{u} - \mathbf{v}) \end{cases} \tag{6.7}$$

where $\varepsilon \ll 1$, $f(\mathbf{u}) = \mathbf{u}(\mathbf{u} - \alpha)(\mathbf{u} - 1)$ ($\alpha \in [0, 1/2]$), and $\kappa_\mathbf{e}$ is electrical conductivity.

The theory of actin–myosin cross-bridge dynamics and the moments [447] applied to the cross-bridge model describe the contraction at the sarcomere and myofiber scale, respectively. The sarcomers of linear stiffness k are in parallel. The resulting stiffness $k_T(t)$ is given by:

$$k_T(t) = k_{\max} \int_{-\infty}^{+\infty} n(\tilde{x}, t) \, d\tilde{x} = k_{\max}M_0(t) \tag{6.8}$$

Each element undergoes a displacement $\tilde{x}$. The total stress $\sigma_\mathbf{c}(t)$ is given by:

$$\sigma_\mathbf{c}(t) = \sigma_{\max} \int_{-\infty}^{+\infty} \tilde{x}n(\tilde{x}, t) \, d\tilde{x} = \sigma_{\max}M_1(t) \tag{6.9}$$

The potential energy is given by:

$$E_c(t) = e_{\max} \int_{-\infty}^{+\infty} \tilde{x}^2 n(\tilde{x}, t) \, d\tilde{x} = e_{\max}M_2(t) \tag{6.10}$$

---

[7] The troponin C sensitivity for Ca^{++} and the cross-bridge availability depend on the sarcomere length.

with $e_{max} = \frac{1}{2}\sigma_{max}$ per unit volume. The resulting sarcomere dynamics, derived by applying the moment-scaling method with the first two moments corresponding to active stiffness and stress, are in agreement with the sliding filament theory. The myofiber activity is controlled by the membrane potential, thus on the ionic currents through various channel types modeled by the action potential ruled by the FitzHugh-Nagumo equations.

Such a model takes into account neither the oxygen consumption nor the link between ATP and Ca^{++}. Ca^{++} overload can induce mitochondrial dysfunction in disease in the presence of a pathological stimulus. Calcium, ATP, and reactive oxygen species are indeed closely connected [448]. The mitochondrion produces ATP, which this synthesis is stimulated by Ca^{++}. The dysregulation of mitochondrial Ca^{++} can lead to elevated ROS concentration and apoptosis.

6.1.3.2 Hill Model

On the macroscopic scale, the cardiac fibers are embedded in a collagen sheath, connected by collagen struts and supported by a ground matrix with elastin fibers. The mesoscopic myofiber constitutive law is incorporated in a *Hill-Maxwell rheological model* [419]. This phenomenological model includes passive elastic and viscoelastic elements and an active component (Fig. 6.2). The sarcomere set of a myofiber is represented by a single contractile element (CE). Isometric deformations are modeled by an elastic serial element (ESE) in series with CE. ESE lengthens when CE shortens at a constant myofiber length. A third elastic element (EPE) in parallel with the CE-ESE branch is introduced. EPE educes the force developed from a certain myofiber length in the absence of stimulation. These three components are not related to the muscle constituents. The activity of CE depends on the action potential u (u > 0 during contraction and u < 0 during active relaxation). The element CE accounts for electrochemically-driven contractions and relaxations according to Eq. 6.6. ESE and EPE are the elastic and viscoelastic elements, respectively [449].

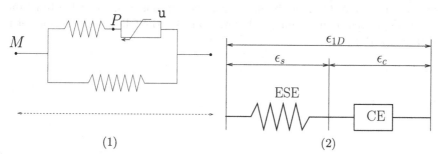

Figure 6.2. (1) Hill-Maxwell rheological model. **(2)** Strains in the active branch. During isovolumic phases, M is fixed and P moves. During systolic ejection and diastolic filling, both M and P move.

The biologically based modeling of the excitation–contraction coupling of the cardiac myofibers [418] suits the myocardium activity more than the current models of active myocardium contraction based on heuristic approaches and experimental data fitting [450–452]. With a wise choice of attachment and detachment rates, the INRIA heart model describes the Hill force–velocity relation [453], as well as viscoelastic passive and active relaxation behaviors.

The cardiac tissue is supposed to be not purely incompressible. With $\mathbf{P}$ denoting the second Piola-Kirchhoff stress tensor [454],

$$\mathbf{P} = -p \det(\mathbf{F}) \mathbf{S}_r^{-1} + \sigma_p^e(\mathbf{G}) + \sigma_p^v(\mathbf{G}, \dot{\mathbf{G}}) + \sigma_{1D} \hat{\mathbf{f}} \otimes \hat{\mathbf{f}} \qquad (6.11)$$

where $p = -B(\det(\mathbf{F}) - 1)$ (B: bulk modulus, used as coefficient of incompressibility penalization), $\mathbf{F}$ is the deformation gradient, $\mathbf{G}$ the Green-Lagrange strain tensor, and $\mathbf{S}_r$ the right Cauchy-Green deformation tensor, $\sigma_p^e(\mathbf{G}) \propto \rho_0 \partial W^e / \partial \mathbf{G}$ (elastic part of EPE, ρ_0: density of the reference state, W^e: elastic strain energy density) and $\sigma_p^v(\mathbf{G}, \dot{\mathbf{G}}) = \partial W^v / \partial \dot{\mathbf{G}}$ (viscoelastic part of EPE, W^v: viscous strain energy pseudo-density) the stresses in the passive materials, σ_{1D} the stress generated by the active element CE, and $\hat{\mathbf{f}}$ the local unit direction vector of the myofiber.

In the preliminary stage, the valves are not incorporated in the model. Constraints on the volume variations of the ventricle are then added:

$$\begin{cases} q \geq 0 & \text{when } p_V = p_a \text{(ejection)} \\ q = 0 & \text{when } p_A < p_V < p_a \text{(isovolumic phases)} \\ q \leq 0 & \text{when } p_V = p_A \text{(filling)}. \end{cases} \qquad (6.12)$$

where $q = -\dot{V}_V$ is the blood flow ejected from the ventricle (V_V: ventricular volume), p_V the ventricular blood pressure, p_A the atrial pressure and p_a the arterial pressure. A regularization[8] is used to overcome numerical failures on flow rate computations. A windkessel or a 1D model of the blood flow provides p_a.

Using Eqs. 6.6 and 6.11, and incorporating the multiple elements of the simplified system: (1) myofibers, (2) passive components, (3) valve functioning, and (4) upstream atrium and exiting artery, the following equation set is obtained [419]:

---

[8] Regularization satisfies the conditions $dq/d(p_V - p_a) \geq 0$, $dq/d(p_A - p_V) \geq 0$ during ventricular ejection and filling, respectively.

$$\begin{cases} \rho\ddot{\mathbf{u}} - \nabla \cdot (\mathbf{F} \cdot \sigma) = 0, \\ \sigma = \sigma_{1D}\,\hat{\mathbf{f}} \otimes \hat{\mathbf{f}} + E_p(\mathbf{G}), \\ \sigma_{1D} = \sigma_c/(1 + \epsilon_s) = \sigma_s/(1 + \epsilon_c), \\ \sigma_c = E_c(\epsilon_c, \mathbf{u}), \\ \sigma_s = E_s((\epsilon_{1D} - \epsilon_c)/(1 + \epsilon_c)), \\ \epsilon_{1D} = \sum_{i,j} G_{ij} f_i f_j, \\ \dot{\mathbf{g}} = \mathcal{G}(\mathbf{g}, t), \\ \text{initial + boundary conditions}, \end{cases} \qquad (6.13)$$

where $E_p(\mathbf{G}) = -p\det(\mathbf{F})\mathbf{S}_r^{-1} + \sigma_p(\mathbf{G}, \dot{\mathbf{G}})$, E_c is a function expressing system 6.6, ϵ_{1D} the deformation in the myofiber direction. From thermomechanical considerations: $1 + \epsilon_{1D} = (1 + \epsilon_c)(1 + \epsilon_s)$. Quantity g stands for either V_V, p_V, p_A, or p_a, the last equation accounting for the set of ordinary differential equations modeling valve opening and closure and arterial pressure changes. Geometric models that contain myofiber directions are used. The complete description of the 3D electromechanical model is given in [419] and numerical solutions in [455].

The action potential can be initiated at the ends of the Purkinje network localized according to the literature data [456]. The model parameters are calibrated according to the available data, which provide: (1) arterial parameters [457, 458], (2) k_c and τ_c [452], from which are computed the respective asymptotic values k_0 and σ_0, and (3) estimation of the passive behavior [459, 460].[9]

6.1.3.3 Blood Flow Incorporation

Patient-specific medical tools combine medical images and signals with a heart model, which involves metabolic (perfusion), electrochemical, and mechanical activities. Such a computer tool can be used to solve inverse problems to estimate parameters and state variables from observations of the cardiac function (data assimilation).[10] The fluid dynamics within the coronary network, which irrigates the myocardium, can be based on a hierarchical approach, taking into account both the large coronary arteries and the intramural vessels. Three-dimensional blood flow model in distensible right and left coronary arteries (proximal part of the heart perfusion network, located on the heart surface) can be coupled to 1D flow model for wave propagation, which can correspond to the first six generations of branches [463, 464], and a poroelastic model of

---

[9] These conflicting literature data are given for any soft tissues.

[10] Data assimilation is aimed at incorporating measurements into a behavioral model to evaluate the state of the explored organ. Local analysis of isochrones could be sufficient for the identification of electrical conductivity. Input isochrone maps can be obtained from different electrophysiological devices, from the invasive sock of electrodes [461] to ECG imaging [462].

the small arteries and the microcirculation that cross the heart wall, using homogenization [465, 466], as wall permeability depends on the wall deformation. This hierarchical flow description can be coupled to oxygen transport.

6.1.3.4 One-Dimensional Heart–Artery Coupling

In a preliminary model, vessel curvatures, wall frictions, axial displacements, flow developments have been neglected [467]:

$$\frac{\partial \mathbf{g}}{\partial t} + \frac{\partial}{\partial z}\mathcal{G}(\mathbf{g}) = 0, \tag{6.14}$$

where

$$\mathbf{g} = \begin{pmatrix} A \\ q \end{pmatrix}, \quad \mathcal{G}(\mathbf{g}) = \begin{bmatrix} q \\ q^2/A + (2\beta/3\rho)A^{3/2} \end{bmatrix}.$$

This hyperbolic system have two eigenvalues $\lambda(A, q) = V_q \pm c$.

The coupling problem then has the following formulation:

$$\begin{cases} \rho \ddot{u} - (C_p \dot{u}_x + K_p(u_x) + E_s(u_x - e_c))_x = 0, \\ \dot{\sigma}_c = k_c \dot{\epsilon}_c - (\alpha|\dot{\epsilon}_c| + |\mathbf{u}|)\sigma_c + \sigma_0|\mathbf{u}|_+, \\ \dot{k}_c = -(\alpha|\dot{\epsilon}_c| + |\mathbf{u}|)k_c + k_0|\mathbf{u}|_+, \\ \sigma_c = E_c(\epsilon_c, \mathbf{u}), \\ \sigma_s = E_s(\epsilon, \epsilon_c), \\ \text{closed valve: } A_{lv}\dot{u}(0, t) = q = 0, \\ \text{open valve: } p_{lv}(t) = p_a(t) = p(0, t) \\ \partial A(z, t)/\partial t + \partial q(z, t)/\partial z = 0, \\ \partial q/\partial t + \partial(q^2/A)/\partial z + (A/\rho)\partial p/\partial z = 0, \\ p = p_0 + \beta(\sqrt{A} - \sqrt{A_0}), \\ q = A_{LV}\dot{u}(0, t), \\ \text{outlet absorbing boundary conditions} \\ \text{initial conditions} \end{cases} \tag{6.15}$$

Data for the arterial network are obtained from [468] (Fig. 6.3). The network is constituted by artery sets 1–5, 12–17, 22, and 27–38 with features given in Table 6.2.

Relatively simple computational models can couple the heart not only to the efferent blood vessels represented by a windkessel model, but also to the afferent compartment (pulmonary circulation and left atrium), to take into account the preload and afterload, using electrical analogues at the boundaries of the 3D deformable left ventricle [469]. Moreover, the baroreflex control of heart rate can be incorporated in such lumped parameter models [470]. Ventricular interactions and pericardium effects on hemodynamics can also be introduced [471].

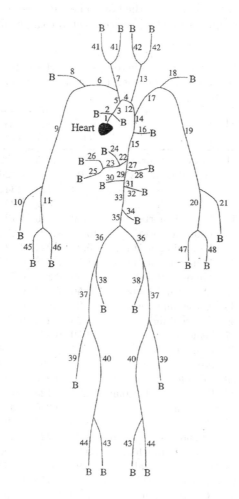

Figure 6.3. Artery network (artery-assigned numbers correspond to labels of Table 6.2).

6.2 Blood Flow in Large Vessels

Detailed investigations aiming at describing fields of physical variables have most often been focused either on blood flow in vessels and heart cavities or in heart wall mechanics. Studies on blood flow in the main arteries coupled to the left ventricle are based on simple 0D/1D models. The vascular impedance expresses the relationship between pressure and pulsatile flow, measured at any given point in the arterial system, and associated with lumped (concentrated) parameter models. It is defined as the ratio of the pressure harmonics and the corresponding harmonics of the flow. The windkessel model represents the arterial system as an elastic chamber connected to a peripheral resistance.

Table 6.2. Lengths, inlet and outlet radius of systemic arteries (numbers correspond to labels of Fig. 6.3; R.: right; l.: left; com. common; ext.: external; int.: internal; post.: posterior; ant.: anterior).

Label	Artery	L (cm)	r_{in} (cm)	r_{out} (cm)
1	Ascending aorta	1.00	1.525	1.502
3	Ascending aorta	3.00	1.502	1.420
4	Aortic arch	3.00	1.420	1.342
12	Aortic arch	4.00	1.342	1.246
14	Thoracic aorta	5.50	1.246	1.124
15	Thoracic aorta	10.50	1.124	0.924
27	Abdominal aorta	5.25	0.924	0.838
29	Abdominal aorta	1.50	0.838	0.814
31	Abdominal aorta	1.50	0.814	0.792
33	Abdominal aorta	12.50	0.792	0.627
35	Abdominal aorta	8.00	0.627	0.550
36	External iliac	5.75	0.400	0.370
37	Femoral	14.50	0.370	0.314
40	Femoral	4425	0.314	0.200
38	Internal iliac	4.50	0.200	0.200
39	Deep femoral	11.25	0.200	0.200
43, 44	Post. and ant. tibial	32.00	0.125	0.125
2	Coronaries	10.00	0.350	0.300
5	Brachiocephalic	3.50	0.950	0.700
6, 17	R. and l. subclavian	3.50	0.425	0.407
9, 19	R. and l. brachial	39.75	0.407	0.250
10, 21	R. and l. radial	22.00	0.15	0.175
11, 20	R. and l. ulnar	22.25	0.175	0.175
46, 47	R. and l. ulnar	17.00	0.200	0.200
45, 48	R. and l. interosseus	7.00	0.100	0.100
6, 18	R. and l. vertebral	13.50	0.200	0.200
7	R. com. carotid	16.75	0.525	0.400
13	L. com. carotid	19.25	0.525	0.400
41,42	Ext. and int. carotid	15.75	0.275	0.200
16	Intercostal	7.25	0.630	0.500
28	Superior mesenteric	5.00	0.400	0.350
22	Celiac trunk	2.00	0.350	0.300
23	Com. hepatic	2.00	0.300	0.250
24	Splenic	6.50	0.275	0.250
25	Gastric	5.75	0.175	0.150
26	Hepatic	5.50	0.200	0.200
30, 32	R. and l. renal	3.00	0.275	0.275
34	Inferior mesenteric	3.75	0.200	0.175

6.2.1 Windkessel Model

Windkessel (der Windkessel: air chamber) model is an example of lumped-parameter model with a single independent variable, the time (t), used to describe: (1) the afterload undergone by the heart which propels blood through the arterial system; and (2) pressure–flow relationships in the artery. Windkessel models were introduced by Frank [472]. The heart and systemic arterial system are modeled by a closed hydraulic circuit with, at least, a compliance[11] and a resistance element (two-element windkessel model). The arterial network, which, in simplest windkessel models, represents the circulatory loop, is considered as a single distensible pipe, with a given time-dependent compliance, connected to the left ventricle, and a peripheral resistance at its distal end, an outlet linear pressure–flow relationship being assumed. Such a model does not take into account wave propagation.

Three-element (time-varying aortic characteristic impedance, arterial compliance and systemic resistance) non-linear windkessel model have been used to analyze aortic-ventricular coupling. The windkessel model provides a reasonable representation of afterload for predicting stroke volume, and systolic and diastolic aortic pressures, but underestimates peak aortic flow and fails to reproduce real impedance as well as realistic aortic pressure cyclic variations [473].

Heart–arterial interaction has been studied using a time-varying elastance model coupled to an arterial model, which is a four-element windkessel models, consisting of total peripheral resistance (R), total arterial compliance (C), total blood inertance (L), and the characteristic impedance of the aorta (Z_0) [457]. Heart function is described by five parameters: slope (E_{max}) and intercept (V_d) of the end-systolic pressure volume relation, the time to reach maximal elastance and heart rate. Left ventricular preload can be expressed as venous filling pressure (p_v) and minimal elastance (E_{min}), or by EDV. In any case, the windkessel model is too simple to accurately describe the coupling between the heart and the main afferent and efferent vascular compartments.

6.2.2 3D Blood Flow

Navier–Stokes equations cannot be solved analytically in most cases. The necessary step is the reduction of the continuum to a discrete finite set of points and hemodynamics quantity values. Computational fluid dynamics (CFD) approximately solves the Navier-Stokes equations. There are several discretization techniques to get numerical solutions of the Navier-Stokes equations. These include boundary elements (BEM), finite difference (FDM), finite element (FEM), finite volume (FVM), spectral methods, and others.

---

[11] When liquid enters in the compliant element, it compresses the air in this circuit component and pushes water out of it. The compressibility of the air in the vat simulates the artery distensibility. The resistance encountered by the fluid leaving the windkessel element models the peripheral resistance.

In FDM, the differential operators are approximated by difference formulae constructed from values of the dependent variable at a number of predefined nodes of a discretization mesh, which approaches the continuum geometry [474]. Navier-Stokes equations are enforced only at these points. The velocity and pressure values at any node are related to the values at neighboring nodes. The nodes are connected in a structured mesh.

In FVM, the fluid volume is divided into a finite number of cells. The local form of the equations balances mass and momentum fluxes across the faces of each individual cell. FVM deals with mean cell values; cell input and ouput depend only on face values. The shape of the cells may be unstructured. However, irregular flow domains can be immersed in regular Cartesian grids [475].

The discrete model used in FEM consists in a set of values of the problem functions at a finite number of domain points associated with piecewise approximations of the functions, using continuous interpolation functions, over a finite number of finite elements [476, 477]. The numerical unknowns are uniquely determined everywhere inside the element by values specified at the element nodes. The algebraic equations are formed as the result of either a variational formulation or the weighted residual method.

All these methods, if properly applied, give better and better approximation to the actual solution of Navier-Stokes equation as the number of nodes is increased, i.e., when the approximate solution converge toward the continuum solution. The mesh node density increase is associated with a reduction in the approximation error and with a computational time growth. A numerical technique can be optimal for a given application and useless in others. However, FEM handles 3D unsteady flows, multiple kinds of boundary conditions [478], and complex deformable geometries. Mass conservation is not always satisfied by the Navier-Stokes solvers based on the finite element method, certain methods being much more robust than others [479].

Lattice Boltzmann Models

Lattice Boltzmann models (LBM) are based on a kinetic approach for the simulations of flow and wave propagation in complex environments [480, 481]. LBM can treat curved boundaries [482], as well as deformable pipe walls [483]. LBM is suitable for parallel and grid computing [484], as shown for 3D flow through rigid artificial heart valve [485]. At the mesoscale of blood flows (arterioles and venules), where the typical length is at most about 10 times the size of the blood cells, the multiphase blood flow must take into account the dense suspensions of deformable cells, conveyed by a surrounding incompressible fluid. Computational methods that simulate the blood cells have a high solving cost. They are mainly used at the vasculature microscale. The lattice Boltzmann model considers momentum distribution functions evolving on a discrete lattice at discrete time steps based on a discrete form of the

Boltzmann equation. The interactions of the flowing cells between them and with the vessel wall can be set by appropriate interface methods.

Fictituous Domain Method

The principle of the fictituous domain method relies on solutions in larger domains, fictituous domains, which contain original domains, but have much simpler configurations [486, 487]. The boundary conditions are imposed by introducing a Lagrange multiplier on the fictituous domain boundaries. The advantage is to work with a structured mesh. Such a method can be used for a body that moves and/or undergoes shape variations, such as a valve, without mesh change, hence in the absence of wall contact. The main drawback of the method is precision loss.

Fictituous domain method and arbitrary Eulerian-Lagrangian technique (Sect. 6.3.3) have been combined in numerical simulations of the behavior of the aortic valve [488]. The fluid domain mesh is hence not modified by immersed leaflet displacement, arbitrary Eulerian-Lagrangian formulation being applied for the interactions between the deformable aortic wall and the flowing blood. Both the arbitrary Eulerian-Lagrangian technique and the fictituous domain method are not adapted in the case of contact between computational domain boundaries, as contacts between the valve cusps or between the leaflets and the wall that are required to close the anatomical conduit to avoid backflows.

6.2.3 Image-Based Detailed Models

Numerical modeling of the cardiovascular system is often carried out to provide a complete description of the blood flow in a specific region. The local fields of hemodynamic variables are obtained using 3D simulations, based on the numerical approximation of the incompressible Navier-Stokes equations, in general, in a rigid domain computed from image processing. In vivo rheological properties of the artery and aneurism walls indeed remains unknown. Moreover, in vivo velocimetry at both vessel ends is not available in mid-size arteries due to limitation in spatial resolution of the scanning device. The 3D modeling deals thus with a frozen domain, an unsteady flow at the inlet, and most often, stress-free conditions at the domain outlets. The procedure, with its sequence of operations, is illustrated by the 3D reconstruction and the meshing of a saccular aneurism (Sect. 7.5.1) located at a terminal embranchment of a branch of the middle cerebral artery.

6.2.4 Space Dimension-Related Multiscale Modeling

> *"tout ce qui est simple est faux, tout ce qui ne l'est pas est inutilisable"* (Valéry) [489]

Space dimension-related multiscale modelings deal with the coupling of flow models from dimension 3 (real configuration modeled by fluid domain; partial differential equations expressed in the 3D space) to O (lumped-parameter models associated with ordinary differential equations). This type of multiscale modeling differs from the modeling aimed at taking into account the various length scale from the cell nanomachinery to the body organs and physiological systems. The coupling of models of different dimension allows us to take into account the entire network of the cardiovascular system and to describe short vasculature segment using 3D Navier-Stokes equations.

The numerical simulations of the blood flow in a segment of the vasculature require the specification of boundary data on artificial boundaries, which limit the explored vascular district. These boundary conditions are influenced by the fluid dynamics upstream from the 3D computational domain entry and downstream from it. When physiological measurements cannot produce suitable boundary data, a mathematical description of the action of the reminder of the circulatory system on the studied region is required. Due to the complexity of the cardiovascular system, simplifications are mandatory to get results in an appropriate time duration, the mathematical description of other parts of the circulatory network not requiring the same level of details. Models of lower dimensions are thus connected to the explored region, leading to the concept of multiscale modeling of the blood circulation [464, 490–493]. The coupling involves models at different scales, 3D, 1D, and 0D, with a decreasing level of accuracy and computational complexity. Reduced models neglect the geometrical complexity of the vascular network.

6.2.4.1 Lumped Parameter Models

Lumped parameter models represent most of the compartments in series and in parallel of the vasculature (Figs. 6.4 and 6.5). These reduced models are described by non-linear ordinary differential equations (ODE) with localized[12] averaged flow quantities. Three quantities are involved in these electrical analogues (Tables 6.3 and 6.4) [458], the compliance (vessel storage ability), the inertance, and the resistance (impedance components associated with inertial forces and viscous effects, respectively). A cardiovascular simulator has been built, composed of an arterial network of compliant tubes of known properties, which allows validation of numerical schemes [494].

Networks of resistances, capacitances, and inductances[13] based on analogies between the transmission of a sinusoidal electrical signal and pulsatile blood flow, are devised to fit blood circulation behavior. The transmission line

---

[12] Vessels are assumed to be defined by uniform representative properties that are set in a single, self-sufficient point.

[13] The resistance R is the impedance component associated with viscous forces, the compliance C is the impedance component associated with vessel storage ability, and the inertance L is the impedance component associated with inertial forces.

analogy leads to the vascular impedance Z concept.[14] The vascular impedances

Table 6.3. Electrical analogues. The reactance $X = C + L$.

Hemodynamical quantity	Electrical variable	Scale (per unit length)
Pressure p_i $[M.L^{-1}.T^{-2}]$	Voltage u	
Flow rate q $[L^3.T^{-1}]$	Current i	
Resistance R $[M.L^{-4}.T^{-1}]$	Resistance R	$8\mu/(\pi R^4)$ (Poiseuille flow)
Inertance L $[M.L^{-4}]$	Inductance L	$\rho/(\pi R^2)$ (pure inertance model)
Compliance C $[M^{-1}.L^4.T^2]$	Capacitance C	$2(A^{3/2}(1 - \nu_P^2))/(\pi^{1/2}Eh)$
		(state law for tube)
		(with fixed ends)

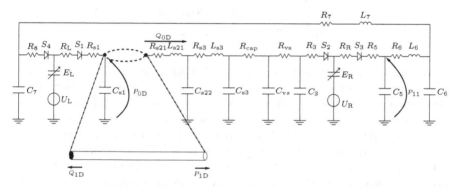

Figure 6.4. An example of 0D–1D coupling. Values of lumped parameters for each selected vascular segment are given in Table 6.4 (from [495]).

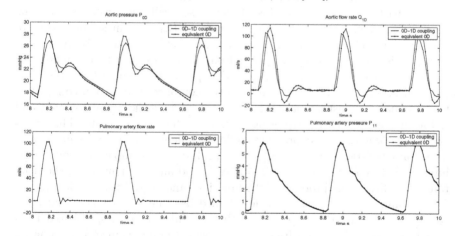

Figure 6.5. Pressure P_{0D}, flow rate Q_{0D}, pulmonary pressure P_{11} and flow rate Q_{12} (from [495]).

---

[14] In a pure resistance model $Z = \Delta p/q$, in a pure inertance model $Z = i\omega L$ $(L = \rho/(AL))$, and in a pure compliance model $Z = dp/q = dV/(qC) = -i/(\omega C)$.

Table 6.4. Coefficients used for 0DM, resistances are given in mmHg s/ml, inductances in mmHg s^2/ml, capacitances in ml/mmHg (labels refer to Fig. 6.4; from [495]).

Body part	Labels	R	L	C
Arteries	s1	3.751×10^{-3}		0.6
	s21	5.47×10^{-4}	3.057×10^{-3}	0.2454
	s22	0.059	1.943-3	0.3546
	s3	0.05	0.6	0.1×10^{-2}
Capillaries	cap	0.83		
veins	vs	0.11		82.5
	3	3.751×10^{-3}		20
Lungs	5	3.751×10^{-3}	-	9×10^{-2}
	6	3.376×10^{-2}	7.5×10^{-4}	2.67
	7	0.1013	3.08×10^{-3}	2.67
	8	3.751×10^{-3}		

are complex quantities that are introduced in unsteady electrical analogues. The vascular impedances expresses the relation of the time-dependent pressure difference generating a flow to the resulting pulsatile blood convection. The vascular impedances include three types, defined below in the case of a sinusoidal input. The *longitudinal impedance* Z_L of a vascular segment per unit length is the pressure gradient-to-flow ratio[15] $Z_L = -(dp/dz)/q$. The *input impedance* Z_z at a given vessel cross-section is the pressure-to-flow ratio $Z_z = (p/q)_z$, considered as input for the downstream vasculature. It is also named the *characteristic impedance* Z_0 for uniform vessels in the absence of wave reflection. The magnitude of the characteristic impedance Z_0 varies directly with the elastic modulus of the vessel wall and inversely with the cross-sectional area, other factors being equal (Z_0 rises in stiffer vessels of smaller bore). The higher the frequency parameter, the smaller Z_0. According to transmission line theory, Z_z differs from Z_0 in presence of wave reflection: $Z_z/Z_0 = (1 + \Gamma^* \exp\{-2\gamma L\})/(1 - \Gamma^* \exp\{-2\gamma L\})$, where Γ^* is the complex reflection coefficient, $\gamma = a + \imath b$ the complex propagation constant, a the attenuation constant ($a = 0$ for inviscid fluid), b the phase constant and L the distance from the exploring station to the reflection site. The *transverse impedance* Z_t is the pressure-to-flow gradient ratio $Z_t = -p/(dq/dz)$. The characteristic impedance $Z_0 = (Z_L Z_t)^{1/2}$. The complex hemodynamic impedance has a real part, the resistance R, and an imaginary part, the reactance X, which is the product of the fluid inertance L and the frequency.

The impedance Z_z has been determined by simultaneous flow and pressure measurements ($Z^* = p^*/q^*$), with their inherent errors, followed by a signal processing based on frequency analysis, the impedance depending on the frequency and not on time. For the harmonic k:

---

[15] By analogy with the electrical impedance $-(dV/dz)/i$.

$$Z(\omega_k) = \frac{p_k(t)}{q_k(t)} = \frac{P_k \exp\{i(\omega_k t + \phi_k)\}}{Q_k \exp\{i(\omega_k t + \psi_k)\}}$$

$$= \frac{P_k}{Q_k} \exp\{i(\phi_k - \psi_k)\} = R_k + iX_k \ . \tag{6.16}$$

Pulmonary Circulation

One-dimensional model of unsteady flow in a four-compartment 40-generation pulmonary vascular network has been developed using a finite difference scheme based on the method of characteristics [496]. The dimensionless tube law $\tilde{A} = -\tilde{p}^{2/3}$ was used for $\tilde{p} < -1.5$ and a third degree polynomial approximation for $\tilde{p} \geq -1.5$. Two-dimensional sheet flow model has been introduced to describe the flow in the pulmonary capillaries [497]. The sheet (sheet thickness of $10.5\,\mu m$) is composed of an aligned circular cross-sections (diameter of $\sim 4\,\mu m$) of posts, connected by a membrane (length of $\sim 8.5\,\mu m$).

6.2.4.2 One-Dimensional Models (Wave Propagation)

When the fluid dynamics is coupled to the vessel wall mechanics, boundary conditions such as stress free for the fluid and free edge for the solid introduce numerical reflection artifacts [498]. The coupling of three- and one-dimensional models, suitable for wave propagation, avoid spurious wave reflections at the ends of the 3D model. However, such distributed parameter models assume that vessels are made of infinitesimal slices without connections between them. As in clinical practice, average data are used. Blood flow phenomena are approximated with 1D model if the streamwise component of the blood velocity is much greater than the secondary flow component. Networks of 1D models are suitable to describe pulse wave propagation along the main arteries and their branches [499, 500].

Detailed 3D FSI models of small CVS regions, 1D wave-propagation models and 0D models of the major part of the circulatory circuit are thus coupled by appropriate BCs. The theoretical analysis of the multiscale coupling has been carried out, providing a local-in-time existence result for the solution [501]. The multiscale coupling must give a well-posed formulation, the reduced models providing averaged values at the model interfaces, whereas the 3D Navier-Stokes equations require pointwise quantities. Numerical experiments show that the imposition of the continuity of all averaged quantities at the interface may causes numerical instabilities, which can be overcome in a straight cylindrical configuration by relaxing the continuity constraint on the cross-sectional area.

The 1D model must a priori take into account heterogeneous velocity distribution in the cross-section. Correction terms are introduced in the energy and momentum equations. The conductance G, momentum coefficient α_m and kinetic energy coefficient α_k are hydraulic parameters deduced from the velocity distribution in the cross-section:

$$G = -\frac{1}{dp_i/dx} \int v\, dA_i, \qquad\qquad [\mathrm{M}^{-1}.\mathrm{L}^4.\mathrm{T}] \qquad\qquad (6.17)$$

$$\alpha_{\mathrm{m}} = \int \frac{v^2}{V_q^2}\frac{dA_i}{A_i}, \qquad\qquad \alpha_{\mathrm{k}} = \int \frac{v^3}{V_q^3}\frac{dA_i}{A_i}, \qquad\qquad (6.18)$$

($V_q = q/A$: cross-sectional average velocity). An uniform straight tube is very crude model for a branching bend of varying geometry and rheology along its length. Nevertheless, it is used to represent the bulk properties of the vessels immediately upstream and downstream form the explored segment. Besides, the hydraulic parameters are generally assumed to be equal to one (uniform velocity distribution or plug flow).

6.3 Fluid–Structure Interaction

Fluid–structure interaction deals with the coupling between the 3D Navier-Stokes equations and the possibly thin[16] 3D vessel wall, which undergoes large displacement (Table 6.5). The blood vessel walls are deformable. The wall compliance has many physiological consequences, blood storage, mainly in veins, wave propagation with a finite speed, flow distribution, etc. Increase in artery wall stiffness induces an overload to the left ventricle. Moreover, arterial wall diseases are susceptible to fissure and rupture, such as stenoses and aneurisms. Interaction between the blood flow and the vessel wall must be taken into account to know whether the high-pressure zones induced by the blood dynamics are correlated to the high-stress concentration areae within the diseased wall. Last, the presence of red blood cells, which are able to agregate and form rouleaux, is responsible of the non-Newtonian behavior of the blood in stagnant blood regions, and particle flow in small vessels. Direct simulations of particule flows consider the flowing RBCs as moving, individual particles, around which the plasma flow is computed [503, 504].

Let Ω^f et Ω^s the fluid and wall domains. The problem must solve an equation set corresponding to: (1) wall mechanics, with its constitutive law and suitable boundary and initial conditions; (2) fluid dynamics, with its constitutive law and appropriate boundary and initial conditions; and (3) coupling conditions, stress and velocity equality at the moving interface.

$$\rho^s \mathbf{u}_{,tt} - \nabla \cdot \mathbf{C}^s = \mathbf{f}(\mathbf{x}, t) \qquad\qquad \text{in } \Omega^s \times [0, T],$$

$$\rho^f [\mathbf{v}_{,t} + (\mathbf{v} \cdot \nabla)\mathbf{v}] = -\nabla p_i + \mu \mathbf{\Delta v} \qquad\qquad \text{in } \Omega^f \times [0, T],$$

---

[16] The ratio between the wall thickness h_{w} and the hydraulic radius of the blood vessel R_h is usually lower than 0.2 ($0.07 \leq h_{\mathrm{w}}/R_h \leq 0.20$ according to the literature data on various arteries, under different pressures). It decreases in peripheral arteries and it is lower in veins than arteries. The most often proposed value is about 0.1. The blood vessel can thus be represented by a thin-walled duct. Whereas 1D model can be used in cylindrical straight pipes, detailed 3D modeling resort to non-linear shell model.

$$\mathbf{u}_{,t}(\mathbf{x}, t) = \mathbf{v}(\mathbf{x}, t) \qquad \text{on } \Gamma^I \times [0, T],$$
$$\mathbf{C}_{ij}^f n_j = \mathbf{C}_{ij}^s n_j.$$

Theoretical aspects of fluid–structure interaction (stability and precision analysis of the coupling scheme, etc.) have been studied [505–507]. The coupling between the Navier-Stokes equations and elasticity ones has been found to be energetically stable, if the convection term in the Navier-Stokes equations is not neglected [508]. In some situations, existence of strong [507, 509] or weak [510–512] solutions has been proved.

Numerical simulations of peculiar features of blood flows must then take into account the wall deformability. The numerical simulations of the fluid–wall interaction raise many issues: (1) the wall displacement being large (artery deformation associated with the radial mode of wave propagation of about 10%, vein collapse), the flow problem must be solved in a moving domain; (2) the densities of the artery wall and the blood, both living tissues full of water, being close, the coupling is strong and must be carefully tackled, using implicit approximations to avoid numerical instabilities [513]; and (3) the boundary conditions on the inlet and outlet(s) of the explored vessel segment must avoid spurious reflections [498]. Certain standard boundary conditions which suit in fixed geometries do not even give rise to a well-posed problems in deformable domains [514, 515]. Suitable boundary conditions for the fluid–structure problem can be provided by the coupling of the detailed model with a simplified 1D hyperbolic system. The results were promising on axisymmetric flows. However, the method is difficult to extend to general flow problems.

6.3.1 Coupling Strategy

The usual strategy splits the entire equation set into FE approximations for each fluid and solid problems. The fluid and solid equations are discretized in time by model-suited methods and to solve the approximated equations [516–518]. Three solvers are then necessary: a fluid, a solid solvers and a coupling one that allows for signaling and mesh updating. Computational cost of the numerical simulations of unsteady flows in compliant vessels

Table 6.5. Current governing principles of blood–vessel wall interactions.

blood	vessel wall
Momentum conservation	
Constitutive law	
Newtonian fluid or	Fiber-reinforced
generalized Newtonian model	hyperelastic material
Interface continuity	

must be reduced. Two strategies can be used, parallel computing [519] and developments of new solving algorithms [520]. Coupled problems with numerical complexities can also be solved using both geometric and algebraic multigrid techniques [521]. Several scenarii and algorithms are thus possible, but research is still active to develop fast solvers based on robust and efficient fluid–structure coupling algorithms. In blood flow modeling, the weak coupling introduced parasite energy and have accuracy and stability problems [522]. Strong implicit coupling needs subiterations and are thus computationally costly [523]. New acceleration techniques for classical fixed-point iteration based on transpiration boundary conditions have been proposed [524]. Newton algorithms based on the exact continuous Jacobian [525] or on approximated Jacobian [526] have been developed. Compared with the common coupling strategies, the computational time has been reduced by one order of magnitude with these new methods. Nevertheless, simulations with a reasonable computational time must use more efficient algorithms.

6.3.2 Wall Models

In addition to shell models, simplified wall modeling deals with the wall radial displacement[17] $\mathbf{u}_r$ and approximations of the Cauchy stress tensor $\mathbf{C}$. The independent-ring model, which neglects the longitudinal deformation is widely used to provide the constitutive equation of the vessel wall:

$$\rho_w \mathbf{u}_{r,tt} = p - \kappa(E, h)u_r. \tag{6.19}$$

The generalized string model has also been employed:

$$\rho_w h \mathbf{u}_{r,tt} - \kappa_T G h \mathbf{u}_{r,zz} + Eh/((1 - \nu_P^2)R_0^2)\mathbf{u}_r - \eta \mathbf{u}_{r,tzz} = \mathbf{f}, \tag{6.20}$$

where κ_T is the Timoschenko shear correction factor, G the shear modulus [498]. The 2D numerical simulation of the left ventricle has been carried out, more to study the numerical aspect of the fluid–wall interaction than to investigate the physiological behavior of the left ventricle [527].

6.3.3 Arbitrary Eulerian-Lagrangian Formulation

The Lagrangian approach is inefficient in complex flows and the Eulerian formulation can not accurately describe the fluid behavior near the moving wall. In the arbitrary Eulerian-Lagrangian formulation (ALE) method, the reference is neither the fluid particle nor the overall flow reference frame, but the moving mesh of the fluid domain surrounded by a deforming wall, with the grid velocity $\mathbf{v}_g$. The time derivative is then defined neither considering a given

---

[17] Both radial and axial wall displacement are produced during wave propagation, but the latter is neglected.

point $\mathbf{x}$ (Eulerian derivative), nor following a fluid particle (Lagrangian derivative), but according to the fluid domain deformation $\mathbf{u}(\mathbf{x}, t) = \mathbf{u}(\varphi(\mathbf{x}_0, t), t)$, where φ is the deformation mapping of the reference configuration Ω_0 to the deformed domain $\Omega(t)$.[18] The Navier-Stokes equation are then reformulated, the convection term including the grid velocity [528–533].

Let Ω_t be the 3D time-dependent computational model of a vessel segment and $\partial \Omega_t$ its boundaries, which are partioned into a vessel inlet Γ_t^i, a vessel outlet Γ_t^o and a wall Γ_t^w.[19] Navier-Stokes equations are derived using Eulerian formulation $\{\mathbf{x}, t\}$. Dealing with deformable vessels, the Lagrangian formulation is used. The configuration at any time Ω_t is mapped from the initial configuration Ω_0. to keep fixed Γ_t^i and Γ_t^o, $\forall t$, the ALE is used. The Navier-Stokes equations becomes:

$$\boldsymbol{\nabla} \cdot \mathbf{v} = 0,$$
$$\rho \left(\mathbf{v}_{,t} + [(\mathbf{v} - \mathbf{v}_g) \cdot \boldsymbol{\nabla}] \mathbf{v} \right) = -\boldsymbol{\nabla} p_i^\star + \mu \boldsymbol{\nabla}^2 \mathbf{v}. \tag{6.21}$$

6.3.4 Immersed Boundary Method

The *immersed boundary method* (IBM) dealts with interactions between a possibly active elastic incompressible flexible solid (fibers and valves) immersed in a Newtonian incompressible fluid, when the equations of the body mechanics ressemble those of the fluid dynamics. The unified description of the composite biofluid–biosolid system is then treated by IBM, the recoil forces differentiating the solid from the fluid [534–536]. The Navier-Stokes equations are solved at the Cartesian grid points. IBM uses a set of control points to represent the fluid–solid interface. The force densities are computed at the control points and spread to the Cartesian grid points by a discrete representation of the delta function. The velocity field is used to update the position of the interface.

The IBM principle is thus to treat, in a fixed uniform cubic lattice, the elastic material as a part of the fluid in which additional forces arising from the elastic stresses are applied. The fluid dynamics is governed by the Navier-Stokes equations, the body mechanics by the 3D linearized elasticity equations [537]. The configuration $\Omega(t)$, at any time t, of a solid body made of an incompressible elastic material is usually described by a Lagrangian variable $\mathbf{X}(\mathbf{x}_0, t)$, which gives the position of the solid particle labeled by its position $\mathbf{x}_0$ in the reference configuration Ω_0. The deformation state of the elastic body is described by a 3×3 matrix $\partial \mathbf{X} / \partial \mathbf{x}_0$.[20] Given the potential $E_p(t)$

---

[18] If the domain deformation is superimposed to the fluid particles trajectories, then the ALE derivative is equal to the Lagrangian derivative.

[19] Both vessel entry and exit, boundaries with upstream (proximal with respect to the heart) and downstream parts of the vasculature, are considered fixed.

[20] Because the material is incompressible, $det(\partial \mathbf{X} / \partial \mathbf{x}_0)$ is time independent. The matrix determines the density of the system potential, which depends on a material function $\mathcal{E}(\partial \mathbf{X} / \partial \mathbf{x}_0, \mathbf{x}_0, t)$ when the material is heterogeneous and active.

and kinetic $E_k(t)$ energy, the action $\mathcal{S} = \int_0^T (E_k(t) - E_p(t)) \, dt$ is introduced. $\mathbf{X}(\mathbf{x}_0, t)$ minimizes $\mathcal{S}$ with the incompressibility constraint and with given initial (and final) configuration(s). The conversion between Eulerian and Lagrangian variables is expressed in terms of the 3D Dirac delta function, the kernel $\delta^3(\mathbf{x} - \mathbf{X}(\mathbf{x}_0, t))$, as well as the mass and force densities, introducing added mass and external force. The kernel δ^3 is replaced for implementation by a discrete kernel δ_h^3 associated with each node of the biosolid, where:

$$\delta_h(x_i) = \begin{cases} (1/4h)(1 + \cos(\pi x/(2h))) & \text{if } |x| \leq 2h \\ 0 & \text{otherwise.} \end{cases}$$

The mesh size must be small enough to track the solid, scouting being led by the recoil forces. The two components of this Eulerian-Lagrangian scheme are thus linked by a smoothed approximation of the Dirac delta function, which is used to apply elastic forces to the fluid, and interpolate the fluid velocity at the representative material points of the elastic tissue.

The immersed boundary method was first introduced to study the fluid dynamics of the heart valves, employing a fiber–fluid representation of the heart. The mathematical formulation of the equations of motion of a viscous incompressible fluid containing an immersed network of elastic fibers forms the foundation of IBM. The massless and volumeless fibers act as pure force generators. The incompressible viscoelastic resulting material is highly anisotropic since the fiber stress points always in the fiber direction. In each time step, forces are generated in elastic and possibly contractile fibers. These forces are then allowed to act on a cubic lattice, on which the equations of fluid dynamics are solved. Finally, the fibers move at the local fluid velocity, updated under the influence of the elastic forces. Neither the fluid motion nor the cardiac tissue motion is assumed known in advance. The first problem deals with the momentum- and mass-conservation equations for flow of incompressible fluid, the body forces (e.g., gravity) being neglected:

$$\rho^f \left(\frac{\partial \mathbf{v}}{\partial t} + (\mathbf{v} \cdot \nabla)\mathbf{v} \right) = \mathbf{f}_E - \nabla \mathbf{p} + \mu \nabla^2 \mathbf{v},$$

$$\nabla \cdot \mathbf{v} = 0.$$

where the applied force density $\mathbf{f}_E$ expresses the effect of the fibers acting in the fluid. The second problem deals with the solid assumed to obey to the linearized 3D elasticity equations:

$$\rho^s \frac{\partial^2 \mathbf{u}}{\partial t^2} - \nabla \cdot \mathbf{C}^s(u) = \mathbf{f}^s$$

$$\mathbf{C}^s(u) = \lambda \operatorname{Tr}(\mathbf{E}(u))\mathbf{I} + 2\mu \mathbf{E}(u)$$

$$\mathbf{E}(u) = \tfrac{1}{2}(\nabla \mathbf{u} + \nabla \mathbf{u}^T)$$

At the fluid–solid interface $\Gamma_I(t)$, the solid must have the same velocity as the fluid and the solid stress must be equal to the force exerted on the solid from the fluid. $\forall \mathbf{x} \in \Gamma_I(t)$,

$$\mathbf{v}(t, \mathbf{x} + \mathbf{u}(t, \mathbf{x})) = \frac{\partial \mathbf{u}}{\partial t}(t, \mathbf{x}),$$

$$\mathbf{C}^f(\mathbf{v}, p)(t, \mathbf{x}) \cdot \hat{\mathbf{n}}^f(t, \mathbf{x}) = \mathbf{C}^s(\mathbf{u})(t, \mathbf{x}) \cdot \hat{\mathbf{n}}^s(t, \mathbf{x}),$$

IBM has the following main features: (1) the solid and fluid are considered to be a composite material, (2) the problem is solved on a steady regular mesh, (3) forces from the solid are calculated in the mesh nodes, (4) the solid is moved at a velocity equal to the local fluid velocity. The main stages of the algorithm are: (1) given the set of the velocities of the wall nodes $\dot{\mathbf{u}}$ and $\mathbf{X}$, compute the local contributions to the wall traction $\mathbf{f}_L$, (2) from $\mathbf{f}_L$, compute $\mathbf{f}_E$ in every mesh node, (3) solve the Navier-Stokes equations, (4) find $\dot{\mathbf{u}}$ at wall points by interpolation from $\mathbf{v}$, and (5) move the boundary.

IBM suffers several drawbacks: (1) the physical process is not well mastered, especially at the different involved length scales; (2) the Reynolds number must be small enough for appropriate treatment of fluid–solid interface displacement; (3) wall leakage occurs if the wall points are not placed sufficiently close; (4) under large deformations, the distance between wall point can become too large; (5) mass balance must be calculated for each cell; and last (6) IBM has stability problems.

The *immersed interface method* (IIM), which deals with an interface between the nodes of an uniform Cartesian mesh, is similar to the immersed boundary method, based on a finite difference scheme, but with a better accuracy near the boundary [538]. The method, developed for Stokes flows, has been afterward extended to Navier-Stokes equations [539].

6.3.5 Image-Based Fluid–Structure Interaction

Due to the role played by the geometry on the flow behavior, the actual vessel shape is of paramount importance for the numerical flow simulation. Because the wall of the large arteries is thin with respect to the vessel radius, shell elements are used to model wall displacements associated with the wave propagation. Quadrilateral shell elements, such as MITC element [540], are more reliable to cope with biological problems than triangular elements, like the DKT element [523]. Quadrilateral shell elements must then be coupled to tetrahedral meshes of the complex geometries of the fluid domains, derived from the facetization associated with the 3D reconstruction from medical images. Rather using interpolation or motar elements [541], the triangular surface mesh, optimized for finite element computations from the facetization, is converted into a quadrilateral mesh, using the INRIA software YAMS. Pairs of triangle produce quadrangles with matched vertices. A Jacobian-free Newton-Krylov algorithm is efficient and robust to tackle blood flows in deformable arteries [542]. This technique was successfully applied to the saccular aneurism (Sect. 7.5.1) used to illustrate a procedure used for image-based computations (Fig. 6.6).

6.4 Shape Optimization and Parameters Identification

Shape optimization and parameters identification belong to optimal control problems of system governed by partial differential equations [543]. In shape optimization [544], the control is on the boundary variations of the domain, and in the inverse problem it is on the physical parameters of the problem. Bypass surgery and mini-invasive stenting (Chap. 7) are commonly used to treat a critically stenosed artery, different varieties of bypass and stent shapes being available [545, 546]. Shape optimization is aimed at minimizing the post-therapeutic complications. *Inverse problems* are aimed at inferring the parameters, which cannot be measured (viscosity, elasticity, etc.) using measurement techniques. Physiological signals can then be assessed by mathematical models of the circulatory system, minimizing the gap between the computed and measured data.

The problem solution must combine: (1) an efficient solver for the state equations, (2) an efficient optimization algorithm, and (3) a fast calculation of gradients of the cost function. Shape optimization has been studied for optimal bypass design, using a Stokes problem as state equation [547]. Solution of inverse problems are currently on lumped parameter or 1D models [548–552]. Further studies are needed to involve the fluid–structure interaction. A difficulty relies on the effective computation of the gradients of the cost function. Some preliminary works have been performed [553, 554]. In addition, hybrid optimization algorithms coupling stochastic and deterministic methods, which have been successfully applied to complex aerodynamic problems, can

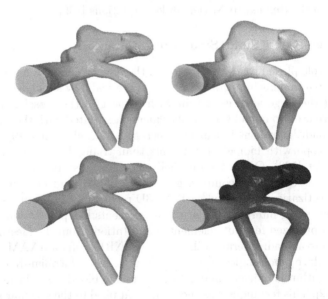

Figure 6.6. Propagation of a pressure wave in a cerebral aneurism (fluid–structure interaction; from [542]).

be used [555]. Genetic algorithms are well adapted ot the case of non-smooth cost function and/or complicated evaluation procedure. This technique has been successfully applied to optimal positioning of electrodes of implanted artificial pacemaker on the wetted surface of the ventricle with a left bundle-branch block (Sect. 7.6.2.2). The cost function based on the depolarization time gives a better criterion than the one defined from ECG features [556].

6.5 Transmural Transport

Three main stages are involved in material transport to and across the vessel wall. The first stage (**stage I**) deals with transport in the vessel lumen. Once conveyed at the possible site for transendothelial migration, the particle must cross the near-wall fluid domain.[21] Certain molecules circulate either more or less freely in the plasma or linked to erythrocytes. Both transport modes can be used by the same molecule, such as albumin. The partition coefficient between plasma and RBC inner fluid depends on molecule diffusivity and RBC membrane permeability. Such process is usually not taken into account in literature models. Besides, the axial dispersion of the tracers in the vessel lumen can be limited by the radial diffusion [557–559].[22] The next stage (**stage II**) corresponds to material uptake by the wall luminal surface and transendothelial transport. Molecules must cross the glycocalyx and either the endothelial cell or the cleft. Adsorption is shear-dependent and can require material conversion. Some lipoprotein conformations, which depend on the wall shear stress, are more convenient for irreversible adsorption, which affects the exchange rate at the interface. The final stage (**stage III**) focuses on transmural transport.

Transport models are currently focused on inner wall layers, neglecting the adventitia and its blood irrigation and drainage. Data on blood adventitial convection to and from the vessel wall remain unknown. Moreover, the studied part of the wall suppose that the wall is composed of a set of layers separated by membranes. Transport through any wall layer is based on convection and diffusion. *Convection* refers to solvent-guided solute motion under a given transtunica pressure. The pressure distribution across the vessel wall remains unknown as well as its dependence with luminal pressure. Increase in blood pressure can induce permeability changes due to porous medium compaction. *Diffusion* is associated with a concentration gradient. The vasomotor tone can also affect the material transfer through the wall poroviscoelastic interstitial

---

[21] Going from intraluminal blood to the vessel wall, the particle cross the diffusion boundary layer. The thickness of the diffusion boundary $\delta \propto (\mathcal{D}z/\dot{\gamma}_w)^{1/2}$ where $\mathcal{D}$ is the species effective diffusion coefficient and z the distance downstream from the inlet.

[22] The longitudinal dispersion of a tracer slug with respect to the cross-sectional average concentration is describable as diffusion in the longitudinal direction with diffusivity $\mathcal{D} + R^2 V_q^2/(48\mathcal{D})$.

matrix. Molecule transport through elastic laminae depends on their fenestrae size, which can vary with applied stresses. In the media, elastic fenestrated lamellae are also perpendicular to the flux axis. Main transport associated phenomena are: (1) hindrance due to friction during solute displacement, (2) solute reflection associated with local partial permeability of the medium to be crossed, and (3) possible reactions.

The solvent (water, subscript w) and solute (subscript s) transport across the vessel wall require transport variables, such as the hydrostatic Δp and osmotic $\Delta\Pi$ pressure difference across the wall layer, as well as physical control parameters such as the *hydraulic conductivity* G_h,[23] layer *permeability* $\mathcal{P}$, *porosity* Ψ,[24] *diffusivity* $\mathcal{D}$, osmotic κ_o and solute drag κ_d *reflection coefficients*,[25] related to the solute selectivity of the wall tunica interfaces, *hindrance coefficient* κ_h[26] (Table 6.6). The wall-layer porosity Ψ is supposed to be time-independent. An example of literature values is given in Table 6.7.

6.5.1 Literature Data

Wall absorption of substances and transport through the vessel wall have been studied by several teams, either to study capillary exchanges or because inadequate efflux of accumulated substances through fenestrated internal elastic lamina and the media to the adventitial blood and lymph vessels may appear. In the second framework, studies were carried out to determine whether convection or diffusion is the major transmural mass transfer process. Diffusion was found to be the dominant transport mechanism rather than convection [564]; but transmural convection, which depends on the wall stress field, influences the macromolecular transport within the arterial wall [565]. Another objective was to determine the most controlling step of the whole process. Stage 1 does not provide a dominant resistance to material motions [566].

---

[23] The hydraulic conductivity depends on the medium density and viscosity as well as on pore size.

[24] The wall-layer porosity is the ratio between the fluid volume and the fibrous matrix volume. Let $\breve{v}_D$ be the volume (V)-averaged filtration speed and $\breve{v}_w$ be the volume (V_w)-averaged solvent velocity, then the porosity $\Psi = \breve{v}_D/\breve{v}_w$ (Dupuit-Forchheimer relation). The continuity equation then states:

$$(\Psi\rho_w)_t + \nabla \cdot (\rho_w \breve{v}_D) = 0.$$

[25] Reflection coefficients depend on membrane permeability for both the solvent and solute.

[26] Collisions between molecules and wall structures reduce the convection and slow diffusion. Volume fraction available for a solute of given radius must thus be taken into account. Hindrance coefficients κ_h are then introduced [560]:

$$c_t + \nabla \cdot (-\mathcal{D}\nabla c + \kappa_h \mathbf{v} c/\Psi) = 0.$$

However, vessel curvature induces local variations in blood–wall mass transfer rate [567]. The endothelium provide a priori a major transport resistance and hence determine the transmural flux. The other wall layers affect the material distribution. The wall porosity is layer-dependent due to layer composition and structure. The layer influence depends on the molecule type. The media that has a greater transport resistance for albumin than the adventitia, acts as a barrier for this molecule [564]. A third goal is to know whether wall ingress-controlled or wall egress-controlled mechanisms are disturbed in wall lesions.

Four factors are involved in disturbed mass transport that leads to atherosclerosis: blood hypoxemia, leaky endothelial junctions, transient intercellular junction remodeling, and convective clearance of the intima and media [568]. High LDL transport is found to be correlated to low O_2 transfer to promote intimal thickening [569]. Proposed models generally do not take into account the different biochemical processes involved in vessel lumen and wall mass transport such as molecule binding, degradation, etc. Neglect of O_2-hemoglobin bonds have been found to lead to large errors in O_2 transfer using a simple one-layer wall model [570].

Table 6.6. Mass transport coefficients and relations. An effective porosity is introduced because collisions between solid parts of the porous medium (length scale L^*; volume V, with solvent volume V_w) and solute slow the transport ($V_{w_{eff}}$: available solvent volume for solute). J_s: solute flux per unit surface area.

Parameter	Relationship
Darcy filtration speed $\breve{v}_D$ [L.T^{-1}]	$J_w \equiv \breve{v}_D(\Delta\Pi = 0) = -(\mathcal{P}_D/\mu_D)/\nabla p = G_h\Delta p$ (Darcy law)
Hydraulic conductivity G_h [M^{-1}.L^2.T]	$\breve{v}_D(\kappa_o = 0) = G_h(\Delta p - \Delta\Pi)$ (Starling law)
Darcy permeability $\mathcal{P}_D$ [L^2]	
Drag reflection coefficient	$\kappa_d(\Delta c = 0) = 1 - J_s/(J_w\breve{c})$
Osmotic reflection coefficient	$\kappa_o(J_w = 0) = \Delta p/\Delta\Pi$
Porosity	$\Psi = V_w/V$
Effective porosity	$\Psi = V_{w_{eff}}/V$
Solute permeability $\mathcal{P}$	$J_s(J_w = 0) = \mathcal{P}\Delta\Pi$
Solute diffusivity $\mathcal{D}$	$J_s(\mathbf{v} = 0) = \mathcal{D}\nabla c$ (first Fick law) $\kappa_h\breve{v}_D \cdot \nabla\breve{c} - \mathcal{D}_{eff}\nabla^2\breve{c} = 0$ (second Fick law)
Slip transport if $\Psi > 0.8$ and $L^* < \mathcal{P}_D^{1/2}$	$\nabla p = -(\mu_D/\mathcal{P}_D)\breve{v}_D + \mu_D\nabla^2\breve{v}_D$ (Brinkman equation)

6.5.2 Transport Model

Dealing with mass transfer in the artery wall, literature models focus on solute dynamics from the vessel lumen across the wall intima and media, separated by internal elastic lamina.[27] The computational domain is divided into three main compartments: (1) the vessel lumen, (2) the subendothelial part of intima and (3) the media. The endothelium and internal elastic lamina define the interfaces between these subdomains. Solute transport between the different subdomains is coupled by mass flux across the interfaces (Table 6.8). Transport models aimed at describing solute transport across the different layers of the artery wall must handle suitable quantitative data for the whole set of physical involved parameters. The advection–diffusion equation (ADE) describes the solute transport in the vessel lumen, the convection term being provided by the local blood velocity field obtained by solving the Navier-Stokes equations:

$$c_t + \mathbf{v} \cdot \nabla c + \nabla \cdot (\mathcal{D} \nabla c) = 0, \tag{6.22}$$

where c is the studied species concentration. Endothelium is usually assumed to be a semipermeable membrane with pores corresponding to junctions and leaky cells. Normal junctions are represented by circular sections around nor-

Table 6.7. Values of wall transport coefficients for albumin (diffusivity of 9×10^{-11} m^2/s and inlet concentration of 572 nmol/cm^3) in a 2D model of an artery (caliber of 6 mm and length of 12 cm) [561]. The inlet velocity profile is parabolic and the lumen, intima, and media Pe are respectively equal to 1.2, 1.7, and 7.1. The wall layers are supposed to be homogeneous. The endothelium and internal elastic lamina are modeled by continuous semipermeable membrane crossed by convective and diffusive fluxes. Permeability coefficients are derived from a pore model [562]. Transport coefficients in the subendothelial space of the intima and the media are based on a fibrous matrix model [563].

	Endothelium	Intima	IEL	Media
		Porous media		
h (μm)		10		300
$\mathcal{P}_D$ (m^2)		9×10^{-17}		5×10^{-20}
$\mathcal{D}_{\text{eff}}$ (m^2/s)		6×10^{-11}		2×10^{-12}
κ_h		1.12		3.08
		Membranes		
G_h (m/Pa/s)	3.6×10^{-12}		3×10^{-10}	
$\mathcal{P}$ (m/s)	5.5×10^{-10}		6.4×10^{-9}	
κ_o	0.75		2×10^{-3}	
κ_d	0.75		2×10^{-3}	

---

[27] The role of EEL and adventitia with its vessels is currently neglected.

mal endothelial cells with a pore set in its central region; leaky junctions are modeled by rings around leaky cells [572].

Blood flows are currently assumed to be laminar, either steady fully-developed in a straight rigid stented[28] artery [561, 573, 574], or unsteady 3D[29] [575], but with questionable parameter values [573]. Mass transport across the subendothelial layer of the intima and the media is modeled by the convection–diffusion–reaction equation. Fluxes J across the endothelium and the internal elastic lamina are computed using *Kedem-Katchalsky equations*:

$$\begin{cases} J_{\mathtt{w}} = G_h(\Delta p - \kappa_o \Delta\Pi), \\ J_s = \mathcal{P}\Delta c + J_{\mathtt{w}}(1 - \kappa_d)\check{c}_* \end{cases} \tag{6.23}$$

($\check{c}_* = \Delta c/\ln(c_2/c_1)$: averaged concentration at interface). The subendothelial space of the intima is assumed to be homogeneous porous media composed of a fibrous matrix with proteoglycans and collagen fibers. In porous media, the pressure and velocity are averaged in representative elementary volume greater than the pore size and smaller than the domain size. The volume-averaged filtration velocity $\check{v}_D$ across any wall layer is computed using the *Darcy law* (Re $\ll$ 1) and a Darcy permeability coefficient $\mathcal{P}_D$, a tensor that is reduced to a scalar in isotropic homogeneous media. The two intima constituents, proteoglycans and collagen fibers, with their respective volume fraction V_{PoG} and V_{Cn}, are supposed to be in series: $\mathcal{P}_D^{-1} = \mathcal{P}_{D,\mathrm{PoG}}^{-1} + \mathcal{P}_{D,\mathrm{Cn}}^{-1}$ [574]. The permeabilities for proteoglycans and collagen fibers are computed using the Karman-Kozeny equation which needs the values of the fiber radius, porosity

Table 6.8. Transluminal and transmural transport (Source : [571]).

Transport layer	Layer thickness (μm)	Governing equations
Vessel lumen	Variable	Advection–diffusion equation (c)
		Navier-Stokes equation ($\mathbf{v}$)
Endothelium	0.5–2	Kedem-Katchalsky equation (J)
Glycolyx	0.1–0.25	Non-modeled
Subendothelium	5–10	Darcy equation ($\check{v}_D$)
(intima)		Advection–diffusion–reaction equation($\check{c}$)
IEL	2–5	Kedem-Katchalsky equation (J)
Media	300–1000	Darcy equation ($\check{v}_D$)
		Advection–diffusion–reaction equation($\check{c}$)
EEL		Usually non-modeled
Adventitia		Uptake by blood and lymph
		Non-modeled

---

[28] A coated stent, which provides drug reservoirs, is implanted in the artery.

[29] The finite element formulation of the Navier-Stokes equations is based on an operator-splitting method and implicit time discretization. The streamline upwind/Petrov-Galerkin (SUPG) method is applied for stabilization.

Table 6.9. Transport in porous media.

Scale	Model	Equation
Macroscopic	Darcy	$-\nabla p = (\mu/\mathcal{P})\breve{v}$
Mesoscopic	Brinkman	$-\nabla p = (\mu/\mathcal{P})\breve{v} - \mu\nabla^2\breve{v}$
Microscopic	Navier-Stokes	$-\nabla p = \rho\,D\mathbf{v}/Dt - \mu\nabla^2\mathbf{v}$

of the fiber matrix, and Kozeny constant. Diffusion transport in porous media obeys to Fick law, the diffusion being caused by a concentration gradient in a moving solvent. The effective diffusivity depends mainly on proteoglycans [40]. The internal elastic lamina is supposed to be a thin semipermeable membrane. The internal elastic lamina is composed of fenestrae of given radius, given density and consequently, given relative surface area. $G_{h,\mathrm{IEL}}$ is derived from the geometry variables of IEL and its fenestrae, as well as $\mathcal{P}_{D,\mathrm{IEL}}$ [562]. κ_o and κ_d may also be given by literature data [562, 576]. The media assumed to be homogeneous porous media composed of a fibrous matrix and smooth muscle cells. Effective coefficients can be obtained from the literature [562, 577, 578] The advection–diffusion–reaction provides the volume-averaged concentration $\breve{c}$. The concentration decrease in the vessel wall is mainly due to the endothelium (82% [561], 68% [575]).

A substance transport model in the microvasculature has been developed [579]. The model is based on basic hexagonal network units composed of a set of parallel capillaries that are periodically arranged in the plane, with cross connections to sources in arteriolar zones and sinks in venular regions. The arteriolar and venular zones belongs also to a periodic structure in the direction normal to the cross-section of the capillary set, the centers of mass of adjacent zones being separated by a distance of 500 μm. Tracer transport based on the convection–diffusion equation is coupled to the flow distribution.

6.5.3 Simulation of the Chemical Coupling

The arterial wall has a selective permeability for blood solutes. Its transfer depends strongly to the local flow behavior. The Navier-Stokes equations provides the velocity and pressure fields, then the convection term for the solute transport. The Navier-Stokes equations are coupled to the convection–diffusion equation for the solute concentration in the vessel lumen. Not only the solute transport within the vessel lumen depends on the velocity field, but also the boundary conditions of the solute transfer on the endothelium are related to wall velocity gradient. The near-wall solute diffusion is associated with a diffusivity tensor, which depends on wall shear stress. The mathematical and numerical modeling of the coupling problem of the lumen transport and the transfer across the rigid endothelium has been analyzed [580].

ADE is coupled to transport equation to model wall mass transfer. The Beavers-Joseph-Saffman condition can be used at the interface between the vessel flow and the porous medium [581]. Inside the arterial wall, a pure diffusive dynamics is commonly considered. The coupling problem has been solved in an axisymmetric stenosis, using the Brinkman model (extension of the Navier-Stokes equations for porous media; Table 6.9), assuming a wall made of a single homogeneous layer [582]. Splitting in the lumen advection–diffusion equation in the wall diffusion equation has been associated with an efficient iterative method based on a Steklov-Poincaré interface equation [583]. Numerical techniques for transport in large artery lumen and wall are presented in [584]. Multi-layered wall models require determination of numerous physical parameters. The parameter set can be obtained from measurement fitting using electric analogues [560]. Numerical results depend on values of transport coefficients in membranes and porous media, which are roughly estimated, generally from in vitro experiments and associated modelings.

Cardiovascular Diseases

Cardiovascular diseases are the leading cause of death in rich countries.[1] Heart failure is due to leakage or narrowing heart valves, heart insufficiency, disturbed electrical activation of the myocardium, etc. Two major arterial pathologies are aneurisms[2] and atherosclerosis, which have been targeted by physical and mathematical modeling. Genetic and environmental risk factors contribute to the variability in disease susceptibility for cardiovascular diseases. Cardiac pathologies remain the number one cause of death from congenital malformations in infancy (45–50% of postnatal deaths due to congenital anomalies). The majority of congenital heart defects arise from abnormal development of the valvuloseptal compartment.

7.1 Risk Factors

The occurrence of cardiovascular diseases depends on the subject pathophysiological ground and habits. Major cardiovascular risk factors include diabetes, hypertension, hypercholesterolemia, obesity (App. B),[3] the familial context (family members with heart diseases), and smoking. Minor cardiovascular risk

---

[1] Cardiovascular diseases cause about 12 million deaths per year world-wide. They account for more than half of all deaths in Europe of people over 65 years old. They are responsible for 33% of all deaths in France. In France, about four people in every 10,000 die prematurely from heart disease, about thirteen in every 100,000 from ischemic cardiopathies, and about seven in every 100,000 from stroke. The rate can be increased by a factor six or more in other european countries (Sources: Cité de la Science in Paris and european health status publication of the European Commission).

[2] The primary spelling aneurism is here used rather than pedantic one aneurysm, although υ was replaced by y in latin and a lot used in the Middle Ages.

[3] The body mass index is the ratio between the weight (kg) and the square height (m). The normal value range is $[18.5 - -25]$. A value between 25 and 30 means moderate overweight; greater than 30, obesity.

factors are age, high-stress job, sedentary habits, lack of daily consumption of fruits and vegetables, and high resting heart rate (beating rate above 75 per minute). The two most important risk factors are cigarette smoking and abnormal ratio of blood lipids (apolipoprotein-B/apolipoprotein-A1). Smoking increases levels of tumor necrosis factor-α and endothelial intercellular adhesion molecule-1, decreases adiponectin level, and enhances production of reactive oxygen species.

Metabolic syndrome, i.e., the set of risk factors (hyperlipidemia, hypertension, and diabetes) of atherosclerosis, can be due to mutation in LRP6, which encodes low-density lipoprotein receptor-related protein-6, a co-receptor in the Wnt pathway, thus impairing Wnt signaling [585].

Obstructive sleep hypopnea–apnea syndrome is characterized by repetitive partial or complete pharyngeal closure during sleep, due to transient collapse of the soft pharyngeal walls.[4] Pharyngeal obstruction leads to arterial oxygen desaturation, persistent diurnal sympathetic activity, which increases during sleep, hypertension, and cardiac tachyarrhythmias or bradyarrhythmias. Obstructive sleep hypopneas and apneas increase the occurrence probability of defects in organ perfusion (angina, myocardial infarction, and stroke). Sleep disordered breathing can elicit not only cardiac and vascular damages, but also resistance to therapies. Obesity, high blood pressure, and aging the are usual features of patients with sleep hypopneas–apneas, although they can occur in patients who are not obese. Obstructive sleep hypopneas and apneas can be associated with other risk factors of cardiovascular diseases, such as insulin resistance syndrome.

7.2 Genetic Background

Many studies are aimed at determining the genetic background of cardiovascular diseases. Increased arterial stiffness rises the cardiovascular morbidity because of an elevation of afterload and altered coronary perfusion. Thirty-five gene expressions[5] are correlated with the pulse wave velocity (PWV), chosen as an index of arterial stiffness in human aortic specimens [586]. Two distinct groups of genes, associated either with cell signaling or with the cytoskeletal–cell membrane–extracellular matrix[6] interactions are involved.

---

[4] The diagnosis of sleep hypopneas and apneas uses nocturnal polysomnography, which records heart rate, respiratory rhythm, electroencephalogram, eye motions, muscle activity, and oxygen saturation. Continuous positive airway pressure is the usual therapy.

[5] Strong correlations are found with the catalytic subunit of the myosin light chain phosphatase, Yotiao, an A-kinase anchoring protein, the regulatory subunit p85-α of the phosphoinositide-3-kinase, protein kinase-Cβ1, and synaptojanin-1.

[6] Integrins α_{2b}, α_6, β_3, and β_5 levels are different between stiff and distensible aortas. Certain proteoglycans, decorin, osteomodulin, aggrecan-1, and chondroitin sulfate proteoglycan-5 (neuroglycan-C), and related proteins, dermatopontin (a

Most common forms of hypertension begin by repeated mild or intermittent elevations in arterial pressure associated with vasoconstriction. As hypertension progresses with age, blood vessels remodel. Hypertension can occur without known cause. Essential hypertension is the most common cardiovascular disease and is a major risk factor. Manifold approaches are used to determine the genetic background of essential hypertension. Certain specific mutations indeed lead to hypertension (others to hypotension), via defective electrolyte transport in the nephron. Many genomic regions are implicated in the regulation of blood pressure. Clusters of blood pressure-related traits are mainly located on chromosomes-1 and -3 [587]. Blood pressure quantitative trait loci can be colocated with those for renin activity and sodium excretion. A cluster of metabolism-related traits, including indices of obesity, are also mapped to chromosome-1. Once QTL regions identified, causal genetic variants must be determined.

Family history is a risk factor for coronary artery disease. Susceptibility genes have been identified, such as Apo gene cluster, gene encoding for myocyte enhancer factor-2A (MEF2A), for 5-lipoxygenase activation protein (ALOX5AP), for leukotriene-A4 hydrolase,[7] for lymphotoxin-α(LTA) gene associated with tumor necrosis factor ligands [588]. Variants of the gene ALOX5AP[8] are risk factors of myocardial infarction, characterized by higher production in leukotriene B4 [589].[9] TNFSF4 genes are markers of the risk of myocardial infarction in humans [590]. The myocardium infarction risk rises in coffee consumers with allele CYP1A2-1F. Susceptibility to stroke has been mapped to chromosome-5q12, in particular to the gene encoding phosphodiesterase-4D [591]. Variants in VAMP8[10] and HNRPUL1[11] genes are associated with early-onset myocardial infarction [592].

Mutations in several Z-disc proteins of the cardiomyocyte sarcomere, which lead to disruption and dysfunction of the contractile apparatus, cause cardiomyopathies [593]. Mutations in SLC2A10 gene, which encodes glucose transporter GLUT10, associated with upregulation of the TGFβ pathway in the arterial wall, cause aortic aneurisms and arterial tortuosity [594].[12]

---

decorin–binding proteoglycan) have different amounts between stiff and distensible aortas.

[7] Leukotriene A4 hydrolase catalyzes the rate-limiting step in LTB4 synthesis.

[8] Gene ALOX5AP on chromosome 13q12-13 is also known as FLAP, coding for arachidonate 5-lipoxygenase (ALOx5) activating protein for synthesis of leukotriene LTA4.

[9] Leukotriene B4 is involved in the 5-lipoxygenase pathway. Leukotriene-B4, which is synthesized from LTA4, activates monocytes, which migrate across the arterial endothelium and differentiate into macrophages, becoming foam cells.

[10] VAMP8 gene is involved in platelet degranulation.

[11] HNRPUL1 gene encodes a ribonuclear protein.

[12] Arterial tortuosity syndrome is an autosomal recessive disorder characterized by tortuosity, elongation, stenosis, and aneurism in the major arteries resulting from disruption of medial elastic fibers in the arterial wall.

Gene mutations of the mitogen-activated protein kinase pathway cause cardio–facio–cutaneous syndrome with cardiac defects among other developmental abnormalities [595]. Atrial septal defect can be associated with a mutation of chromosome-14q12 [596]. The cardiac transcription factor TBX5 regulates expression of the myosin heavy chain MYH6. Mutant forms of TBX5 leads to congenital heart malformation.

Calcification of the aortic valve is one of the leading cause of heart disease in adults. Mutations in the sodium channel gene SCN5A cause cardiac arrhythmia. Mutations in the signaling and transcriptional regulator Notch1 (Part I; App. A.2) cause a spectrum of aortic valve anomalies, from early developmental defect in the aortic valve to later valve calcification [597]. Endothelial dysfunction is associated with hypoalphalipoproteinemia, a genetic disorder with low plasma HDL-C and ApoA1 levels and high coronary heart disease risk.

Heterozygous familial hypercholesterolaemia is due to monogenic disorder associated with one defective allele coding for the low-density lipoprotein receptor. The homozygous disease is defined by both defective alleles, resulting in non-functioning LDL receptors.

7.3 Oxidative stress

Oxidative stresses associated with excessive production of reactive oxygen and nitrogen species lead to cardiovascular diseases. Reactive oxygen and nitrogen species are implicated in hypertension, intimal hyperplasia, and atherosclerosis [598]. Excess superoxide or superoxide-derived substances directly or indirectly promote lipid peroxidation and low-density lipoprotein oxidation. Superoxide inactivates endothelial nitric oxide, hence favoring: (1) vasoconstriction and smooth muscle cell proliferation and migration, and (2) blood cell adhesion. These reactive species can act on inflammatory signaling. Superoxide is produced by NADPH oxidase in endothelial and smooth muscle cells. Vascular NADPH oxidases are regulated by various compounds, such as growth factors (platelet-derived growth factor) and cytokines (tumor necrosis factor-α[13] and interleukin-1), as well as thrombin and oxidized LDLs. Superoxide is metabolized to hydrogen peroxide. These reactive oxygen species serve as second messengers that activate multiple targets, including the epidermal growth factor receptor, cSrc, p38 mitogen-activated protein kinase, Ras, and protein kinase-B [599]. Vascular NADPH oxidase is stimulated by angiotensin-2 via AT1 receptors, in association with EGF receptor, PI3K and GTPase Rac [601]. Moreover, reactive species can activate transcription factor NFκB. Nitric oxide quickly reacts with superoxide to form peroxynitrite (ONOO⁻). Also, nitric oxide hampers NADPH oxidase activity.

---

[13] TNFα recruits subunits to plasmalemmal NADPH oxidase [600]. TNFα also activates eNOS in the plasmalemma nanodomain. This dual activation locally produces superoxide and nitric oxide.

7.4 Dysregulation in MicroRNA Activity

Chronic maladaptive hypertrophic growth of cardiomyocytes due to valvular dysfunction, hypertension, and myocardial infarction leads to heart failure. Such a pathological cardiac remodeling is characterized by an increase in cell size, protein synthesis, and sarcomere assembly, as well as reactivation of fetal genes. During pathological cardiac hypertrophy, several specific microRNAs, especially stress-inducible microRNAs, are dysregulated [602].[14] Overexpression of these microRNAs in cardiomyocytes causes cardiac hypertrophy in vitro.

7.5 Vessel Diseases

The frequency of atherosclerosis[15] is especially high.[16] Millions of patients have dietary and drug treatments. Atherosclerosis is the leading cause of death in developed countries (about half the yearly mortality), and soon in the developing world.[17]

Aneurisms are widenings of the lumen in any artery, most commonly in the aorta for fusiform types, and in the head for saccular types. The arterial fusiform aneurisms and intrinsic stenoses are possible complications of atheroma. Saccular aneurisms can occur as complications of arterial wall trauma or infection. Congenital aneurisms do not have well-defined causes. Intracranial aneurisms, rare in childhood and adolescence, are observed in 3%

---

[14] The expression of miR195, miR23a, miR27a, and miR24.2 is upregulated. miR21, which impedes apoptosis, is also overexpressed during cardiac hypertrophy. miR133, which represses serum response factor, is underexpressed during cardiac hypertrophy.

[15] Atherosclerosis is characterized by endothelial dysfunction, wall inflammation, and intimal plaque formation with deposition of lipids, calcium, and cellular debris. Atherosclerosis is an intimal wall inflammatory plaque in large and medium-sized arteries associated with wall remodeling and luminal obstruction, which reduces oxygen supply to perfused organs and causes blockades after plaque fissuration either by clot at stenosis site or, most often, emboli downstream from the stenosis.

[16] Atherosclerosis begins in childhood; risk factors play a major role (western life style). Due to disease long-term evolution, most cases of atherosclerosis become patent in patients aged 40 to 70 years. The higher prevalence of atherosclerosis in men is thought to be due to the protective effects of estrogens. After menopause, the incidence of coronary heart diseases among women becomes similar to that of men, but women exhibit an approximately 10-year delay in onset of clinical manifestations.

[17] The cost is assessed to be of on the order of hundred of billions of euros a year in Europe or in the United States. Millions myocardial infarctions occur annually in patients with coronary artery stenoses. Cerebrovascular strokes leads to several hundred thousand deaths per year.

to 5% of the population, with a gender ratio of 3 women for 2 men. Abdominal aortic aneurisms most often affect men between 40 and 70 years old (5–7% of people older than 60).

7.5.1 Aneurism

The aneurism ($\alpha\nu\epsilon\upsilon\rho\upsilon\sigma\mu\alpha$: arterial dilation) is the product of multifactorial processes, which lead to a gradual plastic dilation of an arterial segment over a period of years. Aneurism wall stretches and becomes thinner and weaker than normal artery walls. Consequently, an untreated aneurism can rupture. The plastic deformation of the arterial wall is associated with structural changes. Connective tissue repair is deficient.

Unruptured aneurisms may be discovered when they cause symptoms, which depend on the location and size of the aneurism. Aneurisms can also yield symptoms secondary to compression of adjoining structures. Aneurisms often burst. In the case of cerebral aneurisms, the hemorrhage within the skull, less important than for aneurism rupture in any other soft part of the body, is associated with a cerebral vasospasm,[18] which can lead to cerebral ischemia (insufficient blood supply to irrigated tissues), infarction (tissue destruction), and death. Sometimes aneurisms are found when angiography is performed. Within-wall stresses play an obvious role in aneurism rupture. They may also be involved in aneurism development. Gene expression and inflammation in the degenerating wall could be mediated by mechanotransduction. Both flow-induced stresses at the wall and within-wall stress distribution must then be investigated, in particular to estimate the rupture risk.

Two main kinds of aneurisms can occur: fusiform wall dilation and saccular bulging of the artery wall. *Fusiform aneurisms* are often complications of atheroma. They can be located in any artery, but mostly in the aorta. Abdominal aortic aneurisms (AAA) are enlargements of at least 1.5 times the abdominal aorta size. *Saccular aneurisms* are balloon-like defects that protrude from the artery wall. Certain infections of the blood and traumas of the arterial wall can cause saccular aneurism. Congenital aneurisms are located at

---

[18] When a cerebral aneurism ruptures, it bleeds either in the brain or, usually, into the subarachnoid space surrounding the brain. Subarachnoid hemorrhages can lead to vasospasms. Cerebral vasospasm after subarachnoid hemorrhage results from contraction of depolarized vascular smooth muscle cells. The membrane potential of vascular smooth muscle cells is modulated primarily by potassium fluxes, especially those determined by inwardly rectifying potassium channels in small resistance arteries as well as in large conductance arteries of the circle of Willis [603]. Hyperpolarization of the smooth muscle cell membrane caused by opening of K^+ channels close L-type voltage-gated calcium channels, leading to vasorelaxation. Conversely, closing of K^+ channels generate vasoconstriction. Cerebral vascular smooth muscle cells have four main classes of potassium channels: voltage-dependent, large conductance Ca^{++}-activated, adenosine triphosphate sensitive, and inward rectifier channels.

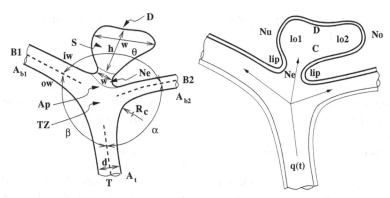

Figure 7.1. Schematic drawing with geometry definitions of a saccular aneurism at the apex Ap of a branching segment of an artery. S: aneurism sac (or cavity C) of width w and height h, Ne: aneurism neck of width w, D: aneurism dome, T: trunk of hydraulic diameter d and cross-sectional area A_t, B: branch of cross-sectional area A_b, TZ: transition zone, ow and iw: outer and inner bend, α, β, θ: branching angles, R_c: local curvature radius. The aneurism sac is here composed of two compartments or loculi lo1 and lo2. No: nose, Nu: nucha.

the branching sites of the brain arterial network, most often, at the branching sites of the arteries afferent to and efferent from the Willis circle. Most of the intracranial saccular aneurisms (ISA) are congenital. Aneurisms can develop from a weakening in the structural layer of an artery, especially with structural discontinuity at branch apex. A mechanically-induced degeneration of the wall internal elastic lamina has thus been proposed as the initiating cause of congenital aneurisms with genetic predisposition (vessel-wall structure deficiency in a frequent familial context), the rheology of the cerebral arteries being different from that of other arteries.

Saccular aneurisms are located either on the vessel edge (*side aneurism*), or at a branching region, and called *lateral* or *terminal* whether the vessel trunk gives birth to a lateral branch or divides into two main daughter vessels. The main region of a saccular aneurism is called the cavity (or pouch or sac, Fig. 7.1). Opposite to the neck is the dome (or fundus) of the aneurism. The cavity can be biloculated with a projecting end, which is called below the nose. The sac wall situated in front of the nose is called hereafter the nucha.

CT and MRI provide aneurism geometry. Saccular aneurisms are classified into three categories according to their largest width (small, large - $12 \leq w_a \leq 25$ - and giant). Other geometrical characteristics are given by the neck size, pouch shape, and angle between the greatest aneurismal length and the local main-flow direction. The geometrical features, especially sac and orifice sizes, are important short-term prognosis factors, because they affect treatment quality. The congenital saccular aneurism wall is fibrous. The media and elastic lamina stop abruptly at the aneurism neck, whereas pads of intimal thickening can be observed in the transition zone of the arterial branching [604].

7.5.1.1 Biochemical Framework

The extracellular matrix is involved in the integrity of the cardiovascular walls. In normal conditions, synthesis and degradation of elastin and collagen fibers balance each other.

Abdominal aortic aneurisms are stiffer than non-aneurismal aortas in the first stage because they occur on atheromatous walls. Secondarily, they become more compliant. Abdominal aortic aneurisms are characterized by chronic wall inflammation, depletion of medial smooth muscle cells and destructive remodeling of elastic fibers. The amount of elastin in aneurisms is significantly lower. In AAAs, the volume fraction of elastin and smooth muscle cells decreases, whereas the combined content of collagen and ground substance increases [605]. However, the ratio of collagen to ground substance between aortas with and without AAA is not significantly different. Moreover, the walls of human abdominal aortas and atherosclerotic aneurisms contain similar amounts of collagen, but the collagen properties are found to differ in AAAs from the normal wall. The total amount of glycosaminoglycans slightly decreases compared with that of normal tissue, but the ratio of particular compounds varies. The percentage of chondroitin sulfate increases and that of heparan sulfate decreases [606].

Imbalance between matrix metalloproteinases[19] and tissue inhibitors of metalloproteinases, are involved in the evolution of abdominal aortic aneurisms. Degradation of interstitial fibers, which is not balanced by synthesis modifies the wall rheology. The lysis of the fibrillar matrix (mainly collagen) is due to an increase in production and/or activation rate of degradation enzyme, like MMPs [607]. Inflammatory cell expression of MMP9 (gelatinase-B) plays a critical role in an experimental model of AAA [608]. The MMP9/TIMP1 ratio is inversely correlated to arterial compliance [609]. Besides, plasma MMP9 and TIMP1 do not differ significantly between AAA and aortoiliac occlusive disease. MMP12 (macrophage elastase) cooperates with MMP9, as mice lacking both proteinases have greater protection from aneurism genesis. MMP2, MMP9, and MMP12 are abundantly expressed in the aneurism wall [610]. The plasminogen system, including the tissue-type (tPA), urokinase-type plasminogen activator (uPA), as well as the plasminogen activator inhibitor PAI1 can be involved in aneurism formation. ProMMPs are indeed activated by plasmin produced from plasminogen. Mice with a deficiency of tPA or of uPA are protected against media destruction and aneurism formation [611]. Abdominal aortic aneurisms are characterized by chronic inflammation, destructive remodeling of the extracellular matrix, and increased

---

[19] Matrix metalloproteinases are secreted as zymogens (proMMPs), which must be matured by activators and are neutralized by inhibitors (Part I).

activity of MMPs [612]. Enhanced expression of MMP2 (gelatinase-A) is a marker of aortic aneurisms, reflecting abnormal flow-induced vessel remodeling. The pathological wall remodeling also includes chronic inflammation with cell recruitment (B and T cells, macrophages, etc.) and possible autoimmunity, triggered by autoantigens such as artery-specific antigenic protein (ASAP), particularly aneurism-associated antigenic protein (AAAP40).

Dipeptidyl peptidase-1 processes for activation granule-associated serine proteases (neutrophil elastase, cathepsin-G, and proteinase-3). These proteases are required for neutrophil recruitment, in particular during the development of elastase-induced experimental abdominal aortic aneurisms [613].

Angiotensin-2 can also be involved in aneurism formation associated with atherosclerosis. Its effect depends on the hyperlipidemic state. Angiotensin-2 promotes an inflammatory reaction in the vessel wall. It induces expression of VCAM1 [614] and monocyte chemoattractant protein-1 by the endothelial and smooth muscle cells. Deficiency in MCP1 receptor CCR2 reduces atherosclerosis in the ApoE$-/-$ mouse [615]. Angiotensin-2 reduces NO production by endothelial cells. Administration of angiotensin-converting enzyme inhibitors improves the endothelial function in patients with coronary artery disease [616]. Angiotensin-2 also stimulates PAI1, which promote thrombosis [617].

In abdominal aortic aneurisms, various stimuli activate c-Jun N-terminal kinase,[20] which stimulates activating protein-1 (AP1), increases the expression of matrix metalloproteinases, and decreases the expression of matrix synthesis enzymes, such as lysyl oxidase and prolyl 4-hydroxylase. Inhibition of c-Jun N-terminal kinase might thereby induce aneurism regression [618].

7.5.1.2 Biomechanical Testing

In clinical practice, the rupture risk must be estimated. Stress distribution in the aneurism is investigated to find regions subjected to high stresses and to plan the treatment accordingly. The geometry of the aneurism can be reconstructed from patient 3D images by manually-assisted automated operation. The surface discretization is then improved to get a suitable mesh for finite-element stress analysis. Although 3D reconstruction from medical imaging, mathematical modeling, and computational techniques have been improved, there is still a lack in suitable input data such as the actual in vivo rheological parameters of the artery wall, both at normal and pathological states. Experiments and numerical simulations have been mostly carried out in abdominal aortic aneurisms and intracranial saccular aneurisms. The abdominal aortic aneurism morphology affects wall deformation and stress distribution, and hence possible rupture. The rupture risk of lateral asymmetric abdominal aortic aneurisms is higher than for anterior-posterior asymmetric ones [619].

---

[20] In AAAs, activated JNK upregulates MMP9 in both vascular smooth muscle cells and macrophages.

During the aneurism development, the risk of rupture must be estimated for a vascular repair decision. Although any criterion to predict the rupture risk does not exist, the wall material is assumed to be subjected to maximum stresses that the wall can withstand. The stress–strain relationships obtained from 1D extension tests exhibit the rheological changes due to wall depletion in interstitial fibers [620]. The typical surgical procedure consists in stitching a vascular graft. However, since the surgery has significant mortality, the surgeon needs to balance rupture risk with surgery risk. Experiments have been carried out in rigid AAA models of various size, conveying a pulsatile flow to study the evolution of the hemodynamic forces at progressive stages of aneurism enlargement, using particle image velocimetry [621]. Even in large AAA models, the flow remains fully attached to the walls during systole, but detaches during diastole. Moreover, the formation of regions of flow stasis was observed even in small AAA models.

Bifurcating stent-graft implantation is used in endovascular repair. However, this mini-invasive procedure has several drawbacks, such as blood seepage into the cavity (endoleaks), stent-graft migration and failure, among others. Bifurcating stent-graft significantly reduces cavity pressure, mechanical stresses, wall motion, and maximal bore change in a computational model of abdominal aortic aneurism [622]. The drag force can exceed 5 N when the end sections are wide, the iliac angle large, and the iliac arteries narrow. The contact of self-expandable or balloon-expandable stent-grafts on the vessel wall hence cannot withstand the blood forces and then requires additional fixation to avoid stent-graft migration and wall repressurization with potential rupture.

7.5.1.3 Flow Simulations

Numerical simulations of flows in simplified geometries have been explored to provide basic knowledge of flow behavior in disease-mimicking vessels with the basic components of the vasculature, bends and branching sites. Moreover, such simulations allow one to easily test the effect of the therapeutics. Nowadays, flow simulations are performed in image-based geometries.

Simplified geometries

> *"C'est une erreur de considérer l'analogie comme une ressemblance imparfaite. La ressemblance n'a pas de valeur dans les sciences"* (Alain) [623]

> *"Sauver les phénomènes, c'est-à-dire formuler une théorie explicative du donné observable."* (Koyré) [624]

Pressure field have been explored in simple pedagogical models of saccular aneurisms located either in lateral or terminal embranchments of the arterial network. The numerical technique to solve classical fluid mechanics equations for incompressible fluids is the finite element method. It is based on a variational formulation of the Navier-Stokes equations on a domain Ω:

$$\mathcal{B}(\mathbf{v}, \mathbf{w}) + \mathcal{T}(\mathbf{v}; \mathbf{v}, \mathbf{w}) + \mathcal{B}'(\mathbf{w}, p) = \langle l, \mathbf{w} \rangle, \quad \forall \mathbf{w} \in V \subset H^1(\Omega)^3, \quad (7.1)$$

$$\mathcal{B}'(\mathbf{v}, q) = 0, \quad \forall q \in Q \subset L^2(\Omega) \quad (7.2)$$

where $\mathcal{B}$ et $\mathcal{B}'$ are bilinear forms, $\mathcal{T}$ a trilinear form, $\langle l, \mathbf{w} \rangle$ the dual product (the quantity l takes into account the non-homogeneous velocity and possible pressure boundary conditions as well as the possible forcing term $\mathbf{f}$), $H^1(\Omega)$ and $L^2(\Omega)$ the Sobolev space of order 1 defined on Ω for vector-valued functions and space of functions that are square integrable in the Lesbegue sense with respect to Ω, respectively. The computational method is suitable for unsteady flow, the time being used as an iterative parameter of the solution. The finite element type is P_1-P_1 bubble element for pressure and velocity approximation respectively [625, 626]. The convective term is approximated by the method of characteristics [627]. The solution is obtained via a generalized Uzawa preconditioned-conjugate gradient method [628]. The initial condition is given by a Stokes problem with the same boundary condition as the unsteady one (period of 1 s).

The neck swirl, the dome slow flow and the branch flow separations define the set of aneurism flow properties (Fig. 7.2). Two-dimensional simple models, characterized by a well-defined geometry, give results in agreement with medical observations, in particular high-pressure zones susceptible to be sources of future development of wall dilation and rupture. The rupture risk and failure of endovascular treatment are associated much more with aneurism geometry than local flow distribution. In every numerical test in the model of a terminal embranchment aneurism, the aneurismal dome, which faces the straight trunk, is subjected to very high pressure. Moreover, the pressure magnitude is much higher than the wall shear stress. When the aneurism is obliterated after coiling, the pressure reaches its highest values at the central part of the new wall, as it does with a normal bifurcation apex. In the model of a lateral embranchment aneurism, the pressure reaches its maximal values at the downstream aneurismal lip and at the outer stem wall, dowstream from the aneurismal neck. The pressure magnitude, here again much higher than the wall shear stress, is much lower than in the terminal aneurism. The downstream lip receives the fluid impact like any bifurcation apex. The highest pressure region in the sac is observed at the aneurismal edge close to the upstream lip. This region of the aneurism sac must be carefully filled during intravascular treatment. When the aneurism is completely blocked, the pressure is maximum at the downstream lip. In a partially filled cavity, the pressure, which remains the highest at the downstream lip, is greater in the fundus region close to the branch with a local high-pressure zone at the boundary between the upstream lip and the adjoining lateral wall. The shape of the post-therapeutic wall has no significant effects on the local flow field.

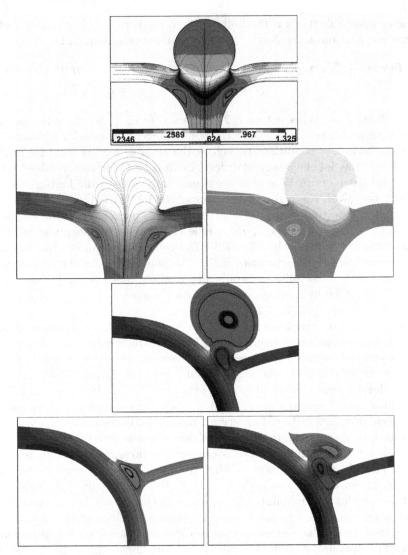

Figure 7.2. Streamlines and pressure field in 2D aneurisms. (**Top**) Terminal branching circular aneurism (aneurism radius to trunk radius ratio of 2, neck length to trunk diameter ratio of 1.8) at the apex of a symmetrical bifurcation of a straight vessel (area ratio of 0.72, curvature ratio at the bifurcation zone of 1/4) (**Top row**) Both symmetric geometry and flow. (**Bottom row, left**) Re = 1000, flow ratio between the left and the right branch $Q_{lb}/Q_{rb} = 0.97$; (**right**) Re = 500, $Q_{lb}/Q_{rb} = 1.5$). (**Bottom**) Lateral branching elliptical aneurism (aneurism ellipticity of 1.2, semi-major axis to trunk diameter ratio of 3.6, neck diameter to trunk diameter ratio of 1.15, neck height to trunk diameter ratio of 0.2) at the apex of an asymmetric branching of a curved trunk (90-degree bend, curvature ratio of 1/10). (**Top row**) untreated aneurism; (**bottom row**) almost completely (**left**) and partially (**right**) filled cavity (steady flow, Re = 500).

Image-Based Aneurism Flow

The fluid domain is composed, at least, of a trunk (inward flow), a transition zone of the branching region, the branches (outward flows), and the aneurism. The flow field usually exhibits persistent or transient vortices in the aneurism cavity, which can also be observed during possible cineangiography [629]. Branch aneurism cavities are commonly irrigated throughout the cardiac cycle, whereas the side-aneurism pouch, especially when the aneurism neck is narrow, can receive blood only during the peak flow period [120].

A terminal aneurism located at the apex of the bifurcation of a branch of the middle cerebral artery has been explored. Its cavity is biloculated with a very long loculus (the nose, on the left in Fig. 7.3). The aneurism cavity is facing the curved stem; the blood thus impacts the aneurism wall. The inlet section of the neck takes the place of both the bifurcation apex and the entrance short segment of the upper edges (with respect to the front view, depicted in Fig. 7.3) of both terminal branches. The neck is badly defined and may be assumed to be axially short and transversally rather broad.

The facetization obtained after extraction of the vessel contour, element reduction and smoothing, is transformed into a FEM-adapted geometrical model to: (1) ensure uniform pressure at the vessel entry and exits, (2) limit the disturbances induced by the boundary conditions (BC) on the flow, and (3) keep upstream and downstream effects of vessel geometry changes on the flow in vessel segments, which are afferent to and efferent from the computational domain. The local direction of the vessel axis is determined at each end of the vessels by a least-squares method from the normal of each selected facet, which belongs to the vessel wall and has an edge on the vessel end. Short straight pipes (two times the local hydraulic diameter only to avoid intersection of the branches, which is not too short when the Reynolds number

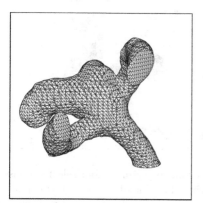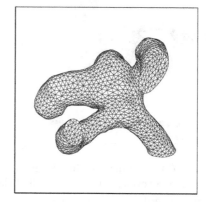

Figure 7.3. Facetization (**left**) and mesh (**right**) of the explored domain with a saccular terminal aneurism (front view)

is not too high and velocity boundary conditions are not applied) are added to the trunk and its bifurcation branches.

The stem peak Reynolds number based on the peak cross-sectional average velocity and trunk radius at the entrance cross-section is equal to about 160. The Stokes number (frequency parameter) based on the trunk radius at the entrance cross-section and the Strouhal number are equal to 1.6 and 0.02, respectively.

Two main regions of high pressure are observed in this aneurism example throughout the cardiac cycle (Fig. 7.4). The first region appears at the dome part facing the neck and the second one at the front face of the neck. The size of high pressure zones varies during the flow cycle. These two regions merge at peak flow and during the beginning of the decelerating phase. The

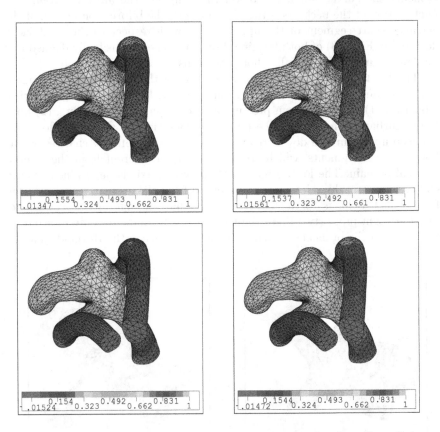

Figure 7.4. Pressure field in a terminal branching aneurism at four selected phases of the cardiac cycle: (1) at about 2/3 of the time duration of the accelerating phase; (2) slightly after peak flow; (3) at the beginning of the decelerating phase, when the instantaneous flow rate is nearly the same as at (1); and (4) at about 3/4 of the time duration of the decelerating phase. Front view.

pressure in these regions rises during the same cycle phases with a relative difference of about 6%. A third high-pressure region occurs at the nose end at peak flow and quickly disappears during the decelerating phase. The dome wall and front face are impacted by the blood flowing from the trunk core region throughout the cycle with more or less strong magnitude.

The high-pressure zone in the neck argues in favor of surgical clipping because the aneurism is superficial with easy surgical access. It is not possible, indeed, to protect efficiently the neck with coiling because coils in this location will always induce emboli. However, if the endovascular treatment is chosen with respect to heavily invasive surgery, frequent angiography control must be made because of the high risk of recanalization.

Stress fields in the wall of saccular and fusiform aneurism has also been computed to point out the site of possible rupture and plan treatment accordingly[21] [630]. However, the models are based on available but unreliable values of the rheological parameters.

7.5.2 Atheroma

Arteriosclerosis refers to a disease family with less-elastic and thicker artery walls. Among these diseases, atherosclerosis (αθηρωμα: fatty substance, σκληρο: hard) is defined by a deposition of fatty materials and then fibrous elements in the intima beneath the endothelium. The artery wall thickens, mainly toward the exterior at the beginning. The lesion then protudes into the lumen and occludes it. Atheroma may be scattered throughout large and medium thick-walled systemic arteries, especially in branching regions.

Atherosclerosis is an inflammatory process characterized by proliferation, migration from the media, and dedifferentiation of vascular smooth muscle cells. The expression of both mitogenic factor PDGF-BB and its receptor increases in smooth muscle cells. The inflammation is composed of four main stages characterized by: (1) *foam cells*,[22] (2) fatty streaks, (3) intermediate lesions, and (4) uncomplicated, then complicated, plaques (atheroma). The atheroma history can also be decomposed into three phases: initiation, progression, and complication.

In most cases, the initial phase of inflammation is silent with a long pre-clinical period. Atherosclerosis produces symptoms when the artery is severely narrowed or a sudden obstruction occurs, locally or downstream from the lesion site. At least 50% of stenotic lesions can lead to complications. The rupture of atheromatous plaques triggers thrombus formation and create embolisms. Emboli, rather than complete lumen obstruction, are responsible for infarction.[23]

---

[21] The surgical solution, AAA grafting, is risky, with significant mortality. The risk of rupture must then be balanced with surgical risk.

[22] Foam cells are characterized by lipid droplet saturation.

[23] The rate of proliferation of cardiomyocytes after acute injury is too low for adequate repair. In the absence of complete myocardial regeneration, fibrosis

The composition of atheroma plaques is more important than the degree of stenosis. Atheroma plaques with large lipid cores, thin fibrous caps and inflammatory cell infiltrates have high rupture and thrombosis risks. Unheralded plaque rupture can occur in small unstable lesions which do not significantly impinge on the arterial lumen and do not reduce the local blood flow rate. Hence, vulnerable atherosclerotic lesions can induce infarction.

Consequently, the characteristics of plaques must be evaluated to identify high-risk patients. Invasive (coronary angiography, intravascular ultrasound imaging, and palpography) and non-invasive (MSSCT) imaging techniques combined with measurements of concentrations of various biomarkers have been proposed to determine subclinical lesions [632].

Prevention of cardiovascular events directly depends on the achieved reduction of LDL concentration down to the level of 2 mmol/l. For each 1 mmol/l reduction in LDL cholesterol, there is a 12% reduction in mortality, 23% reduction in myocardial infarction risk, 24% reduction in need for coronary revascularization, and 17% reduction in stroke rate [633].

7.5.2.1 Atheroma imaging

Nowadays, imaging techniques for atherosclerosis are aimed at improving the diagnosis, enhancing disease characterization, and evaluating therapy effects. Several unconventional imaging techniques, optical coherence tomography, near infra-red spectroscopy, Raman spectroscopy, plaque thermography, and elastography have been recently proposed to study atherosclerotic lesions [634].

High-resolution (< 0.4 mm axial) high-frequency (8-MHz) B-mode ultrasonography with Doppler flow imaging is a useful technique for large- and medium-size peripheral and superficial arteries. Transoesophageal ultrasound imaging is used to investigate carotid and aortic wall thickness and analyze possible plaques. Hypoechoic heterogeneous plaques are associated with internal hemorrhage and lipid core, whereas hyperechoic homogeneous plaques are mostly fibrous. Ultrasonic contrast agents, such as acoustic liposomes with plaque-targeted antibodies, improve image resolution and specificity. Catheter-based intravascular ultrasound (diameter 0.89–1.0 mm) yields direct imaging of the arterial wall.

Magnetic resonance imaging with suitable operating mode, such as T1-, T2-weighted and proton-density-weighted images, and multicontrast techniques can determine plaque morphology and composition. Intravenous administration of paramagnetic gadolinium-containing contrast agents can highlight fibrous caps and lipid-rich areas. Platelet- or fibrin-ligands conjugated to paramagnetic nanoparticles allow thrombus visualization. Gadolinium-associated vitronectin antagonist has been proposed to assess angiogenesis

---

and scarring occur. FGF1 stimulation and p38 inhibition, which are involved in cardiomyocyte proliferation and angiogenesis, enhance cardiac regeneration [631].

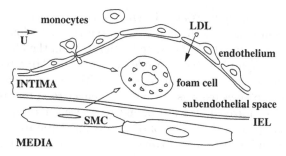

Figure 7.5. Atheroma plaque with migration of LDLs and cells from blood and media into the intima.

in atherosclerotic plaques. Specific antibodies conjugated to paramagnetic liposomes can be used to explore the endothelium expression of adhesion molecules.

Because macrophages are specifically associated with atherosclerotic plaques with high rupture potential, gadolinium-carrying micelles targeting macrophage scavenger receptor can be used to detect in vivo high macrophage content in atherosclerotic plaque using MRI [635]. Positron-emission tomography carried out using labeled derivative of diadenosine-tetraphosphate, an inhibitor of adenosine diphosphate-induced platelet aggregation that accumulates selectively in atherosclerotic lesions, allows one to image macrophage-rich atherosclerotic plaques [636].

7.5.2.2 Plaque Development

When the concentration[24] of circulating low-density lipoproteins is too high, LDLs can accumulate in the inner arterial wall. Endothelial cell injury is also a major stimulus for development of atherosclerosis. Disruptions of the internal elastic lamina are early features of intimal lesions in the apolipoprotein-E knockout mouse (a model of atherosclerosis) [637].

When atherosclerosis begins, low-density lipoproteins, which have crossed the endothelium, travel to the intimal connective layer. They can link to proteoglycans of the extracellular matrix. In the intima, they are modified by enzymes and oxygen radicals to form oxidized LDLs (oxLDL).[25] These modified LDLs induce an inflammatory response.

---

[24] LDL concentration must be less than $1\,g/l$. More than $1.3\,g/l$, especially if there are two or more risk factors for cardiovascular disease, requires treatment.

[25] The earliest detectable event in atherogenesis is the accumulation of oxidized lipoproteins in focal areas of the arterial subendothelium. Lipoprotein aggregation and proteoglycan binding to the extracellular matrix increase particle size and thus impede egress from the arterial wall. The lipoproteins retained in the extracellular matrix trigger inflammation.

Chemokines for monocytes and T lymphocytes are released by the activated endothelium. The endothelium also synthetizes adhesion molecules for these flowing cells to bind to.[26] These cells migrate from the bloodstream into the artery wall. T lymphocytes also release cytokines. Migrated monocytes multiply and differentiate into macrophages with plasmalemmal *scavenger receptors*.[27] Macrophages express not only scavenger receptors but also Toll-like receptors.[28] Oxidized LDLs are bound to scavenger receptors and then ingested by macrophages. Ingestion of oxLDL is unsaturatable. Macrophages then acquire a foamy appearance (foam cell formation). The foam cells secrete inflammatory cytokines and matrix metalloproteinases. The foam cells also release procoagulants into the surrounding environment. They are also involved in migration of smooth muscle cells and in the activity of T lymphocytes.

During the development of atherosclerosis, puringeric receptors P2Y2R are upregulated. Activated P2Y2Rs in endothelial cells induces expression of vascular cell adhesion molecule-1, and subsequent adherence of monocytes, as well as release of pro-inflammatory chemokines.

Blood lipids other than low-density lipoproteins[29] also contribute to atherosclerosis, such as remnant lipoproteins.[30] *Remnant lipoproteins* are cholesterol-enriched particles derived from the lipolysis of intestinal chylomicrons and hepatic very-low-density lipoproteins. Chylomicrons (< 0.95 g/ml),

---

[26] Endothelial cells express leukocyte adhesion molecules at the initial stage of atherosclerosis. Leukocytes bind to these adhesion molecules and migrate into the intima, stimulated by locally produced chemokines, such as CC-chemokine ligand-2 [638]. Cells in the lesion at its initial step also produce the T-cell attractant CCL5, the CXC-chemokine ligands CXCL10 and CXCL11, the mast-cell attractant CCL11, the Janus molecule CXCL16, and the CX3-chemokine ligand-1 (CCL5 is also designated as RANTES, CXCL10 as IP10, CXCL11 as ITAC, CCL11 as eotaxin, and CX3CL1 as fractalkine).

[27] Monocytes differentiate into macrophages in response to various stimuli, particularly macrophage colony-stimulating factor produced by endothelial cells and smooth muscle cells [638]. Scavenger receptors includes CD36, CD68, CXCL16, lectin-type oxidized low-density lipoprotein receptor-1 (LOX1), and scavenger receptors SR-A and SR-B1. These pattern-recognition receptors mediate internalization and lysosomal degradation of modified lipoproteins, lipopolysaccharides, and apoptotic bodies.

[28] Many Toll-like receptors are expressed mainly by macrophages and endothelial cells in atherosclerotic lesions. TLR2 and TLR4 are expressed by endothelial cells in the normal artery wall. Toll-like receptors bind lipopolysaccharides, heat-shock protein-60, oxLDLs and other ligands, which activate the production of pro-inflammatory molecules by macrophages. After ligand binding, TLRs activate nuclear factor-κB and mitogen-activated protein kinase-activator protein-1 signaling pathways.

[29] Most forms of oxidized LDL do not strongly transform cultured macrophages into foam cells.

[30] Remnant lipoproteins activate ACAT and modify cultured macrophages into foam cells.

synthesized by the small intestine, transport triglycerides and cholesterol from the intestinal epithelium. In the blood circulation, the chylomicron triglycerides are hydrolyzed by the lipoprotein lipase on the capillary endothelium, especially in adipose tissue, heart, and skeletal muscle. *Chylomicron remnants* are rapidly cleared from the plasma by the liver. Very-low-density lipoproteins (d < 1.006 g/ml) are synthesized in the liver. VLDL triglycerides are hydrolyzed by lipoprotein lipase in the plasma and liver, generating intermediate-density lipoproteins (1.006–1.019 g/ml) and low-density lipoproteins (1.019–1.063 g/ml). Smaller VLDL and IDL are atherogenic *VLDL remnants*. VLDL remnants are either cleared by the liver using ApoE or converted to ApoB-100-containing LDL, a major cholesterol-transporting lipoprotein in the plasma. LDL bound to the LDL receptor via ApoB-100 are degraded by the liver. Remnant lipoproteins also bind to LDL receptor via ApoE and LDL receptor-related protein (LRP) also via ApoE. Heparan sulfates, abundant in the space of Disse, where processing by hepatic lipases can occur, participate in the uptake of remnant lipoproteins by the liver [639].

Immune mechanisms are involved in the formation and activation of atherosclerosis. Atherosclerotic plaque growth is regulated by multiple inflammatory cellular (monocytes, lymphocytes, platelets, and precursor cells) and molecular (cytokines, chemokines, cell-adhesion molecules, proteases, and signaling pathways)[31] components, from fatty streaks to fibrosis. Atherosclerotic plaques contain inflammatory and immune cells, mainly macrophages,[32] CD3+ T lymphocytes,[33] and also in lesser amounts, mast cells, dendritic cells,

---

[31] Inflammatory cells are recruited at the endothelium by several kinds of inflammatory mediators, such as adhesion molecules and growth factors (especially IL8 and C-reactive protein). The endothelium production of these mediators is stimulated by oxidized phospholipids, such as oxidized (1)palmitoyl(2)arachidonyl-sn(3)glycerophosphorylcholine, which derive from lipoproteins trapped in the intima. However, the human endothelium response to oxidized phospholipids is characterized by genetic-based variability (individual susceptibility to atherosclerosis) [640]. Genes involved in the SREBP pathway or in the UPR pathway are strongly involved in some cases but not others.

[32] Monocytes strongly participate in atherogenesis. Monocytes accumulate continuously during atheroma formation [641]. The monocyte recruitment increases with lesion size and hypercholesterolemia. The expression of gap-junction connexin-37 disappears in the endothelium of advanced atherosclerotic plaques. Connexin-37 hemichannels in monocytes hinder excessive monocyte adhesion and recruitment [642]. ATP-dependent monocyte extravasation is reduced due to ATP release into the extracellular space across connexin-37 hemichannels.

[33] Most T lymphocytes are αβT cells, which coexist with a small proportion of γδT cells [638]. CD4+ T cells react to oxLDLs. T cells interact with antigen-presenting cells, such as macrophages or dendritic cells, to be activated. T-helper-1 differentiation occurs. TH1 cells produce inflammatory cytokines including interferon-γ and tumor-necrosis factor. TNF triggers inflammation via the NF-κB pathway, inducing the production of reactive oxygen and nitrogen species, proteolytic enzymes, and prothrombotic tissue factors by endothelial cells. They

and B lymphocytes. Imbalances between regulatory and pathogenic immunities lead to plaque inflammation and development.

Estrogens attenuate the extravasation of circulating monocytes across the endothelium. The increase of nitric oxide production[34] by endothelial cells induced by estrogens attenuate the cytokine-induced expression of vascular cell adhesion molecule-1 and monocyte chemoattractant protein-1. In contrast, estrogens protect the female brain from stroke by reducing post-ischemic iNOS expression in infiltrating inflammatory cells and cerebral blood vessels within the ischemic territory [645].[35]

Monocytes secrete growth factors for smooth muscle cells. Endothelial cells, macrophages, and intimal smooth muscle cells attracted by cytokine release deliver chemoattractants that stimulate the migration of smooth muscle cells from the media to the intima.

The expression of Cav1.2α1[36] in smooth muscle cells in atherosclerotic plaques is reduced with respect to SMCs in unaffected segments of the same artery. Splice variants of the α1 binding subunit of Cav1.2 Ca^{++} channel undergo molecular remodeling with subsequent alterations in electrical features during atherosclerosis [646]. β-subunits can be another target of Cav1.2 remodeling. Cav1.2 isoform variations could correspond to a transcriptional response to environmental changes associated with cytokine expression and other inflammatory factors in atherosclerosis.

Agglomeration of foam cells, T lymphocytes, and smooth muscle cells, after replication and extracellular matrix synthesis, as well as other intima

---

also secrete interleukins IL12, IL15, and IL18 (IL12 and IL18 are also produced by macrophages and smooth muscle cells in atherosclerotic plaques). TH2 cytokines inhibit TH1-cell responses. Few T helper-2 cells release IL5 and IL10, which have anti-atherogenic effects, and IL4, which promotes atherosclerosis. IL4 stimulates scavenger-receptor expression and MMP12. CD1-mediated NKT cell activation can occur in atherosclerotic plaques. Macrophages and T cells, like endothelial cells, smooth muscle cells and platelets, express CD40 and CD40L, which trigger inflammatory responses. At the opposite, IL10 and transforming growth factor-β, produced by platelets, macrophages, endothelial cells, smooth muscle cells and regulatory T cells, are anti-inflammatory cytokines; thus, they have atheroprotective effects. CD4+CD25+ natural regulatory T cells, which suppress both TH1 and TH2 immune responses against self or foreign antigens, hamper atherosclerosis in several mouse models [643].

[34] L-Arginine is metabolized into nitric oxide and citrulline. Citrulline is converted to L-arginine by arginosuccinate synthetase and arginosuccinate lyase. L-Arginine is also converted by arginase to ornithine, the single source of polyamines putrescine, spermidine, and spermine, which regulate the cell cycle. Arginase can then divert L-arginine metabolism away from nitric oxide synthesis and favor cell proliferation. Estrogens reduce the expression of arginase-2 [644].

[35] Toxicity is attributed to the large amounts of released nitric oxide, which induces oxidative stresses.

[36] Calcium influx, which triggers the contraction of smooth muscle cells, mainly occurs through voltage-gated Cav1.2 channels.

cells leads to *fatty streak* formation. As atherosclerosis progresses, the streaks gradually become larger. Mature plaques (atheromas) have a more complex structure than fatty streaks.

Intermediate lesions evolve into cholesterol-rich plaque characterized by a *fibrous cap* formed by the smooth muscle cell-derived scar tissue. The fibrous cap with smooth muscle cells and collagen fibers surrounds the core of the atherosclerotic plaque. The core contains foam cells, extracellular lipids, and debris from dead cells. Macrophages and T lymphocytes are abundant in the interface between the cap and the core of atherosclerotic plaques. Atheroma first induces an external patchy thickening.

Atheroma continues to grow, and the fibrous cap increases. Matrix metalloproteinases released by macrophages could break the fibrous cap leading to plaque instability. Moreover, smooth muscle cell and macrophage apoptosis can lead to a *necrotic core* with possible plaque disruption.[37] Macrophage apoptosis can be caused by accumulation of free cholesterol associated with: (1) dysfunctioning of proteins of the endoplasmic reticulum membrane, such as endoplasmic reticulum calcium ATPase [648]; and (2) activation of the unfolded protein response in the endoplasmic reticulum with expression of the cell death effector CHOP [649].

The atherosclerotic plaque narrows the artery lumen and collects *calcium deposits*. Episodic fibrous cap ruptures occurring during inflammation stages, stimulated by inflammatory substance release, initiate *thrombus* formation. Indeed, unstable plaques can cracks; the anticoagulant function of the interface disappears once wall structures underneath the endothelium are exposed. The thrombi are often lysed, but any residual thrombus may become a nidus for future coagulation.

In certain cases, the rupture of thin fibrous cap atheroma occurs in the center of the cap, and not at the plaque shoulders. Microcalcifications (bore $10\,\mu m$) in the thin fibrous cap of vulnerable atheromatous plaques are very rare with respect to the numerous calcified macrophages and smooth muscle cells in the necrotic core (calcification size with an order of magnitude of a millimeter, observed by imaging techniques). However, these minute material heterogeneities can affect the rheological behavior of the atheromatous plaques. Certain plaque ruptures could indeed be generated by local stress concentration in cap microcalcifications [650].

---

[37] The apoptosis of vascular smooth muscle cells favors inflammation and vessel wall remodeling. Whereas SMC apoptosis does not cause inflammation in normal arteries, it alone induces plaque vulnerability in atherosclerosis [647]. SMC apoptosis provokes the production of interleukins IL8 and IL19, as well as monocyte-chemoattractant protein-1, leading to monocyte infiltration, with subsequent release of IL6.

7.5.2.3 Mediators

Major risk factors for atherosclerosis, which include acute (infection, immune reaction) and permanent (dyslipemia, obesity, diabetes, hyperhomocysteinemia, hypertension and smoking[38]) factors, induce endothelial dysfunction, cell injury, and an inflammatory environment [651].[39] Artery wall inflammation constitutes a major pathophysiological factor in plaque development, instability, and disruption followed by local thrombosis. Mediators of immunity are involved at various stages of atherosclerosis.

Apolipoprotein-E carries cholesterol to receptors of hepatocyte plasmalemma and regulates cholesterol metabolism. ApoE is the primary ligand for two receptors, the LDL receptor found in different tissues and the ApoE-specific receptor of the liver. Apoε gene and ApoE polymorphism affects lipoprotein metabolism mediated via hepatic binding, uptake, and catabolism of lipid particles [652]. Defects in ApoE can lead to an increase in atherosclerosis risk.

Myeloperoxidase (Mpo) in atherosclerotic lesions generates inflammatory chlorinated reactants that contribute to endothelial dysfunction. The myeloperoxidase-hydrogen peroxide-chloride system of activated phagocytes, associated with the localized inflammatory response in the vessel wall, leads to atherogenic lipoproteins, which are internalized by endothelial cells. The $Mpo–H_2O_2$ complex produces 2-chlorohexadecanal, a chlorinated fatty aldehyde, from HDL-associated plasmalogen, which inhibits the endothelial nitric oxide synthase [653]. LDL oxidation is strongly involved in endothelial injury. OxLDL causes expression of endothelial–leukocyte adhesion molecules on endothelial plasmalemma for adhesion of monocytes and T lymphocytes.

Other *adhesion molecules* are expressed on the surface of endothelial cells, such as selectins, VCAM1, and ICAM1 [654]. Selectins recognize leukocyte cell surface glycoconjugates and substances that interact with leukocyte integrins [655]. Pro-Inflammatory cytokines like TNF and IL1 regulate expression of tissue factor in endothelial cells [656], of adhesion molecules, and induce cell chemotaxis. Leukocyte adhesion is regulated by cytokines such as IL1βand TNFα [657].

Platelets could participate in the initiation of the inflammatory response of the vessel wall [658]. Once adherent, the leucocytes enter the artery wall directed by *macrophage chemoattractant protein* (MCP), the level of which is increased during plaque formation. In endothelial cells, at least three independent pathways regulate MCP1 expression by nuclear factor-κB and activator

---

[38] Smoking affects endothelium functions on the vasomotor tone.

[39] Lifestyle changes thus are recommended for subjects of either sex who have atherosclerosis, especially coronary heart disease: smoking cessation, increase in physical activity, a heart-healthy diet, and weight reduction when necessary. The blood pressure is regularly controlled. Optimum lipid or lipoprotein levels include: LDL cholesterol lower than 100 mg/dl (< 2.59 mmol/l), HDL cholesterol higher than 50 mg/dl (> 1.3 mmol/l), triglycerides lower than 150 mg/dl (< 1.7 mmol/l), and non-HDL–cholesterol lower than 130 mg/dl (< 3.4 mmol/l).

protein-1 [659]. The level of total *tissue factor pathway inhibitor* (TFPI) is higher in hyperlipidemic patients than healthy subjects. TFPI2 has inhibitory activity toward activated factor XI, plasma kallikrein, plasmin, certain matrix metalloproteinases, and the tissue factor–activated factor VII complex. *Matrix metalloproteinases* secreted by macrophages are elevated in patients with unstable angina [660]. Combined determination of MMP9 and IL18 identifies patients at very high risk. In atherosclerotic tissues, TFPI2 expression has been assigned to macrophages, T lymphocytes, endothelial cells, and smooth muscle cells [661].

The blood flow activates NFκB, leading to the production of target adhesion molecules (E-selectin, ICAM1, and VCAM1). High plasma levels of VLDL also activate NFκB, which is involved in inflammation regulation in cultured human endothelial cells [662]. NFκB can then regulate monocyte recruitment. However, the stresses applied by the flowing blood target not only the wetted face of the endothelial cells, but also the abluminal face, via integrin binding with ECM proteins. In normal arteries, these proteins comprise mainly collagen-4 and laminin. In atherosclerosis-prone sites, fibronectin and fibrinogen are deposited. Flow-induced NFκB activation depends on the composition of the extracellular matrix [663]. The binding of integrin on collagen prevents NFκB activation via a p38-dependent pathway activated at adhesion sites, whereas culture supports rich in fibronectin or fibrinogen enhance the activation of NFκB.

The *renin–angiotensin system* can also contribute to atherosclerosis via inflammation [664]. Angiotensin-2 is a regulator of the transcription factor NFκB. It activates inflammatory pathways in monocytes. M-CSF could play a role in monocyte differentiation into foam cells.

Ceramide is implicated in the genesis of atherosclerosis. Ceramide is produced from sphingomyelin by sphingomyelinase and serine palmitoyl CoA. Excess ceramide is removed by ceramidase. Ceramide also exists in circulating LDL, and is upregulated during inflammation. Following uptake by the endothelial cells, a part of the LDL-derived ceramide is converted into sphingosine, whereas another part accumulates inside the cells, with an increased rate of apoptosis [665].

Thrombin activates platelets and several coagulation factors, as well as plaque smooth muscle growth. The release by smooth muscle cells stimulated by thrombin of PDGF-AA (mitogen for fibroblasts and smooth muscle cells) and FGF2 is increased [666].

Interleukins IL8 and IL6 produced by endothelial cells are also found in plaques. IL10 limits atheroma plaque formation in mice [667]. In IL10-deficient mice, microbial pathogens stimulate atheroma plaque formation (the size of the plaque is multiplied by a factor of 10). Moreover, the rupture risk rises. Inhibition of IL18 reduces lesion progression with a decrease of inflammatory cells [668].

Cytokines trigger *C-reactive protein* in the liver, which is involved in the pathogenesis of unstable angina as well as restenosis after intervention [669,

670].[40] Microbes may cause inflammatory plaque activation. Pathogens, such as chlamydia pneumoniae, can act via *Toll-like receptors* to induce smooth muscle cell proliferation and plaque activation [671]. TLR1, TLR2 and TLR4 are markedly increased in macrophages and endothelial cells of atherosclerotic lesions. The hyperplasia suppressor gene blocks device-induced SMC proliferation and restenosis in rat carotid arteries, inhibiting ERK pathway [672].

Oxidative stress,[41] a consequence of hypoxia, is implicated in atherosclerosis. Mitochondrial dysfunction in vascular smooth muscle cells can also trigger the disease. Mitochondrial respiration indeed produces reactive oxygen species (Part I). Oxidative stress can be characterized by increased ROS generation, decreased NO availability, and smooth muscle expression of uncoupling protein UCP1. Uncoupling proteins (inner mitochondrial membrane anion transporters) decrease ATP and ROS synthesis [673]. UCP1 in aortic smooth muscle cells causes hypertension and atherosclerosis without affecting cholesterol levels [674]. UCP1 increases superoxide production and decreases the availability of nitric oxide. Superoxide activates uncoupling proteins for a negative feedback.

During hypoxia, the transcription factor *hypoxia-inducible factor* HIF1 promotes cell survival. Drugs that prevent HIF1 ubiquitination by the von Hipple-Lindau ubiquitin ligase and subsequent hydroxylation could be useful [675].

Molecular chaperone *heat shock protein* Hsp47, expressed within the endoplasmic reticulum, which secretes collagen-1 or -3, is synthesized in atherosclerotic lesions [676]. Hsp47 can be involved in plaque stability. The fibrous cap surrounding the lipid-rich core of advanced atherosclerotic plaques is rich in collagen-1 and Hsp47. In cultured smooth muscle cells, Hsp47 is regulated by the transforming growth factor-β, fibroblast growth factor, and oxidized low-density lipoproteins. Increased Hsp70 levels can reduce infarction size,[42] as well as arrhythmias [677]. Hsp70 can also improve postischemic recovery. αB-crystallin and Hsp27 protect cardiomyocytes against ischemia [678].

Cultured human smooth muscle cells undergo phenotypic changes. They convert into mesenchymal cells able to express markers of smooth muscle cells, osteoblasts, chondrocytes and adipocytes. In atheroma and various other diseases, smooth muscle cells also dedifferentiate, losing their contractile properties. Dysfunctional smooth muscle cells in vivo contribute to lipid accumulation and *calcification* in atherosclerotic plaques. In the normal artery wall, smooth muscle cells express inhibitors of calcification, such as matrix Gla protein, osteonectin, osteoprotegerin, and aggrecan. In atherosclerotic calcification, this expression of inhibitors is reduced, whereas alkaline phosphatase,

---

[40] Restenosis after stenting is mainly due to accute SMC intimal invasion and proliferation. A proliferation peak occurs 5 to 7 days after stent implantation.

[41] Oxidative stress results from an increase in oxidant generation, a decrease in antioxidant protection, or a failure to repair oxidative damage. Cell damage is induced by reactive oxygen species.

[42] The amount of tissue damage after an infarction determines the morbidity.

bone sialoprotein, osteocalcin, and collagen-2 are expressed in the calcified artery wall [679]. Moreover, smooth muscle cells differentiate into chondrocyte-like cells, which surround the necrotic cores [680].

7.5.2.4 Treatment Targets

Several complementary strategies can be used to treat atheroma. Low-density lipoproteins (~ 70 % of total plasma content of cholesterol) are strongly correlated with coronary heart diseases, whereas high-density lipoproteins are inversely correlated with these pathologies. The rate-limiting enzyme in cholesterol synthesis is hydroxy-methylglutaryl-coenzyme A reductase.

Inhibitors of HMG-CoA reductase, the *statins* belong to hypolipidemic drugs used to lower LDL-cholesterol levels in patients at risk for cardiovascular disease because of hypercholesterolemia, but the inhibition of HMG-CoA reductase reduces the sterol pool, which in turn upregulates enzymes of cholesterol biosynthesis and LDL receptors. Statins increases LDL uptake via LDL receptors. These drugs also decrease the liver production of ApoB-containing lipoproteins. Statins also have antioxidant, anti-inflammatory, anticoagulant effects. They inhibit cell proliferation. They regulate NO synthesis by endothelial cells. Moreover, they prevent leukocyte rolling, adherence and transmigration and contribute to the repair capacity of endothelium [681]. However, statins have significant toxicity at high doses. In addition, statins moderately rise HDL concentrations.

High-density lipoproteins have a protective effect. Atherosclerosis treatment can then benefit from increased HDL levels. HDL and its apolipoproteins acts on the efflux of cholesterol from the foam cells, and thus on possible reversal of atherosclerosis. Moreover, inhibition of cholesterol ester transfer protein,[43] which acts on the transfer of cholesterol esters from HDL, increases HDL and lowers LDL. HDL can also act by other mechanisms. HDL hinders the production of cell-adhesion molecules by tumor necrosis factor-α in endothelial cells (Part I), reducing inflammation. Furthermore, HDL binding to scavenger receptor SR-BI activates endothelial nitric oxide synthase (NOS3), inducing vasorelaxation. Components of HDL mediate its anti-inflammatory, antioxidative, anticoagulant, antiaggregatory, and profibrinolytic properties.

Fibrates and *nicotinic acid* could have a greater effect on HDL levels than statins [682]. Fibrates activate the peroxisome proliferative activated receptor PPARα,[44] and then increase the synthesis of ApoA-I by the liver. Fibrates also downregulate hepatic SR-BI, decreasing HDL clearance. Nicotinic acid

---

[43] A significant part of HDL-associated cholesteryl esters is transferred to triglyceride-transporting lipoproteins (VLDL, chylomicrons) by CETP. The triglyceride-transporting lipoproteins can be transformed to be partially used to form LDL.

[44] Peroxisome proliferative activated receptors are nuclear transcription factors with three members: PPARα, PPARγ, and PPARδ (or PPARβ). Many PPAR-regulated genes are involved in fatty-acid metabolism. PPARα and PPARγ are

inhibits adipocyte lipolysis and hepatic synthesis of very-low-density lipoproteins. It acts via Gi-protein-coupled receptors. However, it induces secondary effects. Its efficiency in hampering the progression of atherosclerosis of diverse PPAR or LXR activators, CETP or lipase inhibitors, SR-BI modulators, and cholesterol acceptors (apolipoproteins and synthetic HDLs) is currently being studied.

Because *platelets* are implicated in thrombosis and in inflammation, antiplatelet therapies are proposed. Aspirin inactivates cyclooxygenase COx2 and thus prevents the synthesis by activated platelets of thromboxane-A2, which is involved in platelet aggregation. Thromboxane-A2 synthase inhibitors and thromboxane-A2 receptor antagonists are more specific drugs. Nitric oxide-releasing substances can also be used. Glycoprotein-IIb/IIIa inhibitors have been administered to patients with acute syndroms. In addition to their side and adverse effects, therapy efficiency depends on the administration (intravenous vs. oral) path. Adenosine-diphosphate receptor antagonists can yield an efficient antiplatelet therapy, especially after stenting. CD40 inhibitors might hinder smooth muscle cell mitogenesis and reduce restenosis.

Diets rich in omega-3 fatty acids, such as docosahexaenoic acid and eicosapentaenoic acid, attenuate atherosclerosis. Docosahexaenoate has anti-inflammatory effects, reducing cytokine-stimulated expression of leukocyte adhesion molecules by the vascular endothelium. It reduces IL1 activity and COx2 expression in endothelial cells of saphenous veins subjected to inflammatory stimuli [683]. Docosahexaenoate, after its incorporation into membrane lipids, impedes PKCε translocation to the plasmalemma, attenuates NFκB transcriptional activation of COx2, decreases IL1-induced activation of ERK1/2, and inhibits NADPH oxidase and ROS production,[45] without affecting COx1 expression.

7.5.2.5 Mechanical Background

Various investigations have been performed to discover which blood dynamic factors participate in atherogenesis. Strong correlation has been found between low WSS region and atherosclerotic plaque localization [684]. Intimal thickening occurs where the particule residence time in the near-wall region is long enough and adhesion to the endothelium is enhanced. Moreover, atherosclerotic-prone regions are covered by polygonal endothelial cells, with thinner glycocalix, whereas normal regions are lined with elongated endothelial cells indicative of relatively high wall shear stress.

The stresses applied by blood flow on the vessel wall are involved in pathogenesis as well as complications, such as damage to the fibrous cap and crack-

---

synthesized by macrophages, monocytes, smooth muscle cells, and vascular endothelial cells. PPARα and PPARγ can inhibit NFκB and impede secretion of pro-inflammatory cytokines. They can also hinder adhesion-molecule expression. PPARα and PPARγ- but not PPARδ- activators protect against atherosclerosis.

[45] NADPH oxidase is the main source of reactive oxygen species in the endothelium.

ing of the plaque. The wall shear stress regulate gene transcription, in particular for adhesion molecules and chemoattractants. Intimal thickening can be an adaptive response to imposed local time and space WSS variations. This remodeling, associated with disturbed transmural fluxes of cells and lipoproteins, can secondarily lead to a lesion. Atherosclerosis mainly occurs not only where the wall shear stress is low, but also where the wall shear stress strongly changes both in time and space.

Investigations have been performed mostly in idealized geometry, assuming rigid wall due to atheroma-induced hardening and passive homogeneous plaques. However, evolving plaques are heterogeneous; they are susceptible to break off (fissuring and ulceration), generationg clotting and subsequent blockage of downstream branches.

An ABAQUS-based finite element model of an atherosclerotic plaque has been carried out to find the impact of atherosclerotic plaque geometry on wall stress distribution using an Ogden strain energy function [685]. Three idealized shapes have been explored: (1) an elliptical lumen inside a circular vessel (asymmetrically diseased artery with an atheromatous edge and an opposite normal edge); (2) a circular lumen inside an elliptical vessel (outward expansion of atheroma at a vessel edge); and (3) a circular lumen inside a circular vessel (circumferential expansion of atheroma). Three tissue components have been considered: (1) the artery wall, (2) outer atheromatous plaque, and (3) crescent-shaped lipid core. However, rheological data are questionable.

Other studies were aimed at determining criteria for early diagnosis of plaques and stenosis. In particular, DUS was used to detect flow disturbances distal to a constriction and correlate the perturbation degree to the lesion development via signal processing. In any case, lipid accumulation, which characterizes atherogenesis, is not only secondary to increased influx with changes in endothelial permeability, but is also the consequence of decreased wall efflux. Indeed, the wall transport of compounds is a second pathophysiological factor, the main efflux of substances from the artery wall being provided by the microcirculation of outer half of the vessel wall.

7.5.2.6 Stenosis

Stenoses usually reveal atherosclerotic plaques when lumen narrowing is strong enough (Fig. 7.6). Four types of stenosis can be defined according to the constriction degree (Table 7.1). It induces ischemia and a major accident, tissue infarction (e.g., myocardium infarction and stroke).

The resistance to steady laminar flow through a stenosed artery is increased by surface irregularities at high Reynolds numbers only, and may be largely outweighed by the degree of stenosis asymmetry [686]. The pressure variations through the stenosis can be decomposed into three parts: (1) variations associated with the kinetic energy changes, (2) viscous head losses, and (3) Borda-Carnot term due to the flow separation.

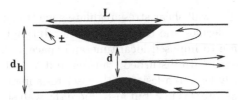

Figure 7.6. Schematic drawing of a stenosis. The usual configuration is asymmetric. The geometry is characterized by the stenosis length, throat (or neck) bore, and upstream and downstream slopes of the converging and diverging segments. According to the Bernoulli principle, the velocity increases at the stenosis throat and the pressure decreases with respect to the values in the upstream convergent. The pressure rises and the velocity decays in the downstream divergent. A strong lumen narrowing induces a jet with flow separation downstream and possibly upstream from it. The greater the divergent slope, the higher the probability of occurrence of an adverse pressure gradient.

The maximal wall shear stress can be estimated by an interactive boundary layer method in the vessel (radius R; conveying blood with cross-sectional average velocity U_q), from the stenosis geometry (convergent length L_c and neck width $\tilde{R}_n = R_n/R$) and an appropriate $\mathrm{Re} = U_q R^2/(L_c \nu)$ by the following formula: $\mathrm{MWSS} = 0.231/(1 - \tilde{R}_n)^{3.311}\mathrm{Re}^{1/2} + 0.718/(1 - \tilde{R}_n)^{2.982}$ [687].

7.5.2.7 Myocardial Infarction

Heart ischemia and reperfusion lead to important irreversible loss in cardiomyocytes by both apoptosis and necrosis (Part I). Bnip3, a mitochondrial member of BH3-only subset of proapoptotic Bcl2 family expressed in the adult myocardium, contributes to autophagy in myocardial ischemia–reperfusion [688]. Hypoxia combined with acidosis can activate Bnip3 in mitochondria. Heart fatty acid binding protein (hFABP), primarily expressed in the heart (but also at low concentrations in other tissues), is released by the myocardium about twenty minutes after the beginning of heart damage. Its detection in blood samples hence allows for rapid diagnosis.

Muscle-specific ring-finger proteins (MuRF) function as ubiquitin ligases in the sarcomere. MuRF3 processes its partners for suitable turnover, and

Table 7.1. Stenosis grades according to the degree of arterial lumen narrowing.

Reduction in vessel bore	Grade
$\leq 39\%$	Mild
40–59%	Moderate
60–79%	Severe
$\geq 80\%$	Critical

thereby contributes to the maintenance of ventricular integrity after myocardial infarction, avoiding cardiac rupture [689].

Additionally, brief pre-infarction episodes protect the heart (and others organs) from severe ischemia (*ischemic preconditioning*), with reduced infarct size, improved ventricular function, and decreased number of cardiac arrhythmias. Several compounds are involved in ischemic tolerance (Table 7.2).[46]

Table 7.2. Examples of mediators of ischemic tolerance, which enhance cell survival and prevent apoptosis (BCL: B-cell leukaemia/lymphoma; Source: [690]).

Cell targets	Cell survival	Cell apoptosis
Plasmalemmal receptors	TNFα, adenosine, Peroxisome proliferator, P2Y2	Death receptors, Nox, Fas ligand
Signaling effectors	Raf, MAPKs, PI3K/PKB, JAK2, STAT5, PKCε, Ceramide, eNOS, MLK3, Growth factors, Chaperones	Caspases, ASK1, GSK3β
Transcription factors	HIF, NFκB, CREB, Activator protein-1, JNK, SRF, MEF2	FOXO
Mitochondrial molecules	BCL2, BCL2-like proteins-1/2, Thioredoxin, Uncoupling protein-2	Cytochrome-C, ROS, BCL2-associated protein X, BCL-associated death promoter
Nuclear compounds	DNA repair enzymes, Survival genes, Inhibitor of apoptosis proteins	Tumor suppressor p53, Pro-apoptotic genes, Poly(ADP-ribose)polymerase

---

[46] The brain also resists to ischemic injury. The ischemic tolerance augments with a non-deleterious preconditioning, such as hyperthermia, prolonged hypoperfusion, oxidative stress, before the ischemic event. During such an adaptive reaction, prosurvival genes can be activated to encode proteins that enhance brain resistance to ischemia. Inflammation and cell apoptosis are reduced. Survival-promoting mechanisms involve mitogen-activated protein kinase pathways, nitric oxide, nuclear factor-κB, hypoxia-inducible factor, neurotrophins, etc. Trophic factors (TGFβ, brain-derived neurotrophic factor, and glial-cell-derived neurotrophic factor) and VEGF might also contribute to the glial and vascular tolerance to ischemia. The primary adaptive response corresponds to a state of low metabolism and perfusion, reduced ion fluxes for stabilization of membrane functions, prevention of anoxic depolarization, efficient clearance of extracellular glutamate, and important glycogen storage [690].

Phosphoinositide 3-kinases are activated either by receptor tyrosine kinases or by G-protein-coupled receptors. Phosphoinositide 3-kinases can have a cell survival role during tissue ischemia, possibly PI3Kα and PI3Kβ isoforms. However, PI3Kγ and PI3Kδ isoforms contribute to reperfusion damaging inflammation in response to mediators, such as vascular endothelial growth factor, platelet-activating factor, cytokines, eicosanoids, histamine, thrombin, and complement factors [691] (Fig. 7.7).

Cyclooxygenase COx2, which catalyzes the conversion of arachidonic acid into prostaglandin-H2, is implicated in ischemia. Prostaglandin-E2 EP1 receptors, effectors of COx2, impair Na^+-Ca^{++} exchanges and lead to neurotoxicity [692].

Hematopoietic precursors migrate into the border zone of the myocardial infarction and differentiate into cardiomyocytes and endothelial cells. Bone marrow-derived precursors give birth to smooth muscle cells, inflammatory cells, and mast cells into the injured heart. Fibrotic ischemia–reperfusion cardiomyopathy arises from daily, brief coronary occlusion. Fabrication and remodeling of connective tissue in the heart participate to cardiac repair. The production of monocyte chemoattractant protein-1 is increased. Bone marrow-derived fibroblast precursors and monocytes are then attracted [693]. Proliferative fibroblasts express collagen-1, α-smooth muscle actin, CD34, and CD45.

Postinfarct myocardial regeneration using stem cells still yields questions concerning the cell types that produce beneficial effect, patient selection, optimal administration, cell number to deliver, and mechanisms of action of transplanted cells (Part I).

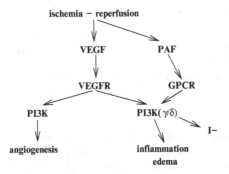

Figure 7.7. Myocardial ischemia–reperfusion injury. Two important mediators, vascular endothelial growth factor (VEGF) and platelet-activating factor (PAF), are produced by ischemic myocardium. Both act directly on endothelial cells to increase the vascular permeability. VEGF via different PI3K isoforms has a proangiogenic (positive) effect and proedema (negative) activity in myocardial infarction. PAF induces edema, promotes leukocyte adhesion to the hypoxic endothelium, activates neutrophils and platelets, and has a negative inotropic effect (I–) via PI3Kγ (Source: [691])

PKB overexpression by mesenchymal stem cells restore the cardiac function after myocardial injury. Secreted frizzled related protein-2 is a major stem cell paracrine factor that mediates myocardium repair after ischemic injury [694].[47]

7.5.2.8 Flow Experiments and Simulations

Flow perturbations induced by a severe lumen narrowing (cross-sectional area reduction > 50%) can be detected by USD techniques. Measurements of the centerline velocity downstream from a 50% stenosis exhibit three types of unsteady poststenotic flow disturbance [695]: (1) a coherent structure associated with the cycle start up, (2) shear layer oscillation, and (3) random structures. Studies have been performed on mild stenoses to find flow features useful for early diagnosis.

The systolic jet through the stenosis is associated with recirculation zones. Flow separations are produced, mainly during the diastole, downstream from the stenosis and sometimes upstream from it, according to the stenosis shape and values of the flow-governing parameters. Blood stagnation or, at least, low-speed transport can more easily trigger clotting on the more or less damaged endothelium of the constricted segment. The stenosis can then induce hypoperfusion of the irrigated tissues by two mechanisms: (1) increased lumen narrowing, and (2) distal flow blockage by emboli. Another mechanical consequence is caused by the jet generated by the stenosis, which can subject the more or less curved vessel wall to additional stresses.

Localized more or less abrupt vessel constrictions and enlargments create transient recirculation zones in pulsatile flows. Localization of the boundary layer separation and reattachment points can be determined in well-defined, usually symmetrical geometries. The stenosis shape of simple models is commonly define by an analytical function.[48] The photochromic tracer method (Sect. 3.4.1) has been used to visualize the velocity profiles in symmetrical (area reduction of 45, 65 and 75%) and asymmetrical (area reduction of 38%) stenoses (upstream and downstream slope angles of 30 and 45 degrees, straight wall of length 1.5 mm in the narrowest section). Flow separation regions are observed with variable positions of separation and reattachment points according to the stenosis degree for a given set of flow parameters,[49] with vortices in severe stenosis [192]. Vortices are generated in 50% stenosis a short

---

[47] Secreted frizzled related proteins compete with frizzled receptor for Wnt ligands by direct binding of Wnt, hence preventing activation of proapoptotoic Wnt signaling. Secreted frizzled related protein-2 increases beta-catenin level within the hypoxic cardiomyocyte, possibly activating antiapoptotic genes.

[48] Either $R(z)/R = 1 - h_s/2R(1 + \cos(\pi z/(L_s/2))))$ where h_s is the stenosis throat thickness, L_s the stenosis length [696, 697], or $R(z)/R = 1 - \exp\{-(2z/R)^2\}/2$ [698].

[49] The pulsatile flow is a nonzero-mean sinusoidal flow with $\overline{\mathrm{Re}} = 575$, $\mathrm{Re}_\sim = 360$ and $\mathrm{Sto} = 7.5$.

time after the peak flow, travel downstream, and decay as time proceeds [699]. The higher the Stokes number, the stronger the vortex. The experimental and numerical tests on simple symmetrical model of arterial stenoses show that the shorter the stenosis and the greater the lumen narrowing, the larger the separation region. The higher the Reynolds number, the longer the distance downstream from the stenosis over which the stenosis affects the flow [700]. This flow disturbance length downstream from the stenosis decreases when the Stokes number rises. Strong differences exist between pulsatile flows in 30% symmetrical and asymmetrical stenoses of the entry segment of the internal carotid artery, the computations being validated by flow visualizations [701].

7.5.3 Chronic Venous Insufficiency

Chronic venous insufficiency (CVI) ranges from varicose veins[50] to venous ulceration and deep vein thrombosis, which prevents venous return. Venous pathology, which can be deep, superficial, or mixed on the one hand and be segmental or impair the whole limb on the other hand, is caused by: (1) inoperative muscle pump; (2) venous obstruction, especially by thrombus formation favored by the Virchow triad (venous stasis, hypercoagulability, and endothelial trauma); or (3) damaged or congenitaly absent valves in the superficial and perforating veins. Valve failure in the perforating veins is responsible for a blood return from deep veins to superficial veins, causing local congestion and secondary failure of superficial vein valves, which lead to varicose veins. During pregnancy, varicose veins can develop due to a rise in progesterone. Long time duration in a standing position is a risk factor (inoperative muscle pump).

Wall thickening, dilatation, and tortuosity of varicose veins demonstrate venous wall remodeling. The collagen content is significantly increased in great saphenous varicose veins with respect to the normal great saphenous veins [702]. Conversely, the elastin content is significantly reduced. Elastase activity is found to be similar in normal and varicose veins. The same observations are made on venous segments of varicose veins without patent lesions. Connective tissue abnormalities occur before valvular insufficiency. In patients with varicose veins, the synthesis of collagen-1 and -3 is dysregulated in smooth muscle cells from varicose veins and similarly in dermal fibroblasts [703, 704]. Collagen-1 level increases with a concomitant reduction in collagen-3 concentration. The production of proMMP2 increases in both cell types.

Sonography analyzes normal and incompetent valve in deep and superficial veins. In particular, abnormalities of ostial valves can be identified, such as insertion defects, valve retraction, valve destruction, as well as the absence of contact between leaflet free edges, valve rigidity and abnormal valve movements, such as valve prolapse and paradoxal motions [705]. VCT (valve, cusp,

---

[50] Varicose veins are swollen and tortuous veins due to loss of normal venous valve function.

and tributary) classification has been proposed from endoscopic images of valvular lesions in the deep and superficial veins [706].

7.6 Heart Diseases

7.6.1 Cardiac Deficiency

Heart failure[51] can be caused by either heart (ischemia, valvular diseases, congenital defects, cardiomyopathies, myocarditis, chronic arrhythmias, etc.) or external factors (hypertension, pulmonary embolism, shunt, etc.). Neurohumoral responses (activation of sympathetic nerves and renin–angiotensin system, increased release of antidiuretic hormone and natriuretic peptides) are compensatory mechanisms, but they can aggravate heart failure by increasing afterload. The heart responds to impairment by progressive hypertrophy[52] to preserve stroke volume at the expense of ejection fraction. Angiotensin and aldosterone stimulate collagen formation and fibroblast proliferation in the heart wall.

Cardiac deficiency has a biochemical background. Mutations in lamins-A/C are associated with dilated cardiomyopathy. Nesprin-2 might be involved in *arrythmogenic right ventricular dystrophy*[53] (ARVD), both mapped to chromosome-14q23. Pregnancy is associated with cardiac hypertrophy, but peripartum and postpartum cardiomyopathies can occur late in pregnancy and a few months after delivery. They are associated with insufficient levels in cardiac manganese superoxide dismutase, hence with excessive levels of reactive oxygen species and cardiac cathepsin-D. Cathepsin-D cleaves prolactin into an anti-angiogenic and pro-apoptotic substance. Cardiomyocyte-specific deletion in signal transducer and activator of transcription STAT3 generates postpartum cardiomyopathy in mice [708].

Certain heart diseases can be linked to air pollution. Ozone and cholesterol interact to form atheronals. Breathing air containing elevated levels of fine particles and ozone can constrict arteries.

7.6.2 Conduction and Rhythm Disorders

Action potential transmission can be delayed at a point of the nodal tissue (sinoatrial, atrioventricular, right or left branch block, of either incomplete or complete type). Action potentials can arise from ectopic foci, in particular

---

[51] Heart fails to supply adequate blood flow to organs, cardiac dysfunction decreasing the cardiac output.

[52] cMyc is not expressed in the adult heart under normal physiological conditions. However, it can be quickly upregulated in the adult myocardium in response to stress, leading to CMC hypertrophy but not myocardium hyperplasia [707].

[53] ARVD is characterized by a substitution of the myocardium by an adipocyte-like tissue.

after ischemia, after toxic administration. The myocardium acquires a new rhythm imposed by the conduction tissue downstream from the pathological zone. Artificial pacemakers are implanted to treat severe block syndromes.

7.6.2.1 Arrhythmias

Heart rate disorders include both increase and decrease in beating frequency. *Tachycardia* (ταχοσ: quickness, καρδια: heart) results from various causes, in particular heart deficiency. *Bradycardia* (βραδοσ: slowness) corresponds to a cardiac frequency lower than 1 Hz. Several kinds of arrhythmias exist.

Heterogeneities in excitability and/or conduction velocities in the heart wall can initiate arrhythmias. Furthermore, increased dispersion of recovery and/or refractoriness (Part I) in nodal cells caused by tissue damage can produce a conduction block. A premature impulse from an area with a short refractory period can propagate and find a region with a longer refractory period, leading to unidirectional propagation.

Electrocardiographic QT-intervals measure cardiac repolarization. Extremely long or short QT-intervals[54] occur in various disorders (long/short-QT syndromes [LQTS/SQTS]). NOS1AP gene for neuronal nitric oxide synthase (NOS1) is associated with QT-interval variation [709].

Long-QT syndromes are caused by an abnormal cardiac excitability characterized by prolonged repolarization and arrhythmias. Eight LQT phenotypes have been identified associated with mutations of genes on chromosomes-3, -4, -7, -11, and -21. LQT1–LQT3 and LQT5–LQT6 loci encode for cardiac ion channels subunits (Table 7.3).

Alterations of the electric properties of cardiomyocytes in heart failure can favor the occurrence of atrial and ventricular arrhythmias by inducing early or delayed *afterdepolarizations*. Early afterdepolarizations can result from a secondary activation of the L-type Ca^{++} channels during the plateau of the action potential, as well as a decreased activity from rapid components of the delayed rectifier K^+ channels in damaged cardiomyocytes. Delayed afterdepolarizations, during phase 3 of the action potential, are caused by spontaneous Ca^{++} release from the sarcoplasmic reticulum. Spontaneous Ca^{++} release from the sarcoplasmic reticulum triggers a premature action potential in failing cardiomyocytes. This Ca^{++} release activates Na^+-Ca^{++} exchangers. The action potential in failing ventricular myocytes is longer than in normal ones. The increased duration of the action potential is induced by enhanced activity of Na^+–Ca^{++} exchangers, slowed decay in Ca^{++} transient fluxes, and reduced activity of inwardly rectifier K^+ channels and Na^+-K^+ pumps.

Re-Entry requires a unidirectional block (the action potential travels in the retrograde but not in the direct direction) within a conducting path,

---

[54] QT-interval measured on the electrocardiogram is the distance from the beginning of the Q-wave to the end of the T-wave. Owing to its sensitivity to the cardiac frequency, this parameter is normalized by the cardiac period (RR-interval).

usually due to partial depolarization, and suitable timing. The action potential traveling into the common distal path, finding the nodal tissue excitable, runs across the block area in the reverse direction. Re-entry can occur either between the atria and ventricles, involving an accessory conduction path, or locally, within a small region of the atrium or the ventricle. Re-entry has been studied by many investigators in 2D and 3D homogeneous, isotropic, excitable media since the 1980s [710–713]. An electromechanical model of a contracting excitable medium has been proposed recently, the non-linear excitation waves yielding contraction and the ensuing deformations exerting a feedback effect on the excitation properties of the medium[55] [714].

MRI tagging can evaluate pacing protocols and locate *ectopic sites*. The temporal evolution of 3D strain fields has been computed from MR tagging of canine hearts for three different pacing sites, the base of the left ventricle free wall, the right ventricular apex, and the right atrium [715]. Right atrium pacing shows rapid synchronous shortening. Left ventricle base pacing exhibits a slowly propagating wave front from the pacing site. RV apex pacing induces regional variations in propagation, with rapid septal activation followed by slower activation than usual. The propagation speed corresponds to the speed of electrochemical propagation in the myocardium.

Table 7.3. Ventricular action potential and ionic currents genes. Locus LQT4 encodes adaptor ankyrin-B, an anchor to the plasmalemma for Na^+, K^+-ATPase, and Na^+/Ca^{++} exchanger. KCNQ1 (LQT1) and KCNE1 (LQT5) genes encode for α (KvLQT1) and β (MinK) subunits of the slow delayed rectifier potassium channel (IKs current), respectively. KCNH2 (LQT2) and KCNE2 (LQT6) genes encode for α (HERG) and β (MiRP1) subunits of rapid delayed rectifier potassium channel (IKr current).

Phase	Ionic fluxes	Genes	LQT types
0	Rapid Na^+ influx (INa)	SCN5A	LQT3
1	Transient K^+ outflux (Ito)		
2	Na^+ influx		
	L-type Ca^{++} influx (ICa,L)	CACNA1C	LQT8
	Delayed rectifying K^+ effluxes (IKr)	KCNH2 (HERG)	LQT2
		KCNE2 (MiRP1)	LQT6
	Delayed rectifying K^+ effluxes (IKs)	KCNQ1 (KvLQT1)	LQT1
		KCNE1 (minK)	LQT5
	Cl^- currents		
3	Rapid K^+ current (IKr)	KCNH2, KCNE2	
	Slow K^+ current (IKs)	KCNQ1, KCNE1	
	Rectifying K^+ influx (IK1, Kir2.1)	KCNJ2	LQT7

---

[55] A three-variable FitzHugh-Nagumo-type excitation-tension finite difference model is coupled to the hyperelasticity finite element model.

During catheterization, using either multi-electrode basket (contact mapping system) or wire mesh, twinned with a second radio-frequency emitting catheter (non-contact mapping system), MR and mobile X-ray imaging can be merged into a common coordinate system using tracking and registration (XMR), to collect cardiac anatomy, motions and electrical activity [716]. In vivo observations can then be combined to numerical simulations of the heart pathological rhythms.

Atrial Arrhythmias

A sinus rhythm[56] with normal, evenly spaced complexes and a rate lower than 60/mn is called *sinus bradycardia*. A sinus rhythm with a rate higher than 100/mn is termed *sinus tachycardia*. Irregular sinus rhythms with the longest RR interval (exceeding the usual value by 160 ms) are called *sinus arrhythmias*.

The origin of atrial contractions can be located elsewhere in the atria than the sinusal node. If the ectopic pacemaker is located close to the atrioventricular node, the atrial depolarization occurs in the opposite direction with respect to the normal one, with inversion of P-wave polarity. Any ectopic pacemaker located in the junction between the atria and ventricles generates a *junctional rhythm* with slow cardiac frequency (40–55/mn, i.e., 0.66–0.92 Hz), normal QRS-complexes, and vanishing P-waves.

Paroxysmal atrial tachycardia, with a frequency of 160 to 220/mn (2.67–3.67 Hz), results from re-entrant activation. P-waves are immediately followed by the QRS-complexes. *Atrial flutter* occurs when the atria contract two or three times for each ventricular contraction. Atrial flutter is characterized by TP interval disappearance.

Atrial fibrillation[57] causes rapid (atrial cells fire at rates of 400–600 times per minute - 6.7–10 Hz) and irregular atrial activity. The atria quiver rather than truly contract. The ventricular frequency is also high (typically of about 150 pulses per minute - 2.5 Hz) and irregular with normal QRS-complexes. The ventricular rate is dictated by the interactions between the atrial rate and the filtering function of the atrioventricular node.

Atrial fibrillation can be caused by an ectopic focus[58], a single small re-entry circuit or multiple re-entries [717]. The refractory period depends on

---

[56] Activations originate at the sinoatrial node.

[57] Atrial fibrillations (AF) are classified into three main types: (1) paroxysmal, intermittent, usually self-terminating AF; (2) persistent AF; and (3) permanent AF. Atrial fibrillation is associated with cardiac and pulmonary diseases. Several cardiac disorders predispose to atrial fibrillations, such as coronary artery diseases, pericarditis, mitral valve diseases, congenital heart diseases, congestive heart failure, thyrotoxic heart disease, and hypertension. Idiopathic atrial fibrillation can be associated with mutations in the gene encoding connexin-40.

[58] Atrial fibrillation is mostly characterized by re-entrant loops in the atrial myocardium initiated by the rapid discharge of ectopic foci, commonly in sleeves of

the action potential duration. Smaller inward currents and higher outward currents reduce the refractory period and promote atrial fibrillation. An additional cause of atrial fibrillation might deal with altered expression of connexin and subsequent abnormal intercellular electrochemical communication. Increased levels of atrial extracellular signal-related kinase and angiotensin-converting enzyme are observed in patients with atrial fibrillation. The densities of angiotensin-2 receptors ATR1 and ATR2 decrease and increase, respectively.

Ventricular Arrhythmias

Premature ventricular contractions have either a supraventricular (QRS duration < 100 ms) or ventricular (QRS duration > 100 ms) origin. *Extrasystoles*, or extra heart beats, correspond to premature ventricular contractions. This type of arrhythmia can be benign, occurring without underlying disease. Nonetheless, they can be associated with abnormalities in electrolyte blood levels, ischemia, or consumption of certain susbstances (smoking, alcohol, caffeine, medications, etc.).

In ventricular arrhythmias, impulses originate from the conduction system of the ventricles. Ventricular activation thus does not originate from the atrioventricular node. QRS complexes are abnormal and last longer than 100 ms. *Idioventricular rhythm* is due to ventricular activation by a ventricular focus with a frequency below $40/\text{mn}$ (< 0.66 Hz). *Accelerated idioventricular rhythm* is defined by short bursts (< 20 s) of ventricular activity at high rates ($40–120/\text{mn}$, i.e., $0.66–2$ Hz). *Ventricular tachycardias* (rate > 2 Hz), with wide QRS complexes, are the consequence of a slower conduction leading to re-entries associated with ischemia and myocardial infarction. Focal transfer using adenoviral vectors of a gene encoding a dominant-negative version of the KCNH2 potassium channel (KCNH2-G628S) to the infarct scar border eliminates ventricular tachyarrhythmias [718].

Ventricular fibrillation, due to multiple re-entry loops, is the most dangerous arrhythmia. Ventricular fibrillation is characterized by irregular asynchronous undulations without QRS complexes. A hereditary factor is involved in the risk of primary ventricular fibrillation. Ventricular fibrillation has been thought to be maintained by unstable re-entry with activation wavelets of changing paths and conduction blocks associated with a heterogeneous dispersion of refractoriness. The multiple-wavelet hypothesis [719] predicts activation patterns different from the observations [720, 721]. High-frequency excitation sources can produce ventricular fibrillations. The *spiral wave* is a possible mechanism leading to a ventricular fibrillation [431]. The spiral wave can break, generating secondary wavelets (spiral breakup) [722, 723]. Rotors are possible organizing centers of fibrillation. During ventricular fibrillation,

---

atrial tissue located within the pulmonary veins. Increased activity of Na^+-Ca^{++} exchangers promotes afterdepolarization-related atrial ectopic firing.

persistent rotor activity has been observed in the anterior left ventricular wall of guinea pig hearts, the fastest activating region of the myocardium, with the shortest refractory period [724]. Action potential duration and activation rates differ between heart wall regions due to heterogeneous distribution of ion carriers, especially repolarizing K^+ channels (Ito, IKr, and IKs; Part I) in the normal guinea pig heart [724, 725].

Ventricular arrrhythmias can be due to channelopathies, resulting from impaired channel function, especially targeting voltage-gated channels. Mutations in human ether-a-go-go-related gene (HERG) associated with hERG potassium channels (Part I), responsible for the rapid delayed rectifier K^+ current (IKr), cause long-QT syndrome, which can lead to arrhythmias, such as *torsade de pointes* and ventricular fibrillation [726]. Mutations in KCNQ1, the α-subunit for the potassium channel that conducts the slow delayed rectifier K^+ current (IKs), or in KCNE1, the β-subunit, also lead to long-QT syndrome. Long-QT syndrome due to small, persistent inward Na^+ current during the plateau phase and membrane hyperexcitability, can be the consequence of mutations in the cardiac sodium channel gene SCN5A that cause slow or impair voltage-sensitive Na^+ channel inactivation. Mutations in KCNJ2, the gene of the inward rectifier K^+ channel, induce ventricular arrhythmias resulting from a delay in the final stage of ventricular repolarization. Mutations in gene CACNA1C for the cardiac L-type Ca^{++} channel prevent channel closure.

7.6.2.2 Conduction Alterations

When PR-interval is nearly constant but shorter than normal (< 120–$200\,\mathrm{ms}$), either the origin of the action potential is closer to the ventricles or the atrioventricular conduction uses a bypass, such as in Wolff-Parkinson-White syndrome.

First-degree atrioventricular block is signed by P-waves always preceding QRS complexes with a prolonged PR-interval (> 200 ms). In *second-degree atrioventricular block*, QRS complexes sometimes do not follow the P-wave.[59] Complete lack of synchronism between P-waves and QRS complexes depicts *third-degree atrioventricular block* (absence of atrioventricular conduction). First-degree and type I second-degree atrioventricular blocks are usually caused by delayed conduction in the atrioventricular node. Type II second-degree atrioventricular block usually is infranodal, especially when QRS complexes are wide. Third-degree atrioventricular block corresponds to a blockage of the two branches of the His bundle, which prevents the propagation front from the atria to reach the ventricles. Third-degree atrioventricular blocks can occur at several anatomical levels (upstream the bundle of His, in the bundle

---

[59] Mobitz type I atrioventricular block is manifested by progressive prolongation of the PR-interval with dropped QRS complex. Usually QRS complexes are narrow. Mobitz type II atrioventricular block is characterized by a constant PR-interval before a dropped QRS. Usually QRS complexes are wide.

of His, or in the upper part of both bundle branches, or even in the right bundle branch and the two fascicles of the left bundle branch).

Bundle-branch blocks denote conduction defects in one of the bundle branches or any downstream fascicle of the left bundle branch. Left or right bundle-branch blocks mean that ventricle activation is initiated by the opposite ventricle. Resulting QRS complexes have abnormal shapes and long duration (> 120 ms). *Right bundle-branch blocks* are characterized by a broad S-wave in leads I and V6, a double R-wave (RSR'-complex) in lead V1, and inversed T-wave in V1 and V2. ECG manifestations of *left bundle-branch blocks* are broad and tall R-wave, with M-shape in leads I, aVL, V5, or V6, inversed T-wave in left precordial leads and in leads I or aVL, deep and large S-wave in right precordial leads.

Wolff-Parkinson-White syndrome is due to action potential passage directly from the atrium to the ventricle via the bundle of Kent, hence bypassing the atrioventricular junctions. QRS complex initially exhibits an early upstroke (δ-wave), which shortens the PQ-interval. Mutations affecting AMP-activated protein kinase lead to Wolff-Parkinson-White syndrome [727]. Mutations affecting MAP3K7 also cause cardiac conduction alterations [728].[60]

7.6.3 Valve Failure

Valvular diseases include two main pathologies: (1) stenosis, when narrowed valves do not open sufficiently, and (2) valvular regurgitation, when leaky valves do not close completely. Valvular diseases are mostly due to either congenital abnormalities or inflammation. Primary valve failure occurs acutely due to leaflet perforation or gradually from leaflet stiffening associated with calcifications and/or thrombus formation.

The mitral cusps are dynamically related to the left ventricular wall via the mitral annulus, chordae tendineae, and papillary muscles. Mitral regurgitation is a complication of myocardium ischemia, which is detected by echocardiography. Defective systolic contraction of the papillary muscles due to acute ischemia and regional dysfunction of the wall of the left ventricle leading to abnormal displacements of the papillary muscles and incomplete leaflet coaptation causes mitral regurgitation.

When the mitral valve fails, LA partial emptying induces increased left p_A and reduced cardiac output. Pulmonary venous congestion and, ultimately, pulmonary edema occur. Valvular regurgitation also leads to a ventricular overload and can induce left ventricle hypertrophy. The compensatory mechanism of increased sympathetic tone rises both heart rate and systemic vascular

---

[60] MAP3K7 (or TGFβ-activated kinase-1), associated with JNK, p38, and IκB kinaseβ, is involved in cytokine signaling and innate immunity. MAP3K7 itself or an associated protein targets kinase LKB1, which phosphorylates (activates) AMPK. MAP4K4 and MAP3K7 triggers AMPK activity; conversely, MAP3K7 is activated by stimulators of AMPK activity.

resistance.[61] Consequently, diastolic filling time decreases and left ventricular flow is impeded. Acute failure of an aortic valve causes a rapidly progressive left ventricular volume overload. Increased left ventricular diastolic pressure engenders pulmonary congestion and edema. Cardiac output is substantially reduced. Increased systolic wall tension causes a rise in myocardial oxygen consumption.

7.6.4 Congenital Heart Defects

Congenital heart defects range from simple to severe malformations, such as complete absence of one or more chambers or valves (7–8 out of 1000 births). Reconstructive procedures for congenital heart defects can create new vessel circuits and connections. For instance, the Fontan operation is performed to treat complex congenital cardiac defects, the serial systemic and pulmonary circulations being driven by the single available (anatomical or functional) ventricle. The surgically created circuit must avoid areas of high stress generated by the fluid, which can damage the wall wetted surface, and provide adequate blood supply. Blood flow models are thus very useful to optimize the treatment [729, 730]. Adequate blood flow distribution into the lungs can be provided by fitting hydraulic design of the cavopulmonary connection, as demonstrated by a 3D finite element model coupled to a lumped-parameter model of the pulmonary circulation [731].

---

[61] Increased sympathetic tone with an increased heart rate and positive inotropic state is aimed at partly maintaining cardiac output. This effect is hampered by an increase in systemic vascular resistance.

8

Treatments of Cardiovascular Diseases

Modeling and simulation of blood flows through reconstructed normal or diseased vascular segments, arterial grafts, and vessel segments with implanted medical devices have been carried out by many biomechanics teams to provide knowledge of flow behavior and applied stress fields. Such investigations are also aimed at optimizing surgical procedures or medical devices. Most of the studies have been carried out in idealized geometries. However, since about 10 years ago, computational domains have been determined by three-dimensional reconstruction of vessels and organs, the anatomy being assessed from medical imaging.[1]

Surgical procedures become less invasive. Furthermore, interventional medicine provides an alternative to surgery. Both techniques require unusual gestures. Applied mathematics and computer sciences are involved in the development of medicosurgical simulators. Image processing coupled to virtual reality on the one hand, and numerical simulations of organ strains and stresses and blood flows on the other hand lead to computer-aided medical diagnosis and treatment planning, as well as improved navigation during image-guided therapies. Finally, telemedicine and telesurgery require robotics and suitable communication technologies.

Two main therapies may be applied to cardiovascular diseases. The first treatment consists of surgical reconstruction to maintain blood flow to the irrigated tissues, which todays attemps to be minimally invasive. The alternative to surgical operation is the image-guided endovascular procedure. This treatment reduces the surgical risks, is less expensive, and can have the same

---

[1] Medical image acquisition provides a model of the anatomy of the explored component of the cardiovascular system. Image acquisition hence is the first modeling stage of a multimodeling investigation leading to numerical simulations (final modeling stage) of the blood circulation. Moreover, the resulting reconstruction of the vascular domain by image processing (second modeling stage) correspond to a frozen state of a physiological system, which is characterized by a strongly varying configuration for adaptation to environmental conditions.

efficiency as surgery. Clinicians, either using catheter-based procedures for implantation of medical devices or mini-invasive surgery, thus require: (1) appropriate visualization of the diseased region and its surroundings; (2) determination of local flow indices and flow disturbances; (3) planning tools for the treatment strategy to minimize the complication risks; (4) a check-up to investigate the effect of the therapy on blood flow and help in prognosis; and (5) simulators to train for these new techniques. Any treatment procedure must limit flow disturbances such as flow stagnation regions, local increased resistances to flow, and strong flow impacts on vascular walls with the risk of wall damage.

8.1 Medical and Surgical Robotics

Interventional computerized tools can be coupled to precise robots. *Robot-supported medical and surgical systems* (RSMCS) are aimed at placing therapeutic devices, possibly from remote locations [732, 733]. To execute the gesture with a robot, the system must accurately determine the position and orientation of the target. Besides, the robot can use surgical manipulators with many degrees of freedom to manipulate target organs on a smaller scale than the current one. Small and tiny robots are aimed: (1) at exploring remote location using vessels and hollow organs to target specific areas, and (2) at controling pre- and post-operation therapeutic effects. Planification of robotic surgery intervention is based on a patient-specific model of the rigid and deformable organs involved in the surgical act, which is fused to the mechanical model of the robotic system. Augmented reality allows the real-time overlay of preoperative data with intraoperative situations, after calibration and correspondence determination [734]. Moreover, the surgeon's motions can be reduced and smoothed to increase precision and avoid hand tremor. Computerized analysis of heart wall motion is necessary during beating heart surgery [735], which is a less invasive method used for grafting in particular. The description of the motions of the heart and coronary arteries provides feeding data for robotic systems [736]. The fast movement of selected points of the left anterior descending coronary artery can range from 0.22 to 0.81 mm.

8.2 Surgery

The cardiovascular surgery: (1) at the heart level, deals with congenital heart defect repairs,[2] valve replacements, heart transplantation, pacemaker place-

---

[2] The ductus arteriosus is a vessel that connects the pulmonary artery and aorta in the fetus to shunt immature lungs, gas exchanges occurring in the placenta. Shortly after birth, when the neonatal lungs carry out blood oxygenation, the

ment,[3] normal rhythm and/or conduction restoration (arrhythmia surgery, i.e., Maze procedure for atrial fibrillation), coronary revascularization; and (2) at the vessel level, with peripheral artery bypass grafting, endarterectomy,[4] with saccular aneurism clipping, fusiform aneurism repair grafting, and mini-invasive stent grafting, etc.

Surgery now uses robotic and imaging systems [738]. Virtual reality-assisted surgery planing (VRASP) has been developed to provide flexible computational support pre- [739] and intraoperatively, which can control large patient-specific datasets in real time.

8.2.1 Grafting

Modeling and simulations have been aimed at understanding possible postoperative complications like *intimal hyperplasia* (IH) and at optimizing anatomical reconstruction. Two surgical procedures can be applied to stenosed arteries: (1) endarterectomy, which removes the plaque once the artery is opened, and (2) grafting. There are two grafting types: (1) vessel replacements, and (2) bypasses (bypass grafting). Replacement or end-to-end anastomosis[5] deals with sutures between ends of both host and graft vessels. Surgical procedures can also be based on end-to-side anastomosis or side-to-end anastomosis. Bypass or side-to-side anastomosis provides an alternate route for the blood to irrigate tissues by bypassing the narrowed or blocked arterial segment (e.g.,

---

ductus arteriosus closes. However, in about 1 in 2000 non-premature infants, the ductus arteriosus remains open (patent ductus arteriosus), causing vascular diseases. There are many other types of heart malformations or congenital heart diseases (ventricular and atrial septal defects, atrioventricular canal, arterial stenosis, coarctation of the aorta, vessel transposition, heart valve atresia, hypoplastic left heart syndrome, tetralogy of Fallot, etc.). Most defects are effectively treated by surgical and heart catheterization procedures.

[3] Cardiac pacemakers are implanted into damaged hearts to correct defects in conduction of the action potential. Sleep apnea is current in adults with pacemakers. Nearly two-thirds of patients with implanted pacemakers have obstructive sleep hypopneas and apneas. Patients with pacemakers thus should be routinely checked, because untreated sleep apnea contributes to cardiovascular deterioration [737].

[4] Endarterectomy used to treat carotid stenosis by removing developed atherosclerotic plaque has several complications, especially peroperative acute cerebral ischemia due to clamping of the internal carotid artery and postoperative hyperperfusion owing to impaired regulatory capacity secondary to chronic ischemia and arterial clamping. Hyperperfusion is defined by an increase in ipsilateral cerebral blood flow above the necessary supply. It can induce cerebral edema and hemorrhage. The risk factors include hypertension, severe stenosis, poor collateral blood flow, and strong contralateral carotid stenosis.

[5] An anastomosis is a natural connection between two vessels. It can refer to a surgical connection.

coronary artery bypass grafting [CABG]).[6] Traditional CABG involves the use of a heart–lung machine and may be required for multivessel grafting. Beating-heart bypass uses a device that stabilizes the operating heart region. Minimally invasive bypass is performed via a set of small incisions, using specially designed instruments.

Vascular grafts can be either: (1) a synthetic tube, made from polyethyleneterphtalate (PET, dacron) with woven or knitted textile structure, or expanded polytetrafluoroethylene (ePTFE);[7] (2) homografts (autografts) using homologous (autologous) veins or arteries (e.g., saphenous vein for femoral artery or internal mammary artery for coronary artery); or (3) heterologous vessel (xenografts, e.g., bovine carotid).[8] Synthetic grafts are porous rugous materials suitable for flowing particule adhesion and transfer as well as pseudointima creation. They have a small (< 6 mm), an intermediate, or a large bore (≥ 10 mm). They are straight, branched, or bifurcated. Taylor patch has been proposed: a vein patch is interposed between the synthetic graft and the artery. Bioresorbable vascular grafts are based on tissue-engineering techniques. Replacement vessels are made from collagen matrices of tubular shapes that have been seeded with autologous smooth muscle cells and endothelial cells. Grafts seeded with bone marrow components, endothelial progenitor cells, or genetically-engineered cells (pseudoendothelial cells) have also been studied recently. Such artificial matrices, biodegradable or not, must: (1) facilitate endothelialization and wall tissue growth, (2) provide mechanosensitive tissue, and (3) have suitable rheological behavior. In particular, their compliance must match vessel distensibility. Moreover, these scaffolding tubes must withstand physiological blood pressures during the initial implantation period.

Complications may arise (dilation, thrombus occlusion, intimal hyperplasia at the artery–graft junction, pseudoaneurism formation at anastomosis, deterioration, and kinking). Interface problems occur when endothelialization is partial along graft lumen. Vascular grafts have specifications to minimize grafting

---

[6] The left internal mammary artery is usually chosen for coronary artery bypass due to better tolerance than the saphenous vein. Saphenous vein grafts develop at a much greater extent intimal hyperplasia and accelerated atherosclerosis. P2X4 receptor expression is elevated in neointimal proliferation, less in arterial grafts than in veinous grafts.

[7] The artificial graft must be made of biocompatible polymers with a coating that provides a smooth wetted surface. Synthetic grafts have been used more or less successfully for large vessels (bore > 6 mm). Problems are encountered for smaller vessels and autogenous saphenous veins are commonly used.

[8] Antigen-presenting plasmacytoid dendritic cells acquire alloantigens in vascularized allografts, then home to peripheral lymph nodes, where they stimulate CCR4+CD4+CD25+Foxp3+ regulatory T cells for tolerance induction [740].

complications.[9] Standard in vitro measurements provide water permeability,[10] usable length,[11] relaxed and pressurized internal diameters, suture retention length and other rheological data[12]. Longitudinal and transverse impedances are measured to fit the lumped parameter models of the regional vasculature.[13] Flow disturbance intensity is computed from ensemble-average data with triggered acquisitions.

Several experiments have investigated flow behavior in bypass grafting. Flow patterns depend strongly on anastomosis configurations (parallel side-to-side, diamond side-to-side, and end-to-side) and graft/host diameter ratio. Optimization of surgical reconstruction of arterial stenoses (endarterectomy or bypass) is based on numerical modeling of the blood flow to minimize flow disturbances and treatment failures (thrombosis or restenosis) [741]. Different geometries of a fluid domain focused on the graft-vessel anastomotic junction have been tested,[14] without dealing with any shape optimization model. Compliance mismatch can lead to increased intramural stresses.

MRI-based model of distal anastomosis after venous bypass surgery have been used to classify bypass geometries [742]. Branching angles between centerlines of each vessel are measured. The spectrum of anastomoses can be reduced to a small subset of cases characterized by two angles: the angle

---

[9] Graft specifications are the following: (1) no physical modification by exposure to biofluids; (2) no foreign body effects (chemical inertness, non-allergenicity, non-carcinogenicity, very low tissue reactivity, non-thrombogenicity, low infectability, biocompatibility, minimal implantation porosity, but maximal tissue permeability); (3) ease of mechanical handling with sutures; (4) good fit at anastomoses, matching quality between graft and host vessel; (5) flexibility without collapsibility and without kinking; (6) stability under continuous mechanical stress and durability; and (7) ease of adequate sterilization and of fabrication with low cost.

[10] Water permeability is associated with an index defined by water flow rate through a given area of sample graft under a given pressure.

[11] The usable length is the length under a prescribed load.

[12] Graft rheological properties are determined from longitudinal and circumferential stress–strain relationships and from pressure-induced dilation. Among rheological quantities, the crip elongation, tensile, circumferential and burst strengths, wall thickness under loading, and compliance are determined.

[13] The impedance $Z = (\rho/A)\,\text{PWV}$ must match the biological values. The pulse wave velocity (PWV) is calculated using the following formulae: $PWV = ((A/\rho)(\Delta p/\Delta A))^{1/2}$ and $\text{PWV} = (Eh/2\rho R)^{1/2}$ for the graft and the host vessel, respectively, to give the host vessel impedance Z_v and the prosthesis impedance Z_p. Local and global reflection coefficients, $\Gamma_l = (Z_v - Z_p)/(Z_v + Z_p)$ and $\Gamma_g = (Z_c - Z_p)/(Z_c + Z_p)$ are estimated (Z_c: characteristic impedance determined from p/q ratio at a certain harmonic).

[14] Several geometric variables of the connecting segment geometry can be tested: the branching angle, toe and heel shape, junction curvature at both the toe and the heel, graft-to-artery bore ratio, taper of the host artery, position of the suture line, etc.

between the graft and the plane of the host artery and the angle between the graft and the proximal branch of the artery.

Shape optimisation methods have been applied to surgical planning [547]. Theoretical investigations based on perturbation analysis and linearized shape design provide results on existence, uniqueness of solution, and well-posedness of the problem [743]. Reduced-basis approximation used for pre-processing can detect essential features of optimisation, like sensitivity analysis [744].

8.2.2 Aneurism Clipping

The surgical treatment of saccular aneurism consists of placing a small metallic clip around the base of the aneurism whereas the normal artery wall is reconstructed to maintain blood flow to the perfused tissues. Aneurism surgery is performed on both ruptured and unruptured aneurisms. Several types of clips exist.

Microsurgery of cerebral aneurisms, performed under microscopic guidance, requires a small opening in the skull. Surgical treatment can lead to partial clipping. Peroperative aneurism rupture can occur. Postoperative complications include infection and damage of neighboring blood vessels. Surgical clipping of unruptured cerebral aneurisms are associated with a higher rate of complications (longer recovery period and persistent symptoms) than coiling [745].

8.2.3 Ventricular Assist Pumps

Because the donor pool for cardiac transplantation is insufficient, ventricular assist devices are used to improve the survival of patients suffering from heart failure and extend the time to transplantation. Ventricular assist devices are mechanical circulatory support systems aimed at unloading the heart and providing adequate body perfusion. They include both portable extracorporeal and implantable pumps for patients with competent lung function. Extracorporeal membrane oxygenation is a short-term support when heart and lung failures are combined. Device implantation can be associated with adjunctive therapy to improve ventricular recovery.

The design of the device configuration and its wall surface must suit hemodynamic and hematological criteria [746]. Left ventricular assist devices[15] have been designed with an optimal size to be permanently implanted in the heart chamber, using minimally invasive procedures. Various cannula designs and diameters are proposed to match patient anatomy. Left ventricular assist devices must minimize hemolysis, thrombosis, as well as heat generation.

---

[15] Left ventricular assist devices (LVAD) target patients with heart failure at a terminal stage waiting for transplantation, individuals with myocarditis and cardiomyopathy expecting complete cardiac recovery, and children in chronic stages of congenital heart disease who develop ventricular failure.

Artificial chambers are composed of two compartments, driving and blood chambers, separated by a flexible membrane [747]. The blood cavity has valved entry and exit orifices. Flow motion in the blood cavity is determined by membrane displacement controlled by the pressure in the second cavity. Numerical experiments have been performed in 2D models of such pumps [748, 749]. Once proved the existence and uniqueness of solutions of an elasticity problem in large displacement and small strain coupled to a viscous fluid, numerical applications in an artificial heart ventricle used either a simplified ALE method or the immersed boundary method. Diaphragme-type ventricular assist devices have also been studied experimentally (LDV) and numerically (using ADINA software) [750, 751].

Such a ventricular assist device is too big to be implanted in the circulatory system. Consequently, micropumps have been devised that can be implanted either in the left ventricle or between the left ventricle and the aorta. However, small pumps designed to be implanted in the left ventricle do not produce pulsatile flow (continuous-flow ventricular assist devices). Centrifugal pumps and pneumatic pulsatile ventricular assist devices have been proposed to be implanted in the body of both adults and children. Pump optimization and design methodologies have been based on 3D numerical simulations combined with design of experiments (DOE) [752], to develop a wholly implantable pump. Scaling has been set for an operating point using the Cordier diagram.

Implantation across any cardiac chamber or vessel lumen of mechanical devices, such as ventricular assist pumps, cardiac valves, filters, etc., induces hemodynamic disturbances and damages flowing cells, causing in particular hemolysis. The implanted devices thus must minimize hemolysis. Hemolysis indices have been proposed, such as a hemolysis index (HI), which measures the plasma-free hemoglobin (Hb) concentration [fHb], a normalized hemolysis index (NHI) relative to total blood [Hb], assuming that the hemolysis rate is small ($[\text{fHb}]/[\text{Hb}] \ll 1$) and varies linearly with time ($\Delta \bullet /\Delta t \equiv d \bullet /dt$), $\text{HI} \propto d[\text{fHb}]/dt \times V_{plasma}/q = (d\text{fHb}/dt)/q$, $\text{NHI} \propto \text{HI}/[\text{Hb}]$ [753]. These authors develop a prediction model derived from the Giersiepen-Wurzinger blood damage relationship using numerical simulations of blood flow field in the vasculature domain with the implanted device, introducing a pointwise function D, the damaged RBC fraction which is supposed to depend on shear

(c_{VM}) magnitude and linearly on exposure time:[16] ($D \propto c_{VM}^{\alpha} t$): $D_t + (\mathbf{v} \cdot \nabla)D = S$, where $S = \kappa_1 \kappa_2 c_{VM}^{\alpha}(1 - D)$.[17]

8.2.4 Valve Prostheses

Malfunctioning valves can be removed and then replaced with prosthetic heart valves. There are two main types of heart valves: (1) mechanical valves and (2) bioprosthetic stentless and stented tissue valves. There are three main designs of mechanical valves: (1) the caged ball valve (e.g., Starr-Edwards valve), (2) the tilting disc valve (e.g., Medtronic Hall and Bjork-Shiley valves), and (3) the bileaflet valve (e.g., St. Jude, CarboMedics, and ATS valves). Valved grafts have also been proposed (e.g., St. Jude aortic valved graft and ATS aortic valve graft prostheses). Bioprosthetic (xenograft) valves are made from porcine valves (e.g., Carpentier-Edwards and Hancock valves) or bovine pericardium (e.g., Ionescu-Shiley and Carpentier-Edwards valves). The prosthetic valve must mimic the static and dynamic characteristics of the natural human valve and the mechanics of flow through it to be successfull. It should not produce turbulences, flow stagnation, or excessive shear stress.

A collapsible trileaflet membrane valve has been developed [754]. It is designed with two cylindrical muffs fixed together, a stiff and a thin flexible, the latter having collapsible cusps. The mechanical behavior must minimize regurgitation and pressure drop across the artificial heart valves.

Several teams have studied in pulsatile flow conditions the mechanical features of the valve prostheses using rigid [755, 756] or deformable [757, 758] test sections, which model either the local valvar region or the heart cavities, possibly with models[18] of existing vessels [759]. The gradual closure of the natural valves during flow deceleration is, in general, not reproduced.

8.2.5 Electrical Stimulation

Electrical stimulation of cardiac tissue is used for cardiac pacing and defibrillation to provide suitable blood flow for nutrient delivery to the tissues.

---

[16] The shear stress is computed using the von Mises criterion:

$$c_{VM}(\mathbf{x}) = \left(1/2\left((c_1(\mathbf{x}) - c_2(\mathbf{x}))^2 + (c_2(\mathbf{x}) - c_3(\mathbf{x}))^2 + (c_3(\mathbf{x}) - c_1(\mathbf{x}))^2\right)\right)^{1/2},$$

where c_i ($i = 1, \ldots, 3$) are the stress vector components computed from the stress tensor components obtained from the Navier-Stokes solution in the pump computational model.

[17] The source term $S = \kappa_1 D(1 - D)/t$ with the hemolysis model $D = \kappa_2 c_{VM}^{\alpha} t$. $\kappa_1 = 0$ if $c_{VM} < c_Y$ and $\kappa_1 = 1$ otherwise (c_Y: stress corresponding to the hemolysis threshold).

[18] The heart geometry used by this researcher team is inversed.

Rate-responsive pacemakers can, indeed, be combined with implantable defibrillators. Any pacemaker is implanted after excluding or correcting electrolyte abnormalities.

Current pacemakers treat bradyarrhythmias and tachyarrhythmias. Implantation of an artificial pacemaker is mandatory in a case of symptomatic bradycardia, such as sinoatrial node dysfunction (or sick sinus syndrome) with very low atrial triggering rate, third-degree atrioventricular conduction block, and certain fascicular blocks. Type II second-degree atrioventricular block can be treated with a pacemaker even in an asymptomatic patient, because it can be a precursor to complete atrioventricular block. Marked first-degree atrioventricular block (PR interval > 300 ms) can benefit from pacing. Pacemakers also are used to treat advanced heart failure with major intraventricular conduction disorders, mainly left bundle-branch block (biventricular pacing for resynchronization).

Automatic calibration leads to adjustments of pacemaker response to normal changes in subject activity. Detection and dynamic analysis of nodal tissue rhythm provide accurate discrimination between simple acceleration and arrhythmias. Expert system associated with the pacemaker provides data for follow-up. Miniaturization of implantable pacemakers allows slightly invasive procedures. Safety features of pacemakers avoid the occurrence of induced currents during magnetic resonance imaging. Transient malfunctions of pacemakers during computed tomography can occur due to X-rays rather than small alternating electrical field ($\sim 150\,\mathrm{V/m}$) and alternating magnetic field ($\sim 15\,\mu\mathrm{T}$) [760].

8.3 Endovascular Therapy

Endovascular techniques are aimed at treating local wall damage in the heart or blood vessels. Endovascular interventions are minimally invasive because they use natural paths. The catheter is inserted into a superficial artery, such as the femoral artery or subclavian artery, and then advanced under image guidance into the diseased segment of the cardiovascular system.

8.3.1 Radiofrequency Ablation

Tachyarrhythmias can be treated by radiofrequency ablation (RFA). A selected small patch (< 5 mm) of heart cells of the conduction paths responsible for the abnormal heart rhythm is destroyed by a radiofrequency wave using an image-guided catheter-based procedure. Cryoablation can also be used, providing cold to freeze and destroy the cells.[19]

---

[19] Small tumors can also be treated by therma ablation, either cryoablation (freezing to at least $-19.4\,\mathrm{C}$ using liquid nitrogen or argon) or radiofrequency ablation. Damage limited to the local vasculature (without repercussion on the organ

Ablation is proposed to treat re-entrant tachycardias due to an extrapath in or adjacent to the atrioventricular node, junctional tachycardias, atria flutter and fibrillation, ventricular fibrillation, and Wolff-Parkinson-White syndrome.

8.3.2 Atherectomy

Catheter-guided atherectomy is used to remove the intimal fatty deposits that block the arterial lumen. Atherectomy uses either a laser catheter, which photodissolves the tissues obstructing the arterial lumen (e.g., percutaneous laser myocardial revascularization [LMR]), or a cutter catheter, which shaves the plaque off. Different cutting devices exist. The first type works like a shaver. The rotational atherectomy is used for hardened plaques and ostial sites. An diamond burr rotates at extremely high speed, breaking up blockages into very small fragments. Besides, ultrasound thrombolysis devices have been used for acute thrombosis with or without infarction.

8.3.3 Intravascular Devices

Endoluminal catheter-based therapy can be aimed at implanting various effective and reliable medical devices. Biomechanical tests can be useful to optimize medical device setting.

8.3.3.1 Stenting

Stents for Arterial Stenosis

Mini-invasive procedures have been developed to treat stenosed arteries. Image-guided balloon angioplasty was introduced by Gruntzig in 1977 for plaque compression and artery dilation. A balloon-tipped catheter is placed from a superficial artery into the stenosed artery, when lumen narrowing is at least equal to 70%. The deflated balloon is positioned across the obstruction and then inflated to the artery lumen physiological size. However, wall dissection and thrombus formation frequently occurred quickly after intervention. After 6 to 12 months, vessel remodeling with neointima characterized by cell proliferation of fibroblasts and smooth muscle cells was often observed.[20]

---

perfusion) can lead to ischemic necrosis, thus contributing to treatment efficacy. At few millimeters from the ice ball edge, the tissue temperature is warmer, thus avoiding lethal tissue damage.

[20] SMC concentration can be given by the reaction-diffusion formula:

$$\partial c_{SMC} = D_{SMC}\nabla^2 c_{SMC} + (r_p - r_d)c_{SMC},$$

where r_p and r_d are SMC production and death rates, both depending on local concentrations of influence factors, either mechanical (WSS) or chemical (NO, ROS, etc.).

Laser angioplasty was used to treat restenosis. The high restenosis rate leads to the development of the permanently implanted stent in 1986 by Puel and Sigwart.

Stents, metallic wire meshes of different configurations, hold the arterial lumen open (thickness of stent strut 70–100 µm; strut height 150–350 µm; between-strut-axis spacing of 0.9–1.8 mm; ratio of stent area to wall area equal to about 0.2). It is squeezed onto a low-radius tube for catheterization and insertion (stent is delivered while collapsed). Premounted balloon stent device is advanced across the constriction and expanded by balloon inflation. Dilated stents undergo plastic deformation without significant recoil. Stent deployment assigns a permanent altered shape to the stenosed arterial segment.

Shape memory alloys can sustain deformations, recovering the initial shape by heat (thermally induced shape recovery). Shape memory alloy stents are inserted into constraining catheters and returned to their original configuration when released from the catheter with body heat. (They are deployed from the catheter end at destination inside the diseased vessel segment, without requiring balloon inflation.) Shape memory alloy reinforced tubular stent-graft, suitable for tapered and bifurcated anatomies, are proposed to treat fusiform aneurisms.

The interface between stents and biological tissues is the focus of desirable and undesirable interactions. The restenosis rate remains significant, although much smaller than for balloon angioplasty.[21] Consequently, coated metallic stents and biodegradable stents have been tested recently. Semiconductor coating is used to reduce electron transfer from stent to circulatory fibrinogen. Other antithrombotic or antiproliferative coatings have been proposed. The administrated dose must be high enough for efficiency, but avoid toxicity.

Several types of drug-eluting stents have been tested in clinical trials to prevent, or at least to limit SMC migration and proliferation into the intima, which lead to restenosis. There are two main types of drug-eluting stents: (1) a coating by drug-polymer matrix, with a possible screen to reduce the transport speed and loss in blood flow, and (2) a cubic (size < 100 µm) reservoir set in the metallic structure. Rapamycin and taxol targets the cell division G1- and M-phase, respectively. HSG (or mitofusin-2) has an antiproliferative effect by inhibition of Ras–ERK signaling and subsequent cell-cycle arrest at the G0–G1 transition [761]. SMC proliferation can also be hindered by the inhibition of fibronectin matrix assembly [762].

---

[21] The response to stent injury can lead to uncontrolled healing. Smooth muscle cells migrate from the media and myofibroblasts from the adventitia across the external elastic lamina to be transformed into smooth muscle cells. The smooth muscle cells then secrete the extracellular matrix. The neointimal proliferation (NIP) can be sufficient to heal the implanted stent site, leaving the lumen opened, but can be extensive and yield vessel restenosis.

Endothelium-like tissue coated stents are contrived to impede the consequences of damaged endothelium during stent implantation. The endothelial cells are fragile and do not adhere strongly to the stent material, which bears very large deformations, in particular during stent implantation. Genetically engineered chondrocytes, which have the main endothelial cell functions and better adhesion properties than endothelial cells, can form a stent coating.

Several properties are assigned to any stent apart from its geometrical properties.[22] The implantable device shares the biological properties of protheses. It must be biocompatible, without any toxicity and allergy. Although associated with thromboresistant guide wires, antithrombogenicity is not fully fullfilled because of endothelium damage during the procedure. Any implantable device must be characterized by: (1) antiproliferative capability, (2) corrosion resistance, and, if possible, (3) repair functionality (particularly endothelial cell growth ability). Last but not least, it can be used for local drug delivery to control wall remodeling, especially for cell-migration and platelet-aggregation inhibitors. Stent mechanical properties[23] include: (1) high flexibility for safe and easy introduction (because of tortuous vessel anatomy); (2) high expandibility (the stent must be deployable safely in various vessel configurations avoiding strut twist); (3) plastic ductility (low elastic recoil after expansion); and (4) spatial stability in both axial and radial directions (absence of significant longitudinal shortening after expansion). It must sustain vessel wall stresses. For a suitable implantation, the stent must be visible by common imaging techniques to check: (1) advance to and across the stenosis, and (2) immediate and future device behavior. However, stent buckling, kinking, and migration have been observed. Optimal patient anatomy-dependent stent design and stent anchoring are thus needed.

The mechanical properties (structure expansion, recoil, and long-term behavior) of a balloon-expandable stent have been evaluated using the finite element method to assess stress and strain fields in the dilated stent wall [763]. FEM-based design of medical devices also allows the reduction of fabrication costs. Mathematical models of coated stents, which are used to prevent restenosis, have been used to study the effect of stent design on drug release and transport in the diseased wall, in particular the strut number and the ratio between the coated strut area and vessel area [764]. The delivered dose[24] depends on both the strut number and the strut surface density. Dose spikes are observed in front of the strut areas along the outer wall edge and dose

---

[22] Geometrical properties include: (1) structure, (2) surface, (3) length and diameter (a small outer diameter with a suitable profile is necessary for safe introduction in small arteries of diameter $< 1\,mm$), and (4) strut thickness and spacing (the strut-to-strut gap affects the local flow behavior).

[23] Mechanical tests determine, in particular, expanded diameter for a selected pressure range ($0 \leq p \leq 800\,kPa$), recoil, radial stiffness (pressure exerted by the vessel wall at which the intravascular device do not resist), fatigue, crimp (stent-balloon adherence), bending stiffness, and shortening due to expansion.

[24] The dose and not the drug concentration is the quantity of interest.

drops in the between-strut regions for a given strut surface density. The dose is smaller at both stent ends. The higher the strut number,[25] the lower the dose maximum, and the more uniform the dose distribution in the wall for a given radial coordinate, the asymptotic stent delivering the lowest dose.

Valvate Stents for Heart Valve Replacements

Aortic valves can be percutaneously replaced by valve-containing stents in the beating heart under image guidance, without strong damage of the aortic vessel wall and obstruction of the coronary ostia. First percutaneous aortic valvuloplasty was proposed using balloon-expandable stents [765, 766]. Catheterized implantation uses a valve prosthesis that can be sutured into an expandable stent and keep its property after crimping and re-expansion. Bovine jugular venous valves are sutured inside the stent [767]. The valve-stent assembly is deployed by balloon inflation. Due to heavy periodic loading, self-expanding stents are preferred to balloon-expandable stents [768]. A pulmonary metal stent with a valve was very recently implanted using a catheter-based method in a child with a congenital heart defect[26].

Stent Grafts for Abdominal Aortic Aneurisms

The risk of rupture of fusiform abdominal aortic aneurisms is currently assessed by its maximum caliber. Simulations can be carried out to display the stress field at the aneurism wall as well as within the aneurism wall to assist the clinicians. When the aneurism transverse size exceeds the critical value of 5 cm, abdominal aortic aneurism repairs use either open surgery or an endovascular graft technique (stent graft, such as bifurcated endografts, aortomonoiliac grafts). Endovascular aneurism repairs (EVAR) of large abdominal aortic aneurisms is another example of mini-invasive treatment (MIT). EVARs lead to a greater number of complications and reinterventions, but slightly increase survival in comparison with open repairs. The benefit related to early mortality is then limited by a costly strict follow-up and possible reintervention. Hence, open repair can be recommended for patients with long life expectancy. Besides, EVAR has a huge short-time operative mortality in patients already unfit for open repair.

Deployment complications can occur. However, the major drawback of endovascular techniques is stent-graft failure associated with stent-graft migration, endoleaks and aneurism redevelopment. Endoleaks are classified into several types (Table 8.1; Fig. 8.1). Fixation defaults generate not only endoleak but also migration of the device, as well as any attachment cuff.

---

[25] Tested strut numbers are equal to 24, 48, 96, 192, 384, and ∞.

[26] 20% of congenital heart defects involve the pulmonary valve. The mini-invasive cardiac catheterization carried out in the Hospital for Sick Children in Toronto takes 90 minutes and requires a single overnight stay.

Variable-Mesh Stents for Branching Saccular Aneurisms

Special stents are designed to limit blood impact in branching aneurisms, with little disturbance in branch flows[27] (Fig. 8.2). Numerical experiments can be carried out, an interface of variable permeability without any tangential velocity component replacing the stent mesh of variable size. Usual stent effects

Table 8.1. Classification of endoleak types occurring in stent grafts according of their origin (Fig. 8.1).

Grade	Flow source
Type I	Persistent perigraft flow due to inadequate sealing of stent-graft ends or suture breakage
Type II	Retrograde flow from collateral branches (i.e., from lumbar and inferior mesenteric arteries)
Type III	Flow due to disconnection or degradation of components (graft wall seam separation, polyester tears, metal frame fractures)
Type IV	Flow through stretched graft (with abnormal high porosity)

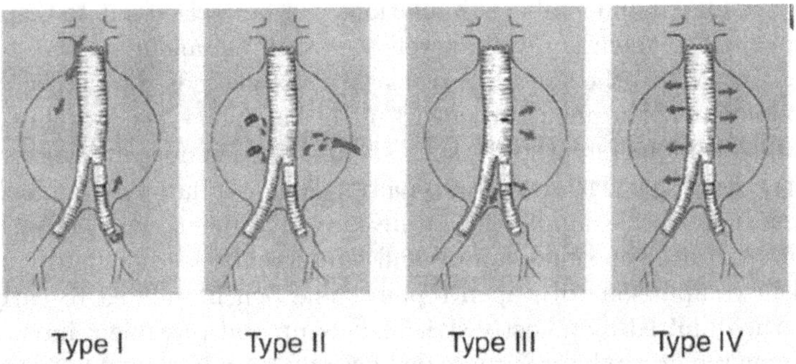

Type I   Type II   Type III   Type IV

Figure 8.1. Stent graft can be used to treat fusiform aneurisms of the abdominal aorta. The main complication of the procedure is endoleaks. There are several kinds of endoleaks according to the source (sutures, branches of the dilated segments, or degraded or permeable stent-graft wall; from A. Choong, St Mary's Hospital, Imperial College of London).

---

[27] Such stents have dense mesh regions to obstruct the aneurism neck as much as possible and stabilize the stent in the afferent vessel trunk on the one hand and large free areas in front of the branches to allow downstream irrigation on the other hand. A stent coating is necessary to limit, and possibly avoid, clotting and hemolysis.

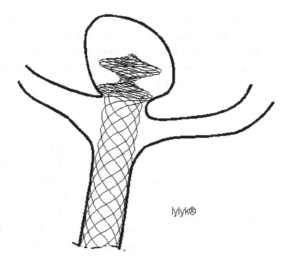

Figure 8.2. Stent for saccular branching aneurisms (courtesy of Cardiatis company).

on flows in the aneurismal cavity have been studied in side aneurisms. Flow disturbances depends on the stent type [769].[28]

8.3.3.2 Coiling

Closure of ruptured and unruptured aneurisms can be accomplished through coiling using platinum detachable coils (caliber of $\sim 400\,\mu m$). A catheter is threaded from a peripheral artery up to the aneurism neck under image guidance. From there a microcatheter is then advanced within the main catheter until its tip lies within the aneurism. Very thin coils are then advanced into the aneurism cavity, avoiding both rupture and embolism. After applying a very low voltage electric current, the coil is detached. Several coils can be sequentially used until the cavity is correctly filled with coils. Once coils are released, the blood in the aneurism clots. In numerical simulations, coils are represented by connected segments of overlapping spheres [770].

Endovascular treatment is necessary in aneurisms at high risk for surgery, but coiling can be inefficient. Coils, which have a circular memory, do not necessarily enter in the whole aneurismal cavity during coiling. Furthermore, coil compaction occurs.[29] Aneurism rupture and recanalization depends on the aneurism type in relation to the local blood flow. Unsatisfactory therapeutic results occur when: (1) the aneurism volume is great, (2) the aneurism

---

[28] Two stent models, helix and mesh, have been used in a model of a side aneurism in a straight duct. Velocities were measured by PIV (Stokes number of 3.9, peak duct Reynolds number of 425).

[29] Coils are pushed toward the dome by the arterial flow.

neck is large (important neck size-to-largest cavity length ratio), and (3) high blood inflow due to favorable aneurism angulation with respect to the stem axis [771]. Stents are also used in aneurism treatment in association with coiling when the aneurism compresses the adjoining artery. Stents have also been proposed to heal side saccular aneurisms.

8.3.3.3 Embolization

Catheter-based transarterial chemoembolization (TACE) is a catheter-based procedure to cure cancers. TACE is similar to intra-arterial infusion of chemotherapy but embolizing materials are added to block the small blood vessels. This treatment is used to destroy liver tumors. However, such a procedure commonly induces regional side effects, the boli not being injected very close to the target. Uterine fibroid embolization is a minimally invasive image-guided technique aimed at blocking the arteries that supply blood to the fibroids. A catheter is placed into the uterine artery branch that irrigates the tumor and small particles are injected to cause fibroid degeneration.

A two fluid phase flow computational model has been carried out in a 3D patient-specific model of the liver with a tumor and its arterial network [772]. The catheter length varies inside the artery to estimate the drug flow rate at each arterial branch.

8.3.3.4 Vein Devices

Apart from postural advices and elastic stockings,[30] superficial vein surgery and injection sclerotherapy may be performed. Endovascular devices may also be implanted in veins.

A *vena cava filter* is a device inserted into a major vein to prevent multiple pulmonary embolisms (PE) due to migration of thrombi into the pulmonary circulation. Since the majority of pulmonary emboli originate from the damaged vein wall in the lower limbs or pelvis, filters to block passage of emboli are mainly placed into the inferior vena cava, commonly in its infrarenal segment. Vena cava filters can be constructed from shape memory alloy, which, once released, form an umbrella shape.

The usual elliptical cross-section shape insures both appropriate venous return and valvular efficiency. These features are lost in dilated varicose veins. *V-shaped clips* can be put into varicose veins to restore the usual vein lumen configuration, which is defined by a preferential flattening according to its cross major axis.

---

[30] Elastic stockings induce redistribution of the venous flow toward the deep veins embedded in muscles, which behave like stiffer thick-walled compliant vessels, while superficial thin-walled diseased veins are collapsed by compression.

8.3.3.5 Drawbacks of Medical Devices

Any implant device may induce several biological and mechanical disturbances when device design is not appropriate and its rheology does not match the wall properties. When a part of the device is located in the flow core, hemolysis may be observed. It can, however, be supposed that the device material is selected to avoid sentitization, cytotoxicity, and carcinogenicity. In any case, a medical device, like any implanted foreign body, provides a matrix for thrombus formation, especially when endothelial cells are damaged or removed during the intravascular implantation. A thrombogenic response occurs after platelet deposition and clotting activation. Vessel wall damage also generates intimal proliferative response and quick restenosis. Growth factors can be released from platelets and fibroblasts. An inflammatory reaction due to interactions between leukocytes and the device can occur once leukocytes and complement have been activated. Finally, the medical device disturbs the blood flow. Locally, the added interface rugosity caused by the implanted device can generate local flow separations. Moreover, the abrupt transition at junctions of native vessel and stented segment disturbs the flow. The device very often induces a compliance mismatch. By means of wave reflection, it can affect remotely the flow.

8.3.3.6 Design of Medical Devices

Computer-aided manufacturing allows one to easily test different device geometry, structure, and material properties, with an important requirement: the material matching of the mechanical properties of the biological tissues. It needs to model accurately the physical processes at the interface, because the interface between the implants and the biological tissues, blood and vascular wall, is the focus of desirable and undesirable interactions. Moreover, it must be adaptable to unsteady loading induced by cardiac contractions.

8.4 Nanotechnology-Based Therapy

Nanotechnology can be applied to lesion treatment. Biocompatible nanoparticles filled with suitable drugs can target diseased cells. Once inside the lesion, they can deliver their content. Such compounds can bind to asialoglycoprotein receptors on hepatocytes [773]. Certain macromolecular vectors, such as dendrimers, are aimed at delivering their content into determined intracellular compartments [774].[31] Vectors can carry manifold substances, such as anti-cancer drugs, imaging agents, and cell receptor ligand. Because cancer cells need more folic acid than normal cells, and display a higher number of folate

---

[31] Dendrimers have a tree-like structure with many branches on which can be attached various molecules.

receptors on their plasmalemma, folate can be added to dendrimers [775]. Another solution for homing to specific body sites, such as tumors, is to target clotted plasma proteins. Tumor blood vessels are lined with a mesh of clotted plasma proteins not found in other tissues. Tumor-homing peptides that bind to this mesh can be fixed to nanoparticles to selectively target tumors. Tumor-homing peptide accumulation in tumor vessels induces additional binding sites for more particles, mimicking platelets [776]. These nanoparticles must avoid fast clearance by the liver and spleen, except in the case of quick efficient delivery from nanoparticles.

Another aspect of nanoparticles under investigation at the SCCS Laboratory of the National Taiwan University is related to physical destruction rather than chemical degradation of targeted cells. With selective ligands for targeted cell receptors, these nanoparticles concentrate the energy delivered by ultrasound to kill malignant cells. Lipid-coated perfluorocarbon nanodroplets can be used for therapy and ultrasound imaging. Subjected to ultrasound with acoustic energy at usual power levels, these nanoparticle enhance drug delivery. The displacement of perfluorocarbon nanoparticles can be forced in the direction of US propagation. Ultrasound (peak pressure on the order of MPa, frequency on the order of MHz) produces a particle velocity proportional to acoustic intensity, which also increases with rising center frequency [777]. Within a vessel (bore of hundreds of micrometers), a fluid motion is produced with a velocity of hundreds $\mu m/s$, which conveys the nanoparticles. Furthermore, ultrasound generated by conventional ultrasound imaging devices can enhance interactions between nanoparticle lipid layer and targeted cell plasmalemma via appropriate integrins, thus enhancing lipophilic drug delivery [778]. Ultrasound imaging can be combined with nanoparticle-based chemotherapy. Mixtures of drug-loaded polymeric micelles and perfluoropentane nanobubbles stabilized by the same biodegradable block copolymer are transported selectively into tumor interstitium, where the nanobubbles coalesce to produce microbubbles with ultrasound contrast [779]. Drugs in polymeric micelles are released under the effect of tumor-directed ultrasound. Such nanoparticles have many functions: drug carriers, ultrasound contrast agents, and enhancers of ultrasound-mediated drug delivery.

8.5 Medical and Surgical Simulators

Various minimally invasive vascular procedures exist. Mini-invasive techniques are still being improved to limit complications and mortality. Minimally invasive techniques are characterized by indirect visualization of the operation field on video monitors. Mini-invasive techniques uses specially designed instruments that are manipulated either directly or via mechanical linkage. Direct manipulations must be operated over the patient, whereas robot-assisted surgery can be done from a remote location. In the operating room, video assistance can be based on the cheaper and lighter 3D ultrasound imaging, the images being merged with preoperative data, such as MRI.

Minimally invasive procedures are also used to treat cancers. Percutaneous image-guided cryoablation and radiofrequency ablation[32] are aimed at destroying small tumors rather than surgically removing them. These less invasive techniques must be carefully monitored using an imaging technique and suitable navigation tools.

Mini-invasive therapies in medicine and surgery can be practiced with much less risk after learning and training, using virtual-reality based simulators. Any real-time medical simulator must be extensible, scalable, maintainable and flexible to be handled by multiple users. Any user must easily interact with the simulator. The first element deals with detection of tool contact and both mechanical (tissue deformation) and possible associated biological responses. Such contact generates haptic feedback useful for depth sensation. The second element corresponds to action and its desired and unwanted consequences (tissue traction, cutting, bleeding, gaseous emboli, etc.). Computer-generated scenes are then aimed at simulating in real time complex therapy procedures and tissue reactions. Medical and surgical simulators are associated with subject-specific images, virtual reality hardware, and haptic devices. The training system for MIT is based on infographics and data banks of recorded forces from actual tasks. Computed-aided training must, indeed, reproduce visual and haptic senses experienced during a minimally invasive procedure. Reaction forces, calculated with a frequency on the order of 1000 Hz, are sent to sensors to differentiate tissues and suitably navigate across the working field (depth sensing particularly).

Thoracoscopy and laparoscopy consist of performing surgery by introducing an endoscope and different instruments into the patient body through small incisions [781, 782]. MIT is beneficial because it reduces the surgical trauma and hospital duration, and therefore, the care cost on the one hand, and the morbidity for well-mastered procedures on the other hand. However, it brings new constraints on surgical practice. First, it significantly limits the surgeon's access to the organs of interest. Furthermore, this technique requires specific hand–eye coordination, which must be acquired after a training period. Training methods for medical and surgical interventions use either "endotrainers"[33] or living animals.[34] Limitations of such procedures stimulate development of computerized gesture-training systems. Much less sophisticated techniques can also be learned with simulators. For example, ultrasound

---

[32] Mobile phones in operation emit a radiofrequency electromagnetic field. The wave energy is partially absorbed by head organs and affects the brain electrical activity. It also induces a decrease in regional cerebral blood flow [780].

[33] Endotrainers use mannequins, inside which are placed plastic organs. The whole anatomy, organ interactions, and influence factors (cardiac pulsations and respiratory motions) are not taken into account. Consequently, these mechanical devices are of limited interest.

[34] Between-species anatomical differences are a limitation factor of animal testing.

imaging, which is a cheap, quick, and noninvasive technology, is commonly used for diagnosing many medical situations. Ultrasound-scan simulators are developed for training on patients, not only for sonography education, but also for evaluations of skills with a variety of normal and abnormal cases.

The development of MIT simulators raises important technical and scientific issues: (1) the geometry and mechanical behavior of the anatomical structures must be modeled and (2) the simulator must provide an advanced user interface including visual and force feedback.[35] Five main elements must then be introduced in a medical simulator: (1) geometric modeling,[36] (2) physical modeling,[37] (3) instrument interaction,[38] and (4) visual[39] and haptic[40] feedback [733].

---

[35] This interface can be decomposed into three distinct modules. The first module must model the interaction between surgical instruments and virtual organs. In particular, this task includes the detection and processing of the contacts that occur during the simulation. The second module aims at displaying the operating field on a video monitor in the most realistic manner. The third module must control a force-feedback device so the user can feel the applied forces that provide a 3D sense to the operator.

[36] Anatomical structures are now extracted from medical imaging. However, the automatic delineation of structures is still considered an unsolved problem. Many human interactions are required for 3DR.

[37] The mechanical behavior of organs is defined by non-linearity, poroviscoelasticity, plasticity, and fatigue phenomena. The usual constitutive laws must be simplified and optimized for real-time computation before implementation in a surgical or medical simulator.

[38] The hardware interface driving the virtual instrument is essentially composed of one or several force-feedback systems having the same degrees of freedom and appearance than actual instruments used in minimally invasive therapy. In general, these systems are force-controlled, sending the instrument position to the simulation software and receiving force targets. Once the position of the virtual instrument is known, contacts between two instruments or between an instrument and an organ must be detected. When a contact is detected, a set of constraints is applied to soft-tissue models. However, modeling the physics of contacts can lead to complex algorithms and therefore purely geometric approaches are often preferred.

[39] Visual feedback is important in video therapy because it gives a 3D perception of the environment. In particular, the effects of shading, shadows, and textures are important clues that must be reproduced in a simulator.

[40] The touch experienced by an operator when manipulating an instrument gives 3D information. The coupling between visual feedback and force feedback produces the sense of immersion. Haptic feedback requires a greater bandwidth than visual feedback. For simulating the contact with a soft object, a refresh rate of 300 Hz should be sufficient, whereas for a hard object a refresh rate greater than 1000 Hz should be used.

Conclusion

"Il y a deux sortes d'esprits : l'une de pénétrer vivement et profondément les conséquences des principes, et c'est là l'esprit de justesse ; l'autre, de comprendre un grand nombre de principes sans les confondre et c'est là l'esprit de géométrie. L'un est force et droiture d'esprit, l'autre est amplitude d'esprit. Or, l'un peut bien être sans l'autre, l'esprit pouvant être fort et étroit, et pouvant être aussi ample et faible. [Because it is much better to know something of everything than to know anything of a thing.]" (B. Pascal) [783]

" To want to get on two mounts, poem and philosophy, one runs the risk to ever become neither poet, nor philosopher " (G. Pico della Mirandola) [784].

The blood is pumped by the automatic heart, under the control of the central nervous system through two serial networks, the high-pressure systemic and low-pressure pulmonary circulations. Both circulatory circuits are composed of an arterial tree, capillary set, and venous bed. Most studies are performed in large blood vessels, either in normal or pathological conditions, especially in arteries. Analysis of the blood flow in large vessels shows that: (1) the vessel network is characterized by a succession of geometry singularities; (2) the anchored vessel has a compliant, composite, multilayered, thin ($h/R_h \sim 0.1$) wall, a small length in general (entry flow), a curved axis in various directions, and a possibly varying cross-section both in size and shape (environment prints); (3) the incompressible blood can have a non-Newtonian behavior in stagnant blood regions of diseased vessels. Consequently, the quasiperiodic (due to choatic heart behavior), three-dimensional (characterized by fluid-particle helical trajectories), most often laminar flow develops in vessels (entry lengths). The blood flow is governed by a set of dimensionless parameters, which vary along the vascular bed. Blood flow dynamics are more or less coupled to vessel wall mechanics according to the distance from the heart and the vasculature environment.

Explored parts of the cardiovascular system are currently isolated. Numerical simulations use artificial boundary conditions at vessel inlet and outlets (generally several exits due to branchings). Model hierarchy is proposed to take into account interactions with the bulk circulation. Different model scales (different equation dimension orders) are considered for: (1) wave propagation resulting from the coupling between the flowing blood and deformable vessel wall; and (2) main segments of the vasculature. The former is associated with a first order non-linear hyperbolic equation set for cross-sectional averaged velocity and pressure with varying variables due to rheological changes along the vessel length, and the latter with pointwise non-linear ordinary algebraic-differential equations.

Simulations of blood flows in large vessels are based on Navier-Stokes equations for an unsteady, three-dimensional, laminar flow, which can be recast in arbitrary Lagrangian-Eulerian formulation to model deformable vessel walls. However, this computational technique is not suited for very large deformations undergone by immersed valves of the heart, veins, and lymph vessels, as the valve cusps form a more or less broad contact region to prevent back flow.

Mathematical analysis and numerical simulations of the concentrated suspension of flowing cells in the microcirculation are required to develop suitable models of tissue perfusion. Modeling of the microvasculature must incorporate several features, such as trains of deformable erythrocytes surrounded by a lubrification layer, which trap plasma boli, and plasma skimming, as the vessel lumen and blood cells have the same order of size magnitude.

Blood circulation is aimed at conveying cells and various substances; therefore, blood flows need to be coupled to transport equations. Mass transfer across the vascular wall is required not only for tissue nutrition but also for inflammation, healing, and tissue growth and remodeling. Transported materials, after adhesion on and possible activation at the vascular endothelium, migrate across the multilayered composite vessel wall, where they can react with other substances and interact with biochemical pathways.

Rheological behavior of biological tissues and cells depends on their microscopic features as well as interactions with the vicinity. Biorheology modeling needs a complete description of the microstructure of the cell and tissues to handle the influence factors. However, constitutive equations are derived using the continuum concept. Therefore, most of the components characterized by a small size are neglected, but their influence can be integrated using effective rheology quantities and appropriate functions.

Mechanotransduction is investigated to clarify the manifold stress-induced processes from sensing to processing. This research field is an example of cooperation between biochemists and biomechanicians. At the cell scale, modeling on the continuum level may be questionable. However, it is used to estimate stress distribution in wall layers and interactions among large cell components, nucleus, cytosol and plasmalemma, as cytoplasmic organelles are neglected.

Such interactions can affect the local flow and, consequently, mass transport and cell responses. Genes involved in mechanotransduction, coding for ion channels, receptors, or responsive substances, are targets for additional studies.

The computational domains as well as experimental test sections are constructed from imaging data. However, the time and space resolutions of the imaging techniques most often are not good enough, and the data too noisy, to get a perfect definition of the inner and outer wall of the vascular region of interest, as well as ossible connections with the surrounding organs. Nevertheless, these connections yield important boundary conditions. Three-dimensional reconstruction strongly depends on the quality of input images. The fields of the hemodynamical variables is affected not only by the mesh density[1] but also by the investigated domain design [120]. Moreover, further works remain to be done to reconstruct and to mesh the vessel domain and to carry out numerical blood flow experiments in a suitable time for the medical practice. The computer-aided diagnosis of vessel pathologies requires the coupling between the blood flow and the vessel wall. The mathematical and numerical processing of this coupling was very recently successful. However, appropriate rheological data are still lacking.

Biomechanics also contribute to the development of new diagnosis methods; new measurement techniques, from signal acquisition to processing; new surgical or medical implantable devices; and new therapeutic strategies. Nanotechnology is implicated in drug delivery and medical devices. Nanomaterials[2] can be used in medicine for their ability to cross biological barriers and target specific tissues. Nanomaterials can thus be used to develop new therapies, such as nanoparticle-based ultrasound or magnetic hyperthermia for the treatment of cancer.[3]

The next generation of medical tools is based upon experience in sensor fusion, computer vision, robotics, virtual reality, and image and signal processing. They include, in particular, navigation, guidance, and positioning tools prior to and during the procedures. Navigation systems, indeed, allow one to determine the optimal patient-specific location, and guide the operator to achieve the desired placement. Integration of robotics, medical image processing, and infographics give birth to cybermedicine with its *computed-aided procedures* (CAP) and *image-guided therapy* (IGT). Telemedicine is based on systems of electronically communicating data from one site to a distant site

---

[1] Grid-convergent results (mesh independency) must be provided, numerical solutions being given for a range of significantly different mesh resolutions.

[2] Nanomaterials usually correspond to objects with dimensions in the range of 1 to 100 nm. In the medical field, they include objects up to 1 μm in size.

[3] Nanoparticles coated with aminosilane are taken up faster by tumor cells than by normal cells and subsequently heated by a magnetic field. Similarly, investigations are focused on nanoparticles used to concentrate the energy of ultrasound beams in tumors to be destroyed by the heat. Moreover, the treatment can be repeated, as nanoparticles form stable deposits within the tumor.

with data fusion by superimposing patient-specific data. Telepresence operation procedures have two major components: (1) a remote site with a 3D camera system and responsive manipulators with sensory input, and (2) an operating work station with a 3D monitor and dexterous handles with force feedback. A robot will thus be capable of executing the procedure at the actual site of the operation. Teletaction sensors will react depending on the type of material with which the operator is dealing, and imitation tools at the work station correspond to actual tools on the robotic arms at the site of the operation. Thereby, cardiovascular diseases provide an opportunity for multidisciplinary research aimed at developing computer-aided diagnosis and therapy.

A

Anatomical, Biological, Medical Glossaries

A.1 Anatomy Terminology

Anterior: nearer to the front of the body (ventral).
Contralateral: on the opposite side of the structure.
Coronal Plane: (frontal plane) separates the structure into ventral (anterior) and dorsal (posterior) regions.
Distal: farther from the entry.
Inferior: away from the head.
Ipsilateral: on the same side of the structure.
Lateral: farther from the midline of the body.
Medial: nearer to the midline of the body.
Posterior: nearer to or at the back of the body (dorsal).
Proximal: nearer to the face respiratory orifices.
Sagittal Plane: separates the structure into right and left compartments.
Superior: toward the head.
Transverse Plane: separates the structure into upper (cranial) and lower (caudal) parts. The anatomical position of the human body is upright; the transverse plane is horizontal. Lying on the bed of an imaging apparatus, the plane is vertical.

A.2 Biochemistry Terminology

A-Kinase Anchoring Protein (AKAP): scaffold protein that regulates in space and time signaling cascades. AKAPs interact with diverse enzymes.
Aaa Proteins: protein superfamily that uses ATP-dependent conformational changes to drive various cellular processes.
ABC Transporter: proteins coupled to ATP hydrolysis that transfer peptides and small molecules across the cell membranes.

Acetylcholine: a neurotransmitter and a hormone particularly involved in contraction of smooth muscle cells and cardiomyocytes. Acetylcholine receptors include the *nicotinic acetylcholine receptor*, a transmitter-gated ion channel, and the *muscarinic acetylcholine receptor*, a G-protein-coupled receptor. Acetylcholine, once bound to cardiac muscarinic M2-receptors, reduces the heart rate, activating Gi/o proteins and the related G-protein-gated inward rectifier K^+ channels.

Actin: a major constituent of the cytoskeleton of every cell and of the sarcomere of myocytes. Three main actin types exist α, β, and γ. Monomeric actin is a ATPase which interacts with one molecule of ATP or ADP. It contains one high-affinity and several low-affinity binding sites for divalent cations. Globular actin (G-actin monomer) polymerizes with the concomitant hydrolysis of ATP to form polarized filamentous actin (F-actin polymer), a double-helical filament. The helical actin filament has a barbed end and a pointed end. Actin stress fibers are bundles of actin filaments that cross the cell and are linked to the extracellular matrix via focal adhesions. In vitro, actin polymerizes from both ends, rapidly at the plus end (barbed end) and slowly at the minus end (pointed end). Actin polymerization depends on affinity filament ends (actin-monomer concentration must be higher than a critical concentration, which depends on ionic levels). During cell displacements, plus ends are often oriented toward the motion direction (cell leading edge), as filamentous actin gives rise to filopodia and lamellipodia. Actin interacts with manifold proteins, actin-binding proteins (ABP), and actin-severing proteins (ASP). Actin is involved in interactions between the cytoskeleton and plasmalemma, in association with proteins, to determine the cell shape, stabilize the cell membrane, and construct specialized membrane domains in particular (membrane–skeleton proteins of the erythrocyte particularly involve spectrin, ankyrin, band-3, and protein-4.1). Dystrophin of the sarcolemma links the membrane to actin filament bundles. Actin binds to ATP. When actin hydrolyzes ATP (ADP bound actin), it is then able to bind to myosin.

Actin-Binding Protein: associates with either actin monomers or actin filaments in cells and subsequently modifies actin properties.

α-*Actinin*: actin cross-linking protein of the spectrin family. α-Actinin forms homodimers in a rod-like structure with one actin-binding domain on each side of the rod, thus cross-linking two actin filaments. α-Actinin also binds various cytoskeletal proteins (titin, zyxin, vinculin, and α-catenin), hence associating actin to focal adhesions. α-Actinin binds phosphatidylinositol 3-kinase, Rho kinase, plasmalemmal receptors, and β-integrins, thereby bridging signaling molecules to the cytoskeleton. The α-actinin family includes four members. α-Actinin-1 and α-actinin-4 are involved in the organization of the actin cytoskeleton in non-muscle tissues. α-Actinin-2 and α-actinin-3 are specific isoforms of the striated muscle. α-Actinin-2 is the single cardiac isoform.

A Disintegrin and Metalloprotease (ADAM): molecule of the plasmalemma with cell adhesion and protease activities. A disintegrin and metalloprotease with thrombospondin motifs (ADAMTS) forms a family of extracellular proteases. Certain members of the ADAMTS family are implicated in embryonic development and angiogenesis.

Adaptor Protein (AP): is involved in molecular interactions of signaling pathways. It is particularly required in the regulation of signal transduction initiated by the commitment of plasmalemmal receptors. Certain adapters are expressed in all cells, whereas others are restricted to specific celles.

Growth factor receptor binding proteins (Grb) are adapters associated with receptor tyrosine kinases for growth factors. Adaptor proteins (AP1–AP4) are implicated in endocytosis, connecting transported molecules with components of the vesicle coat. Adapters are heterotetramers formed with two large subunits (a first subunit α, γ, δ, and ϵ, and a second subunit $\beta2$, $\beta1$, $\beta3$, and $\beta4$ in AP2, AP1, AP3, and AP4, respectively), a medium-sized subunit ($\mu1$–$\mu4$ in AP1–AP4, respectively) and a small subunit ($\sigma1$–$\sigma4$ in AP1–AP4, respectively). AP2 is the main clathrin adapter in the plasmalemma. AP2 is recruited by phosphatidylinositol(4,5)bisphosphate or phosphatidylinositol(3,4,5)trisphosphate. AP1, AP3 and AP4 are located in endosomes and the trans-Golgi network. AP1 binds to phosphatidylinositol(4,5)bisphosphate and ADP-ribosylation factor Arf1.

Adducin: belongs to a family of cytoskeletal proteins. α- and γ-adducins are ubiquitously expressed, whereas β-adducin is expressed in brain and hematopoietic tissues. Adducin binds to calcium–calmodulin. It serves as a substrate for protein kinases-A and -C.

Adenosine Diphosphate (ADP): nucleotide produced by ATP hydrolysis by ATPase. It regenerates ATP when phosphorylated particularly by oxidative phosphorylation.

Adenosine Monophosphate (AMP): one of the four nucleotides (i.e., nucleosides, a purine or pyrimidine base linked to a ribose or deoxyribose sugar, with at least one phosphate group; Table A.1) of RNA molecules.

AMP-activated Protein Kinase (AMPK): regulator of cellular metabolism and cell structures activated by serine/threonine kinase LKB1 in response to energy deprivation.

Adenosine Triphosphate (ATP): nucleoside provider of chemical energy in cells. ATP synthase catalyzes the formation of ATP from ADP and inorganic phosphate during oxidative phosphorylation. Mitochondria produce adenosine triphosphate by transferring electrons from organic substrates through a set of respiratory enzyme complexes to molecular oxygen. In energy-generating oxidative phosphorylation, proteic complexes and cofactors are required in the chain of redox reactions. Protons (H^+) are transported across the mitochondrial inner membrane. The resulting transmembrane proton gradient is used to generate ATP. ATP is consumed to generate cellular processes that require energy (molecule synthesis and active transport of molecules across membranes).

ATP is not only the basic fuel molecule of the cell but also a messenger (i.e., between endothelial cells, smooth muscle cells, and perivascular nerves), particularly a paracrine messenger using gap junctions for adjacent cell communication.

Adenylyl Cyclase (or adenylate cyclase): a membrane-bound enzyme that catalyzes the formation from ATP of cAMP, a second messenger controlling numerous intracellular activities.

Adipokines: produced by adipose tissue, have cytokine-like properties. Epicardial adipose tissue manufactures chemokines such as monocyte chemotactic protein-1 and cytokines, such interleukins IL1 and IL6, and tumor necrosis factor-α.

Adiponectin: protects the heart from ischemia. Adiponectin overexpression promotes insulin sensitivity and angiogenesis and hinders cardiac hypertrophy. Adiponectin inhibits apoptosis and TNFα production. Adiponectin stimulates COx2-dependent synthesis of prostaglandin-E2 in CMCs [785]. It also has antiapoptotic effects via AMP-activated protein kinase.

Adrenaline (or epinephrine; Ad): catecholamine secreted by chromaffin cells in the adrenal medulla which serves as hormone and neurotransmitter. Adrenaline acts in short-term stress reactions (vasoconstriction and positive chronotropy particularly). Each chromaffin cell contains 1 to $3\,10^4$ vesicles, each vesicle 3 to $5\,10^6$ adrenaline molecules.

Affinity: tendency of a molecule to bind to its biological target.

Agonist: substance that interacts with a receptor and initiates the receptor-mediated response.

Amino Acid: building blocks of peptides (less than about 50 amino acids) and proteins. They contain both amino (NH_2) and carboxyl (COOH) groups attached to the same carbon. Standard amino acids (alanine, argi-

Table A.1. Nomenclature of bases, nucleosides, and nucleotides. Deoxyribonucleotides, DNA components, are obtained from the corresponding ribonucleotides by reduction catalyzed by ribonucleotide reductase.

Bases	Ribonucleosides	Ribonucleotides
Adenine	Adenosine	Adenylate (AMP)
Guanine	Guanosine	Guanylate (GMP)
Uracil	Uridine	Uridylate (UMP)
Cytosine	Cytidine	Cytidylate (CMP)
Bases	Deoxyribonucleosides	Deoxyribonucleotides
Adenine	Deoxyadenosine	Deoxyadenylate (dAMP)
Guanine	Deoxyguanosine	Deoxyguanylate (dGMP)
Thymine	Deoxythymidine	Deoxythymidylate (dTMP)
Cytosine	Deoxycytidine	Deoxycytidylate (dCMP)

nine, asparagine, aspartic acid, cysteine, glutamic acid, glutamine, glycine, histidine, proline, serine, tyrosine, etc.) are encoded by the genetic code. Non-standard amino acids (taurine, dopamine, etc.) are not incorporated into proteins. Non-standard amino acids (ornithine and citrulline) often act in metabolism of standard amino acids. In humans, essential amino acids (isoleucine, leucine, lysine, methionine, phenylalanine, threonine, tryptophan, and valine) are supplied by food.

Amphiphysin: implicated in T-tubule formation in muscle and clathrin-coated vesicle formation. Amphiphysins recruit GTPase dynamin, which is responsible for vesicle scission, and synaptojanin, which participates in vesicle uncoating.

Anabolism: edification of new large compounds from simpler substances (constructive metabolism).

Angiotensin-2 (ATn2): molecule that regulates the expression in vascular endothelium of soluble epoxide hydrolase. The latter degrades antihypertensive and antiatherosclerotic mediators epoxyeicosatrienoic acids. Once bound to its receptor AT2R1, ATn2 activates sEH promoter via cJun, which stimulates transcription factor AP1 [786].

Ankyrin: attaches the spectrin cytoskeleton to the erythrocyte membrane.

Antagonist: agent that, at the receptor level, opposes the receptor-associated responses induced by agonists.

Apoptosis: programmed cell death triggered by multiple agents (receptor stimulation, mutations in mitochondrial DNA, production of reactive oxygen species, mitochondrial release of cytochrome-C, etc.). Apoptosis is generated by two main proteolytic pathways, the intrinsic and extrinsic pathways, both involving initiator caspases and executioner caspases.

Apoptosome: molecular complex formed by cytochrome-C, Apaf1, and caspase-9. Apaf1 binds both cytochrome-C and caspase-9 and contains an ATPase region. In the absence of apoptotic signal, Apaf1 is an autoinhibited monomer. Autoinhibition loss is triggered by recruitment of cytochrome-C released from mitochondria (other Apaf1 regions remain in autoinhibited conformation). ATPase activity of Apaf1 leads to oligomerization needed for apoptosome formation. Pro-caspase-9 are then recruited, the apoptosome providing a platform for caspase-9 dimerization (activation).

Arachidonic Acid: molecule derived from phospholipids of cellular membranes involved in vascular inflammation. It can be converted to eicosanoids by cyclooxygenase, lipoxygenase, and CYP450 epoxygenase. Endothelial cells express both CYP2C and CYP2J.

Arf Protein: ADP-ribosylating factor of the GTPase family that regulates both CoP1 coat and clathrin coat assembly at Golgi membranes.

ARP Complex: complex of proteins involved in actin filament growth from the minus end.

β-Arrestin: specialized proteins which interact with a single type of molecules. β-Arrestins stop GPCR signaling by binding to GPCRs, which are thereby not able to interact with G proteins. Moreover, β-arrestins, which bind to both clathrins and receptors, bring GPCRs to clathrin-coated vesicles for endocytosis.

Autocrine Signaling: secreted signal molecules act on the producing cell after secretion (αυτο: in itself, or on other adjacent cells of the same type).

Autophagy: degradation of defective and damaged cell components, proteins, and organelles, which can injure the cell. This process can also occur in response to stresses (high temperature or microbial invasion) and starvation. Proteins and organelles are transported in autophagosomes through the cytoplasm to lysosomes, then catabolized in lysosomes to provide cells with amino acids and energy during starvation and stress. In opposition to apoptosis which leads to cell death, autophagy contributes to cell survival. However, co-occurrence of autophagy and apoptosis generates more or less delayed cell death. In certain infections, autophagy is required for the onset of apoptosis.

Bag Proteins: proteins (Bag1–Bag6) which bind and modulate the activity of heat-shock proteins 70 (Hsp70 chaperones). Several Bag proteins control cell apoptosis. Bag proteins are also involved in the transcriptional activity of steroid hormone receptors. Bag1 forms a complex with both B-Raf and PKB.

B-Raf: serine/threonine protein kinase of the Raf family. It can be activated by GTPase Ras and Rap1. cAMP activates mitogen activated protein kinase via B-Raf.

Bcr Protein: contains a serine/threonine protein kinase, a guanine nucleotide-exchange factor, a GTPase-activating protein, and a ligand for a PDZ domain. Therefore, Bcr is a negative regulator of the small GTPase Rac. It blocks p21-activating kinase. It also negatively regulates the Ras–Raf–MAPK–ERK signaling pathway as well as β-catenin, and thus the expression of proliferation-promoting genes [787].

Bone Morphogenetic Protein (BMP): member of the TGFβ family. It binds complexes of BMP receptors BMPR1 and BMPR2. Once bound by BMP, BMPR2 phosphorylates BMPR1. The BMPR complex phosphorylates a set of Smad proteins (Smad1, Smad5, and Smad8). The Smad set then forms a complex with Smad4, which translocates to the nucleus to regulate the transcription of specific target genes.

Bradykinin: molecule formed by action of proteases on kininogens. It stimulates nitric oxide formation by the vascular endothelium (vasodilation), as well as prostacyclin formation. Bradykinin also increases vascular permeability. It is a component of the kallikrein–kinin system.

C-Reactive Protein (CRP): acute-phase protein that binds to ligands exposed in damaged tissue and activates complement. In particular, C-reactive protein binds to phosphocholine in the presence of calcium. It is produced by hepatocytes stimulated mainly by interleukins IL1β and IL6, and by tumor necrosis factor-α.

Cadherin: calcium-dependent adhesion protein implicated in tight cell–cell contacts.

Calcineurin: Ca^{++}-dependent serine/threonine protein phosphatase (PP2B or PP3). It binds to a Ca^{++}-sensititive transcriptional factor, the nuclear factor of activated T cells (NFAT). Calcineurin dephosphorylates NFAT, which translocates to the nucleus. A sustained rise in calcium level is necessary for NFAT-dependent transcription. A decay in nuclear calcium level, indeed, deactivates calcineurin and allows NFAT kinases to rephosphorylate NFAT, which then leaves the nucleus. Calmodulin transports calcium ions to the nucleus.

Calcitonin Family of Peptides: a molecule set that includes calcitonin, amylin, two calcitonin gene-related peptides (αCGRP and βCGRP), and adrenomedullin.

Calmodulin (Cam): calcium-binding protein that binds to other proteins according to the intracellular calcium concentration. The binding affects the activity of many targets, such as enzymes and transport proteins. Calcium–calmodulin-dependent protein kinase, once stimulated by calcium-activated calmodulin, mediates the effects of calcium by phosphorylation of other proteins. Calmodulin bound to calcium can inactivate voltage-gated calcium channels. Calmodulin can also facilitate the ion motion across certain voltage-gated calcium channels. In the heart, calcium–calmodulin interacts with plasma membrane Ca^{++} ATPase, L-type Ca^{++} channel, and ryanodine channel-2. Calmodulin is the calcium sensor for both small and intermediate conductance calcium-gated potassium channels. Calmodulin affects the activity of voltage-gated sodium channels. Calcium–calmodulin regulatory activities are mediated by Cam-dependent kinases.

Calmodulin-Dependent Kinase (CamK): serine/threonine protein kinase that has a calcium–calmodulin-dependent activity and by autophosphorylation, a calcium–calmodulin-independent function. CamKs are phosphorylated by Cam-binding or CamK kinase (CaM2K). Several CamK types are multisubstrate kinases, such as multimeric CamK2, whereas others are specific for a given protein, such as the myosin light chain kinase. Monomeric skeletal and cardiac MLCKs regulate the light chain of cardiac myosin-2, increasing sarcomere sensitivity to calcium (positive inotropy). PKC also phosphorylates MLC.

Calponin: actin-binding protein, which composes a protein family, with Neutral calponin-2, basic calponin-1, and acidic calponin-3. The two latter members can inhibit actomyosin Mg^{++}-ATPase.

Calsequestrin: protein of the region of the sarcoplasmic reticulum near the sarcolemma.

CAP: adapter encoded by the SORBS1 gene, which has a sorbin peptide homology (SoHo) domain. CAP regulates diverse cell processes (transport, transcription activation, and ubiquitination).

Caspase (Cys-dependent Asp-specific protease): intracellular protease with a catalytic activity governed by a Cys side chain of the protease and one or two proteic cleavages after Asp residues of substrates. Caspases are involved in apoptosis (proteolytic cascade). The activator caspases include initial caspase-8 and -10 (activated via receptors Fas, which recruit FADD and RIP1, and TNFα receptors [plasmalemma receptor pathway]) and caspase-9 (activated by the mitochondrial release of cytochrome-C and interaction with Apaf1 [mitochondrion pathway]). The effector caspases, caspase-3, -6 and -7 cleave various cell targets. Caspases particularly cut components of the nuclear pore complex, then disturb nucleocytoplasmic exchange of molecules. Caspase inhibitors bind to caspases.

Catabolism: reaction chains involved in degradation of complex substances in simpler molecules, with possible release of catabolites and energy (destructive metabolism). The process can lead to excretion.

Catecholamines: adrenaline and noradrenaline, act on adrenergic receptors. α1-Adrenergic receptors mediate catecholamine effects, particularly on smooth muscle cell growth and contraction, myocardial function, and synthesis of the atrial natriuretic factor.

Catenin: cytoplasmic protein involved in cell–cell adhesion, linking cadherins to the actin cytoskeleton. α-Catenins link cadherins to the actin cytoskeleton. β-Catenin is an adherens junction protein. Zonula adherens, or adherens junctions, serve for signal transmission between neighboring cells and anchor the actin cytoskeleton.

Caveola: invagination of the plasmalemma that buds off internally to form vesicles.

Caveolin: transmembrane protein. The family includes caveolin-1, caveolin-2, and muscle-specific caveolin-3. Caveolins are constituents of caveolae. They can associate with inactive signaling molecules, such as Src family kinases and Ras family GTPases, acting as a scaffold for signaling complex assembly. Caveolin-1 also binds integrins.

Cbl Protein: ubiquitin ligase and adapter involved in the regulation of signal transduction in various cell types in response to different stimuli.

Cdc42: Rho family GTPase activated by integrins and bradykinin. Cdc42 acts on actin by binding WASP, an activator of Arp2/3 and actin polymerization. During cell migration, Cdc42 is involved in cell polarization and reorientation of the Golgi apparatus toward the leading edge. Cdc42 is also implicated in endocytosis.

(Molecular) Chaperone: a protein that controls the folding of nascent polypeptides into the suitable 3D structure so that the targeted polypeptides become functional (it avoids misfolding, hence inactive polypeptides). It also maintains polypeptides in an inactive state until they have been transported to their destinations to be assembled into functional compounds. Molecular chaperones also participate in the assembly into higher-order structures and disassembly of polypeptides, without being a component of the final molecules. In addition, molecular chaperones are

implicated in the repair of denatured proteins or their degradation after stress (denatured and misfolded proteins that cannot be refolded by the molecular chaperones undergo proteolysis). They also prevent protein aggregation, which is observed in many diseases, such as hypertrophic cardiomyopathy. Molecular chaperones were originally identified as "heat-shock proteins" (Hsp), because they have been observed in cells subjected to thermal stresses. Members of the Hsp family are identified by their molecular mass (Hsp70, Hsp90, etc.). Hsps have cytoprotective functions when the cell undergoes different types of stresses, either environmental or metabolic (not only heat, but also hypoxemia, toxic elements, oxidative stress, etc.). They are then better defined as stress proteins. Many molecular chaperones are expressed in all cell types, whereas other are restricted to certain cell types.

Chemoattractant (cytotaxin, chemotaxin): induces a chemotactic response.

Chemokine: chemoattractant cytokine for leukocytes via interactions with G-protein-coupled receptors. Chemokines intervene in the functioning of the immune system, as well as on endothelial cells involved in angiogenesis. Chemokines are big peptides (or small proteins) made of 60 to 100 amino acids with 20 to 90% homology in sequences. Chemokines are divided into four families: CXC, CC, C, and CX3C. CXC and CC chemokines belong to the two major subsets. CXC chemokines attract neutrophils, T lymphocytes, B lymphocytes and natural killer cells. CC chemokines attract monocytes, macrophages, basophils, T lymphocytes and eosinophils. Chemokine-receptor nomenclature follows that of chemokines: CXCR, CCR, XCR, and CX3CR. Chemokines also act as communication molecules between neurons and glia, neurons and endothelial cells, and in neuron–neuron interactions, leading to calcium influx. Chemokines might also function as neurotransmitters and/or neuromodulators in the central nervous system.

Cholesterol (Cs): component of cell membranes and precursor in the synthesis of steroid hormones, vitamin-D, and bile acids. Cs metabolism is tightly regulated to prevent abnormal deposition. A balance between cell influx and efflux is necessary for Cs homeostasis. Cs flux is mediated by receptors and involves lipoproteins. Endocellular Cs inhibits both Cs synthesis and membrane receptors. Atherosclerosis risk markers include ApoA-I and ApoB with respective concentrations - measured once the Cs upper limit has been reached - less than $2.65\,g/l$ and $1.2\,g/l$. Efflux from plasma membrane involves several possible mechanisms: (1) simple desorption and diffusion to phospholipid-containing lipoproteins; (2) removal by apolipoproteins to form HDL, involving cAMP; and (3) scavenger receptor SR-BI mediated exchange associated with calveolae (Part I), in particular in macrophages and the liver [788]. Cholesterol acceptors used for uni- or bidirectional flux include: (1) cyclodextrins, bidirectional shuttles between the cell membrane and extracellular Cs acceptors (blood lipoproteins); (2) phospholipid vesicles, for unidirectional Cs motion from the cell; (3) Apos, which participates in cellular Cs and phospholipid efflux; and (4) HDL responsible for efflux efficiency, and blood proteins, such as LCAT, phospholipid transfer protein PLTP, and CETP. Multiple membrane proteins are involved in cholesterol transport, MLN64 in the endosome; NPC1, NPC2, STAR, at the mitochondrion; SCP2 in the cytosol; and peroxisomes and oxysterol binding protein at the Golgi apparatus [789].

Chromatin: complex of DNA, histones, and other proteins of the cell nucleus of a eukaryotic cell, chromosome material. Heterochromatin corresponds to regions of condensed chromatin. Both histones and DNA undergo modifications, especially during transcription. Direct modifications are due to either a change in electrostatic charges or internucleosome contacts, allowing opening or closing of DNA. Indirect modifications are related to alterations in the nucleosome surface that promote the association of chromatin-binding proteins.

Chromosome: single, long DNA macromolecule, that contains many genes, regulatory elements, and other nucleotide sequences. The chromosome represents the largest unit of genome organization. Chromosomes are packed by proteins, particularly histones, into chromatin. Each chromosome has a centromere and arms projecting from the centromere. There is a network of communications within and between chromosomes. The chromosome territory is a unit of nuclear organization separated by an interchromatin compartment between adjacent chromosomes. Within the nucleus, each chromosome has its dynamic location (inner or outer nucleus, with possible relocation); and within the chromosome, each gene has its site (interior or exterior of the chromosome territory, with possible relocation outside of chromosome territory).

Clathrin: assembles on the cytosolic side of the plasmalemma to form a clathrin-coated pit, which buds off to form a clathrin-coated vesicle containing extracellular fluid and solutes dissolved in this solvent. Clathrin is the predominant protein present in the coat of clathrin-coated vesicles. A clathrin coat is a three-dimensional array of triskelia. Each triskelion is made of three clathrin heavy chains and three clathrin light chains.

Clathrin Adaptors: bind directly to both the clathrin lattice and plasmalemmal lipids and proteins, thus connecting the clathrin scaffold to a membrane component. Interactions with clathrin are involved in vesicle formation. Clathrin adaptors are required in assembling of clathrin triskelia. Identified clathrin adaptors are the protein-binding and phospholipid-binding adaptor protein complexes AP1 and AP2, as well as AP180.

Claudin: binds to claudins on adjacent cells to form the primary seal of the tight junction. Distinct species of claudins can interact within and between tight junction strands, except in some combinations.

Clone: population of genetically identical cells produced from a common ancestor.

Coactivator: protein that regulates gene transcription, generally in a tissue-specific fashion.

Coenzyme: substance participating in enzymatic reactions as acceptor or donor of chemical groups or electrons, such as NAD^+ and coenzyme-A. The latter is used in the enzymatic transfer of acyl groups.

Cofilin: depolymerizes actin when phosphorylated by LIM-kinase.

Collagen: a major component of the extracellular matrix. Collagen fibrils are formed by self-assembly of secreted fibrillar collagen subunits.

Colony-Stimulating Factor (CSF): signaling molecule that controls the differentiation of blood cells in the bone marrow.

Complement System: set of serum proteins, once activated by antibody–antigen complexes or microorganisms, implicated in biochemical cascades of the innate and adaptive immunity, leading to phagocytosis, cytolysis, chemotaxis, opsonization (opsonin coating of a particle), and inflammation. The proteins of the complement system are either involved in the complement pathways or have regulatory functions. The pivotal component C3 is activated in the three complement pathways (classical, lectin and alternative). The complement system generates anaphylatoxins, such as C3a, for chemotaxis and cell activation. The C3-convertase initiates the membrane attack pathway, which results in the membrane attack complex (MAC), consisting of C5b, C6, C7, C8, and polymeric C9. The *classical pathway* is triggered by activation of the C1-complex (C1q, C1r, C1s) by C1q binding to its target. The C1-complex splits C2 and C4 into C2b and C4b, whichs bind together to form C3-convertase. In the *alternative pathway*, C3 is split into C3a and C3b. C3b binds to factor-B, and C3b-fB is cleaved by factor-D into Ba and the alternative pathway C3-convertase. The *lectin pathway* is similar to the classical pathway, but the opsonin mannan-binding lectin (MBL) replaces C1q (MBL complex is formed by MBL and MBL associated serine protease MASP-1 and -2).

Connexon: plasmalemmal water-filled pore formed by a ring of six proteic subunits. Connexons from two adjoining cells join to build a continuous between-cell channel, the gap junction.

Cooperativity: process by which binding of a ligand to receptor or enzyme influences binding at a second site.

Coronin: protein involved in cell cycle progression, signal transduction, apoptosis, and gene regulation.

Cortactin: actin-binding protein implicated in endocytosis. Cortactin organizes the structure of the cytoskeleton and cell adhesions. It binds directly to GTPase dynamin-2, which mediates the formation of vesicles from the plasmalemma and Golgi stack.

Cotransmission: synthesis, storage, and release of at least two transmitters by a single nerve.

Cell Cortex: cytoplasm actin-rich layer on the inner face of the plasmalemma, responsible for motions of the cell surface.

Chicken Tumor Virus Regulator of Kinase (Crk): adapter promoting the assembly of signaling complexes. It recruits other adapters and tyrosine kinases. It activates Ras GTPases. Adapters Crks regulate transcription and cytoskeletal organization during cell growth, proliferation, adhesion, differentiation, apoptosis, and motility, by linking tyrosine kinases to small GTPases. Crk1 and Crk2 are Crk splicing isoforms. Oncoprotein Crks are involved in the localization and activation of several different effectors (guanine nucleotide-releasing proteins C3G, protein kinases of the Abl and GCK families, and small GTPases such as Rap1 and Rac).

Cyclic Adenosine MonoPhosphate (cAMP): nucleotide generated from ATP by adenylyl cyclase stimulated by ligand-bound plasmalemmal receptors. cAMP mediates intracellular signaling. It mainly acts via protein kinase-A, which phosphorylates many intracellular enzymes. PKA can be complexed with A-kinase anchoring proteins. It also activates certain ion channels and guanine nucleotide-exchange factors. It is catalyzed by phosphodiesterases, such as PDE4D3. The cAMP level depends on the balance between adenylyl cyclases and phosphodiesterases.

Cyclic Guanosine MonoPhosphate (cGMP): intracellular signaling molecule formed from GTP by guanylyl cyclase (or guanylate cyclase) in response to receptor stimulation.

Cyclooxygenase (COx): forms a family of three isoforms of cyclooxygenases (COx1, COx2, and COx3), which catalyzes the conversion of arachidonic acid into prostaglandins.

Cytochrome: electron carrier that transfers an electron from an electron donor (becoming oxidized) to an acceptor molecule during cellular respiration. Cytochromes are either monomeric proteins or components of enzymatic complexes, with heme groups, generally bound to membrane (mitochondrial inner membrane and endoplasmic reticulum). In many cytochromes, the metal ion surrounded by the heme group is iron, which evolves between reduced (Fe^{++}) and oxidized (Fe^{+++}) states.

There are several kinds of cytochromes, such as cytochromes-A, cytochromes-B, cytochrome-C. Cytochrome-B-C1 complex is one of the three electron-driven proton pumps of the respiratory chain, accepting electrons from ubiquinone. Cytochrome oxidase complex, another electron-driven proton pump of the respiratory chain, accepts electrons from cytochrome-C and generates water using oxygen as an electron acceptor (NADH dehydrogenase complex is the third electron-driven proton pump). Electron-transport chains of the inner mitochondrial membrane generate a proton gradient across the membrane used for ATP synthesis. Cytochrome-C oxidase reduces oxygen to water to power the proton pump, generating only a small quantity of damaging oxidation products (peroxide, superoxide). Cytochrome-P450 (P450 refers to the "pigment at 450 nm," due to absorbance of light at wavelengths near 450 nm when heme iron is reduced) is a plasmalemmal oxidase of a multienzyme complex, which also includes NADPH–cytochrome-P450 reductase and cytochrome-B5.

Cytochrome-P450 targets (with oxygen and NADPH) cholesterol, vitamins, and steroids, among other substances. Certain cytochrome-P450 isoforms are specific, but many catalyze numerous chemical reactions. Cytochrome-P450 particularly metabolizes arachidonic acid to active products, thereby, determining a third pathway of arachidonic acid metabolism, in addition to cyclooxygenases and lipoxygenases. Epoxygenases of the arachidonic acid metabolism in the vasculature belong to cytochrome P450-2B,-2C, and -2J. ω-Hydroxylases are members of the family of cytochrome-P450-4A.

Cytokine: extracellular regulator of the innate and adaptive immune systems. Cytokines are pleiotropic, acting on different types of cells. Cytokines are redundant, different cytokines performing the same function. Cytokines are multifunctional, regulating different functions. Cytokines such as interferons acts via the cytokine receptor and the JaK–STAT pathway.

Cytoskeleton: dynamic structure of the cell that ensures cell stability and motility. It is responsible for the internal organization of the cell. It allows the cell to adapt to physical forces applied by its environment, transport substance in its cytoplasma, give birth to daughter cells, and move.

Dalton: unit of molecular mass, approximately equal to the mass of a hydrogen atom $(1.66 \cdot 10^{-24}\,\text{g})$.

Death Receptor: member the tumor necrosis factor receptor family. Ligand binding provokes death receptor aggregation and initiates the signaling cascade, leading to either inflammation or cell death.

Decorin: small cellular or pericellular matrix proteoglycan. It binds to collagen-1.

Deoxyribonucleic Acid (DNA): polynucleotide formed from deoxyribonucleotides, and made of four bases: adenine, cytosine, guanosine, and thymine. DNA stores hereditary information.

DNA Operon: organization of linked genes intervening in the same pathway for rapid temporal and spatial coordination of their expression (rare in eukaryotic genomes).

Desmin: myocyte-specific intermediate filament. It connects myofibrils to each other and to the plasmalemma.

Destrin: actin depolymerizing factor, which binds F-actin in a pH-dependent manner. It also binds and sequesters G-actin, hindering actin polymerization.

Diacylglycerol: signaling lipid produced by the cleavage of phosphoinositides, which activates protein kinase C.

Diacylglycerol Receptors: include protein kinase-C (PKC) and -D (PKD), chimerins, Ras guanyl-releasing proteins (RasGRPs), Munc13, and DAG kinases. PKCs are subdivided into isozymes: a first group, PKCα, PKCβI, PKCβII and PKCγ (calcium, phospholipid, and DAG-activated); a second one, PKCδ, PKCε, PKCη, PKCθ (calcium-insensitive, phospholipid-dependent, and DAG-responsive), and a third one PKCζ and PKCλ (calcium-insensitive and DAG-unresponsive).

Dicer: ribonuclease that cleaves double-stranded RNAs (dsRNA) and RNA hairpins into small inhibiting RNAs, either exogenous small interfering RNAs (siRNAs) associated with gene silencing triggered by foreign dsRNA, or endogenous microRNAs (miRNAs).

Dimerization: chemical association of two molecules. Heterodimerization is related to a substance interacting selectively with a non-identical protein to form a heterodimer.

Docking protein: moorage protein responsible for attaching other substances to molecular complexes (protein–protein docking) for protein–protein interactions, or membrane constituents involved in transport or motion.

Dopamine: neurotransmitter and neurohormone released by the hypothalamus (inhibit prolactin release from the anterior pituitary). As a member of the catecholamine family, dopamine is a precursor to adrenaline and noradrenaline. Dopamine is synthesized mainly by nervous cells and adrenal medulla. Dopamine is involved in behavior, sociability, salience, motivation, cognition, sleep, and motor activity.

Dynamin: cytosolic GTPase implicated in budding of clathrin-coated vesicles.

Dynein: large multimeric nanomotor protein that undergoes ATP-dependent motions along microtubules. Dynein is approximately four times bigger than myosin-2 and about 10 times larger than kinesin. Its motor domain belongs to the heavy chain, which comprises a ring of six ATPases associated with diverse cellular activities (Aaa) and two protruding domains, the tail and the stalk.

Dystrobrevin: forms the dystrophin-associated protein complex at the sarcolemma with dystrophin, dystroglycan, sarcoglycan, and syntrophin. Dystrobrevin-α binds syntrophin, which recruits signaling proteins. Dystrobrevin-β could interact with syntrophin and the short form of dystrophin.

Dystroglycan: laminin-binding component of the dystrophin–glycoprotein complex that bridges the subsarcolemmal cytoskeleton and extracellular matrix. Dystroglycan also acts as a receptor for agrin and laminin-2.

Dystrophin: cytoskeletal protein at the inner surface of myocytes. Dystrophin belongs to the dystrophin–glycoprotein complex, which links the F-actin cytoskeleton and extracellular matrix.

Eicosanoid: member of the family that includes the prostaglandins, thromboxanes, and leukotrienes. Eicosanoids are divided into two subgroups, the prostanoids, including prostaglandins and thromboxanes, which are synthesized by cyclooxygenase, and the leukotrienes by the lipooxygenase. Numerous stimuli activate phospholipase-A2, which hydrolyzes arachidonic acid, from which derive the eicosanoids.

Elastin: protein that forms extracellular extensible fibers.

Emilin: extracellular matrix glycoprotein acting in elastin deposition.

Endothelin (ET): a vasoconstrictor synthesized by the endothelial cell. ET binds to specific G-protein-coupled receptors on the vascular smooth muscle cell.

Enhanceosome: stable multimolecular protein–DNA complex including a set of partner transcription factors. Each transcription factor of the enhanceosome binds to target sites in the genome during a more or less short duration (transient or stable binding).

Enlargeosome: small cytoplasmic vesicle coated by annexin-2, which upon stimulation undergoes calcium-dependent fast transport.

Enzyme: a molecule, usually a protein, that commonly catalyzes a single reaction (reaction selectivity), operating on a single substance (substrate selectivity). Selective enzymes facilitate specific reaction paths. Enzymes are specialized proteins that drive specific chemical reactions. They convert specific substrates to the corresponding products. Enzymes catalyze chemical reactions by lowering the energy barriers between substrate and product, acting on transient structures formed in rate-limiting steps of chemical reactions, and thus increasing the rate of reaction.

Enzymes are able to change their conformation to properly function (the enzyme can be more active [positive allosterism] or less active [negative allosterism]). Each enzyme has a specific three-dimensional shape with an active site where the biochemical transformation takes place. The active site has two subsites, the binding site (enzyme reversibly binds to its substrate and forms an enzyme–substrate complex) and the catalytic site. In living cells, metalloproteins bind their metal enzymatic cofactors, such as zinc, copper, iron, and manganese. Most metalloproteins appear to be flexible. Decay in local concentration of specific metal or increase in level of another metal that can bind to the metalloprotein leads to metabolic disorders. Metal specificity is achieved via metallochaperone proteins [790]. Toxic Cu, a cofactor for mitochondrial, cytosolic, and vesicular (e.g., ATPase cation transporter of secretory vesicles) enzymes, is available without damage by Cu-handling protein Hah1.

The cell regulates the concentration of active enzyme by: (1) synthesis and degradation according to the cell need, or (2) phosphorylation/dephosphorylation at a specific amino acid residue, switching the enzyme activity on or off. Some enzymes are synthesized in an inactive form, the zymogen, which must be processed by a protease to become active. Isozymes are multiple types of an enzyme for the same catalytic reaction with different properties (affinity or conversion rate). Enzymes operate at an optimal temperature as well as at a tissue-dependent pH optimum. Enzyme inhibitors interfere, either reversibly or irreversibly, with the enzymatic reaction. In competitive inhibition, the inhibitor competes with the substrate for the active site and forms an enzyme–inhibitor complex (occupied active site). Non-competitive inhibition deals with reversible interactions between inhibitor and enzyme that do not occurs at the active site. Feedback inhibition from an effector (a downstream product of a reaction chain) can downregulate the enzyme involved in a rate limiting step (slowest part of the pathway) of the biochemical cascade.

There are many classes of enzymes. Oxidoreductases catalyze oxidation or reduction reactions (aldehyde dehydrogenase converts acetaldehyde to acetyl CoA). Oxidoreductases often require cofactors, such as NAD+, which accept hydrogens released during oxidation. Oxidases use oxygen as an acceptor molecule. Transferases move a functional group from a donor to an acceptor. Hydrolases (lipases, phosphatases, acetylcholinesterase, and proteases) break bonds and add water across the broken bond. Ligases and lyases catalyze the formation and cleavage of C-C, C-O and C-N bonds. Isomerases is involved in intramolecular rearrangements.

Enzymes are released during cell destruction. After myocardial infarctions, creatine phosphokinase, lactate dehydrogenase, α-hydroxybutyrate dehydrogenase and serum glutamic oxaloacetic transaminase are elevated.

Erythropoietin-Producing Hepatoma Receptors (Eph): form a family of receptor tyrosine kinases that associate with their plasmalemmal ligands, the ephrins. They mediate cell-to-cell signaling and regulate cell adhesion and migration, especially in vasculogenesis.

Eph Receptor Interactor (ephrin): transmembrane signaling protein with bidirectional role. Ephrins send forward signals via the activation of its receptor on another cell. Reverse signals use several pathways such as tyrosine phosphorylation of its cytoplasmic domain by Eph receptors, FGF receptors, or Src kinases, and recruitment of adapter growth factor-receptor-bound protein Grb4/Nck2, which increases focal adhesion kinase activity, GTP exchange factor PDZ-RGS3, and scaffold Disheveled.

Epoxyeicosatrienoic Acids (EET): are generated by CyP450 epoxygenase from arachidonic acid. Four types of epoxyeicosatrienoic acids exist: (5,6)EET, (8,9)EET, (11,12)EET, (14,15)EET. They are hydrolyzed by epoxide hydrolases (soluble epoxide hydrolase, mainly in the cytosol and peroxisomes, and microsomal epoxide hydrolase, which binds to the intracellular membranes) to dihydroxyeicosatrienoic acids. They exert autocrine effects on vascular endothelial cells, as well as paracrine activities on neighboring cells, such as vascular smooth muscle cells. They act on proliferation of endothelial cells, and thus on angiogenesis. These endothelium-derived hyperpolarizing factors increase the intracellular calcium concentration, subsequently activating Ca^{++}-activated K^+ channels of the smooth muscle cells, inducing cell hyperpolarization and vasodilation. They inhibit cytokine-induced inflammation by hindering leukocyte adhesion to the vessel wall.

Extracellular Signal-Regulated Kinase (ERK): member of the mitogen-activated protein kinase family of serine/threonine kinases. Effector of the Ras signaling pathway. ERK regulates cell adhesion and cytoskeletal dynamics. It is also involved in cell proliferation and differentiation.

Ezrin Radixin Moesin (ERM): linkers of the plasmalemma to the actin cytoskeleton. This scaffold proteic complex of adapters is involved in the localization and coordinated activity of apical transporters. Cytokine-induced adhesion molecules lead to the formation of the ezrin–radixin–moesin complex.

Fascin: actin-bundling protein.

Fibroblast Growth Factor (FGF): growth factor involved in embryogenesis, cell growth, tissue repair, and angiogenesis.

Fibronectin: extracellular matrix protein involved in cell–matrix adhesion (plasmalemmal integrins are receptors for fibronectin) and guidance of migrating cells. This glycoprotein is present in a soluble dimeric form in plasma.

Filamin: actin-binding protein that anchors membrane proteins for the actin cytoskeleton.

Filopodium: thin protrusion with an actin filament core generated on the leading edge of a migrating cell.

Flavin Adenine Dinucleotide (FAD): and reduced flavin adenine dinucleotide (FADH2) are involved in the tricarboxylic acid cycle.

Focal Adhesion Kinase (FAK): cytoplasmic tyrosine kinase at focal adhesions, which are associated with integrins.

Feline Yes Related Protein (Fyn): member of the Src family of protein tyrosine kinases. Activated Fyn binds to and phosphorylates Shc, thereby recruiting Grb2 and Sos for ERK activation.

Gap Junction: channels permeable to large hydrophilic solutes used for between-cell communication. They are formed by end attachment of two connexin hexamers (hemichannels or connexons) from each neighboring cells.

Gelsolin: protein that affects actin structures in response to external stimuli, in a calcium- and phosphoinositide-regulated manner.

Gene: specific sequences of bases (DNA segment linked to the chromosomes) yielding the basic unit of heredity, which contains the information necessary to produce RNA copy (transcription). It is made of a promoter, which controls the gene activity, and a coding sequence (exon), as well as noncoding regions (intron) that regulate gene expression. Introns and exons are determined during splicing. The gene must be transcribed from DNA to messenger RNA, then be translated from mRNA to protein (gene expression). A single gene can lead to the synthesis of various proteins by alternative splicing (pre-mRNAs are arranged in alternative ways.). An active gene forms a cluster with other active genes. A cis-regulatory element (cis-element; cis: on this side) is a region of DNA or RNA that regulates the expression of genes located on that same strand. A trans-regulatory element (trans-element; trans: across) regulates the expression of genes distant from the gene which produces it.

Growth Arrest-Specific Protein (Gas): component of the microfilament network. Gas2 colocalizes with actin fibers at the cell cortex and along the stress fibers.

Glycan: glycosylation provides a post-translational modification of proteins. Protein glycans have multiple roles. The glycan function varies according to its ligand and attachment site. Achievement of functions of certain proteins requires glycans. Glycans also allows recognition by lectins (complementary binding proteins), involved in homeostasis and body defense. Glycans promote protein folding into suitable tertiary and quaternary structures. Certain forms of glycans, such as glycolipids and proteoglycans, have structural roles.

Glycoprotein: composed of a protein and a carbohydrate (or glycan, such as glucose, glucosamine, galactose, galactosamine, mannose, fucose, or sialic acid). Carbohydrates are monosaccharides, disaccharides, linear or branched oligosaccharides, polysaccharides, or sulfo- or phospho-derivatives. A single, few, or many carbohydrate units can be covalently linked to the protein. Proteoglycans are a subclass of glycoproteins in which carbohydrates are polysaccharides (glycosaminoglycans) that contain aminosugars.

Glycosaminoglycan (GAG): highly charged polysaccharide mainly linked to a protein core of proteoglycans in the extracellular matrix.

Glycosylation: addition of saccharides to protein

Glycosylphosphatidylinositol Anchor (GPI): lipid linkage of membrane-bound proteins.

Glypican: plasmalemmal heparan sulfate proteoglycan.

Graf: GTPase-activating protein associated with focal adhesion kinase. Graf is located in focal adhesions and actin stress fibers. Graf targets Rho GTPases.

Growth Factor (GF): extracellular signaling protein that stimulates cell proliferation and other activities.

Growth Factor Receptor Binding Protein (Grb2): adapter that binds to receptors and other adapters, such as Shc, Sos, Gab1, and Gab2. Grb2 activates Ras GTPases.

Guanosine DiPhosphate (GDP): nucleotide that produces guanosine triphosphate by phosphorylation.

Guanosine TriPhosphate (GTP): nucleotide that releases free energy by GTPase hydrolysis of GDP. GTP is involved in microtubule assembly, protein synthesis, cell signaling.

G Protein: protein family of the intracellular portion of the plasma membrane that binds to activated receptor complexes. After conformational changes, binding, and hydrolysis of GTP, G proteins act in channel gating and/or in receptor pathways. Among G proteins, there are either stimulatory G (Gs) or inhibitory G (Gi) proteins. G-protein-coupled receptors mediate the cellular responses of signaling molecules (hormones, neurotransmitters, and local mediators). These receptors send messages of the extracellular ligands via their functional relationship to G proteins.

Gi protein regulates ion channels and inhibit adenylyl cyclase, whereas Gs protein activates adenylyl cyclase. Gq is another type that activates phospholipase-C and the phosphoinositide signaling pathway.

G-Protein-Coupled Receptor (GPCR): receptor coupled to an intracellular trimeric GTP-binding protein. Theses plasmalemmal receptors transmit signals to the cell and are thus involved in many physiological processes, after ligand binding and GPCR activation. Signaling is controlled by the regulator of G-protein signaling (RGS) proteins, which act on GPCRs to confer signaling specificity, a ligand being able to bind both RGS and GPCR. Subtle changes in either the structure of the ligand or receptor can switch the receptor on or off [791]. An amino acid component has been identified in the receptor binding site that determines whether a ligand behaves as an agonist or an antagonist.

GTP-Binding Proteins (Small GTPases): cycle between GTP-bound active and GDP-bound inactive states. GTP/GDP exchange is modulated by GTPase activating proteins (GAP; inactivation), by guanosine nucleotide-exchange factors (GEF; activation), such as Ras-GRF and Ras-GRP, and guanosine nucleotide dissociation inhibitors (GDI). Once activated for signal transduction, they interact with their downstream targets, which include the protein kinase Raf and the lipid kinase phosphatidylinositol kinase (PI3K).

Small GTPases include isoforms of Ras, Rap, Rho, Rab, Ral, Arf, and Ran. There are three Ras species: H-, N-, and K-Ras. Ras GTPases are activated by several guanine nucleotide exchange factors. Ras GTPases then interact with several downstream targets, such as Raf kinase and phosphoinositide 3-kinase. Rap GTPases are homologues of Ras. Rap1 interacts with the set of Ras effectors. Ras-related nuclear proteins Ran regulate nucleo-cytoplasmic transport across nuclear pore complexes. Ran-GTP is only generated in the nucleus. The Rho family has six different classes: Rho (RhoA, RhoB, and RhoC), Rac (Rac1, Rac2, and Rac3, or Rac1B or RhoG), Cdc42 (Cdc42Hs, Chp, G25K, and TC10), Rnd (Rnd1 or Rho6, Rnd2/Rho7, and Rnd3 or RhoE), RhoD, and TTF. Rho GTPases, RhoA, RhoB, and RhoC, although playing distinct roles, regulate the actin cytoskeleton, endocytosis, and transcription stimulated by extracellular signaling. Rnd proteins have antagonistic effects with respect to Rho and Rac. Effectors of Rho, Rac, and Cdc42 are Ser/Thr kinases, Rho kinases and p21-activated kinases, or function as scaffold proteins (WASP). Rho is activated by extracellular ligands such as lysophosphatidic acid and bombesin, inducing formation of actin stress fibers. Rac activated by growth factors and active Ras leads to formation of lamellipodia and, afterward, actin stress fiber. Cdc42 stimulates formation of filopodia (protrusions containing tight actin bundles along the lamellipodium). Rac and Cdc42 together induce the assembly of focal complexes. Rho kinases link to myosin light-chain phosphatase. Phosphorylation of MLC induces a conformational change in myosin increasing its affinity for actin filaments. Rac and Cdc42 effectors include PI3K, MAPKs, and p21-activated kinases. Phosphorylation of MLCK by p21-activated kinases inhibits MLCK and reduces actin stress fiber content. p21-Activated kinases phosphorylate (activate) LIM kinase. LIMKs phosphorylate and inactivate cofilin, which depolymerizes actin.

The pathway from Ras through Raf, mitogen-activated protein kinase and extracellular signal-regulated kinase regulates, together with: (1) many scaffolding proteins (which stimulate the pathway, such as kinase suppressor of Ras KSR1 and MAPK partner-1, or mediate crosstalk with other pathways); (2) adapters (which drive effectors to suitable cell locations, such as connector enhancer of KSR) for signaling distribution; and (3) inhibitors (e.g., Raf kinase inhibitor protein) is involved in many fundamental cellular processes. The Ras–ERK cascade recruits Raf from the cytosol to the plasmalemma to be activated. Downstream kinases and scaffold proteins translocate to the cell membrane also for activation.

The Rnd family, a subgroup of the Rho family, consists of three proteins: Rnd1 (Rho6), Rnd2 (Rho7), and Rnd3 (Rho8 or RhoE). Rnd proteins are always bound to GTP. Rnd proteins are expressed in the brain, Rnd1 in the liver as well, and Rnd2 in the testis. Rnd3 is found ubiquitously. In fibroblasts and epithelial cells, Rnd1 and Rnd3 proteins colocalize with cadherins in adherens junctions. Rnd proteins, interacting with RhoGAPs, inhibit Rho. Rnd1 inhibits the formation of stress fibers. Rnd3 can induce cell migration.

GTPase-Activating Protein (GAP): binds to a GTP-binding protein and stimulates its GTPase activity, hence inactivating it.

Guanine Nucleotide Exchange Factor (GEF): links a GTP-binding protein and allows it to bind GTP rather than GDP.

Harvey Rat Sarcoma Viral Oncogene Homologue (Hras): growth factor effector.

Heat-Shock Protein (Hsp): prevents irreversible protein aggregation. Once formed, complexes between substrates and Hsps are very stable at physiological temperatures. Release of proteins bound to Hsps requires ATP-dependent chaperones. Many Hsp families are inactive or partially active under physiological conditions [792]. They are specifically activated by stresses, such as elevated temperatures, helping the cell to survive the stress. Although members of the Hsp superfamily are diverse in composition and size, most share the same main features. Hsps assist the folding of proteins in the cytosol, maintain protein conformation suitable for refolding, and can refold denatured proteins. Therefore, they counteract any effect that can induce protein aggregation or instability.

Hedgehog (Hh): with its three human homologous compounds, SHh, Dhh, Ihh, are plasmalemmal morphogens. Hh proteins serve as ligands for Patched receptors (Ptc1 and Ptc2) of targeted cells which cause, via ac-

tivation of transcription factors Gli1, Gli2, and Gli3, either short range
signaling when tethered to the signaling cell plasmalemma, or long range
signaling when released from the signaling cell. The release is regulated by Dispatched protein. In the absence of Hh, Ptc1 inhibits effector
Smoothened. Protein kinase-A, glycogen synthase kinase-3, and casein
kinase-1 phosphorylate Gli, which becomes a repressor.

Hemojuvelin: molecule that belongs to the repulsive guidance molecule
(RGM) family, which also includes the bone morphogenetic protein core-
ceptors RGMA and RGMB. Hemojuvelin can act as a bone morphogenetic
protein coreceptor. The bone morphogenetic protein (BMP) upregulates
hepatocyte expression of hepcidin. Hemojuvelin mutations associated with
hemochromatosis lead to defective BMP signaling and decreased hepcidin
expression.

Heparan Sulfate Proteoglycans (HSPG): molecules composed of a protein and at least one heparan sulfate glycosaminoglycan chain. There are
three HSPG subsets: (1) the membrane-spanning proteoglycans (syndecans, β-glycan and CD44v3), (2) the glycophosphatidylinositol-linked proteoglycans (glypicans), and (3) the proteoglycans secreted in the extracellular matrix (agrin, collagen-18, and perlecan).

Heparan sulfate proteoglycans act as co-receptors for growth factor receptors. Heparan sulfates bind to multiple proteins, such as fibroblast
growth factors and their receptors, transforming growth factors, bone morphogenetic proteins, Wnt proteins, chemokines and interleukins, lipases,
apolipoproteins, proteins of the extracellular matrix, and plasma proteins.
HSPGs favor cell adhesion to the extracellular matrix. They transport
chemokines across cells. HSPGs collaborate with integrins to modulate
cell adhesion.

In the circulatory network, HSPGs modulate lipid metabolism functioning as lipase anchors. Endothelial lipases hydrolyze triglycerides of circulating chylomicrons and very-low-density lipoproteins. HSPGs cooperate
with LDL receptors. HSPGs of the smooth-muscle cells can transactivate
VEGF signaling in adjoining endothelial cells, enhancing the formation of
ligand–VEGFR2 complexes on endothelial cells.

High-Density Lipoprotein (HDL): a class of lipoproteins that have protective effects, removing cholesterol from the arterial wall. HDLs counteract the effects of oxidized LDL. HDLs also inhibit blood cell adhesion
to vascular endothelium. HDLs reduce platelet aggregability and coagulation.

Histamine: released by mast cells in response to inflammation and allergy
(after specific antigen binding to immunoglobulin IgE attached to plasmalemma receptors on mast cells). It causes arteriolar vasodilation, venous
constriction in certain regions, and increased capillary permeability. Histamine induces phosphorylation of VE-cadherin, creating gaps between
the endothelial cells. It also causes the contraction of smooth muscles of
the airway walls.

Hormone: substance produced by endocrine (ενδο-κρινω: inside secrete) glands and released into the blood circulation to exert remote effects on specific tissues.

Hydrolysis: cleavage of a molecular bond in association with water (a hydrogen atom H being added to one product and a hydroxyl group OH to the other).

Hydrophilicity: tendency of a molecule to be solvated by water (antonym: hydrophobicity).

Hypoxia-Inducible Transcription Factor (HIF): its two isoforms HIF1 and HIF2 adapt cells to hypoxia, increasing the gene expression for glucose metabolism, angiogenesis, hematopoiesis, vascular tone and cell survival. HIF1 and HIF2 non-redundantly bind to both the transcriptional coactivators CREB-binding protein and p300. CREB-binding protein and p300 interact with transcriptional regulatory proteins. They have protein and histone acetyltransferase activities. HIF also acts by another pathway involving histone deacetylase inhibitor trichostatin A. An interaction between coactivators CREB-binding protein and p300 and deacetylases is required for full HIF response [793].

Immunoglobulin Superfamily: family of proteins that contain immunoglobulin domains (a characteristic domain of a hundred amino acids) or immunoglobulin-like domains. Most are involved in cell–cell interactions or antigen recognition (five classes of immunoglobulin - IgA, IgD, IgE, IgG, and IgM - are implicated in the immune response).

Inflammasome: cytosolic proteic complex, particularly formed by adapter ASC and cryopyrin, which activate caspase-1 to process interleukins IL1β and IL18.

Integrin: heterodimeric plasmalemmal molecules involved in both cell–cell and cell–matrix interactions. This family of adhesion molecules is composed of multiple classes. β_1 Integrins bind to extracellular elements (collagen, laminin, fibronectin, and VCAM1); β_2 integrins to ICAMs, clotting factors, and C3b; (α_M), β_3 integrins to clotting factors in particular, β_4 integrins to laminin, β_5 integrins to vitronectin, β_6 integrins to fibronectin, and β_7 integrins to fibronectin and VCAM1. Cytoplasmic domains bind protein tyrosine kinases (FAK and Syk) and cytoskeletal proteins (talin and paxillin).

Integrin-Linked Kinase (ILK): serine/threonine kinase located in focal adhesions. It binds LIM-domain protein PINCH, phosphatidylinositol-(3,4,5)trisphosphate, paxillin, parvins, and β-integrins. ILK can phosphorylate glycogen synthase kinase-3, myosin light chain kinase, and protein kinase-B.

Interferon: antiviral molecule. There are different classes of such cytokines. The IFNs-I include IFNα (with subtypes, IFNα1, α2, α4, α5, α6, α7, α8, α10, α13, α14, α16, α17 and α21), IFNβ, IFNε, IFNκ, and IFNω in humans. IFN-I genes belong to chromosome-9. IFNs-I bind a common IFN-I receptor. IFN-I receptor is composed of two subunits, IFNAR1, associated with the Janus activated kinases (JAK), and IFNAR2, associated with tyrosine kinase-2 (TyK2) and JAK1. Activation of JAKs associated with IFN-I receptor results in phosphorylation of signal transducer and activator of transcription STAT1 and STAT2, leading to the formation of complexes with IFN-regulatory factor-9 (STAT1–STAT2–IRF9): the IFN-stimulated gene (ISG) factor-3 complexes (ISGF3). These complexes translocate to the nucleus and bind to IFN-stimulated response elements (ISREs) in DNA to initiate gene transcription. STAT5 is associated with TyK2 bound to IFNAR1. CrkL, a member of the Crk family of adapters associated with the guanine nucleotide-exchange factor (GEF) C3G, is activated by the engagement of IFN-I. The activated form of CrkL forms a signaling complex with STAT5, which also undergoes a phosphorylation. The CrkL–STAT5 complex translocates to the nucleus and binds to the IFNγ-activated site (GAS) for gene transcription. The phosphorylation of CrkL also stimulates C3G. C3G subsequently activates RAP1. RAP1 regulates the activation of p38 cascade.

There is a single type of IFN-II, IFNγ. The gene encoding IFN-II is located on chromosome-12. IFNγ binds the IFN-II receptor. This receptor is composed of two subunits, IFNGR1 and IFNGR2, which are associated with JAK1 and JAK2, respectively. Activated JAK1 and JAK2 phosphorylate STAT1. Both IFN-I and IFN-II induce the formation of STAT1–STAT1 homodimers which translocate to the nucleus and bind to IFNγ-activated site (GAS), initiating the transcription of ISGs. The IFNγ-activated JAKs also activate phosphatidylinositol 3-kinase. The activated PI3K activates protein kinase-Cδ, which in turn regulates an additional phosphorylation of STAT1, which is required for full transcriptional activation.

IFNλ1, IFNλ2 and IFNλ3 are also known as IL29, IL28A, and IL28B, respectively. IFNλ bind a different receptor, which is composed of two chains, IFNLR1 (IL28Rα) and IL10Rβ.

Interleukin (IL): cytokine expressed particularly by leukocytes, involved in cell communication, immune system function, and hematopoiesis.

Intermediate-Density Lipoprotein (IDL): produced from VLDLs. A part of its triglyceride content is hydrolyzed by hepatic lipase, producing LDLs.

Intracrin: compounds act on the producing cell without extracellular release (intra: within).

Intron: noncoding component of genes that interrupt protein-coding sequences. Introns are removed from the pre-mRNA transcript by RNA splicing.

JaK–STAT Signaling Pathway: biochemical reaction cascade that involves plasmalemmal receptors, cytoplasmic Janus kinases (JaK) and activators of transcription (STATs) to activate gene expression.

c-Jun N-Terminal Kinase (JNK): belongs to a subset of mitogen activated protein kinase family. There are three members, JNK1, JNK2, and JNK3, with isoforms. JNK1 and JNK2 are ubiquitiously expressed, whereas JNK3 is restricted to the brain, heart, and testis. JNK binds and phosphorylates the N-terminal activation domain of transcription factor cJun. JNK can be activated by cytokines and environmental stresses.

Kaptin: ATP-sensitive actin-binding protein

Kinesin: motor protein that undergoes ATP-dependent motions along microtubules. It transports organelles within cells and moves chromosomes during cell division.

Kinetochore: protein complex that connects the chromosome to the microtubules of the spindle; it is used during chromosome segregation.

Lamellipodium: wide protrusion supported by a mesh of actin filaments at the leading edge of a migrating cell.

Laminin: protein of basal laminae.

Leukotriene: lipid effector arising from oxidative metabolism of arachidonic acid through the action of the lipoxygenase, leading to the unstable leukotriene-A4 (LktA4). This intermediate is the substrate for leukotriene-A4 hydrolase and leukotriene-C4 synthase, producing LktB4 and cysteinyl leukotrienes, respectively. The leukotrienes can then be divided into two different classes: (1) the cysteinyl leukotrienes (CysLkt or peptidoleukotrienes), leukotriene-C4 (LktC4), -D4 (LktD4), and -E4 (LktE4); and (2) leukotriene-B4 (LktB4). LktB4 is a chemotactic substance for neutrophils. LktC4 is converted into LktD4, and then into LktE4. LktC4 and LktD4 are contracting agents of vascular smooth muscle cells. CysLkts promote plasmatic exudation in the microcirculation. CysLkts act via receptors (CysLktR1 and CysLktR2) coupled to G proteins, inducing calcium influx.

(Cell) Lipids: Triacylglycerols and cholesteryl esters are found in cytosolic lipid droplets and traveling lipoproteins [789]. The main part of cellular lipids serve as membrane materials. Glycerophospholipids include phosphatidylcholine (a major membrane constituent), phosphatidylethanolamine, phosphatidylserine, and phosphatidylinositol. Glycerophospholipids are synthesized in the endoplasmic reticulum and Golgi complex. Phosphatidylethanolamine is also manufactured in mitochondria. The endoplasmic reticulum and Golgi apparatus are nearly exclusively made of glycerophospholipids. Sphingolipids, such as sphingomyelin, form a solid gel phase at body temperature, which is fluidized by cholesterol, a third type of membrane component. Glycosphingolipids are synthesized in the Golgi complex. Glucosylceramide is produced on the cytosolic surface and others on the lumenal part of the Golgi apparatus. The ceramide transfer protein is required for sphingomyelin synthesis. Sphingolipids do not translocate across the membrane, whereas cholesterol moves between and across membranes. Cholesterol preferentially interacts with sphingolipids, especially sphingomyelin. The trans-Golgi network, its secretory organelles, and endosomes contain sphingolipids and cholesterol. Late endosomes and lysosomes are composed of lysobisphosphatidic acid. Cells use lipids not only for its membrane structure and for carriers, but also for relocation of membrane proteins during signal transduction. Specialized lipids include lysophosphatidic acid, lysobiphosphatidic acid, ceramide, diacylglycerol, and sphingosine-1-phosphate. Phosphoinositides regulate binding of cytosolic proteins to the membrane.

Lipid Translocators: such as ATP-binding cassette (ABC) transporters, translocate lipids across the plasma membrane.

Lipophilicity: affinity of a molecule for a lipidic environment.

Lipopolysaccharides (LPS): molecule made of a lipid and a polysaccharide. They are major components of the outer membrane of Gram-negative bacteria. They promote the synthesis of pro-IL1β, IL1RA, IL6, TNFα, COx2, NOS2, and matrix metaloprotease ADAMTS.

Low-Density Lipoprotein (LDL): large molecular complex, composed of a single protein, many cholesterols, and other lipids, aimed at transporting cholesterol through the blood to cells. This lipoprotein subfamily is richer in cholesterol and its esters than IDL. About 75% of LDL is taken up by the liver with its Apo B acting as the ligand to the receptor; 24% of LDL is delivered to other tissues. About 1% of LDL with oxidized apolipoprotein B is removed from the circulation by scavengers after spending abnormally long time in the circulation.

Matrix Metalloproteases (matrixins, MMP): secreted and plasmalemmal proteases involved in embryogenesis, tissue remodeling, wound healing, and angiogenesis.

Membrane Raft: regional component of the plasmalemma. Differences in lipidic composition of the plasmalemma can explain transport specificity [789]. Mixtures of sphingomyelin, phosphatidylcholine, and cholesterol can spontaneously segregate either into a liquid-ordered or -disordered phase. On the internal surface of the Golgi apparatus of polarized cells, phosphatidylcholine domains serve in the retrograde pathway to the endoplasmic reticulum, phosphatidylcholine–cholesterol domains in the basolateral route, and sphingolipid–cholesterol domains in an apical surface destination in polarized cells or for different types of transport. Different lipid environments can recruit a specific set of transport proteins. GPI-anchored proteins are assigned to sphingolipid–cholesterol rafts.

Metabolism: chemical processes involved in the cell life to generate energy and new molecules.

Microtubule: component of the cytoskeleton.

Microtubule-Associated Protein (MAP): protein that binds to microtubules and modifies their properties.

Mixed Lineage Kinase (MLK): serine/threonine kinase that activates the JNK and p38 pathways. It binds to Cdc42 and Rac GTPases.

Mitogen-Activated Protein Kinase (MAPK): set of dual-specificity kinases, which relay signals from the plasma membrane to the nucleus. MAPKs are classified into three main groups: (1) the extracellular signal-regulated kinases, ERK1 and ERK2; (2) the p38 family, p38α, p38β, p38γ and p38δ; (3) and the JUN amino-terminal kinases, JNK1, JNK2 and JNK3. Additional MAPKs, such as ERK3, ERK5, ERK7 and ERK8, are not classified into these sets. The activation of the different MAPKs is regulated by MAPK kinases (MAP2K). They phosphorylate MAPKs on both threonine and tyrosine sites which are specific for the distinct MAPK families. The activation of MAP2Ks is regulated by other upstream kinases, MAP2K kinases (MAP3K). Activation of MAP3Ks or MAP2Ks usually occurs downstream of small GTPases, which are regulated by guanine nucleotide-exchange factors (GEF), which in turn are substrates for receptor or non-receptor tyrosine kinases.

Multivesicular Body: transport vesicles of the stage between endosome and lysosome.

cMyc: oncogene encoding a transcription factor that is able to transform cells and, with Trail, promotes apoptosis. Epithelial organization prevents oncogenic transformation as well as apoptosis.

Myosin: motor protein family. Myosin-2, the sarcomeric myosin, have two heavy chains, each with a globular domain characterized by an ATP binding site and an actin-interacting region. Light chains (MLC), calmodulin or calmodulin-like proteins, wrap around the neck region of each heavy chain (MyHC). Each myosin-2 monomer binds two distinct light chains, essential and regulatory MLC. Tails of heavy chains associate in a helical coiled-coil rod. Intertwining heavy chains determine two distinct regions: (1) a pair of free parts (15 nm long and 9 nm wide) with globular heads (S1) and rod-like necks (S2); and (2) a coiled-coil domain ($\sim$ 150 nm long, 2 nm in bore). S1 has binding sites for ATP and Ca^{++}, S2 for MLCs. Proteolytic cleavage generates two fragments $\sim$ 80 nm from the end (C terminus) of MyHC: (1) the heavy meromyosin (HMM), which contains S1 motor domain and S2 neck associated with MLCs; and (2) the light meromyosin (LMM), two-thirds of the coiled-coil domain, aggregation part of MyHC. The tail domains can interact to form complexes of many myosins. Myosin monomers assemble into parallel myosin dimers, which associate to form a minifilament with an antiparallel arrangement.

Myosin-1 and -6 have only one heavy chain, with a tail ending in a globular domain. Myosin-5 has two heavy chains like myosin-2, but with a longer neck region, and a shorter coil region with a globular domain at the end of each heavy chain tail. The neck domain binds three couples of two distinct light chains.

The filament sliding during contraction requires a reaction cycle with successive steps. In the absence of calcium ions, tropomyosin locks the myosin binding site of actin. When calcium binds to troponin, troponin and tropomyosin are moved on actin and free the myosin binding site of actin. The myosin head then binds to actin forming a cross-bridge. Afterward, the myosin binds ATP and unbinds. The myosin binding site of actin is blocked. The myosin hydrolyzes ATP and undergoes a conformational change. ADP and inorganic phosphate dissociate from myosin. The myosin head rotates and can then slide along the actin filament to bind to a new actin site.

Myosins-1, -5, and -6 bind to membranes or macromolecules via their globular tail domains, myosins-1 and -5 with membranes of Golgi apparatus and Golgi-derived vesicles, myosin-6 with actin filaments. These myosin types are involved in motions of organelles and plasmalemma.

Myosins can be regulated by: (1) Ca^{++} and calmodulin; (2) phosphorylation of myosin light chains by either Ca^{++}–calmodulin-dependent myosin light chain kinase or Rho kinase; and (3) caldesmon in smooth muscle cells.

The Rho proteins regulate the actin cytoskeleton. Rac1 and Cdc42 activate WASP, which in turn activates Arp2/3 acting on actin polymerization. Rho promote formation of stress fibers.

Myosin Binding Protein-C (MBP-C): member of the family of immunoglobulins expressed in isoforms specific for cardiac and skeletal muscle. The cardiac isoform of myosin-binding protein-C can be phosphorylated by protein kinase-A, but slow and fast skeletal muscle isoforms do not depend on protein kinase-A. S2 segment of myosin of striated muscle binds myosin-binding proteins-C.

Myosin Light Chain Kinase (MLCK): Ser/Thr kinase that phosphorylates the regulatory light chain of myosin for contraction or cell motility. MLCK is stimulated by ERK and inhibited by PAK.

Natriuretic Peptides: include not only the cardiac hormones, the atrial natriuretic peptide (ANP), and the brain natriuretic peptide (BNP; isolated in the porcine brain and also synthesized by the heart), but also the C-type natriuretic peptide (CNP; isolated in the porcine brain), which is produced by the endothelial cells, and the urodilatin (renal natriuretic peptide). ANP and BNP act on guanylyl cyclase-A (GCAR) and CNP on guanylyl cyclase-B (GCBR) receptors.

Neuromodulator: substance that alone has no effect under basal condition, but that modulates neuronal activity. A neuromodulator can convey information to a neuron set. It either enhances or dampens the neuron activity. Certain neuromodulators can remain during a noticeable period in the cerebrospinal fluid; they then are able to modulate the activity of large cerebral regions. Some neurotransmitters, such as acetylcholine and serotonin, also are neuromodulators. Several neuropeptides and chemokines act as neuromodulators. Certain neuropeptide and chemokine receptors are located at non-synaptic sites. Neuromodulators can affect the induced activity of neurons without disturbing their basal function.

Neuropeptide: peptide made of 3 to 40 amino acids that is synthesized in the neuron of both the central and peripheral nervous system and transported through the axon. Neuropeptides serve as neurotransmitters with excitatory or inhibitory effect or neuromodulators. Many neuropeptides also operate as hormones. Neuropeptides are generally stored in large, dense-core vesicles in all cell regions (soma, dendrites, axon, and nerve endings). Some neurons can produce several types of neuropeptides. Neuropeptides have multiple effects, modifying information transmission, gene expression, local blood flow, and glial cell activity. Generally, neuropeptides act as specific signals between one neuron population and another. Neuropeptides can have prolonged actions. Neuropeptide receptors can be located outside synapses. The neuropeptide family (about 100 members) includes opioids (endorphins, enkephalins, dynorphins, substance-P, and octopamine), secretins (secretin, motilin, glucagon, vasoactive intestinal peptide, and growth-hormone releasing hormone), gastrins (gastrin and cholecystokinin), somatostatin, and other neurohormones (i.e., neurophysin, vasopressin, oxytocin, and melanin-concentrating hormone), galanin, ghrelin, agouti-related protein, cocaine and amphetamine regulated transcript, bradykinin, neuropeptide-Y, neurokinins, neurotensin, neuromedin-U, etc. (Table A.2).

Neurotransmitter: substance that conveys and modulates signals either between two neurons or between a neuron and another cell, such as a muscular, glandular, and sensory receptor cell. Neurotransmitter has the following features: (1) existence of precursors and synthesis enzymes in the presynaptic neuron; (2) location in nerve terminals; (3) abundant storage possibly with other neurotransmitters in nerve-terminal vesicles; (4) release or corelease after membrane depolarization; (5) electrophysiological effects; (6) binding to cognate pre- and postsynaptic receptors; and (7) inactivation after action. Classical neurotransmitters include acetylcholine, biogenic amines (catecholamines such as adrenaline, noradrenaline, and dopamine, indolamines, such as serotonin and melatonin), histamine, amino acids (excitatory glutamate, inhibitory γ-aminobutyric acid, aspartate, and glycine), purines (adenosine, ATP, GTP and their derivatives), neuropeptides (vasopressin, somatostatin, neurotensin, etc.), nitric oxide, and ions (zinc; Table A.3). Most neurons use a single molecule as neurotransmitter. Some neurotransmitters (e.g., noradrenaline) also function as hormones. Certain neurotransmitters, such as zinc, not only modulates the sensitivity to other neurotransmitters, but also enters into the post-synaptic cells by specific carriers. When action potential occurs, the rapid depolarization primes calcium influx which provokes vesicle transport to the synaptic membrane and release in the synaptic cleft. Neurotransmitters then diffuse across the synaptic cleft to bind to cognate receptor clusters. The receptors can be classified into ionotropic (ligand-gated ion channels) and metabotropic receptors (signal transduction via secondary messenger and effectors). Neurotransmitters generally affect the excitability of postsynaptic neurons by depolarization or by hyperpolarization (excitatory or inhibitory states). In the central nervous system, combined input from several synapses usually initiate an action potential. Neurotransmitters are removed from the synaptic cleft either by reuptake or degradation into a catabolite by a cognate enzyme.

Nicotine Adenine Dinucleotide (NAD^+): electron acceptor in oxidation–reduction reaction, whereas NADH is an electron donor that carries electrons for oxidative phosphorylation. NADH dehydrogenase complex, an electron-driven proton pump of the mitochondrial respiratory chain, accepts electrons from NADH.

Nicotine Adenine Dinucleotide Phosphate (NADP$^+$): carrier involved, with its reduced form NADPH, in biosynthesis.

NADPH Oxidase: enzyme made of plasmalemmal cytochrome-B558, which contains gp91phox and p22phox, and cytosolic regulatory subunits p47-phox and p67phox and Rac1 GTPase.

Netrin: member of a family of secreted proteins stored in the extracellular matrix. Three netrins (netrin-1–netrin-3) are structurally related to the

Table A.2. Selected neuropeptides. The hypothalamus secretes several releasing hormones (CRH, GHRH, GnRH, TRH, somatostatin, etc.), the hypophysis various hormones (ACTH, MSH, GH, FSH, LH, TSH, prolactin, endorphin, oxytocin, vasopressin). Other bioactive neuropeptides include gastrin, cholecystokinin, gastrin releasing peptide, motilin, tachykinins (substances-P and -K), natriuretic peptides, calcitonin gene-related peptide, vasoactive intestinal peptide, glucagon, pancreatic polypeptide, calcitonin, insulin, secretin, parathyroid hormone, melanin-concentrating hormone, etc.

	Main functions
Angiotensin-2	Release of aldosterone
	Vasoconstriction of arterioles
Bradykinin	Nitric oxide-induced vasodilation
	Increase of vascular permaeability
Cocaine- and amphetamine related transcript	Modulation of activity of leptin and neuropeptide-Y
Dynorphin	Inhibition of oxytocin secretion
Endorphin	Inhibition of substance-P
	Control of body's response to stress
Enkephalin	Suppression of substance-P2 release
Galanin	Regulation of neurotransmitter and hormone release (inhibition of acetylcholine and glutamate release)
Ghrelin	NOS3 activation
Neuromedin-B	Smooth muscle contraction
Neuropeptide-Y	Enhancement of effect of noradrenaline
	Regulation of circadian rhythms
	Regulation of peripheral vascular resistance
	Regulation of heart contractility
Neurotensin	Elevation in ileal blood flow
Nociceptin	Pain transmission
Nocistatin	Inhibition of nociceptin
Orexin	Trigger eating
Substance-P	Nitric oxide-induced vasodilation
	Stimulation of intestine motility, saliva production
Urocortin	Potent and long-lasting hypotensiion
	Increase of coronary blood flow

γ-chain of laminins, and netrin-4 to the β-chain of laminins. A subset of netrins includes netrin-G1 and netrin-G2, with glycosylphosphatidylinositol cell anchorage.

Nitric Oxide (NO): is produced by the endothelium by NO synthase (NOS) in response to the wall shear stress and ligands of G-protein-coupled receptors. NO activates the guanylyl cyclase in the vascular smooth muscle cells and induces SMC relaxation. NO inhibits platelet aggregation, leukocyte extravasation, and proliferation of endothelial and vascular smooth muscles cells. In the heart, NO enhances both perfusion and electromechanical coupling.

At low (nanomolar) concentrations, nitric oxide regulates iron-binding proteins, particularly soluble guanylyl cyclase. sGC Activation generates cGMP, which controls the expression of many processes such as vascu-

Table A.3. Selected neurotransmitters and their activity on blood circulation and respiration.

	Main functions
Acetylcholine	Transmission of nervous cues Decrease in heart frequency Release of NO and vasodilation Contraction of skeletal muscles Bronchoconstriction, increase in bronchial secretion
Adenosine	Vasodilation
Dopamine	Regulation of muscle tone Control of venous return
Gamma-aminobutyrate acid	Inhibitory neurotransmitter
Glutamic Acid (Glutamate)	Stimulatory neurotransmitter Conversion into GABA Precursor of proline, ornithine, arginine, and polyamines
Glycine	Inhibitory neurotransmitter Synthesis of many compounds
Neuropeptide-Y	Regulation of energy Augmentation of noradrenaline-induced vasoconstriction
Noradrenaline	Smooth muscle contraction Stimulation of cardiac activity
Serotonin	Vasoconstriction Temperature regulation Sensory perception

lar smooth muscle relaxation, angiogenesis, wound healing, inflammation, and cancer. At higher (micromolar) concentrations, NO generates reactive nitrogen species. NO and RNS target zinc ions, as well as cysteine residues, particularly, leading to MMP activation.

Noradrenaline (NAd; or norepinephrine): catecholamine that acts as both a neurotransmitter and a hormone released from the adrenal glands. It acts on both α and β adrenoreceptors.

Notch Pathway: intercellular signaling cascade in which both ligand and receptor are plasmalemmal molecules. Signal transmission is thus restricted to adjacent cells, requiring cell–cell contact via Notch receptors and their DSL ligands. Notch pathway regulates cell differentiation. Four Notch receptors (Notch1–Notch4) have been identified in mammals. Ligands include Jagged (Jag1 and Jag2), and delta-like (Dll1, Dll3, and Dll4). Delta-like-4 is implicated in vasculogenesis/angiogenesis, Notch pathway regulates sprouting and branching, influencing the formation of specialized endothelial tip cells at the leading edge of vascular sprouts.

Nuclear Factor-κB (NFκB): transcription factor inactive in resting cells, due to sequestration in the cytosol by its inhibitor IκB, is activated particularly by the tumor necrosis factor and interleukin-1. The nuclear factor-κB cascade especially acts in inflammatory responses.

Nuclear Hormone Receptor: transcription factor mostly working as a partner.

Nuclear Receptor: transcription factor that positively and negatively regulate development and homeostasis. Three classes can be identified according to the binding properties: (1) the steroid- and thyroid-hormone receptors, (2) the orphan receptors, and (3) the adopted orphan receptors, with known ligands. Adopted orphan receptors, which are regulated by paracrine or autocrine ligands, include peroxisome-proliferator-activated receptors (PPARα, PPARβ, and PPARγ) and liver X receptors (LXRα and LXRβ). Retinoic-acid receptors, PPARs and LXRs bind to DNA with retinoid X receptor RXRα, RXRβ, or RXRγ. PPARs can be activated by fatty-acid metabolites produced during inflammation (leukotriene B4 and 8S-hydroxyeicosatetraenoic acid bind PPARα, 15-hydroxyeicosatetraenoic acid, 13-hydroxyoctadecadienoic acid bind PPARγ.) [795]. LXRs are stimulated by oxysterols (24S-hydroxycholesterol, 22R-hydroxycholesterol). PPARs are expressed by macrophages, B and T lymphocytes, dendritic cells, endothelial cells among others. LXRs work with the sterol-regulatory-element-binding protein family of sterol-regulated transcription factors to control the cellular cholesterol level. LXRs stimulate the production of ATP-binding cassette transporters ABCA1 and ABCG1 for efflux of cholesterol to extracellular apolipoprotein-A1 and high-density lipoproteins. Peroxisome-proliferator-activated receptors and liver X receptors control in a combinatorial manner inflammatory programs of gene expression in implicated leukocytes. LXRs control macrophage survival.

Nuclear Reprogramming: four main strategies used for nuclear reprogramming include: (1) nuclear transfer (injection of a somatic nucleus into an enucleated oocyte, which can lead to a reproductive cloning), (2) cell fusion of differentiated cells with pluripotent embryonic stem cells, (3) exposure of somatic cells or nuclei to cell extracts from oocytes or embryonic stem cells, (4) explantation of germ cells in culture cells that have regained pluripotency [796].

Nucleoside: glycosylamine made of a nucleobase, or base, and ribose. Nucleosides include adenosine, cytidine, guanosine, thymidine, and uridine in particular. Deoxynucleosides contain deoxyribose rather than ribose.

Nucleosome: package of DNA and histones. Histone modifications, such as methylation, control DNA transcription, replication, and repair. During transcription, the transit of RNA polymerase across the transcription unit is preceded by histone post-translational modifications to open the chromatin by transient displacements of nucleosomes.

Nucleotide: modified nucleoside, which contains a base, a sugar, and one or more phosphate groups (mono-, di-, or triphosphate compound). Nucleosides are phosphorylated by specific kinases in the cell. The sugar is deoxyribose or ribose.

Oligomer: small polymer composed of amino acids (oligopeptides), glucids (oligosaccharides), or nucleotides (oligonucleotides).

Osteopontin (Opn): phosphoprotein implicated in bone resorption and ventricular remodeling, regulating motility of various cells (osteoblasts, osteoclasts, fibroblasts, and macrophages), but also mediates TLR9-dependent interferon synthesis by plasmacytoid dendritic cells and TH1 immune responses [797]. Osteopontin is either secreted as a cytokine or kept in the synthesizing cells (T lymphocytes, dendritic cells, and macrophages) for nuclear translocation and gene expression, in an intracellular location different from the secreted Opn one. Plasmacytoid dendritic cells, innate immune system cells different from conventional dendritic cells according to plasmalemmal markers, release interferon-1 under the control of Toll-like receptors TLR7 and TLR9. The engagement of TLR9 on plasmacytoid dendritic cells produces interferons that activate other cells. The interferon synthesis by plasmacytoid dendritic cells particularly increases cross-presentation by conventional dendritic cells. Although TLR9 is expressed in both plasmacytoid and conventional dendritic cells, upregulation of osteopontin expression is only observed in plasmacytoid dendritic cells.

Oxidation: loss of electrons from an atom, as during oxygen addition or hydrogen removal.

Oxidative Phosphorylation: ATP formation in mitochondria driven by the transfer of electrons from metabolites to oxygen.

p21-Activated Kinase (PAK): serine/threonine kinase that binds to and is activated by Rho GTPases, leading to autophosphorylation and activation. PAK is involved in actin polymerization and cytoskeletal dynamics.

Palmitoylation: attachment of palmitoleic acid to cysteine residues of membrane proteins.

Paracrin: substances act on adjacent cells ($\pi\alpha\rho\alpha$: close to). Hence, paracrin signaling corresponds to a short-range cell–cell communication.

Pathway: reaction set, in which a stable form of a protein is activated to become an enzyme, which then catalyzes the next reaction of the cascade. The location of signaling effectors affects signaling magnitude and time constant (Table A.4). A-kinase anchoring proteins form signaling complexes for both spatial regulation at specific sites and temporal control by uploading or downloading enzymes for signal transduction or termination. Pathways leading to all-or-none responses impart qualitative information, i.e., the presence or absence of a signal. Simple spatial and temporal nature of the biochemical cascade is then required. At the opposite, signaling pathways must be precisely tuned to encode information on quantitative features of the stimulus which can occur over a wide range of magnitudes and time scales (both the dose and duration of a cue must lead to the adapted cell response). The pathway architecture must then be characterized by different spatial and temporal patterns, possibly associated with several branches of biochemical cascades. Also, the plasmalemmal concentration of cognate signaling effectors organized in nanodomains in association with scaffold proteins increase the efficiency and specificity of signaling. The higher the ligand concentration, the larger the number of transient effector clusters (linear relationship). The regulated motion of scaffold proteins can control the extent of clusters and hence modulate signaling magnitude.[1]

Pathway Specificity: cell strategy to prevent unwanted cross-talks between a set of signaling pathways that share common components (especially mitogen-activated protein kinase modules) in the cytosol and prime distinct cellular responses. Cross-activation followed by downstream mutual inhibition (cross-inhibition), between-pathway insulation by different proteic complexes, sequestration by scaffold proteins, feedback loops, and

Table A.4. Signal sensitivity and subcellular location of effector module (Source: [799]). Raf-MAP2K-ERK cascade serves as pathway hub that governs cell fate.

	Plasmalemma	Cytosol
Activation threshold	Low	High
Complex lifetime	Long	Short

---

[1] Ras GTPase activation in association with scaffold proteins (galectins) triggers the formation of signaling clusters that activate Raf kinase and the MAP2K–ERK module [798]. The regulated motion of scaffold proteins and signaling effectors can control the extent of Ras clusters and hence modulate MAPK signaling magnitude.

localization of signaling effectors (such as plasmalemmal nanodomains of Ras GTPases) allow pathway specificity. Dynamical behavior, i.e., signaling activation and deactivation rates and signaling duration downstream from common components, can also achieve signal specificity.

Pattern-Recognition Receptor (PRR): microbial sensor. The family of pattern-recognition receptors includes plasmalemmal and cytosolic receptors. Components of bacteria and viruses can enter into the cytoplasm. They are recognized by cytosolic receptors. The cytosolic receptors are classified into two main families: the nucleotide-binding oligomerization domain-like receptor family (NLR; with NALP, IPAF, and NOD proteins) and the RIG-I-like receptor family (RLR; with retinoic acid-inducible gene RIG-I and melanoma differentiation-associated gene MDA5). Plasmalemmal receptors include the set of Toll-like receptors, lectin-like family molecules (mannose receptors and β-glucan receptors), immunoreceptor tyrosine-based activation motif-associated receptors, Fc receptors (FcR), triggering receptors expressed on myeloid cells (TREM), and osteoclast-associated receptors (OSCAR). Immune response are efficiently triggered when sets of pattern-recognition receptors made of Toll-like receptors and other receptor types are stimulated [800]. The combined activation of different receptors leads to complementary responses that modulate innate and adaptive immunity.

Paxillin: protein of focal adhesions. It binds β-integrin, protein tyrosine kinases (Src and FAK), structural proteins (vinculin and actopaxin), regulators of actin organization (PIX and PKL), and, once phosphorylated, adapter Crk.

Paxillin Kinase Linker (PKL): (*G-Protein-Coupled Receptor Kinase Interacting Protein* [GIT]) member of a set of Arf GTPase activating proteins. It binds to paxillin and G-protein-coupled receptor kinase. It also can bind focal adhesion kinase.

Paxillin Kinase Linker-Interactive Exchange Factor (PIX): (*Cloned-Out Of Library* [COOL]) guanine nucleotide exchange factor of focal adhesions.

Perlecan: heparan sulfate proteoglycan of the basement membrane involved in blood vessel growth.

Peroxisome: membrane-bounded intracellular organelle that carries out oxidative reactions in lipid metabolism.

Phosphatidylinositol 3-Kinase (PI3K): enzyme involved in intracellular signaling pathways that phosphorylates phosphoinositides.

Phosphodiesterase (PDE): enzyme family that hydrolyzes second messengers cAMP and cGMP. Twenty-one PDE genes are classified into eleven families.

Phosphoinositides (PI; inositol phospholipids): contain phosphorylated inositol derivatives. Phosphoinositides are implicated in cell signaling. They also serve as labels to define organelle identity, as recruiters and regulators of cytoskeletal and membrane dynamics. Phosphoinositides are thus involved in cell transport, cell adhesion, and motility. Phosphoinositide metabolites, inositolpolyphosphate 3 (IP3), diacylglycerol (DAG), and arachidonic acid, which derive from degradation of phosphatidylinositol(4,5)biphosphate by phospholipase-C and phospholipase-A2, respectively, are second messengers. IP3 acts on Ca^{++} dynamics. PI(4)P and PI(4,5)P2 are the most abundant phosphoinositides [801]. PI(4)P is the predominant phosphoinositide in membranes of the Golgi complex and its secreted vesicles. PI(3)P and PI(3,5)P2 are selectively concentrated on early and late endosomes respectively. PI(4,5)P2 and PI(3,4,5)P3 are located in the plasma membrane. PI(4,5)P2 and DAG are involved in exocytosis. Phosphoinositides bind to endocytic proteins, like clathrin adapters of the plasma membrane AP2 and AP180, epsin, Hip1/Hip1R and ARH/Dab. Clathrin adapters that do not function at the plasma membrane bind other phosphoinositides. AP1 binds PI(4)P on Golgi membranes. Dynamin also binds to PI(4,5)P2. Furthermore, phosphoinositides, particularly PI(4,5)P2 and PI(3,4,5)P3, regulate actin polymerization.

Phospholamban: protein of cardiomyocytes and, at lower levels, of slow-twitch skeletal muscle and smooth muscle cells. Its phosphorylation by different PKs might be responsible for lusitropic and inotropic effects of β-agonists. Phospholamban prevents SERCA2 functioning. It is phosphorylated by protein kinase-A, which is regulated by G-protein-coupled receptors, and calcium–calmodulin-dependent kinase-2. Its inhibition of SERCA2 is subsequently released.

PhosphoLipase A2 (PLA2): hydrolyzes membrane phospholipids and then produces several phospholipid metabolites such as lysophospholipids, arachidonic acid, eicosanoids, and platelet-activating factor.

PhosphoLipase C (PLC): enzyme bound to the cytoplasmic edge of the plasmalemma that forms inositol(1,4,5)trisphosphate (IP3) and diacylglycerol (DAG) from plasmalemmal phosphatidylinositol(4,5)bisphosphate.

Phospholipids: formed from four components, fatty acids, a negatively charged phosphate group, alcohol and either glycerol (glycerophospholipid or phosphoglyceride), or sphingosine (sphingomyelin).

Plakoglobin: protein of desmosomes and intermediate junctions. It forms complexes with cadherins.

Plasmerosome: plasmalemma microdomains and the adjoining junctional endoplasmic reticulum, separated by a tiny junctional space, form plasmerosomes, which communicate by Ca^{++} signals. Ion concentrations in the junctional space can transiently differ from those in the remaining cytosol. The plasmerosomes facilitate Ca^{++} transfer between the stores and cytosol. In cardiomyocytes, sarcolemmal microdomains contain dihydropyridine receptors and the adjacent sarcoplasmic reticulum ryanodine receptors, which interact for excitation–contraction coupling. Structural and functional relationships between plasmalemma and junctional endoplasmic reticulum exist in other cell types, such as arterial smooth muscle cells. These buffer barriers control the release of Ca^{++} from endoplasmic reticulum stores, as well as the refilling of stores via plasmalemmal store-operated Ca^{++} channels. The plasmerosome serves to carry Ca^{++} between the extracellular fluid and endoplasmic reticulum. It can regulate Ca^{++} concentration in the stores, Na^+/Ca^{++} exchange modulating the process.

Pleckstrin Homology Domain (PH domain): allows intracellular signaling proteins to bind to phosphoinositides phosphorylated by PI3K.

Plexins: have intracellular domains homologous to Ras GTPase-activating proteins. There are four plexin types (A–D), which include four A-type, three B-type, one C-type, and one D-type.

Porin: the most abundant protein of the outer mitochondrial membrane. Porins are transmembrane proteins that allow passive diffusion, forming non-selective pores through the outer membrane.

Profilin: actin monomer-binding protein. Profilin binds to numerous profilin binding proteins. It can interact with phosphoinositides, such as phosphatidylinositol bisphosphate (PIP2). It can thus also be involved in the cell signaling cascade.

Prostacyclin (PGI2): an endothelium-derived vasodilator that binds to IP receptors (Gs-coupled prostaglandin receptors for prostacyclin) on the vascular smooth muscle cells to induce SMC relaxation. PGI2 is synthesized from arachidonic acid by cyclooxygenase (COx), which has different isoforms, a c COx1 and an inducible COx2. PGI2 inhibits platelet aggregation.

Prostaglandin (PG): member of the eicosanoid family. Their formation from arachidonic acid is catalyzed by cyclooxygenase (PGG/H synthase and COx). There are several isoforms of COxs. Constitutive COx1 is expressed in almost all tissues. Inducible COx2 is produced in response to cytokines, growth factors, etc., at inflammation sites, and in monocytes, macrophages, neutrophils, and endothelial cells. Prostaglandins were first discovered in seminal fluid and assumed to be produced in the prostate (hence the name). They are generally quickly degraded. Prostaglandins have various functions. Prostaglandins are mediators of inflammation and thrombosis. PGD2, PGE2, and PGI2 induce vasodilation, whereas PGF2, PGH2 (a precursor to thromboxanes A2 and B2), TXA2, and TXB2 generate smooth muscle contraction. PGE2, PGH2, and TXA2 induce and PGD2 inhibits platelet aggregation. The four Gs-coupled prostaglandin receptors (EP2, EP4, IP for prostacyclin, DP for PGD2) and the four Gq/Gi-coupled receptors (EP1, EP3, FP for PGF2, and TP for TXA2) are homologous.

Protein: polymers made of amino acids. The number of amino acids can range from two to about 50 (peptides), or more, up to several thousands. Proteins are characterized by their structure and configuration. The primary protein structure is defined by the sequence of amino acids. The secondary structure is built by hydrogen bonding, usually leading to α helix and β sheet. The tertiary structure corresponds to molecule folding. Is is stabilized by hydrogen and disulfide bonds. The quartenary structure is observed in multi-subunit compounds by the stable association of two or more folded polypeptides. Proteins act as structural components, molecular motors, enzymes, carriers, receptors, messengers, transcription factors, antibodies (immunoglobulines), growth factors, clotting factors, etc.

Protein Kinase: adds phosphate groups to target proteins, leading to a functional change of the proteic substrate associated with modifications in enzyme activity, cellular location, or association with other proteins, especially during signal transmission. Such phosphorylation occurs by removing a phosphate group from ATP, which is converted to ADP. Their activity is regulated, being switched on by phosphorylation, either by the kinase itself (autophosphorylation) or by activators. Most kinases act on both serine and threonine, others act on tyrosine, and some on the three residues. Protein kinases are thus divided into three families: protein tyrosine (Tyr) kinases (PTK), protein serine/threonine (Ser/Thr) kinases (PSTK), and dual-specificity protein kinases (TTK), located in centrosomes and kinetochores, which phosphorylate all three amino acids. (TTK is required for mitosis arrest in response to spindle defects and kinetochore defects [794].) MAP2K is a mixed Ser/Thr and Tyr kinase.

Serine/threonine protein kinases phosphorylate proteins. They can be regulated by numerous signals, such as cAMP, cGMP, diacylglycerol, and calcium–calmodulin. The Ser/Thr protein kinase family includes phosphorylase kinase, protein kinase-A, protein kinase-B,[2] protein kinase-C, calcium–calmodulin-dependent protein kinases, mitogen-activated protein kinases, and Raf kinases.

Protein Kinase A (PKA): cAMP-dependent protein kinase that phosphorylates target proteins in response to a rising level in intracellular cAMP.

Protein Kinase B (PKB): a serine/threonine kinase that has three isoforms: PKB1, PKB2, and PKB3. PKB is activated by growth factor receptor. PKB is involved in cell survival and proliferation signaling via a phosphatidylinositol 3-kinase pathway. PKB blocks the activity of the nuclear factor of activated T cells (NFAT), promoting its ubiquitination. PKB suppresses cell motility, especially tumor cell migration.

Protein kinase C (PKC): calcium-dependent protein kinase activated by diacylglycerol that phosphorylates target proteins on specific serine and threonine residues.

Proteoglycan: element of the extracellular matrix composed of a core protein and one or more (up to about 100) glycosaminoglycans. The glycosaminoglycans of the proteoglycans are linear polymers of repeating disaccharide units (up to about 200). Proteoglycans can also contain one or more oligosaccharides. Proteoglycans are either extracellular matrix proteins or

---

[2] Protein kinase-B is also known as AKT kinase.

integral transmembrane proteins. The proteoglycans include chondroitin sulfates, dermatan sulfate, heparin, heparan sulfate, and keratan sulfate. Members of the three families of heparan sulfate proteoglycans - syndecan, perlecan, and glypican - can serve as storage of growth factors in the basement membrane.

Proteolysis: process executed by proteases that cuts peptide bonds, hence creating irreversible protein modifications. Proteolytic cascades occur during blood coagulation, fibrinolysis, gastrulation, and apoptosis. At each stages, an activated protease stimulates the zymogen of the next protease acting in series. Such a set of serial reactions can amplify the triggering signal and add regulatory nodes throughout the pathway.

Purinergic (Nucleotide) Receptor: ubiquitously expressed receptors for either adenosine (P1 receptor) or adenosine and adenosine/uridine di/triphosphate (P2 receptor).

Rab: small GTPase of the plasmalemma and organelle membranes involved in specific vesicle docking. Rab GTPases control transport fluxes. They can be regulated by guanine-exchange factor complex TRAPP.

Rac: small GTPase of the Ras family. Guanine nucleotide exchange factors, GTPase activating proteins, and guanine nucleotide dissociation inhibitors determine its activity by regulating of the ratio of GTP/GDP bound form.

Raf: mitogen-activated protein kinase kinase kinase (MAP3K) that acts downstream from Ras GTPases. Raf1 kinase phosphorylates (activates) MAP2K1 and MAP2K2, which in turn phosphorylate (activate) ERK1 and ERK2.

Free Radical: an unstable reactive molecule having unpaired electrons. Free radicals can be endogenous, from the mitochondrial respiratory chain (oxygen forming superoxide) or exogenous. Free radicals react with lipids, proteins, and DNA.

Ral: small GTPase that is activated by the guanine nucleotide-exchange factor RalGEF and interacts with filamin. Ral regulates vesicle transport.

Ran: GTPase of the cytosol and nucleus required for transport through nuclear pores.

Rap: small GTPase activated by the calcium and cAMP pathway.

Ras: GTPase relaying signals from plasmalemmal receptors to the nucleus. The family is composed of N-Ras, K-Ras, and H-Ras. Ras is anchored at the plasma membrane. Ras is activated by Son-of-sevenless or Ras GDP releasing factor. It binds to phosphoinositide 3-kinase.

R-Ras subset of the Ras family includes R-Ras1, R-Ras2, and R-Ras3. R-Ras are attached to the plasmalemma and stimulated by the activated integrin–FAK–p130Cas–Crk pathway, which recruits R-Ras guanine nucleotide-exchange factors. R-Ras modulates JNK.

Reactive Oxygen Species (ROS): radicals (superoxide) and nonradical reactive oxygen derivatives. Some are highly toxic. Reactive oxygen species are major regulators of cell senescence. Many of the reactive oxidative species are formed during normal metabolic activity, but their concentrations rise under oxidative stress. An oxidative stress occurs when ROS cannot be eliminated. The overproduction of ROS leads to the oxidation of cellular macromolecules, leading to lipid peroxidation, protein damage, and DNA lesions (mutations). Oxidative stresses induce degenerative diseases. Antioxidants regulate oxidative reactions. Preventive antioxidant proteins (albumin, transferrin, myoglobin, ferritin, etc.) hamper ROS formation. Scavenger lipid-soluble and water-soluble antioxidants (vitamin C, E, α- and β-carotene, superoxide dismutase, hydrogen peroxide, and glutathione peroxidase) remove ROS.

Reactive Nitrogen Species (RNS): radical nitrogen-based molecules facilitating nitrosylation reactions. Non-neutralized RNSs induce nitrative stresses.

Receptor: molecule on the cell surface that specifically binds a ligand (first messenger).

Receptor Interacting Protein Kinase (RIP): regulator of cell survival and death that constitutes a family of seven members (RIP1–RIP7). They are serine/threonine kinases. RIP1 has a death domain and RIP2 a caspase recruitment domain. RIP3 interacts with RIP1.

Receptor Tyrosine Kinase (RTK): transmembrane protein acting as receptor for growth factors and hormone (insulin). Ligand binding leads to multimerization and autophosphorylation (activation). RTKs bind enzymes (phosphoinositide 3-kinase and phospholipase-Cγ) or adapters (Shc, Grb2, IRS1, and GAB1), which recruit enzymes involved in signal transduction.

Reduction: electron addition to an atom, as during hydrogen addition or oxygen removal.

Regulator of G-Protein Signaling (RGS): protein that accelerates GTP hydrolysis (inactivating G proteins and GPCR-induced signaling). Negative regulators of G-protein signaling form a family of proteins, which have different structures (but share a RGS domain or RGS-like domain, with GTPase activity), expression, and function [802]. Certain RhoGEFs contain a RGS-like domain. Many different RGSs are expressed by cardiomyocytes. RGS4 have antihypertrophic effects. RGS4 regulates G-protein-gated inward rectifier K^+ channels. RGS4 also inhibits angiogenesis. RGS5 is expressed in vascular endothelial cells. RGSs are substrates of different protein kinases for fine-tuned signaling. Certain RGSs inhibit endothelin-1-stimulated PLC activity via Gq- and Gq/11-mediated contractile effects. Phosphatidylinositol(1,4,5)trisphosphate inhibits RGS. PIP3 binds to RGS4 and RGS16 and inhibits their GAP activity at low $[Ca^{++}]_i$, whereas Ca^{++}–calmodulin, which binds to the same site on RGS4 as PIP3, restores its GAP activity.

Remnant Lipoproteins: are produced by the catabolism of triglyceride-rich lipoproteins made by enterocytes and hepatocytes, with subsequent relative cholesterol enrichment by the loss of triglycerides.

Renin–Angiotensin–Aldosterone System (RAAS): regulates blood volume and, subsequently, arterial pressure. Sympathetic stimulation, acting via β1-receptors, decreased renal arterial pressure, and reduced sodium delivery to the distal tubules, stimulates renin release by the kidney. Renin acts on *angiotensinogen* to form *angiotensin-1*. The vascular endothelium, particularly in the lungs, has the *angiotensin converting enzyme*, which yields *angiotensin-2*. Angiotensin-2 is a vasoconstricor via specific receptors. It releases aldosterone from the adrenal cortex, which in turn acts upon the kidneys to increase sodium and fluid retention. It stimulates the release of vasopressin from the posterior pituitary, which increases fluid retention in the kidneys. It facilitates noradrenaline release from sympathetic nerve endings and inhibits NAd reuptake by nerve endings.

Residue: unit of a polymer

Respiration: cellular oxidative process that produces CO_2 and H_2O. The respiratory electron-transport chain of the inner mitochondrial membrane with its respiratory enzyme complex receives electrons from the Krebs cycle and generates the proton gradient across the membrane for ATP synthesis.

Rho: small GTPase. The family includes Rho, Cdc42, and Rac1. Rho GTPase regulates actin cytoskeleton organization and assembly, in particular actin stress fibers and focal adhesion formation.

Rho Kinase (RoK): serine/threonine kinase. Its activation, which depends on integrins, induces interaction with Rho and translocation of Rho kinase to the plasmalemma. Rho kinase enhances the formation of focal adhesions and the assembly of actin stress fiber. Rho kinase binds with and phosphorylates (inhibits) myosin phosphatase.

Ribonucleic Acid (RNA): polymer made of ribonucleotides, composed of four bases, adenine, cytosine, guanine, and uracil. RNA is the DNA precursor during evolution. RNAs can be single-stranded molecule with a much shorter chain of nucleotides than DNA. However, most biologically active RNAs (transfer RNA, ribosomal RNA, small nuclear RNA, and other non-coding RNAs) are not single-stranded. These structured molecules then achieve chemical catalysis. RNA can catalyze its replication and the synthesis of other RNAs. Riboenzyme is a RNA-enzyme. Hairpin loops of large RNAs can be sources of RNA folding and be used as binding sites for interactions with other nucleic acids and proteins. Furthermore, hairpin loops confer stability to RNA.

RNA-Binding Proteins (RNABP): co-regulators of mRNA splicing, export, stability, location, and translation.

RNA Interference: gene activity reduction or suppression due to the processing of double-stranded RNA. RNA interference also stabilizes the genome by keeping transposons (mobile elements of the genome) silent. This RNA-dependent gene silencing process is initiated by short double-stranded RNAs which form RNA-induced silencing complex (RISC).

RNA Regulon: (RNA operon) coordinator of mRNA post-transcriptional events.

RNA Splicing: factor of gene regulation that removes introns from coding mRNA sequences. RNA splicing is either favored or impeded by spliceosome components when the cell is subjected to stress.

Double-Stranded RNA (dsRNA): RNA with two complementary strands. Trimmed into double-stranded small interfering RNAs, it initiates RNA interference.

Guide RNA (gRNA): RNA that guides mRNA modifications via RNA-protein complex.

Messenger RNA (mRNA): RNA synthesized from a gene sequence that contains the information on amino acid sequence for protein synthesis. Messenger RNAs carry the information to the ribosome. Each codon of three bases encodes for a specific amino acid, except the stop codon, which terminates protein synthesis. Coding regions begin with the start codon and end with the stop codon. Segments of pre-mRNA coding regions can serve as enhancers or silencers of exon splicing. Untranslated regions of a given transcript are RNA sequences located upstream from the start codon and downstream from the stop codon for mRNA stability and location. Moreover, proteins that bind to untranslated regions affect translation by interfering with the ribosome binding to mRNA or modifying mRNA structure.

MicroRNA (miRNA): small inhibitory RNA (single-stranded) that undergoes post-transcriptional modification from single-stranded precursor pre-miRNA to regulate gene expression by base-pairing to mRNA, leading to either mRNA degradation or suppression of translation.

Non-Coding RNA (ncRNA): RNA that is not translated into a protein, such as transfer RNA (tRNA) and ribosomal RNA (rRNA).

Ribosomal RNA (rRNA): constituent of ribosomes, synthesized in the nucleolus, which participates in protein synthesis. There are two mitochondrial rRNAs (16S and 23S) and four cytoplasmic rRNAs (5S, 5.8S, 18S, and 28S). rRNA decodes mRNA into amino acids in small ribosomal subunit and interacts with tRNAs. Ribosomal RNA catalyzes the formation of peptide bonds during translation.

Small Interfering RNA (siRNA): exogenous double-stranded RNA produced by trimming of double-stranded RNA.

Small Nuclear RNA (snRNA): short RNA of the nucleus transcribed by RNA polymerase-2 or -3. They are associated with specific proteins to form the so-called small nuclear ribonucleoproteins (snRNP), for RNA processing, such as splicing and polyadenylation (addition of adenines). Polyadenylation, addition of pre-mRNA tail, and protein binding protect mRNA from degradation by exonucleases. Polyadenylation is also used for transcription termination, mRNA export from the nucleus, and translation. snRNAs are also involved in the regulation of transcription factors or RNA polymerase-2, and the maintenance of telomeres.

Small Nucleolar RNA (snoRNA): RNA synthesized by RNA polymerase-2. They form with specific proteins small nucleolar ribonucleoproteins (snoRNP). snoRNPs modify various types of RNAs, such as rRNAs and spliceosome snRNAs.

Transfer RNA (tRNA): RNA that transports amino acids toward ribosomes for protein assembling on ribosomal RNA sites in an order determined by the messenger RNA. tRNA thus reads the code and transfers the appropriate amino acid to be incorporated into a growing polypeptide chain.

S100 Protein: members of a protein family that have manifold tissue-specific intra- and extracellular functions. They are involved in contraction, motility, growth, differentiation, cell cycle, transcription, and secretion.

Sarcomere: contractile unit with special arrangement. It is composed of (1) A band (length between free ends of adjacent myosins), with M line (attachment of adjacent myosins), with H zone (myosin structure, absence of actin), and with a region of intercalated actin and myosin filaments, and (2) I band (actin lattice), with Z disc (sarcomere ends).

Scaffold Proteins: bind and organize proteins in a signaling pathway. Scaffold (or scaffolding) proteins not only function as an assembly element, but also can serve as regulatory nodes, tuning pathway fluxes.

Scavenger Receptor-B1: HDL receptor both for cell uptake of cholesteryl esters from HDL and cellular cholesterol efflux to HDL. SR-B1 can bind both apolipoproteins and, with a greater affinity, HDLs.

Second Messenger: intracellular macromolecule activated by ligand-bound receptor.

Serotonin (or 5-HydroxyTryptamine [5HT]): has multiple effects. It affects, in particular, the cardiac frequency. Released by platelets, it triggers vasoconstriction (its name comes from its discovery as serum agent affecting vascular tone). Serotonin has several receptors (5HTR) [803].

5HT1 Gi/Go-coupled receptors, with different subtypes (5HTR1A, 1D, 1E and 1F), when they are linked to 5HT, activate G protein, which induces K^+ channel opening and hinders the synthesis of cyclic adenosine monophosphate. Conversely, complexes 5HT–5HTR2 and 5HT–5HTR4 yield K^+ channel closure. Ligand-stimulated 5HT2 Gq/G11-coupled receptors, with three subtypes (5HT2A, 5HT2B, and 5HT2C) promotes the formation of inositol trisphosphate and diacylglycerol. 5HT–5HTR2 do not interact with cAMP. 5HT3 is a ligand-gated Na^+ and K^+ cation channel. 5HT4 and 5HT7 Gs-coupled receptor augments cAMP production after ligand binding.

Serotonin acts distantly from its production sites. It is released from almost all brain regions, especially by serotonergic neurons of the raphe nucleus in the brainstem. Extracellular 5HT is inactivated mainly by carriage across the cell membrane of presynaptic axons by serotonin transporter SERT.

Serum Amyloid-A (SAA): cardiovascular risk factor weaker than C-reactive protein. Serum amyloid-A belongs to a family of apolipoproteins of high-density lipoprotein. The synthesis of certain SAA members of the family strongly rises during inflammation.

Src-Homologous and Collagen-like Substrate (Shc): compound required in signaling pathways that connects to certain activated receptors for Tyr-phosphorylation. Activated adapter Shc binds Grb2 and stimulates Ras GTPase and phosphoinositide 3-kinase.

Signalosome: proteic complex that comprises activated receptor.

Gene Silencing: inhibition of the expression of homologous sequences by RNAs, either at the transcriptional level (transcriptional gene silencing), or the post-transcriptional level (post-transcriptional gene silencing or RNA interference).

Sirtuin: nicotinamide adenine dinucleotide (NAD^+)-dependent deacetylase silent information regulator (SIR) enzymes.

SNAP Receptor (SNARE): protein of organelle membranes and derived vesicles guiding the vesicles to their correct destinations. They work in pairs, a vSNARE of the vesicle membrane specifically binding to a tSNARE of the target membrane.

Son-of-Sevenless (Sos): guanine-exchange factor that stimulates GDP release from Ras and GTP binding (Ras activation). Sos binds to Grb2. Activated (phosphorylated) receptors or adapters recruit Grb2–Sos complex to the plasmalemma at Ras loci.

Spectrin: forms a rigid network that supports the plasmalemma of the erythrocyte. It thus determines the plasmalemma features (shape and deformability). This heterodimer, composed of subunits α and β, forms tetramers or polymers linked by actin and protein-4.1.

Splicing: pre-mRNA removal of non-coding introns, keeping protein-coding exons. Pre-mRNA can be spliced by the spliceosome, an RNA-protein complex, or by ribozyme in different ways. Therefore, protein isoforms are encoded from a single gene (alternative splicing). There are several splicing modes: (1) alternative selection of promoters in the case of different promoters, (2) alternative selection of cleavage and polyadenylation sites, (3) intron retaining in the case of coding intron, and (4) exon cassette by splicing out exon to alter the sequence of amino acids.

Schmidt-Ruppin viral oncogene homologue (Src): cytoplasmic tyrosine kinase activated by Gαi/o, which particularly phosphorylates STAT3. Cytosolic c-Src is often self-restrained to an inactive conformation. In the plasmalemma, c-Src phosphorylates a truncated, relocalized estrogen receptor-α, leading to NOS3 activation.

The Src family of non-receptor tyrosine kinases are signaling proteins associated with cell adhesion, differentiation, survival, migration, and proliferation, especially during angiogenesis. The Src family includes at least eight members, which are reversibly coupled to the inner leaflet of the plasmalemma: c-Src, Yes, Fgr, Fyn, Lck, Lyn, Blk, and Hck. c-Src, Fyn, and Yes are ubiquitously expressed. Blk, Hck, Fgr, Lck, and Lyn have a restricted expression pattern.

Store-Operated Calcium Channel (SOC): channel that functions during several cell processes, such as cell transport, cell displacement, and gene expression. The Ca^{++} release-activated Ca^{++} channel belongs to the set of store-operated calcium channels. Activated phospholipase-C produces inositol(1,4,5)trisphosphate, which opens IP3 receptors of the endoplasmic reticulum and produces an initial transient rise in intracellular calcium concentration resulting from the release of calcium from its intracellular stores. A subsequent sustained calcium entry across the plasmalemma from extracellular sources is carried out by Ca^{++} release-activated Ca^{++} channels of the cell membrane which are activated by the reduction in calcium ions in the endoplasmic reticulum. In cells with repleted calcium stores, calcium sensor of the endoplasmic reticulum, the stromal interaction molecule STIM1 and Ca^{++} release-activated Ca^{++} channel of the plasmalemma Orai1/CRACM1 are dispersed along the membrane of the endoplasmic reticulum and the cell, respectively. STIM1 and Orai1 migrate independently to closely apposed sites of interaction, separated by a narrow gap of cytosol. Store depletion induces (1) STIM1 accumulation at junctional endoplasmic reticulum, STIM1 then forming multimers; and (2) Orai1 displacement in regions of the plasmalemma opposite to the STIM1 clusters. Orai1 slowly transports calcium ions when the concentration of extracellular calcium is high enough and inactivated when intracellular calcium level rises.

Stress Fiber: contractile actomyosin bundle that forms when the cell interacts with its environment. These bundles assemble either in adherent patches at focal adhesions or in arcs. They undergo continuous remodeling during cell motion.

Sumoylation: post-translational modification involved in various cellular processes (protein stability, molecular transport, transcriptional regulation [mainly transcription repression], apoptosis, stress response, and cell cycle progression). Small ubiquitin-related modifier (SUMO), indeed, attaches to and detaches from targeted proteins to modify their function. There are four known human isoforms: SUMO1, SUMO2, SUMO3, and SUMO4. Sumoylation can increase protein lifetime and can change protein location. Attachment of a small ubiquitin-like modifier is reversible; several SUMO-specific proteases remove SUMO from proteic substrates.

Antisense Suppression of Gene Expression: due to fragments of single-stranded RNA bound to mRNA which block translation without mRNA degradation.

Syk: protein tyrosine kinase that phosphorylates scaffolding proteins to recruit signaling effectors and factors, such as phosphatidylinositol 3-kinase, phospholipase-Cγ, adapter Grb, guanine nucleotide-exchange factor Sos, etc.

Syndecan: plasmalemmal heparan sulfate proteoglycan. Syndecans interact with cytoskeletal proteins, kinases (Src, calcium–calmodulin-dependent serine protein kinase CASK), and phosphatidylinositol(4,5)bisphosphate.

Talin: component of the integrin adhesion complex. This actin-binding protein is located in focal adhesions. It interacts with F-actin, vinculin, β-integrins, phospholipids, and focal adhesion kinases. This scaffolding protein thus connects β-integrins to the actin cytoskeleton. Talin is required for focal adhesion assembly. Talin also regulates transcription.

Telomere: chromosome end used during replication to prevent chromosome shortening.

Tenascin: extracellular matrix glycoprotein. The tenascin family includes tenascin-C, -R, -X, and -W. Tenascin-C is found in developing tendons, bone, and cartilage; tenascin-R in the nervous system; tenascin-X in connective tissues; tenascin-W in the kidney and developing bone.

Tensin: protein of the focal adhesions. It cross-links actin filaments. The titin features, especially elasticity, vary among its I-band, M-line and Z-disc regions.

Tetraspanin: plasmalemmal protein that forms complexes with integrins, EGF receptor, and PKC. Tetraspanins are found at the edges of lamellipodia.

Thrombospondin: extracellular protein that participates in cell-to-cell and cell-to-matrix communications. Thrombospondin-1 binds to fibrinogen bound to a GPIIb/IIIa receptor. Platelets are thus linked via several fibrinogens bound by thrombospondin-1.

Thyroid Hormone: hormone family that affects growth, metabolism, and function of nearly all organs. Thyroid hormones activate nuclear hormone receptors, which modulate the expression of target genes. In particular, they lower the concentration of atherogenic lipoproteins.

Thyroid Hormone Receptors: set of four different isoforms (α1, α2, β1, and β2) with differential expression among anatomical tissues. Thyroid hormone receptors α and β are predominant in the heart and in the liver, respectively.

Titin: protein with kinase activity that spans the hemisarcomere. This scaffold protein binds to myofibrils and other sarcomeric proteins.

Titin-Cap/Telethonin (T-Cap): protein of the sarcomere of both striated and cardiac myocytes. It is a substrate of titin kinase.

Toll-Like receptor (TLR): receptor expressed on immune cells (dendritic cells and macrophages), which sense pathogen associated molecular patterns produced by bacteria, viruses, fungi, and protozoa. Toll-like receptors recognize lipids, carbohydrates, peptides, and nucleic acids expressed by microorganisms. Many pathogens yield ligands for several TLRs. Toll-like receptors have homology with Toll, a molecule that stimulates the production of antimicrobial proteins in Drosophila melanogaster. The family of Toll-like receptors includes 12 known plasmalemmal proteins that trigger innate immune responses. Toll-like receptors can activate NFκB (except TLR3), interferon-regulatory factor, phosphatidylinositol 3-kinase and MAPK pathways. Toll-like receptors need adapters. Toll-like receptors, interleukin receptors IL1R and IL18R induce cytokine production via NFκB, using adapter myeloid differentiation primary-response gene 88 (MyD88). TLR3 and TLR4 require adapter TRIF; TLR2 and TLR4 adapter TIRAP; and TLR4 adapter TRAM.

Signaling by Toll-like receptors involves several adapters [804]. Myeloid differentiation primary-response gene MyD88, MyD88-like protein (MAL, associated with TLR2 and TLR4), TIR domain-containing adapter protein inducing interferon-β (TRIF, associated with TLR3), TRIF-related adapter molecule (TRAM, associated with TLR4) are recruited to initiate signaling. Sterile α- and armadillo motif-containing protein (SARM) inhibits TRIF.

Transcription Factor: protein that binds to corresponding promoter on DNA to regulate the gene activity. It either suppresses or triggers gene transcription into mRNA, promoting the binding of RNA polymerase to DNA. A given transcription factor often regulates multiple genes. Conversely, the expression of a given gene can be regulated by different transcription factors.

Transcriptome: set of messenger RNA produced in a cell at a given time in a given context.

Transduction (signaling pathway): process that conveys received informations from the cell environment to the nucleus. Transduction is triggered by signaling ligand binding to its plasmalemmal receptor, which undergoes a subsequent conformational change and possible phosphorylation to recruit cytoplasmic signaling components. The latter initiates a cascade of events, the signal being transferred from the membrane through the cytoplasm and into the nucleus, to express various genes involved in the cell response.

Transforming Growth Factor-β (TGFβ): a family of proteins that includes bone morphogenetic proteins, growth factors, inhibins, and activins. They are involved in cell proliferation and differentiation. They bind to receptors-1 (seven known types) and -2 (five known types).

Transient Receptor Potential Cation Channels (TRP): carriers of cell membrane responsible for calcium influx into non-excitable cells. They include seven families: canonical (TRPC), vanilloid (TRPV), melastatin (TRPM), no mechanoreceptor potential-C (TRPN), ankyrin-like (TRPA), mucolipin (TRPML), and polycystin (TRPP).

Translocation: molecule transit within the cell or ion transfer across its membrane carrier.

Translocon: membrane-embedded proteic assembly used for protein transport into and across cell membranes.

(Active) Transport: carriage of a solute across a biological membrane against the concentration gradient associated with energy dissipation.

Tricarboxylic Acid Cycle (TCA, Krebs cycle, citric acid cycle): closed loop of biochemical reactions involved in the metabolic aerobic pathway which oxidizes acetyl groups of metabolites to CO_2 and H_2O in the mitochondria.

Tropomyosin (TMy): protein associated with actin, which yields structural stability and modulates filament function. Various isoforms exist (TMy1, TMy2, TMy3, and TMy4), which can be filament specific. In myocytes, tropomyosin moves its position on actin during contraction, which is regulated by calcium.

Troponin (TN): proteic complex formed by troponin-C, -T, and -I. Troponin-C expression is restricted to skeletal and cardiac muscles, where it regulates contraction. After calcium binding to troponin-C, actin undergoes changes that allow ATP hydrolysis and its interaction with myosin. Troponin-I is the inhibitory subunit of troponin. Troponin-T binds the complex to tropomyosin.

Tubulin: globular protein. The tubulin is a stable heterodimer of α and β subunits, which bind guanine nucleotides and make up microtubules. The intradimer interface contains a non-exchangeable GTP, whereas the interdimer interface contains GDP. GTP favors polymerization and GDP depolymerization. γ-Tubulin, another member of the tubulin family, is involved in the nucleation (onset of a state transition) of microtubule assembly and polar orientation of microtubules.

Tumor Necrosis Factor (TNF): cytokine involved in inflammation. TNF major role is the regulation of immune cells. TNFα signaling activates NFκB and MAPK either for cell survival or apoptosis. The soluble homotrimeric cytokine is released from transmembrane homotrimeric proteins cleaved by the metalloprotease TNFα-converting enzyme. TNF is mainly produced by macrophages, but also by lymphoid cells, mast cells, endothelial cells, fibroblasts, and cells of the nervous system.

Ubiquitin (Ub): the ubiquitin-proteasome set is used for protein degradation. Attachment of ubiquitins to a target protein (ubiquitination) tags for degradation by the 26S proteasome (non-lysosomal intracellular proteolysis). The ubiquitin–proteasome system is responsible for the lysis of short-lived proteins, like those having a regulatory function, and proteins with abnormal composition or structure. The ubiquitination is a three-step process involving Ub-activating (E1), Ub-conjugating (E2), and Ub-ligating (E3) enzymes. Ubiquitin is activated by a ubiquitin-activating enzyme E1. E1 hydrolyzes ATP and forms a complex with the resulting ubiquitin adenylate. The attachment of ubiquitin, as a single molecule or as a multi-ubiquitin chain, to the substrate protein is then catalyzed by an ubiquitin-conjugating enzyme E2. Finally, the ubiquitinated protein is transported to the proteasome. Ubiquitination of certain proteins can require an ubiquitin-protein ligase or recognin E3. Attachment of Ub to a target protein is not only a cue for protein destruction but also a localization signal. Ubiquitination also control DNA repair and replication, in embryogenesis, gene transcription regulation, the cell cycle (targeting cyclin during the G1 phase), apoptosis, and intracellular trafficking. Attachment of: (1) a single Ub molecule to a single lysine (Lys) residue leads to protein monoubiquitination; (2) several single Ub to different Lys residues to multiple monoubiquitination; and (3) a chain of Ub molecules to one or more Lys residues to polyubiquitination. Many cytoplasmic and nuclear proteins become ubiquitinated after phosphorylation. More precisely, the addition of a single ubiquitin to target proteins is involved in transcription control, receptor internalization, and endosome sorting. Monoubiquitination might be implicated in protein location, conformation, and interactions. The addition of multiple ubiquitins is implicated in proteasome degradation, DNA repair, ribosome function, and signal transduction. Several Ub-like proteins are conjugated to target proteins, using the same conjugation chemistry. The small Ub-related modifier (SUMO) modulates protein activity. The interferon stimulated gene product-15 (ISG15) is attached to proteins in response to interferon and lipopolysaccharides, among others. Ubiquitin corresponds to an inducible and reversible signal. Deubiquitinating enzymes rapidly remove ubiquitin and Ub-like proteins from proteins (regulation tuning of vesicular transport, cellular signaling, DNA repair, etc.).

E3 Ubiquitin Ligase: component of the ubiquitin proteasome system for the degradation of proteins. There are three groups: ring-finger, HECT-domain (homologous to E6AP carboxyl terminus), and SCF-complex (Skp1–Cul1–F-box protein) E3 ligases. The ring-finger E3 ubiquitin ligases include muscle-specific ring-finger proteins (MuRF1, MuRF2, MuRF3) expressed specifically in cardiac and skeletal muscle. MuRF1 targets titin and troponin-I. MuRF2 binds to titin. MuRF3 interacts with Four-and-a-half LIM domain protein (FHL2) and γ-filamin.

Vascular Endothelial Growth Factor (VEGF): regulates the growth of blood vessels and the vascular permeability. It regulates cell proliferation via JNK and transcriptional factor AP1. Ligand-bound receptor dimerization causes autophosphorylation and triggers the signaling cascade. The VEGF family includes VEGF-A, -B, -C and -D. VEGF binds to its cell-surface receptors VEGFR1 (or Flt1), VEGFR2 (or Flk1/KDR) and neuropilin-1/2 (NP1/NP2). VEGFR2 is associated with vascular endothelial cadherins and certain integrins α_v at between-cell junctions. VEGF-A is also a vascular permeability factor, inducing vascular leaks. VEGF-mediated vascular permeability normally leads to a fibrin impenetrable barrier. However, in cancer and ischemic tissues, VEGF-A induces edema, tumor cell extravasation, etc. It indeed deletes junctions between endothelial cells. VEGF alters gap junction by connexin-43 phosphorylation via VEGFR2 and Src kinase activation [805]. VEGF disrupts tight junctions via Src-dependent phosphorylation of zonula occludens-1 and occludin. VEGF acts via Src on adherens junction proteins (VE-cadherin, β-catenin, γ-catenin–plakoglobin and p120-catenin), inducing tyrosine phosphorylation. Plasma VEGF released from circulating cells acts on the luminal surface of the vascular endothelium, whereas VEGF secreted from pericytes or other cells plays on the abluminal surface. VEGF signaling uses PI3K–PKB, Ras–Raf–ERK, Src, and PLCγ–NOS.

VAV: protein of the Dbl family of guanine nucleotide-exchange factors for the Rho GTPase family. It is involved in hematopoiesis, angiogenesis, and T- and B-lymphocyte development and activation

Very-Low-Density Lipoprotein (VLDL): transport triglycerides from the intestine and liver to other tissues, particularly adipose and muscle tissues. Nascent VLDLs synthesized and secreted by hepatocytes contain triglycerides, apolipoprotein-B, and cholesterol esters. Mature functional VLDLs acquire apolipoproteins-E and -C2 from HDLs. VLDLs are cleaved by endothelial lipoprotein lipase, which is activated by apolipoprotein-C2. VLDLs then lose apolipoproteins and transfer triglycerides and give birth to intermediate-density lipoproteins.

Villin: actin regulatory protein of the gelsolin/villin family, which cap, nucleate, and/or sever actin filaments. Supervillin associates actin filaments to the plasmalemma.

Vimentin: protein that can recept protein kinase, synapsin, glycogen synthase, myosin light-chains, and microtubule-associated τ-protein.

Vinculin: protein of focal adhesions, which binds talin, α-actinin, actin, paxillin, and VASP. Vinculin stabilizes interactions between talin and actin and recruits profilin–G-actin to actin polymerization sites.

Vitronectin: protein of the pexin family, in plasma and tissues. It promotes cell adhesion and spreading.

von Hippel-Lindau Protein (vHL): degrades the transcription factor hypoxia-inducible factor (HIF), after oxygen-dependent hydroxylation by prolyl hydroxylases and cosubstrate 2-oxoglutarate, generating succinate. Reduction in fumarate hydratase or succinate dehydrogenase, both belonging to the tricarboxylic acid cycle, lead to activation of HIF. Conversely, during hypoxemia, increased fumarate concentrations and augmented levels of succinate inhibiting prolyl hydroxylases, HIF is stabilized and modifies the gene expression for tissue adaptation. In particular, HIF upregulates vascular endothelial growth factor (VEGF) to improve vascularization and oxygen supply. Activated HIF also increases glucose uptake and glycolysis.

von Willebrand Factor (vWF): mediator between platelets and blood vessel wall during blood coagulation.

Wiskott-Aldrich Syndrome Protein (WASP): form a set with WASP-family verprolin-homologous protein (WAVE). This protein set includes five known members: WASP, Neural WASP (N-WASP), WAVE1, WAVE2, and WAVE3. These scaffolds lead to a burst of actin polymerization via WASP-activated ARP2/3 complexes [806]. WASP and N-WASP bind to several adapters, such as Grb2, Nck, and Crk. WASPs also interact with phosphoinositides and small GTPases Cdc42. WASP and N-WASP are phosphorylated by Src.

Neural Wiskott-Aldrich Syndrome Protein (N-WASP): protein of cell leading edges, whereas WAVE, another set of the WASP family (WAVE is also known as SCARs) is implicated at the front of migrating cells. These proteins use the Arp2/3 complex to stimulate actin polymerization. New actin filaments are initiated at the membrane by filament branching catalyzed by the N-WASP or WAVE–Arp2/3 complexes, which are activated by Cdc42 and Rac, respectively. N-WASP acts on actin-based vesicular transport. WAVE generates Rac-dependent lamellipodial protrusions during cell migration and macropinocytosis, Abi1 acting downstream from Rac.

Wnt Signaling (Wingless in drosophila): a signaling pathway that acts during embryogenesis as well as in adult development. Wnt proteins control cell proliferation, morphology, motility, and cell fate, as well as stem cell maintenance. Aberrant Wnt signaling is implicated in developmental disorders, degenerative diseases, and cancers. The Wnt signaling pathway activates transcription of genes via β-catenin. β-Catenin acts as a transcriptional coactivator with members of the LEF–TCF family. Without stimulation, β-catenin is phosphorylated by a protein complex, which includes Apc, axin, and GSK3. Wnt signal activates two membrane receptors, Frizzled and LRP6, which form a complex and trigger signaling, stopping the degradation of β-catenin.

LDL receptor-related proteins-5 and -6 are required for transmission of Wnt/β-catenin signaling. These Wnt signaling receptors are phosphorylated by the membrane-bound casein kinase-1γ to be tranduced [807]. Activated LDLR-related protein-6 then promotes the recruitment of the scaffold protein axin.

Wnt signaling is also done via various factors, such as RhoA, JNK, PP3, PKC, and casein kinase-2. Bcr is a negative regulator of Wnt signaling.

Zonula occludens proteins (ZO): interact with occludin. ZOs serve as recognition proteins for tight junctional placement, and support structure for signal transduction proteins. ZOs belong to the membrane associated guanylate kinase-like proteins (MAGUK) family of proteins.

Zymogen: proenzyme (inactive precursor) of enzyme.

A.3 Physiology/Pathophysiology Terminology

Action Potential: electrochemical wave that quickly and transiently propagates in the plasmalemma of a neuron or a myocyte. It results from ion fluxes across the plasmalemma.

Actuator: a type of transducer that converts energy into mechanical force (motion). Actuators and sensors are hardware elements of instrumentation used in computer-aided medicine and surgery and medical simulators.

Adipocyte: cell of adipose tissues and component of the endocrine system and the immune system by its secreted molecules: adipokines, cytokines (tumor-necrosis factor, interleukin-1, interleukin-6), CC-chemokine ligand-2, clotting substances (plasminogen-activator inhibitor-1), and certain complement factors.

Afterload: load to be overcome by the left ventricle to eject a blood volume during every heart beat.

Aging: process that involves numerous pathways, as well as genetic and environmental factors. The genetic control of longevity is associated with survival processes (maintenance and repair of damaged DNA). DNA damage accumulates with aging.

Anastomosis: direct connection between arteries or veins.

Anatomic Atlas: data set used to differentiate tissues of similar composition, then of similar image intensity.

Aneurism: localized dilation of the wall of a blood vessel.

Angiogenesis: blood vessels development from existing capillaries by sprouting, pruning, and/or splitting (intussusception). Primitive plexi of capillaries remodel and grow by angiogenesis. Angiogenesis occurs in adults, particularly during wound healing (physiological angiogenesis). Abnormal angiogenesis appears in certain chronic diseases and tumors.

Angiogenesis is regulated by endothelial growth factors (vascular endothelial growth factor, fibroblast growth factor, platelet-derived growth factor, etc.) and angiogenesis inhibitors (angiostatin, endostatin, and thrombospondin-1). The angiogenic factors activate vascular and circulating endothelial cells, as well as endothelial progenitor cells from the bone marrow. Activated endothelial cells, as well as local stromal cells, secrete matrix metalloproteinases so that endothelial cells can invade and proliferate. The process is mediated by plasmin generated by the urokinase plasminogen activator in association with its receptor and the tissue-type plasminogen activator inhibitor-1. The migration of endothelial cells is regulated by adhesion molecules, such as integrins $\alpha_v\beta_3$ and $\alpha_v\beta_5$.

The expression of the vascular endothelial growth factor is regulated by several factors, such as hypoxia, oxidative and mechanical stresses, glucose deprivation, acidosis, mutation of oncogenes and tumor suppressor genes, and cytokines. In normal conditions, hypoxia inducible factor-1α is degraded by the ubiquitin-proteasome complex in coordination with the von Hippel-Lindau protein. When hypoxia occurs, hypoxia inducible factor-1α forms a complex with hypoxia inducible factor-1β, which translocates to the nucleus and binds to the VEGF promoter. The VEGF pathway up-regulates integrin expression, stimulates urokinase plasminogen activator (uPA) and uPA-receptor, activates phosphatidylinositol 3-kinase and matrix metalloproteinases, enhances the production of nitric oxide. VEGF release hence facilitates the invasion of endothelial cells using matrix metalloproteinases, urokinase plasminogen activators, and tissue-type plasminogen activator inhibitor-1. VEGF leads to migration of endothelial cells by activation of p38 mitogen-activated protein kinase, nitric oxide, and focal adhesion kinase. It favors survival of endothelial cells by action of phosphatidylinositol 3-kinase and protein kinase-B, by inhibition of caspases, and upregulation of the apoptosis inhibitors. VEGF increases the endothelium permeability and proliferation by activation of extracellular signaling-regulated kinase-1 and -2, c-Jun-NH2-terminal kinase–stress-activated protein kinases, and protein kinase-C.

Atherosclerosis: inflammatory process of large artery walls characterized by proliferation, migration from the media, and dedifferentiation of vascular smooth muscle cells. Atherosclerosis is characterized by progressive lipid accumulation inflammation, fissuration, thrombosis, calcification, and lumen stenosis.

Autonomic Nervous System: part of the nervous system composed of the sympathetic and parasympathetic systems. It modulates physiological functions, without conscious control. The sympathetic and parasympathetic systems function in opposition to each other (sympathetic accelerator and parasympathetic brake). Moreover, the sympathetic system is involved when quick responses are required, whereas the parasympathetic system intervenes when immediate reactions are not needed. The autonomic nervous system also causes the release of hormones, especially those involved in blood circulation functions (renin, vasopressin). The coordinated activity of the sympathetic and parasympathetic systems particularly regulates the blood flow (Table A.5), acting on heart functioning, vasculature resistances, breathing frequency (venous return), micturition (blood volume), etc. (the sympathetic system quickly adjusts the cardiovascular system to the body needs, increasing heart rate and cardiomyocyte contractility, and inducing vasoconstriction.) In addition to its involuntary actions, the autonomic nervous system can cooperate with the consciousness.

Both sympathetic and parasympathetic systems include afferent and efferent fibers. The sensory fibers lead signal to the nervous centers. The area postrema serves as a chemosensory center. The nucleus of the solitary tract integrates visceral information. Leaving these centers, efferent fibers can relay in ganglia. Preganglionc neurons belong to the central nervous system. Preganglionc sympathetic neurons originates in the spinal chord (lateral horns of thoracic and lumbar segments of the spinal cord), pregan-

Table A.5. Predominance of sympathetic and parasympathetic tone at effector sites of blood circulation

Site	Predominant tone
Heart	Parasympathetic (cholinergic)
Arteries	Sympathetic (adrenergic)
Veins	Sympathetic (adrenergic)

glionic parasympathetic neurons in the medulla oblongata and the sacral spinal chord. The preganglionic neurons modulate the activity of autonomic ganglionic neurons.[3] The sympathetic ganglia are located near the spine column (close to the spinal chord), the parasympathetic ganglia close to organs. The sympathetic ganglia is subdivided into two groups, paravertebral[4] and prevertebral (or preaortic) ganglia, forming the plexi of the digestive tract. The preganglionic neurotransmitter for both sympathetic and parasympathetic systems is acetycholine (cholinergic neurons). Postganglionic neurons, traveling with blood vessels, belong to the peripheral nervous system. Most postganglionic sympathetic neurons release noradrenaline, most often with cotransmittors (ATP, neuropeptide-Y, nitric oxide, etc.), which in most cases acts on adrenergic receptors. Postganglionic parasympathetic neurons release acetylcholine to stimulate either muscarinic (peripheral tissues) or nicotinic (sympathetic and parasympathetic postganglionic neurons, and adrenal chromaffin cells) receptors.

Autoregulation: intrinsic ability of an organ to maintain a constant blood flow despite changes in perfusion pressure.

Bathmotropy: refers to the exitability of the myocardium.

Blood-Brain Barrier (BBB): is formed by the endothelial cells of the cerebral microvasculature. It maintains a suitable environment for neuronal signaling.

Body Surface Area (BSA): estimated for adults as BSA = ([height(cm) + weight(kg)] − 60)/100.

Bradycardia: $f_c < 1\,\text{Hz}$.

Cancer: a population of malignant cells, either clustered as tumors or dispersed, as in leukemias.

Although mutations occur in differentiated cells, cancers contain stem cells. Cancers can thus be derived from stem cells. Stem cell alterations explain, at least partially (dedifferentiated cells mimic stem cell behavior), heterogeneity of most tumors, with different degrees of cell transformation. Cancer growth depend on the activity of these stem cells, which are targeted by effective therapy.

Six main alterations in cell physiology characterize a tumor cell: (1) self-sufficiency in growth signals, (2) insensitivity to growth-inhibition, (3) resistance to apoptosis, (4) limitless replicative potential, (5) sustained angiogenesis, and (6) tissue invasion and metastasis.

Carcinogenesis can be divided into three phases: (1) initiation, with stable genomic alterations (DNA is mutated by chemical or physical carcinogens, with subsequent activation of oncogenes and/or inactivation of

---

[3] Axons of sympathetic neurons make synapse on either sympathetic ganglion cells or chromaffin cells in the adrenal gland.

[4] 21 to 22 pairs of sympathetic ganglia exist: 3 cervical (superior, middle, and stellate ganglia), 10 to 11 thoracic, 4 lumbar, 4 sacral, and a single ganglion in front of the coccyx. Paravertebral ganglia provide sympathetic innervation to blood vessels.

tumor-suppressor genes and proteins), (2) promotion, with proliferation of genetically altered cells, and (3) progression, with tumor spreading. During the progression stage, additional mutations can be acquired.

Many biological processes are involved in tumorigenesis. Inflammation and innate immunity could exert pro-tumorigenic effects, whereas adaptive immunity could have antitumorigenic effects. Inflammation acts via leukocytes (macrophages, tumor-associated macrophages,dendritic cells, neutrophils, mast cells, and T lymphocytes), which are recruited to the tumor microenvironment. These leukocytes produce cytokines, growth and angiogenic factors (tumor-necrosis factor, interleukin-6, and vascular endothelial growth factor), as well as matrix metalloproteinases and their inhibitors. Necrosis of cancer cells in the hypoxic tumor core leads to the activation of tumor-associated macrophages. Interferons, interleukin-4, IL12, IL13, and granulocyte/macrophage colony stimulating factor might increase tumor rejection, inducing maturation of dendritic cells and promoting initiation of an adaptive immune response. Colony stimulating factor-1 (M-CSF) and interleukin-6, produced by tumor-associated macrophages, can inhibit maturation of dendritic cells. In association with IL1 and tumor-necrosis factor, M-CSF and IL6 can promote tumor progression. Viral and bacterial infection (human papillomaviruses, hepatitis B and hepatitis C viruses, and Epstein-Barr virus are risk factors for malignancies), subsequent inflammation and cancer can be connected. Inflammation can induce tumor growth by the activation of nuclear factor-κB [808]. Tumors are hypoxic at some stages because of high oxygen consumption and inadequate blood supply. The transcription factor hypoxia-inducible factor (HIF) is activated for angiogenic signaling and glycolysis. A decay in tumoral HIF activity can then reduce tumor growth.

Cardiac Index: CI = CO/BSA (l/mn/m^2).

Cardiac Output (CO): blood amount pumped by the heart in one minute. This hemodynamic variable is continuously regulated by the nervous system.

Central Venous Pressure (CVP): pressure measured in the right atrium.

Central Nervous System: part of the nervous system, including the brain and spinal cord, which analyzes sensory data and emits nerve impulses.

Chronotropy: refers to the cardiac frequency (positive chronotrope effect (C+): increase, negative (C−): decrease in heart rate).

Coagulation: transformation from a liquid to a solid form, generally applied to the clotting of blood.

Compliance: usually in physiology dV/dp, the reverse of the elastance (dp/dV). Total peripheral compliance is a hemodynamic index currently estimated using a windkessel model. The vessel compliance depends on the geometry and rheology of the blood vessel for a given pressure range. A bulk quantity does not have any mechanical meaning. However, it is used in medical practice.

Composite Material: material composed of more or less stiff fibers surrounded by a matrix.

Computer-Aided Therapy: integration of computer technologies into care, procedures being planned and guided by a computer model.

Conductance: refers to the rate of the ion flux across ion channel or to the flow rate for a given pressure drop.

Correlation: measures the interdependence between two variables, most often assuming a linear relationship between them and the absence of discontinuity.

Datum Fusion: merged informations from multiple sources.

Diastole: the period of cardiac relaxation (in opposition to systole).

Display Device: small, high-resolution CCD cameras introduced in the body to display target images on a video-monitor. Consequently, more people in the operating room follow the performance of the operation and can anticipate the needs.

Distal: further away (in opposition to proximal).

Dromotropy: refers to the conduction velocity of the nodal system.

Edema: manifestation of a strong imbalance between hydrostatic and osmotic pressures at the capillary wall, with excess accumulation of water in the interstitium without reabsorption into the capillaries or filtration into the lymphatics. Edema can be caused by lymphatic or venous obstruction, permeability changes, such as in inflammation, reduction in plasma protein concentration, such as in liver dysfunction.

Efferent: away from (in opposition to afferent).

Ejection Fraction (EF): ratio of the stroke volume to the end-diastolic ventricular volume (EDV), used as index of contractility.

Electrocardiogram (ECG): the electrical signal caused by depolarization and repolarization of the myocardium during the cardiac cycle. The electrical activity of the heart reflects the time- and voltage-dependent behavior of ion channels of the cardiac wall (nodal cells and cardiomyocytes) and the resulting transmembrane ionic gradients.

Embolus: air bubble, fat particle, or blood clot conveyed in the blood.

Endocrine: pertaining to glands excreting hormones.

Endothelium: monolayer of cells lining blood vessels.

Epigenetics: study of heritable changes in gene expression without alteration in the nucleotide sequence due to chemical modifications of nucleic acids or of histones (DNA methylation and chromatin remodeling). Epigenetics thus describes interactions of genes with their environment, especially the effects of chromatin.

Extravasation: process during which cell leaves a blood vessel and migrates within the vessel wall.

Fascia: sheet of connective tissue, largely collagen, surrounding or separating muscle fibers.

Fatty Streak: subendothelial accumulation of foam cells (cholesterol-engorged macrophages).

Fibrillation: atrial or ventricular rate greater than 5 Hz.

Field of View (FOV): selected area for acquisition and display.

Flutter: rapid but regular contractions of atrial or ventricular myocardium caused by re-entry ($f_c > 4\,Hz$).

Focal: occurring at specific locations.

Force Feedback (FFB): virtual force simulation, which requires a device generating the reaction to the action, for example, haptic feedback in surgical simulator. Haptic devices must retrieve both the applied force and the tactile quality of the biological tissues (texture and shape [smoothness or sharpness]) back to the fingers.

Gating: opening and closure processes of ion channels. Voltage-gated ion channels open and close according to the electrical potential of the plasmalemma. Ligand-gated ion channels respond to binding of either extracellular or intracellular molecules. Gating also refers to image acquisitions and measurements triggered by ECG.

Gene Expression: process by which a gene DNA sequence is converted into the structures and functions of a cell.

Glycocalyx: small covering on a cell surface.

Haptic Feedback: interface that gives the sense of touch (απτο: to touch, to come into contact; αφη: sense of touch) to the user via applied forces and/or displacements. Haptic devices[5] assess forces exerted by the user. Haptic interfaces are required for medical simulators aimed at training for minimally invasive and/or remote procedures.

Heart Failure: results from defects in myocardial metabolism, especially in calcium handling, alterations in myosin isoforms, abnormalities in cardiomyocyte cytoskeleton, and excessive cardiomyocyte apoptosis.

Homeostasis: maintenance of the internal equilibrium. The body continuously experiences modifications of its environment and must keep a internal stability, the physiological data evolving within given ranges. Homeostasis disturbances occur following diverse kinds of stresses, which trigger adaptive responses of the body. The response type depends on the stress duration. The body response is based on various systems, including the hypothalamus–pituitary–adrenal axis, the autonomic (vegetative) nervous system with its parasympathetic and sympathetic components, the limbic system, and the immune system.

The biophysical and biochemical quantities, which described the internal equilibrium, evolve in a given range centered around an average. The averaged quantities are mathematical data which serve as indices to measure the bulk activity of the physiological systems. The precise aspects of the activity are usually much more difficult to predict.

---

[5] Haptic devices were also called dactylokinesthetic because they are related to both tactile (δακτυλος: finger) and kinesthetic (kinestesic) representations, sensing contact with a targeted organ or another tool in association with tool positioning.

Humoral: related to body fluids, particularly serum.

Image Acquisition: Process of measuring, reconstructing (for quick visualization), and storing image data. Images are collected with a given acquisition time, and afterwards (after required reconstruction time) display on a screen.

Image-Guided Therapy: spatial informations (target volume and localization) in relation to the surrounding structures used to plan access trajectories and approach the target using suitable instruments with minimal risks. The peroperative imaging is combined to the preoperative geometrical model of the target organ with periodic updating to guide the therapeutic gesture. The peroperative image guidance must provide the accurate definition of the target margins for full efficiency of the procedure. Several sensors (optical, electromagnetic, ultrasonic, etc.) can be attached to the instruments to provide a continuous spatial localization during the intervention (instrument tracking).

Image Registration: transformation of different image sets acquired at different times and from different perspectives, and hence in different coordinate systems, into a single coordinate system. Registration allows to compare and integrate patient images obtained from different acquisition techniques. Elastic registration copes with deformations of body's organs. Non-rigid registration of medical images are also used to register patient's images to anatomical atlases.

Image Segmentation: separation of mutually exclusive regions of an image to retrieve the organ contours.

Adaptive Immunity: immune reponses created by foreign body invasion. Adaptive immune responses to a previously experienced stimulus are enhanced with respect to the first encounter. The main cells of adaptive immune responses are lymphocytes, which express a large repertoire of antigen-specific receptors. The adaptive response indeed is characterized by its diversity due to the random recombination of antigen receptor gene segments during lymphocyte development, which gives rise to numerous T and B lymphocytes, each with a unique antigenic specificity. MHC class II molecules on the surface of antigen-presenting cells are displayed for recognition by CD4+ T helper cells and adaptive immune responses. Naive T lymphocytes are targeted for a given antigen after interaction with specialized antigen-presenting cells in secondary lymphoid organs, which triggers effector T-lymphocyte proliferation. Effector T lymphocytes give rise to long-lived memory T lymphocytes that protect against reinfection.

Innate immune responses can determine both the type and intensity of adaptive immune responses. The recognition phase of innate responses depends on receptors, such as Toll-like receptors expressed by dendritic cells, which detect molecules carried by bacteria and viruses. Toll-like receptors prime the synthesis of interferons that regulate the development of both innate and adaptive immunity.

Innate Immunity: pre-existing, antigen-independent immune mechanism that defends against infection and cellular transformation. Innate immune responses involve both soluble and plasmalemmal molecules that recognize a finite set of substances associated with tissue damage and certain pathogens. Repeated exposure to the same stimulus does not substantially alter the nature of the immune response. Innate immunity involves natural killer cells, $\gamma\delta T$ lymphocytes, phagocytes, epithelial cells, and dendritic cells. These cells are activated by plasmalemmal sensors.

Impedance: opposite of *admittance*, determines the pressure to be generated by the heart pump to create the blood flow.

Incidence: number of individuals who have newly contracted a given disease during a specified period in a given population.

Inflammation: response of the immune system to irritation, injury, or infection.

Inotropy: refers to the myocardial contractility.

Interface Technology: deals with interactions between a person and an electromechanical device, with inputs, outputs, and the interaction environment, which can be partially real, partially virtual (from images of the operating field).

Intimal Hyperplasia: cell proliferation in intima associated with abnormal hemodynamic stresses, oxidative stresses, and endothelial injury, observed in atherosclerosis, stented vessel restenosis, and grafting.

Kinesthesia: sense that detects position and motion of body's parts (κινησι: to generate a motion; αισθησις: sensation). Afferent information comes from skin, bones, tendons, joints, and muscles. The sense of the position of the components of the locomotor system become automatic by training (paramount importance of medical simulators). Kinesthesia-associated haptic perception strongly relies on fine sensation of pressure experienced during touch. In addition, kinesthesia is involved in hand-eye coordination used in particular during therapeutic procedures.

Kinesthetic Feedback: device that prevents forbidden trajectories.

Lesion: localized abnormality in tissue organization.

Limbic System: nervous system involved in emotion formation and processing, motivation, learning, and memory. The limbic system (limbus: edge, boundary, margin) contains many brain structures, such as the amygdala, hippocampus, hypothalamus, mammillary bodies, fornix (which connects the hippocampus to the hypothalamus), fornicate gyrus (which consists of cingulate and parahippocampal gyri), dentate gyrus (a part of the hippocampal region), thalamus, and basal ganglia (caudate nucleus, putamen, globus pallidus, and substantia nigra). The limbic system is connected with the cerebral cortex. The orbitofrontal cortex plays a role in decision making. The sense of smell is related to all parts of the limbic system, especially to the olfactory cortex. The limbic system acts on both the autonomic nervous system and the endocrine system.

The amygdala is involved in signaling related to emotional responses (aggression and fear). The hippocampus and the parahippocampal gyrus are required for storage of short-term memory, formation of long-term memory,[6] and spatial orientation. The mammillary bodies are implicated in emotion, sexual arousal, and memory processing. The cingulate gyrus regulates the blood flow and cognition. It coordinates sensory inputs associated with emotions. It is also involved in memories related to odors and pain. The dentate gyrus plays a role in formation of new memories. The nucleus accumbens acts in various processes, such as pleasure and addiction. The thalamus is implicated in motor control. It relays sensory signals to the cerebral cortex. The hypothalamus controls organ functioning by hormone release, especially the blood circulation. The basal ganglia are responsible for repetitive behaviors, reward experiences, and attention.

(Vessel) Lumen: space inside any blood vessel in which blood flows. The lumen (lumen, lumina: light, shine, glow) is bound by the wetted surface of the endothelium with the glycocalyx, which senses the stresses applied to it by the moving blood (mechanotransduction).

Lusitropy: ability of cardiomyocytes to relax.

Extracellular Matrix: ground substance of connective tissue.

Mean Arterial Pressure (mAP): time-mean pressure during the cardiac cycle (virtual pressure that would exist without arterial pulsations). Its approximate value is mAP = (systolic AP − diastolic AP)/3 + diastolic AP.

Metadata: collection of physio/pathophysiological process data. A knowledge of the methodology used to compile datasets is important for assessing and interpreting data. Data quality depends on geographical coverage, sampled population, and size of data collection. Data collected a long time ago can be no longer relevant due to changed situations. Data collection is also affected by exploration times (physiological cycles) and possible breaks in collecting observations. Data representativity depends on the selection technique for sampling of the population. Compilation method is worth knowing, because definitions of concepts and physio/pathophysiological parameters can vary, as well as, in the case of diseases, incidence and prevalence of factors can be not very clearly defined.

Metabolic Syndrome: collection of risk factors of atherosclerosis, including hyperlipidemia, hypertension, and diabete.

Microelectromechanical Systems (MEMS): microelectronics-based devices used to sense, to act and to compute.

Minimally Invasive Therapy: minimal access to the target organ that implies limited field of view and depth perception. The usual straight hand-eye axis disappears. The instruments are remote manipulators, needing sensory output back to the hand to feel the organs and sense the depth.

---

[6] The hippocampus is not involved in procedural memory, i.e. learning by repetition.

Mitochondrium: organelle in the cytoplasm of cells containing genetic material and many enzymes important for cell metabolism, including those responsible for the conversion of nutrient to usable energy.

Mucosa: tissue containing an epithelium and excretory cells.

Myofiber: small set of myocytes.

Necrosis: process characterized by disruption of the plasmalemma and release of intracellular components into the surrounding tissue, which then yield an inflammatory response and can damage neighboring cells.

Nervous System: body command system divided into central and peripheral nervous systems.

Ostium: opening or passage.

Pacemaker: collection of specialist myocytes in the sinoatrial node (the so-called natural pacemaker, because of its highest emission frequency, although reduced by the parasympathetic system, among nodal tissue), which oscillate electrically to initiate the cardiac cycle.

Parasympathetic: part of the autonomic nervous system generally responsible for conserving energy and reducing metabolism (in opposition to sympathetic).

Perfusion Pressure: arterial pressure minus venous pressure.

Peripheral Nervous System: part of the nervous system, which carries nerve impulses from (afferents) and to (efferents) the body organs. It includes cranial and spinal nerves with centers in the brain stem and the spinal cord, respectively. It is made of sensory (afferent) and motor (efferent) limbs; neurons form reflex arcs. Certain sensory neurons monitor the concentrations in blood gases (chemosensors for carbon dioxide and oxygen), the arterial pressure (baroreceptors), etc.

Pitch: angular displacement according to the anterior-posterior direction. In MSSCT, slice thickness-to-table translation ratio.

Preload: load undergone by the left ventricle associated with diastolic filling. Preload involves stored mechanical energy.

Prevalence: number of cases of a given disease in a specified population at a peculiar time interval (small or large), regardless of the date of contraction. Prevalence is distinct from incidence, which is a measure of the number of new cases.

Prognosis: expected course of a disease, either without therapy (natural prognosis) or after treatment (clinical prognosis). Any prognosis is defined by certain criteria, such as the absence of pain or recurrence in case of malignancies. Prognosis results from clinical studies, statistical evaluations and, last but not least clinician experience.

Prolapse: abnormal protrusion of a part of an organ, in particular the movement of a cardiac valve into its upstream chamber.

Proprioception: sense (proprius: which belongs to me; perceptio: to gather, to collect) of the relative position and possible motion of neighboring body's parts. Proprioception corresponds to the sensory information imparted from the locomotor system (bones, muscles, ligaments) and skin to know the position of body's elements and motion amplitude. Proprioception is then a sensory modality, which differs from the six exteroceptive senses (exteroception: sight, hearing, balance, smell, taste, and touch; exteroceptors are the eyes, ears, nose, mouth, and skin) and interoceptive perceptions of the state of body's organ (interoceptors send information from stressed organs). The proprioceptive information comes from sensory neurons connected to the inner ear (motion and orientation) and stretch receptors of muscles and ligaments. During the exploration of objects, the force information is more important than proprioception.

Proximal: closer (in opposition to distal).

Pulsatility Index: peak flow-to-mean flow ratio ($\hat{q}/\bar{q}$).

Three-Dimensional Reconstruction: generation of a volumetric representation of the region of interest. The 3D reconstruction can combine the anatomy (size, shape, and possibly color, texture, etc.) to functional data (organ segmentation, vascularization, etc.), physical characteristics (flow rate, rheology), and biochemical activity.

Re-Entry: electrical impulse return into a recently activated myocardium area.

Remodeling: long-term changes in cell proliferation, differentiation, migration and apoptosis occuring in hypertension, atherosclerosis, ischemia, and restenosis.

Biological Rhythm: functional process variable in time. A circadian rhythm is a cycle endogenously generated of about (circa: around) one day, then including both diurnal and nocturnal variations. The circadian rhythm can be modulated by external factors (light and temperature). Circadian rhythms are defined by three criteria: persistancy in constant conditions, resetting by light/darkness exposure, and temperature compensation with the same rate within a temperature range. The governing circadian clock in mammals is located in the suprachiasmatic nucleus of the hypothalamus. The suprachiasmatic nucleus secretes melatonin in response to cues from the retina. Melatonin release rises during night and falls during day. However, the circadian rhythm is also found in many body cell types. Ultradian and infradian rhythms designate short (i.e., heart and respiration rates) and long cycle periods. The latter include circaseptan or weekly rhythm (immunity), circavigintan (about 20 days; production of certain hormones), circatrigintan (about 30 days; menstrual cycle and seasonal variations with winter/summer differences in lipid levels and in thyroid functions), and circannual rhythm (organ sensitivity to environmental stimuli). In opposition to homeostasis that supposes a relatively constant internal state, chronobiology explores periodic processes in living organisms and exhibits predictable temporal variability. Chronomics refers to molecular mechanisms involved in chronobiology.

Sarcoplasm: cytoplasm of striated myocytes.

Sarcoplasmic Reticulum: set of convoluted tubules and flattened sacs throughout the sarcoplasm of cardiomyocytes.

Sensor: a type of transducer. Biological sensors are sensitive to: (1) mechanical contact stresses such as those generated by blood motion (pressure and shear), as well as body forces such as gravity and magnetic and electrical fields; (2) physical factors (light, temperature, humidity, etc.); (3) internal body's state (position and motion, i.e. proprioception); (4) chemical environment (levels of oxygen, nutrients and possibly toxins); and (5) signaling molecules such as hormones, neurotransmitters, and cytokines.

Septum: partition of an organ. In the heart, the myocardium dividing the left chambers from the right ones.

Serous Membrane: tissue composed of epithelium and underlying loose connective tissue, particularly the lining of the pericardial, pleural, and peritoneal cavities.

Signal Averaging: signal processing method for improving the signal-to-noise ratio by taking the average of several samples of a given signal acquired under similar conditions.

Spindle: cell structure made of chromosomes, microtubules and associated proteins, which segregates the duplicated chromosomes during mitosis, a part of the cell cycle.

Standard Deviation: square root of the sample variance.

Stem Cell: Certain stem cell descendants produce new stem cells, keeping the stem cell properties (self-renewal). Others differentiate, entering into a division set to give birth to functional cells and leading to a loss of stem cells. Stem cells can be labeled according to descendant function or development stage (adult or embryonic). Totipotent stem cells have unlimited differentiation potential, whereas pluri- (zygote), multi-, and bipotent (adult) stem cells give rise to a limited number of differentiated descendant types.

Stem cells with their potential to develop into specialized cells can serve in regenerating failing organs, avoiding the creation of tumors or aberrant tissues. Stem cells can, indeed, become malignant, especially when they move away from their niche. Adult stem cells are multipotent, giving birth to cell types that belong to the lineage of their embryonic layer, rather than pluripotent like embryonic stem cells, which come from very early-stage embryos and produce cell types of all three embryonic lineages (ectoderm, endoderm, and mesoderm). Stem cells derived from amniotic fluid are pluripotent, being able to differentiate into various tissue cells, representing each embryonic germ layer (adipogenic, osteogenic, myogenic, endothelial, neuronal, and hepatic lineages) [809].[7] Stem cells exist in tissues, especially those with high cell turnover (skin, gut, and blood). The liver can regenerate more than 50% of its initial mass within weeks.

The pool of hematopoietic stem cells represents less than 10^{-4} of bone marrow cells in adults, but each of these cells gives birth to a large population of intermediately differentiated progenitor cells. These progenitor

---

[7] Amniotic fluid contains multiple cell types derived from the fetus.

cells divide and differentiate after several stages to form mature cells. The stem cells simultaneously self-renew and remain undifferentiated, whereas progenitor cells lose this property. Stem cell fate and function are controlled by signaling pathways that involve Notch, sonic hedgehog, and Wnt genes. The identity of molecules that confers the ability to self-renew is not fully known, but the stem cell environment, the so-called stem cell niche, plays a fundamental role. However, hematopoietic stem cells can migrate to distant body regions. Progenitor cells move away from the niche.

Smooth Muscle Cell: non-striated, involuntary myocyte in vessel walls.

Stasis: unchanging state.

Stenosis: constriction or narrowing of a vessel.

Stent: short narrow tube, often in the form of a mesh, which is inserted into the lumen of a anatomical conduit such as an artery, bronchus, or bile duct to keep a previously blocked passageway open.

Stroke Index (SI): blood volume expelled by the left ventricle per heart beat indexed by BSA.

Stroke Systemic Vascular Resistance Index (SSVRI): per-beat measurement of afterload.

Stroke Volume (SV): amount of blood ejected by the left ventricle into the vasculature per heart beat.

Suprachiasmatic Nucleus: anatomical center of the circadian clock in the anterior hypothalamus.

Synapse: junction (συν: together; απτο: to attach) usually between a neuron and its target cell. It consists of axon transmitting ending (terminal), tiny synaptic cleft between the cells, and dendrite (receiving ending) of the downstream neuron or the effector cell. In particular, synapses lead to neural circuits and neuromuscular junctions. The enlarged tip of the presynaptic nerve terminal (synaptic button, bouton, or knob) is separated from the post-synaptic target cell surface (a dendrite or a cell body) by the synaptic cleft (width 20 nm). Plasmalemmas of the two adjacent cells are connected by cellular adhesions. Proteins of the postsynaptic membrane anchor and convey neurotransmitter receptors, as well as modulators of receptor activity.

Immunological Synapse: interface between an antigen-presenting cell and a responding T lymphocyte that is similar to synapses of the nervous system. Both antibody-mediated (humoral immunity; antibodies in body fluids bind cognate antigens and trigger a response) and cell-mediated (T lymphocytes bind to the surface of antigen-presenting cells and trigger a response, which can involve other types of leukocytes) immune processes indeed involve close contact between a T lymphocyte and an antigen-presenting cell (i.e., helper T lymphocyte Th1 with dendritic cell or macrophage; helper T lymphocyte Th2 with B lymphocyte; cytotoxic T lymphocyte with cell infected by virus, bacterial components, or parasites, allograft, and tumor cell; NK lymphocyte with virus-infected or cancerous cell). Several types of molecules intervene: (1) T-cell receptors and the major histocompatibility complex on the antigen-presenting cell; (2) adhesion molecules; and (3) cytokines (such as interleukins and perforin) and clusters of cytokine receptors.

Ventricular Stroke Work: work applied to the blood at each ejection by the ventricle.

Left Stroke Work Index: normalized amount of work the heart expends over one heart beat interval.

Sympathetic: part of the autonomic nervous system that is generally responsible for elevating metabolism and increasing alertness (in opposition to parasympathetic).

Systemic: pertaining to the cardiovascular system with the exception of vessels perfusing the lungs (in opposition to pulmonary).

Systole: period of cardiac contraction (in opposition to diastole).

Tachycardia: $f_c > 1.6\,\text{Hz}$.

Telepresence: remote manipulation of instruments via controlled communications and robotic arms.

Test Sensitivity: ratio between the number of true positive and total number of abnormals (sum of true positive and false negative) $\text{TP}/(\text{TP} + \text{FN})$.

Test Specificity: ratio between the number of true negative and total number of normals $\text{TN}/(\text{TN} + \text{FP})$.

Thermoablation: therapeutic technique, either based on heat, created at a targeted zone by electromagnetic wave (radiofrequency ablation) or pressure wave (ultrasound destruction), or cold (cryoablation), aimed at removing dysfunctional cells or added to partial resection of tumors.

Thrombus: blood clot.

Tip Cell: specialized endothelial cell at the leading edge of vascular sprouts that make filopodia for guiding vascular sprout growth according to VEGF concentrations. VEGF-A induces endothelial tip cells. Delta-like ligand-4/Notch signaling, which occurs downstream from VEGF, decreases the number of tip cells and vessel branchings.

Total Peripheral Resistance: determinant of the cardiac load controlled by the nervous system, which is commonly estimated either by the aortic flow-to-pressure ratio or using a windkessel model.

Variance: average value of the square of the difference between the random variable and its mean.

Vasoconstriction: reduction in bore of blood vessels, particularly the arterioles, which control the distribution of blood in the body (in opposition to vasodilation).

Vasculogenesis: development of blood vessels that involves the differentiation of angioblasts into endothelial cells and their assembly into primary vascular plexi. Vasculogenesis is regulated by interactions of growth factors such as vascular endothelial growth factor, fibroblast growth factor, transforming growth factor-β, and platelet-derived growth factor. The vessel growth direction is driven by leading endothelial tip cells in response to guidance molecules, whereas the lagging endothelial cells form vascular lumens. During maturation, pericytes and smooth muscle cells are recruited to capillaries and large blood vessels, respectively. The vascular network remodels with body's growth by pruning and branching in response to growth factors, hypoxia, and blood flow to form a structure of arteries, capillaries, and veins.

Vasodilation: increase in caliber of blood vessels, particularly the arterioles (in opposition to vasoconstriction).

Venous Pump: local external forces acting on valvular veins that improve venous return.

Venous Return: blood flow reaching the right atrium.

Virtual Reality: computer modeling associated with real-time simulations.

Windkessel effect: arteries swell and shrink during each heart beat, especially elastic arteries near the heart, during blood ejection by myocardium contraction and the absence of cardiac output during myocardium relaxation, respectively. The phase lag of the artery wall displacement with respect to the beginning of myocardium contraction/relaxation depends on pressure wave propagation, which is finite in deformable arteries. The arterial buffer, with this windkessel effect, allows the transformation of input pressure, which varies from almost zero to its maximum, into a pressure changing from a diastolic minimum (much greater than in the left ventricle) to a systolic maximum. The starting-stopping flow at the cardiac outlet is transformed into a pulsatile flow (time-dependent but continuous throughout the cardiac cycle) in the arteries.

B

Adipocyte

Adipocytes (the latin root adeps means fat) store lipids for restriction periods. Adipose tissues have an important growth potential, then requiring angiogenesis (part I). However, excess inputs develop adipose tissues, increasing cell size and number, and augment the occurrence of cardiovascular diseases. Lipid accumulation in adipocytes disturbs the adipokine secretion (Table B.1), impairs insulin signaling, and dysregulates the cell functioning.

The adipokines (or adipocytokines) belong to a group of cytokines (between-cell communication, i.e., paracrine function) secreted by adipose tissues and other organs. They can also function as hormones (endocrine function). The adipokine family include leptin, adiponectin, resistin, visfatin, and retinol binding protein-4. Adipokines reduce fatty acids in non-adipose tissue cells.

Table B.1. Adipocyte production. Adipokins regulate food intake, and thereby energy homeostasis.

Adipokines	Leptin, adiponectin, resistin
Cytokines	Tumor necrosis factor-α, interleukin-6
Inflammatory reactants	Serum amyloid A, pentraxin, lipocalin, ceruloplasmin, macrophage migration inhibitory factor
Angiogenic factors	Vascular endothelium growth factor, monobutyrin
Lipogenic factors	Acylation-stimulating protein
Matrix components	Collagen-4
Proteases	Adipsin
Miscellaneous	Osteonectin, stromolysin

Adipocytes also secrete angiotensinogen, adipocyte differentiation factor, interleukin-6, tumor necrosis factor-α, and plasminogen activator inhibitor-1, as well as others mediators, such as nitric oxide, prostaglandins, acylation-stimulating protein, and adipsin (complement-D). Visceral adipose tissues seem to have up to five times the number of PAI1-producing stromal cells compared with subcutaneous adipose tissues [810].[1] Circulating PAI1 level is correlated with an accumulation of visceral fat.

Adipocytes and resident macrophages synergistically secrete TNFα and IL6, particularly in obesity. Obesity and insulin resistance increase cardiovascular risk by dyslipidemia, hypertension, glucose dysmetabolism, and mechanisms that implicate adipokines, cytokines, and hypofibrinolytic factors [811]. Adipocytes release certain compouds that alter glucose and lipid metabolism, blood pressure, coagulation, and fibrinolysis, and lead to inflammation.

Dyslipidemia in obesity is characterized by increased concentrations of VLDLs and of LDLs, and decreased levels of HDLs. Hepatic overproduction of VLDL is a consequence of hepatic steatosis. In insulin-resistant states of obesity,[2] the dyslipidemia is characterized by an increased concentration of smaller, denser LDLs, after increased lipolysis by hepatic lipase [811]. These LDLs are more exposed to oxidation. They are mostly targeted by macrophage scavenger receptors rather than the normal LDL receptor.

B.1 Adiponectin

Adiponectin (necto: to bind) is synthesized mainly by adipocytes. It is also expressed by skeletal myocytes, cardiomyocytes, and endothelial cells.

---

[1] Plasminogen activator inhibitor-1 is a marker of hypofibrinolysis.

[2] Insulin resistance leads to: (1) impaired glucose uptake, particularly in myocytes, in hepatocytes, and in adipocytes, (2) impaired LDL receptor activity, with delayed VLDL clearance, and (3) inability to suppress hepatic glucose production and release of non-esterified fatty acids from hypertrophic adipocytes. The level of non-esterified fatty acids increases owing to decreased lipolysis, fatty acid oxidation and low levels of adiponectin (the latter favoring fatty acid oxidation), stress-induced adrenergic stimulation, and to inflammation. Increased levels of non-esterified fatty acids cause lipotoxicity, impair endothelium-dependent regulation of the vasomotor tone, increase oxidative stress and have cardiotoxic effects. Reduced lipoprotein lipase activity decays the clearance of triacylglycerol-rich lipoproteins. Impaired lipolysis of triacylglycerol-rich lipoproteins decreases the transfer of apolipoproteins and phospholipids from triacylglycerol-rich lipoproteins to HDL, thus reducing HDL concentration. Furthermore, delayed clearance of triacylglycerol-rich lipoproteins facilitates CETP-mediated exchange between cholesterol esters in HDL and triacylglycerols in VLDL. Also, the degradation rate of ApoB100, which regulates VLDL secretion, is decreased in insulin resistance. In the cardiovascular system, insulin resistance is associated with inhibition of the PI3K pathway and overstimulation of the growth factor-like pathway.

Adiponectin circulates in blood at high concentrations (5–10 mg/ml). Adiponectins exist as a low-molecular-weight full-length trimers and globular cleavage fragments. The full-length trimer can dimerize to form a middle-molecular-weight hexamer, which can oligomerize to form a polymer. Full-length adiponectin stimulates AMPK phosphorylation (activation) in the liver, whereas globular adiponectin yields this effect in skeletal myocytes, in cardiomyocytes, and hepatocytes.

Adiponectin binds to its G-protein-coupled receptors, adipoR1 and adipoR2 (Table B.2). T-cadherin could act as a co-receptor for the middle/high-molecular-weight adiponectin on endothelial cells and smooth muscle cells, but not for the low-molecular-weight trimeric and globular forms [812, 813]. Activation of adipoR1 and adipoR2 by adiponectin stimulates the activation of peroxisome-proliferator-activated receptor-α (PPARα), AMP-activated protein kinase (AMPK) and p38 mitogen-activated protein kinase. Stimulation of AMP-activated protein kinase in the liver and skeletal muscle strongly affects fatty acid oxidation and insulin sensitivity.

Adiponectin affects gluconeogenesis and lipid catabolism. Adiponectin hinders atherosclerosis. Adiponectin favors insulin activity in the muscles and liver via activated AMPKs (Table B.3). PPARγ upregulates adiponectin expression and reduces the plasmatic TNFα concentration. TNFα, produced in adipose tissues, prevents adiponectin synthesis. TNFα phosphorylates insulin receptors, and hence desensitizes insulin signaling.

Adiponectin has dominant anti-inflammatory features, and thus anti-atherogenic and antidiabetic properties (Table B.4). Adiponectin regulates the expression of both pro- and anti-inflammatory cytokines. It suppresses the synthesis of tumor-necrosis factor-α and interferon-γ and favors the production of anti-inflammatory cytokines, such as interleukin IL10 and IL1-receptor antagonist by monocytes, macrophages, and dendritic cells. Adiponectin

Table B.2. Receptors of adipocyte hormones and their main transducers. Adiponectin avoids G-protein (Source: [814])

Type	Main transducer	Ligand
AdipoR1	AMPK, MAPK	Adiponectin
AdipoR2	AMPK, MAPK	

Table B.3. Effects of adipokines on glucose level.

Decrease	Increase
Adiponectin	Resistin
Leptin	RBP4
Omentin	TNFα, IL6
Visfatin	

increases the synthesis of tissue inhibitor of metalloproteinase in macrophages via IL10 [815].

Adiponectin inhibits the expression of adhesion molecules via inhibition of TNF and NFκB. Adiponectin thus also impedes endothelial-cell proliferation and migration [813]. Adiponectin suppresses endothelial cell apoptosis [816]. Adiponectin also hampers foam cell formation.

B.2 Leptin

Leptin (λεπτοσ: thin), mainly produced by adipocytes in response to high lipid levels, regulates satiety. Leptin, indeed, represses food intake and promotes energy consumption.[3] Leptin can be co-expressed with growth hormone in somatotropes of the anterior pituitary.[4] Leptin circulates in the blood (at

Table B.4. Adipocytokines and their activity in inflammation and immunity (Source: [813]).

Adipokin	Inflammatory and immune effects
Adiponectin	Anti-inflammatory ($\downarrow$ endothelial adhesion molecules, $\downarrow$ phagocytosis, T-cell responses, $\downarrow$ B-cell lymphopoiesis, $\downarrow$ NFκB, TNFα, IL6, IFNγ, $\uparrow$ IL1RA, IL10) Pro-inflammatory ($\uparrow$ CXCL8 in presence of lipopolysaccharide)
Leptin	Pro-inflammatory ($\uparrow$ TNFα, ROS, IL6, IL12, chemotaxis, neutrophil activation, thymocyte survival, lymphopoiesis, T-cell proliferation, NK-cell function, $\uparrow$ TH1 response, $\downarrow$ TH2 activity)
Resistin	Pro-inflammatory ($\uparrow$ endothelial adhesion molecules, $\uparrow$ NFκB, TNFα, IL1β, IL6, IL12)

---

[3] Adipose tissues are aimed at storing energy. Adipocytes saturated with lipids can lead to lipid accumulation in other tissues, reducing their functioning. Adipose tissues act as endocrine organs, secreting adipokines. Leptin is detected by the arcuate nucleus in the hypothalamus. Increased arcuate nucleus activity inhibits the production of neuropeptide-Y in the paraventricular nucleus, thereby reducing feeding.

[4] Somatotropes have leptin receptors. Leptin can thus be an autocrine or paracrine regulator.

a concentration of a few ng/ml) and in the cerebrospinal fluid, crossing the blood–brain barrier in order to regulate food intake by the hypothalamus.

Leptin interacts with six types of receptors (LepRa–LepRf). LepRb is found in the hypothalamus, especially in the satiety center.[5] Leptin hampers the activity of neurons expressing neuropeptide-Y and agouti-related peptide (AgRP), and favors the activity of neurons expressing α-melanocyte-stimulating hormone.

Leptin receptors are widely distributed on endothelial cells and vascular smooth muscle cells. Leptin stimulates mitogen-activated protein kinases and phosphatidylinositol 3-kinase. Leptin induces SMC proliferation and migration [817]. Leptin also favors platelet aggregation and promotes angiogenesis [818]. Leptin intervenes not only in angiogenesis but also in hematopoiesis, upregulating the expression of vascular endothelial growth factor via activation of NFκB and PI3K [812]. Leptin also favors the production of nitric oxide synthase-2, and thereby of reactive oxygen species.

Leptin stimulates AMP-activated protein kinase which decreases ATP-consuming anabolism and increases ATP-manufacturing catabolism. Leptin decreases insulin levels by inhibition of proinsulin synthesis and reduction of secretion. In myocytes, leptin improves insulin sensitivity and reduces intracellular lipid levels by direct activation of AMP-activated protein kinase combined with indirect inputs to the central nervous system. In the liver, leptin also enhances insulin sensitivity.

Leptin has dominant pro-inflammatory effects (Table B.4). Leptin binds to its receptor OBRb and activates: (1) the mitogen-activated protein kinase pathway (p38 and ERK) and (2) signal transducer and activator of transcription STAT3, thus producing pro-inflammatory cytokines TNFα and interleukins IL6 and IL12 in monocytes and macrophages. Leptin favors the activities of monocytes, macrophages, and natural killer cells [813]. Leptin stimulates neutrophil chemotaxis and the production by neutrophils of reactive oxygen species. Leptin stimulates the production of IgG2a by B lymphocytes. Leptin increases IL2 secretion by T lymphocytes.

B.3 Resistin

Resistin is synthesized by adipocytes and other cells of adipose tissues, as well as by myocytes, pancreatic cells, and macrophages. Resistin circulates in high-molecular-weight hexamers and low-molecular-weight complexes [813]. The resistin synthesis is affected by pituitary, steroid and thyroid hormones;

---

[5] Leptin receptor is expressed at low levels in manifold tissues and at high levels in the mediobasal hypothalamus, particularly in the arcuate nucleus, ventromedial nucleus, and dorsomedial nucleus. Activation of leptin receptors in the hypothalamus represses orexigenic pathways involving neuropeptide Y and agouti-related peptide and stimulates anorexigenic pathways involving pro-opiomelanocortin and cocaine and amphetamine-regulated transcript [812].

adrenaline; endothelin-1; and insulin. Resistin could reduce glucose uptake by muscles, adipose tissues, and the liver, affecting insulin sensitivity.

Resistin activates phosphatidylinositol 3-kinase and members of the mitogen-activated protein kinase family, p38 and ERK. Resistin has dominant pro-inflammatory features. Resistin increases the production of tumor-necrosis factor and interleukins IL1β, IL6, and IL12 by various cell types via NFκB-dependent process. IL1, IL6, and TNF upregulate the resistin expression. Resistin upregulates the expression of adhesion molecules, such as vascular cell adhesion molecule-1 and intercellular adhesion molecule-1, as well as CCL2 by endothelial cells. It also favors the release of endothelin-1 by endothelial cells.

B.4 Other Adipokines

Visfatin has an insulin-like activity because it binds to and activates insulin receptor at binding site different from that of insulin [819]. It thus favors glucose uptake. Also, visfatin inhibits neutrophil apoptosis.

Visceral adipose tissue-derived serine protease inhibitor (vaspin) suppresses the production of tumor-necrosis factor, leptin, and resistin [813].

Omentin, an insulin sensitizer made by stromal-vascular cells within the fat pads, enhances glucose uptake [812].

Retinol-binding protein-4 (RBP4) impairs insulin action on the liver and muscles [812]. Retinol binding protein-4 contributes to insulin resistance.

Insulin resistance is also associated with lipolysis and release of non-esterified fatty acids into the circulation [812]. Circulating non-esterified fatty acids reduce glucose uptake by adipocytes and myocytes, and promote glucose release by hepatocytes. Transient increases in non-esterified fatty acid levels, such as acute changes after a meal, enhance insulin secretion, whereas chronic elevations associated with insulin resistance reduce insulin secretion by the pancreas.

C

Basic Aspects in Mechanics

C.1 Dimensionless Parameters

Dimensional analysis groups together the influence factors in dimensionless ratios. The formulation of the dimensionless equations depends on the choice of variable scale $(\cdot^{\star})$. The dimensionless equations exhibit a set of governing dimensionless parameters, which have a suitable physical meaning for the problem (Table C.1). The phenomenological analysis gives the order of magnitude of the different terms of dimensionless equations. Scaling takes place in dimensionless coefficients allocated to the corresponding terms. Models are devised using geometrical and dynamical similarity.

In a vessel cross-section that is not circular, the hydraulic diameter is introduced: $d_h = 4A/\chi$ where A is the cross-sectional area and χ the wetted perimeter. Different types of forces act on every fluid particle (ρ: fluid density, μ: fluid dynamic viscosity). (1) Remote body forces generated by a potential like the gravity ($\rho\mathbf{g}$) or an electromagnetic field such as in MRI, are currently neglected. The other forces are fluid-particle surface forces. (2) Pressure forces (p) result from the pressure gradient in the streamwise direction (downstream (D) and upstream (U) directions, then applied to the D and U faces of the hexahedral infinitesimal control element taken as the fluid particle). Adjacent north and south (N/S) and west and east (W/E) particles in the two other directions normal to the local flow direction induce pressure forces on the fluid particle, which prevent particle rotation, although torque results from particle

Table C.1. Examples of flow dimensionless ratios.

Length ratios	L/d_h, R/R_c, δ/R, $V_q/(R\omega)$, $\widehat{V_q}^2/(R_c R\omega^2)$, $u_*/(R\omega)$, $u_*^2/(\omega\nu)$, R/λ,...
Time ratios	$R\omega/\overline{V_q}$, $\delta\omega/\widehat{V_q}$,...
Velocity ratios	$\widehat{V_q}/\overline{V_q}$, $V_q(t)/c(p(t))$,...
Force ratios	Re, De, St, Sto, $L(dV_c/dt)/V_c^2$, $\mu L/(a_0^2(K\rho)^{1/2})$,...

Table C.2. Dimensionless parameter values in the arteries at rest with $f = 1\,\text{Hz}$.

Blood vessel	Radius (mm)	V_q (cm/s)			Re			St		Sto
		Mean	Peak	Min	Mean	Peak	Min	Mean	Peak	Min
Ascending aorta	10	20	70	−20	500	1750	500	0.31	0.09	0.31 12.5
Descending aorta	10	20	60	−10	500	1500	250	0.31	0.10	0.62 12.5
	3	10	50		75	375		0.19	0.04	3.8
	2	7	30		35	150		0.18	0.04	2.5
	1	7	20		18	50		0.09	0.03	1.3

shearing, which provides vorticity. (3) Shear forces ($\propto \mu V^\star / L^{\star^2}$ are caused by the friction between adjacent particles and the wall for the particle close to it, acting in the flow opposite directions when they are caused by slower moving particles. (4) Inertia forces, with their temporal ($\propto \rho U^\star \omega$) and convective ($\propto \rho V^{\star^2}/L^\star$) components are reactions to the fluid motion. The values of the main dimensionless parameters in the arteries are given in Table C.2.

C.1.1 List of Main Dimensionless Flow Parameters

Amplitude Ratio (or modulation rate): $\gamma_v = \widehat{V_q}/\overline{V_q}$. The pulsatile flow is less unsteady when the modulation rate is lower.

Archimedes Number: determines the motion of fluids due to density differences $gL^3\rho(\rho_b - \rho)/\mu$ (g: gravity acceleration (9.81 m/s^2), ρ_b: body density).

Courant Number (CFL number): provides the discretization step used in numerical solution of PDEs to fet a stable scheme. There are the *propagation Courant number* $c\Delta t/h$ (c: wave speed, Δt: time step, h: mesh element size) and the *convection Courant number* $v\Delta t/h$ (v: fluid velocity).

Dean Number: $De = (R_h/R_c)^{1/2}Re$ in laminar flow through curved vessels is the product of the square root of the vessel curvature ratio $\kappa_c = R_h/R_c$ by the Reynolds number (R_h: hydraulic radius, R_c: curvature radius). De is then proportional to the ratio of the square root of the product of convection inertia by centrifugal forces to viscous forces. The dynamical similarity of a steady laminar motion of an incompressible fluid through a rigid smooth bend, with a small uniform planar curvature, is found to depend upon De, introduced by Dean (1927, 1928). For fully developed turbulent flow, the friction factor depends on the Ito number Ito $= (R_h/R_c)^2 Re$ (Ito, 1959). The Dean number De usually cannot be calculated because of the complex curvature of the vessel axis, which varies continually in the three spatial directions.

Deborah Number: ratio of the polymer characteristic relaxation time and the flow time scale. The smaller the Deborah number, the less gel-like structure there is in the material.

Froude Number: $V^*/(gL^*)^{1/2}$ quantifies the relative influence of gravity. The *reduced Froude number*, or densimetric Froude number, is a Froude number in which gravity g is replaced by Archimedes thrust.

Grashof Number: combination of Reynolds and reduced Froude numbers $\text{Gr} = (\text{Re}/\text{Fr})^2$.

Kàrmàn Number: dimensionless parameter used to specify the non-zero mean sinusoidal flow $(dp/dz)R^3/(\rho\nu^2)$.[1] In turbulent flow, it is also defined as $\delta u_*/\nu$ (u_*: friction velocity), which is a Reynolds number representing the ratio of the large eddy scale (δ: boundary layer thickness) to the small motion scale ν/u_*

Knudsen Number: ratio of the mean free path of molecules to the length scale.

Mach Number: ratio of the fluid speed to the propagation speed $\text{Ma} = V_q/c$. The wave speed c depends on the involved deformable parts, fluid compressibility χ and vessel distensibility D ($\chi \ll D$ even for inhaled air). The usual speed is the speed of sound. In blood circulation, Ma is related to the propagation speed of pressure waves. The Mach number defines convection regimes associated with the vessel compliance (collapsible vessels).

Nusselt Number: $\text{Nu} = \text{h}L^*/l g_T$ is involved in heat transfer between a solid or a fluid and a moving mono- ou polyphasic fluid (h: convection coefficient, λ_T: thermal conductivity).

Péclet Number (Pe): involved in convection exchanges, is the ratio of the convection mass transport to diffusion mass transport $\text{Pe} = L^*V^*/D = \text{ReSc} = L^*V^*/\alpha_T = \text{Re} \times \text{Pr}$ (D: molecular diffusivity, α_T: thermal diffusivity). Pe is a Reynolds-like number based on molecular or thermal diffusivity rather than momentum diffusivity.

Prandtl Number: ratio of kinematic viscosity to thermal diffusivity ν/α_T.

Reynolds Number: $\text{Re} = V^*L^*/\nu$ ($\nu = \mu/\rho$, $V^* \equiv V_q$: cross-sectional average velocity, $L^* \equiv R$: vessel radius) is the ratio between convective inertia and viscous effects applied on a unit of fluid volume. Re is also the ratio between the momentum diffusion time scale and the convection characteristic time $\text{Re} = (R^2/\nu)/(R/U)$. In pulsatile flows, both mean $\overline{\text{Re}} = \text{Re}(\overline{V_q})$ and peak Reynolds numbers $\widehat{\text{Re}} = \text{Re}(\widehat{V_q})$, proportional to the mean and the peak cross-sectional average velocity, respectively, can be calculated. Re controls flow pattern transition. $\text{Re}_\delta = \text{Re}/\text{Sto}$ is used for flow stability study (δ: boundary layer thickness). Branching pulsatile flows are currently based on stem peak Reynolds number $\widehat{\text{Re}}$ calculated from $\widehat{V_q}$, on

---

[1] $\text{Ka}_\sim = R^3/(\rho\nu^2)G_{p\sim}$ ($G_{p\sim} = (dp/dz)_\sim \propto \rho\omega V_\sim$) can be expressed as $\text{Ka}_\sim = \text{Re}_\sim \times \text{Sto}$. $\overline{\text{Ka}} = R^3/(\rho\nu^2)\overline{G_p}$ ($\overline{G_p} = \overline{(dp/dz)} \propto \mu\overline{V}/R^2$) is the Reynolds number: $\overline{\text{Ka}} \equiv \text{Re}$.

the trunk radius R and on the blood kinematic viscosity.[2] In bend flows, a secondary motion-associated Reynolds number can be introduced, using the velocity scale $V_2^\star$, when centrifugal forces $\rho V^2/R_c$ are balanced by local inertia effects $\rho \omega V_2^\star$) $Re_2 = V^2 R/(\omega \nu R_c) = V^2/(\omega \nu) \times \kappa_c = Re/St$. The wall shear Reynolds number is defined by $u_\star \delta/\nu$

Schmidt Number: ratio of the kinematic viscosity to the molecular diffusivity of the specy $Sc = \nu/D$. It provides the ratio of the viscous boundary layer to the concentration boundary layer.

Stokes Number: $Sto = R(\omega/\nu)^{1/2}$ (also called Witzig-Womersley number) is the frequency parameter of pulsatile flows (ω: pulsation of flow oscillation). Sto is the square root of the ratio between time inertia and viscous effects. The Stokes number is a Reynolds-like number for periodic flow (local acceleration replaces convective acceleration). The Stokes number is proportional to the ratio between the vessel hydraulic radius and the Stokes boundary layer thickness ($Sto \propto R_h/\delta_S$) and the ratio between momentum diffusion time and the cycle period ($Sto \propto ((R^2/\nu)/(1/\omega)) \equiv (T_{\mathrm{diff}}/T)^{1/2}$).

Strouhal Number: $St = \omega L^\star/V^\star$ is the ratio between time inertia and convective inertia ($St = Sto^2/Re$). In quasi-periodic flow in a branching vessel region, the peak Strouhal number is based on the trunk peak cross-sectional average velocity: $St = \omega R/\widehat{V_q}$. The Strouhal number is proportional to the ratio between the steady and the Stokes boundary layer thicknesses ($St = \delta/\delta_S$). The dimensionless stroke length $(\widehat{V_q}/(R_h\omega))$, which is also the ratio of the length scale of the axial displacement of a fluid particle to the vessel radius or the ratio between the flow cycle and the convection time scale, is the inverse of the Strouhal number. In the aorta at rest, $\widehat{V_q}/R_h\omega = 11.1$ with the following value set: $f = 1\,Hz$, $R_h = 10\,mm$, $\widehat{U_q} = 0.7\,m/s$. Bend flows depend, for small values of the frequency parameter, on the Strouhal number for the secondary motion (5elocity scale $V_2^\star$), when centrifugal forces $\rho V^2/R_c$ are balanced by local inertia effects $\rho \omega V_2^\star$, $St_2 = (\omega^2 R R_c)/V^2$. In turbulent periodic flows, $St = R\omega/u'$ is the ratio of the time scale of turbulent fluctuations R/u' (u': turbulent intensity) to the flow period. The turbulent Strouhal number $St = \omega R/\overline{u}_\star$ can also be defined by the time mean friction velocity $\overline{u}_\star$.

The *Helmholtz Number* $He = \omega L^\star/c$, used to estimate whether or not the fluid compressibility must be taken into account ($He \ll 1$: the compressibility is ignored), is a Strouhal-type number.

Taylor Number: based on the Stokes boundary-layer thickness δ_S,

$$Ta = V^2 \delta_S^3/(R_c\nu^2) = V^2/(\omega^{3/2}\nu^{1/2}R_c)$$

is used as a stability parameter in bends. Ta is also equal to $V/(R_c\omega)^2 \times R_c(\omega/\nu)^{1/2} = Sto \times St_2^{-1}$ or to $Re_\delta^2 \times \delta_S/R_c$.

---

[2] The blood kinematic viscosity $\nu = 4 \times 10^{-6}\,m^2/s$ when it is assumed to be Newtonian.

Weissenberg Number: dimensionless number for viscoelastic flows, defined as the product of relaxation time and shear rate.

Complex flows have been analyzed by flow decomposition into an inviscid core and a viscous boundary layer considered to be the Stokes type in periodic flows. The parameter $L(d\tilde{V}_\infty/d_t)/\tilde{V}_\infty^2$ (L: distance from vessel entry, $\tilde{V}_\infty$: dimensionless free-stream velocity) must be introduced when the flow rate is close to zero, time inertia forces being greater than convection inertia forces. The quasi-steadiness assumption is only valid when $L^\star \tilde{\tilde{V}}^\star / \tilde{V}^{\star 2} \ll 1$ [820].

C.1.2 Scaling of Time-Dependent Flow in Planar Uniform Bends

The bend is the basic simple model of anatomical ducts. Phenomenologic analysis of time-dependent flows leads to the following classification of flows in curved pipes, from which the velocity scale of the virtual secondary motion can be determined [821].

- Unsteady inertia dominated secondary motion
  - time inertia is greater than centrifugal forces,

$$\frac{V^2}{R_c} \sim V_2 \omega \implies V_2 \sim \frac{V^2}{\omega R_c} \; ;$$

  - time inertia is greater than convective inertia:

$$V_2 \omega \gg \frac{V V_2}{L} \implies \text{St} \gg 1 \, , \quad V_2 \omega \gg \frac{V_2^2}{R} \implies \text{St}_2 \gg 1 \, ;$$

  - time inertia is greater than viscosity:

$$V_2 \omega \gg \frac{\nu V_2}{R^2} \implies \text{Sto} \gg 1 \, .$$

- The dominant term of the secondary motion is the viscosity:

$$\frac{V^2}{R_c} \sim \frac{\nu V_2}{R^2} \implies V_2 \sim V^2 \frac{R^2}{\nu R_c} = \text{Sto}^2 \frac{V^2 \omega}{R_c} \, ,$$

$$\frac{\nu V_2}{R^2} \gg \frac{V^2}{R} \implies \frac{V_2 R}{\nu} = \text{Re}_2 \ll 1 \, ,$$

$$\frac{\nu V_2}{R^2} \gg \frac{V V_2}{L} \implies \frac{V R^2}{\nu L} \ll 1 \, ,$$

$$\frac{\nu V_2}{R^2} \gg \omega V_2 \implies \text{Sto}^2 \ll 1 \, .$$

- The dominant term of the secondary motion is the convective inertia:

$$\frac{V^2}{R_c} \sim \frac{V_2^2}{R} \implies V_2^2 \sim V^2 \kappa_c \, ,$$

$$\frac{\nu V_2}{R_c^2} \ll \frac{V_2^2}{R} \implies \mathrm{Re}_2 \gg 1 \, ,$$

$$\omega V_2 \ll \frac{V_2^2}{R} \implies \mathrm{St}_2 \ll 1 \, .$$

C.2 Poiseuille Flow

Poiseuille flow is used as a reference for comparison as well as a simplification in many models dealing with relationships between the flow rate and the pressure drop in a network composed of several vessel generations. This flow type cannot be observed in physiological vessels because it corresponds to a fully-developed, steady, laminar flow of a homogeneous, incompressible, Newtonian fluid in a long, straight, cylindrical, pipe with a rigid, smooth wall and a uniform circular cross-section [822]. The fluid particle flows with straight paths in concentric layers parallel to the pipe wall. The velocity profile is invariant. The pressure drop, balanced by the viscous effects, varies linearly with the distance along the duct. This conditions allows an analytical solutions of the Navier-Stokes equation.

$$u = \frac{1}{4\mu} \frac{\Delta p}{l}(r^2 - R^2) = u_M \left(1 - \left(\frac{r}{R} \right)^2 \right) = 2, V_q (1 - \tilde{r}^2) \, ,$$

$$q = -\frac{\pi}{8\mu} \frac{\Delta p}{l} R^4 \, , \quad V_q = q/(\pi R^2) = \frac{1}{8\mu} \frac{\Delta p}{l} R^2, \quad u_M = 2V_q \, ,$$

$$\frac{du}{dr} = -2u_M r / R^2 = -\frac{\Delta p}{2\mu l} r \, , \quad \Delta p = \mathrm{R} q \, ,$$

$$\mathrm{R} = \frac{8\mu l}{\pi R^4} G = \frac{\Delta p}{l} = \frac{32\mu}{d^2} V_q \, , \quad \Lambda = \frac{64}{\mathrm{Re}} \, ,$$

$$\tau_w = -\mu \left(\frac{du}{dr} \right) \bigg|_{r=R} = -\mu \left(\frac{\Delta p}{2\mu l} r \right) \bigg|_{r=R} = -\frac{R \Delta p}{2l}, \quad \tau_w = -4\mu \frac{V_q}{R} \, ,$$

$$\tau_w = C_f \rho \frac{V_q^2}{2} \, , \quad C_f = \Lambda \frac{l}{d} = \frac{\Lambda}{4} = \frac{16}{\mathrm{Re}} \, .$$

For an elliptical cross-section,

$$q = -\frac{\pi}{4\mu} \frac{\Delta p}{l} \frac{a^3 b^3}{a^2 + b^2} \, .$$

C.3 Womersley Flow

Consider a fully developed laminar flow of a homogeneous incompressible Newtonian fluid in a horizontal, cylindrical, uniform, straight pipe of circular cross-section, of smooth rigid wall [823]. The pulsatile flow is composed of a steady component $(\overline{V_q})$ and a sinusoidal modulation, with a given circular frequency (ω) and amplitude $(V_{q\sim})$: i.e., a nonzero-mean sinusoidal flow $(\widehat{V_q} = \overline{V_q} + V_{q\sim} = \overline{V_q}(1 + \gamma_u), \gamma_u = V_{q\sim}/\overline{V_q}$: amplitude ratio or modulation rate). This an example of analytical solution of the Navier-Stokes equation.

The velocity field is $\mathbf{v} = \mathbf{v}(r,t)$ $(v \equiv v_z)$.

$$v_{,t} = -\frac{1}{\rho}p_{,z} + \nu \frac{1}{r}\partial_r(ru_{,r})$$

let $G_p = -p_{,z}$ be the constant pressure gradient,

$$G_p = \overline{G_p} + G_{p\sim}\exp\{\imath\omega t\}.$$

With the dimensionless quantities $\tilde{v} = v/V$ $(V = R/T)$, $\tilde{r} = r/R$ and $\tilde{p} = p/(\rho V^2)$, the equation becomes:

$$\tilde{v}_{,\tilde{t}} = \frac{V}{R}\widetilde{G}_p + \frac{\nu}{R^2}\left(\tilde{u}_{,\tilde{r}\tilde{r}} + \frac{1}{\tilde{r}}\tilde{u}_{,\tilde{r}}\right).$$

The decomposition of the equation in a real steady part and an imaginary unsteady part leads to the following system $(V/(R\omega) = 1)$:

$$\overline{\tilde{v}}_{,\tilde{t}} = \frac{V}{R}\overline{\widetilde{G}_p} + \frac{\nu}{R^2}\left(\overline{\tilde{u}}_{,\tilde{r}\tilde{r}} + \frac{1}{\tilde{r}}\overline{\tilde{u}}_{,\tilde{r}}\right),$$

$$\imath\tilde{v}_{\sim,\tilde{t}} = \widetilde{G}_{p\sim} + \mathrm{Sto}^{-2}\left(\tilde{u}_{\sim,\tilde{r}\tilde{r}} + \frac{1}{\tilde{r}}\tilde{u}_{\sim,\tilde{r}}\right).$$

With the variable change $w = \hat{\tilde{v}} + \imath\widehat{G}_p$, the equation of imaginary part becomes:

$$w_{,\tilde{r}\tilde{r}} + \frac{1}{\tilde{r}}w_{,\tilde{r}} - \imath\mathrm{Sto}^2 w = 0.$$

The term $(-\imath\mathrm{Sto}^2)^{1/2}\tilde{r}$ represents a new variable, a Bessel equation is obtained, which leads to the solution[3]:

$$\tilde{v}_\sim = \imath\widehat{G}_{p\sim}\left(\frac{J_0(\imath^{3/2}\tilde{r}(\omega/\nu)^{1/2})}{J_0(\imath^{3/2}\mathrm{Sto})} - 1\right).$$

The higher the frequency parameter, the greater the distorsion of the velocity profile (Fig. C.1).

---

[3] $J_0(\imath^{3/2}\tilde{r}(\omega/\nu)^{1/2}) = 1 + (\imath/2^2)(\tilde{r}^2\omega/\nu) - 1/(2^24^2)(\tilde{r}^2\omega/\nu)^2 + \imath/(2^24^26^2)(\tilde{r}^2\omega/\nu)^3 + \mathcal{O}[(\tilde{r}^2\omega/\nu)^4]$.

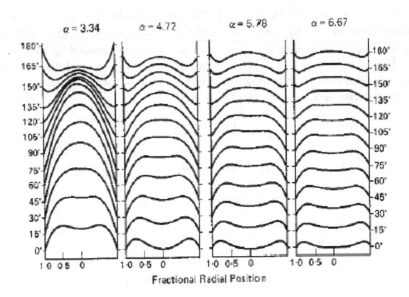

Figure C.1. Velocity profiles of the Womersley solution for different values of the Stokes number. The phase lag rises and the amplitude decays when the frequency increases (from [374]).

C.4 Entry Steady Flow in a Straight Pipe

The entry, or entrance, length Le has been first defined, for a steady laminar flow in a long straight cylindrical conduit of circular cross-section and smooth rigid impermeable wall, with uniform injection velocity, for instance, as the pipe length from which the deviation of the velocity distribution from the Poiseuille (subscript P) distribution[4] is less than 1% [824]. The viscous effects have then pervaded the whole tube lumen. The inlet length has also been defined, for an easier record, as the distance through which the developing maximum velocity, which is the centerline velocity in a long straight tube, reaches 99%[5] of the peak velocity of the fully developed flow: $v_{max}/v_{Pmax} = 0.99$ ($v_{Pmax} = 2V_q$: maximum of the Poiseuille velocity distribution). There is a huge between-author variability in the value of the entry length for a single and simple flow that leads to the Poiseuille flow.

$\widetilde{Le}$ through which the velocity is redistributed approximately into a parabolic

---

[4] $u_P(r) = 2V_q(1 - (r/R)^2)$ (V_q: cross-sectional average velocity, r: radial distance from the tube axis, R: tube radius).

[5] The same threshold is used to define the boundary layer thickness δ, as the distance from the wall where $u(z,\delta) = 0.99\,V_\infty$.

profile:[6]

$$\widetilde{\mathrm{Le}} = \frac{\mathrm{Le}}{d\mathrm{Re}} = 0.065, \qquad (C.1)$$

where Re is the Reynolds number based on the tube hydraulic diameter d. However, this value is overestimated, and the following value is proposed:[7]

$$\widetilde{\mathrm{Le}} = 0.015 , \qquad (C.2)$$

when $200 < \mathrm{Re} < 2500$. When $\mathrm{Re} < 200$, $\mathrm{Le}^+ = \mathrm{Le}/d = 1.2$.[8]

Equations are also proposed, as the following ones:

$$\mathrm{Le}^+ = \kappa_1 \mathrm{Re} + \frac{\kappa_2}{\kappa_3 \mathrm{Re} + 1} , \qquad (C.3)$$

where κ_1, κ_2, κ_3 are constant (e.g., [840, 841]),[9]

$$\mathrm{Le}^+ = \kappa_1 + \kappa_2 \mathrm{Re} , \qquad (C.4)$$

where κ_1, κ_2 are constant (e.g., [842]).[10]

C.5 Rheology Glossary

Let us consider a body elementary particle[11] that undergoes a displacement **u**. This position change of material points can induce a configuration variation

---

[6] An approximate analysis from the estimation of the boundary layer thickness gives a similar result: $\delta \sim 2(\nu z/V_q)^{1/2}$ (ν: fluid kinematic viscosity); therefore, $\mathrm{Le}/(d\mathrm{Re}) = 0.0625$. A value of 0.06 is given in many textbooks (e.g., [825–827]). Normalization quantities $R\mathrm{Re}_d$ ($\mathrm{Re}_d = V_q d/\nu$) and $R\mathrm{Re}_R$ ($\mathrm{Re}_R = V_q R/\nu$) rather than $d\mathrm{Re}_d$ have also been used. Besides, the boundary layer thickness is estimated by a function of the characteristic length $R\mathrm{Re}_R$. The boundary layer is the flow region where the friction forces $\propto \mu V_\infty/\delta^2$ are balanced by the inertia forces $\propto \rho V_\infty^2/R$ (ρ: fluid density). The balance yields the approximative formula for the boundary layer thickness $\delta \sim R\mathrm{Re}_R^{-1/2}$.

[7] The fluid mechanics literature shows a great between-author data variability, e.g., 0.01 [374], 0.028–0.030 [828–834] and 0.04 [835, 836]. The last value is equal to the inlet length in a straight channel with flat parallel walls of width w: $\mathrm{Le}/(w\mathrm{Re}_w) = 0.04$ [837, 838]. A range is sometimes provided to take into account between-author variability of the dimensionless entry length (e.g., [839]: $\widetilde{\mathrm{Le}} \in$ [0.03–0.06], $100 < \mathrm{Re} < 2000$).

[8] The Reynolds number threshold of 200 is author-dependent, a value of 100 is often found.

[9] [840] gives the following values: $\kappa_1 = 0.056$, $\kappa_2 = 0.6$, $\kappa_3 = 0.035$ and [841] - for Le and not Le^+ - $\kappa_1 = 0.061$, $\kappa_2 = 0.72$, $\kappa_3 = 0.04$.

[10] $\kappa_1 = 0.59$, $\kappa_2 = 0.056$.

[11] This infinitesimal volume element of the continuum is assimilated to a material point, which is also called a control volume, the smallest analysis volume, which contains a large number of material molecules. There is a one-to-one relation between the material point and its spatial position.

measured by the strain tensor[12] **E**, which expresses translations and rotations. The displacement **u** is generated by forces, the effects (compression, elongation, and shear) of which are measured by the stress tensor[13] **C**.

For a given loading distribution, one can define: (1) five variables (1a) the displacement vector **u**, of components $\{u_i\}_1^3$, (1b) the stress vector **c** ($c_i = C_{ij}\hat{n}_j$, $f_i = c_i\,dA$), of components $\{c_i\}_1^3$ and the stress tensor **C**, of components $\{C_{ij}\}_{i,j=1}^3$ ($C_{ij} = C_{ji}$), (1c) the strain vector **e**, of components $\{e_i\}_1^3$ and the strain tensor **E**, of components $\{E_{ij}\}_{i,j=1}^3$ ($E_{ij} = E_{ji}$), E_{ii} representing the elongations (extensions) or normal strain, E_{ij}, $i \neq j$, the angular displacements or shear strains; and (2) three relations (2a) the deformation-displacement relation **E**(**u**), (2b) the stress-strain relation **C**(**E**), (2c) the equilibrium and motion equations.

Breaking strength: load corresponding to a maximum extension required to produce material rupture (1D test framework).

Bulk modulus: parameter that quantifies the reaction of a material to a volume change when it is subjected to a given load ($B = p/(dV/V)$). Its physical magnitude is homogeneous to a pressure.

Complex viscoelastic moduli: parameter related to a sinusoidal shear applied on a material:

$$e^* = \hat{e}\exp\{\imath(\omega t)\} = C^*(\omega)c^*(t) \ ,$$
$$c^* = \hat{c}\exp\{\imath(\omega t + \varphi)\} = G^*(\omega)e^*(t) \ ,$$
$$G^*(\omega) = \Re[G(\omega)] + \imath\Im m[G(\omega))] = G'(\omega) + \imath G''(\omega) \ ,$$

where $G^*(\omega)$ is the complex shear modulus, $G'(\omega)$ the *storage modulus* and $G''(\omega)$ the *loss modulus*.

In the case of 1D sinusoidal traction tests, the analysis is based on a Voigt phenomenological model, $E(\omega) = E_{\mathrm{dyn}} + \imath\omega\eta$, where E_{dyn} is the dynamic

---

[12] The deformation is a measure of change in size, shape, and volume. The elastic and plastic deformations are reversible and irreversible, respectively. There are: (1) lineic deformation, change in length per unit length; (2) shear deformation, angular shift with shape change due to tangential stresses; and (3) volumic deformation, volume change per unit volume. A loading applied at time t on an unstressed body produces either instantaneous or delayed deformation. The residual deformation is a deformation that persists after loading withdrawal. The permanent deformation is a limit toward which the residual deformation tends when $t \to \infty$.

[13] The stress is a measure of forces resulting from internal reactions between body elementary particles due to sliding, separation, and compaction induced by external promptings. The internal resistance forces to the deformation are resultants of normal and tangential forces, continuously distributed, with variable magnitude and directions, which act on elementary surfaces across the entire material. When the loading is quickly applied, it can affect the process via stress and strain wave propagation.

elastic modulus and $\omega\eta$ the loss modulus. A complex incremental elastic modulus can be defined:

$$E^*_{\text{dyn}} = E_{\text{inc}}(R_e)\exp\{i\varphi\}\,,$$

where φ is the phase lag between the imposed sinusoidal pressure of amplitude Δp and the resulting radial excursions of magnitude ΔR_e. Therefore, $E_{\text{dyn}} = E_{\text{inc}}(R_e)\cos\varphi$ and $\omega\eta = E_{\text{inc}}(R_e)\sin\varphi$.

Compliance: refers to cross sectional area variations due to pressure changes at a given vessel station $C = (\partial A/\partial p)$.

Constitutive Law: relation between stress and strain tensors. It must agree with experimental data for a large loading range. It must contain a minimal number of independent constants, which have a physical meaning and can be easily calculated.

Creep at constant stress: when a stress is suddenly applied and maintained constant for a long time (step function), the strain gradually increases. When the stress is removed, the strain either does not come back to or goes back slowly to its original value.

Displacement Decomposition: Any displacement at any time can be decomposed into a uniform translation, rigid rotation and deformation, with respect to the reference frame (Cauchy-Stokes theorem):

$$\mathbf{u}_{P_2} = \mathbf{u}_{P_1} + \mathbf{du} = \mathbf{u}_{P_1} + \mathbf{\Omega} \times \mathbf{dr} + \mathbf{E}\cdot\mathbf{dr}.$$

Distensibility (specific compliance): refers to cross-section deformation

$$D = (\partial A/\partial p)/A = C/A.$$

Elasticity: property that enables the material to resist deformation by the development of a resisting force.

Elastic Modulus: ratio of the applied stress, or reacting stress, to the resulting deformation. A material is linearly elastic in a given loading range if the elastic modulus remains constant, the stress being proportional to the strain ("*ut tenso, sic vis*", Hooke law: $E = C_{ii}/E_{ii}$).

Extensibility: refers to 1D loading.

Generalized Newtonian Model: in shear-thinning fluid flows, the extra-stress tensor, characterized by $\mathbf{T} = 2\mu(T,\dot{\gamma})\mathbf{D}$ $(\mathbf{D} = (\nabla\mathbf{v} + \nabla\mathbf{v}^T)/2)$ is given by $\mathbf{T} = 2\mu(T,i_2(\mathbf{D}))\mathbf{D}$ $(i_2(\mathbf{D}) = (\text{tr}(\mathbf{D})^2 - \text{tr}(\mathbf{D}^2))/2)$.

Hysteresis: successive loading-unloading cycles show a loop, with ascending and descending branches not superimposed. Sinusoidal inputs are currently used $\varepsilon(t) = \bar{\varepsilon} + \varepsilon_\sim\sin\omega t$, $c(t) = \bar{c} + c_\sim\sin(\omega t + \varphi)$. The loop shape depends on: (1) the mean loading value $\bar{\varepsilon}$, (2) loading amplitude $\Delta\varepsilon$, (3) loading rate $\dot{\varepsilon}$, and (4) loading history $\mathcal{H}(\varepsilon)$: $\int_{-\infty}^{t}\varepsilon(t-\tau)\,d\tau$ $(E = E(\bar{\varepsilon}, \Delta\varepsilon, \dot{\varepsilon}, \mathcal{H}(\varepsilon))$.

Incremental Elastic Modulus: the elastic modulus is considered constant in small loading range, as the non-linear stress-strain relationship is decomposed into small intervals (piecewise constant elastic modulus). For a point of the outer surface of the vessel wall, which is easy to observed experimentally, Bergel (1961) proposed the following formula:

$$E_{\mathrm{inc}}(R_i) = \frac{2(1 - \nu_P)R_i^2 R_e}{R_e^2 - R_i^2} \frac{\Delta p}{\Delta R_e},$$

and for a point of the wetted surface, easy to target by medical imaging,

$$E_{\mathrm{inc}}(R_e) = \frac{(1 + \nu_P)R_i}{R_e^2 - R_i^2}\left((1 - 2\nu_P)R_i^2 + R_e^2\right)\frac{\Delta p}{\Delta R_e}.$$

Isotropic Material: the material properties are independent of direction.

$$E = 2G(1 + \nu_P), \qquad \nu_P = (3B - 2G)/(2(3B + G)),$$
$$B = E/(3(1 - 2\nu_P)), \qquad G = E/(2(1 + \nu_P)).$$

Memory Effect: the behavior of certain material depends not only on loading applied at the observation time t, but also on the previously imposed stresses. The history of a physical variable $\mathbf{g}$ is the set of values taken by $\mathbf{g}$ during previous times: $\mathbf{g}(t)_{-\infty}^t \to \int_{-\infty}^t \mathbf{g}\, d\tau$. An influence function can be introduced ($\mathcal{H}$). Soft biological materials undergo at any instant stresses (deformations) that depend on the stress (strain) magnitude at that time, on the loading rate and on the loading history.

Orthotropic Material: a material which has at least two orthogonal planes of symmetry, within which material properties are independent of direction.

Poisson Ratio: ratio of the relative contraction in the transverse direction j to the relative deformation in the direction i of the applied load $\nu_P = E_{jj}/E_{ii}$ ($i \neq j$). 1D extension is characterized by: (1) a longitudinal lengthening $e_\ell = \Delta L/L$ and (2) a transverse shortening $e_t = \Delta d/d = -(\nu_P/E)c$ (transverse strain-to-axial strain ratio).

Preconditioning: initial period of adjustment to loading. The cyclic loading response reaches a quasi-steady state after several succeeding cycles (adaptation period), probably due to matrix reorganization.

Prestress: biological tissues in the physiological state are not unstressed. Once excised, they shrink (tethering effect of the surrounding tissues). Once axially cut, blood vessels widen.

Pseudoelasticity: an approximative splitting description of the stress–strain relationship associated with cyclic loadings.

Relaxation Function: incorporates the response to stress history. Y.C. Fung proposed decomposition into a reduced relaxation function, a normalized function of time, and an elastic response that depends on the strain.

Reference State: state that is commonly determined according to the physiological requirements. Consequently, it does not correspond to an unstressed state. For an artery, it is defined by $p_i = 13.3\,\mathrm{kPa}$, knowing that in an artery deconnected from the vessel network $L/L_{\text{in vivo}} \sim 0.9$, and in an excised artery $L/L_{\text{in vivo}} \sim 0.6$–$0.7$.

Shear Modulus: quantifies the resistance of a material to a shape change caused by a shear, keeping a constant volume (shear stress-to-shear strain ratio $G = C_{ij}/E_{ij}$, $i \neq j$). Application of equal and opposite tangential surface stresses **c** at opposite faces of a control volume induces sliding with an angle α, without rotation due to normal stresses applied by adjacent particles ($G = |\mathbf{c}|/\tan\alpha$).

Strain: There are several definitions of strains. The *stretch ratio* in the direction of the applied 1D stress is the relative displacement, i.e., the ratio of the length change to the unstressed length $\lambda = L/L_0$. The *engineering strain* for an uniaxial loading is the stretch ratio minus one $\varepsilon = \Delta L/L_0 = \lambda - 1$. Another strain measure refers to the deformed configuration $\varepsilon' = \Delta L/L = 1 - \lambda^{-1}$. Quadratic strains can be easily incorporated in strain energy densities. The *Green-St Venant strain* is defined by $\varepsilon_G = (L^2 - L_0^2)/(2L_0^2) = (\lambda^2 - 1)/2$ and the *Almansi-Hamel strain* by $\varepsilon_A = L^2 - L_0^2/(2L^2) = (1 - \lambda^{-2})/2$. The *natural strain* $\varepsilon = \ln\lambda$ allows the easily handling of successive loadings because the resulting strain is the sum of the constitutive strain measures.

The static deformation of cylindrical orthotropic vessels generated by internal pressure and uniform axial extension is described by $\lambda_z = L/L_0$, $\lambda_r = R/R_0$, $\lambda_\theta = \overline{\chi}/\overline{\chi}_0 = \lambda_r{}^{14}$ (perimeter associated with the wall neutral line).

Let $\mathbf{F} = \partial\mathbf{x}/\partial\mathbf{x}_0$ be the deformation gradient tensor ($\imath_1, \imath_2, \imath_3 = det\mathbf{F}^2$). Using the polar decomposition theorem, $\mathbf{F}$ can be expressed by the product of the right $\mathbf{U}$, or left $\mathbf{V}$, stretch tensor and the rotation tensor $\mathbf{R}$ ($\mathbf{R}^T\mathbf{R} = \mathbf{R}\mathbf{R}^T = \mathbf{I}$): $\mathbf{F} = \mathbf{R}\mathbf{U} = \mathbf{V}\mathbf{R}$.

The right and left *Cauchy-Green deformation tensors* are associated with dilation/compression and shearing actions: $\mathbf{S}_r = \mathbf{F}^T\mathbf{F} = \mathbf{U}^T\mathbf{U}$ and $\mathbf{S}_l = \mathbf{F}\mathbf{F}^T = \mathbf{V}\mathbf{V}^T$, respectively. The *Biot-Finger strain tensor* $\mathbf{B} = \mathbf{S}_l^{-1} = \mathbf{F}^{T^{-1}}\mathbf{F}^{-1}$. The *Green-Lagrange strain tensor* and *Almansi strain tensor* are given by: $\mathbf{G} = (\mathbf{S}_r - \mathbf{I})/2 = \mathbf{F}^T\mathbf{D}\mathbf{F}$ and $\mathbf{A} = (\mathbf{I} - \mathbf{B})/2$.

Strain Energy Density (elastic potential): a function of the strain invariants if the elastic material is homogeneous and isotropic.

Stress: force per unit surface area producing a deformation. The stress tensor components C_{ij} are conveniently expressed in an orthogonal basis $\{\hat{\mathbf{e}}_i\}_{i=1}^3$. In any material point $P \in \Omega$, a second order stress tensor $\mathbf{C}(P)$ exist (combination of the stresses acting on the faces of the infinitesimal control volume), such that the local force per unit surface area $\mathbf{c}(P, \hat{\mathbf{n}}) = \mathbf{C}(P)\hat{\mathbf{n}}$

---

[14] λ_θ can be defined as: $\lambda_\theta = (\pi/(\pi - \theta))\lambda_r$ when the opening angle θ is known [249].

[Cauchy theorem, 1822].[15] The diagonal elements (C_{ii}) represent normal (tensile) stresses, and others $(C_{ij}$ $(i \neq j))$ the tangential (shearing) stresses.

$$\mathbf{c} = \mathbf{C} \cdot \hat{\mathbf{n}}, \quad c_i = C_{ij} n_j \quad \forall i, \forall j, \; i,j = 1,2,3. \qquad \text{(Cauchy formula)}$$

Stress is defined with respect to either the reference (*Lagrange-Piola stress* $C_{Lii} = f_i/A_0$) or the deformed (*Cauchy-Euler* stress $C_{Cii} = f_i/A$) configuration. *Kirchhoff* stress is defined by $C_{Kii} = C_{Lii}/\lambda_i$ or $C_{Kii} = (\rho_0/\rho)C_{Cii}/\lambda_i^2$.

Stress Relaxation at constant strain: when a strain is suddenly applied and maintained constant for a long time (step function), the induced stress decreases after reaching its maximum.

Tensile Strength (yield point): tension at which a stretched material cannot go back to its original configuration and undergoes an irreversible plastic deformation (the material yields, with breaking of links between its constituents).

Tensor Decomposition: The velocity gradient tensor $\mathbf{L} = \boldsymbol{\nabla}\mathbf{v}$ ($\dot{\mathbf{F}} = \mathbf{LF}$, $(\boldsymbol{\nabla}\mathbf{v})_{ij} = \partial v_i/\partial x_j$) can be decomposed into a symmetrical tensor $\mathbf{D}$, the *deformation rate tensor*, and a antisymmetrical tensor $\mathbf{W}$, the *rotation rate tensor* or the *vorticity tensor*:

$$\boldsymbol{\nabla}\mathbf{u} = \mathbf{D} + \mathbf{W} = 1/2(\boldsymbol{\nabla}\mathbf{u} + \boldsymbol{\nabla}\mathbf{u}^T) + 1/2(\boldsymbol{\nabla}\mathbf{u} - \boldsymbol{\nabla}\mathbf{u}^T).$$

Thixotropy: the response of a thixotropic material depends on body structure changes, and consequently on the loading rate, duration of the unstressed period, and loading duration with respect to the body-kinetic time scale.

Viscocity: material property dealing with resistance to deformation and motion.

Relative Viscocity of a suspension: ration of the suspension viscocity ot the suspending fluid (plasma) viscocity.

Volume Dilation: $dV/V = \boldsymbol{\nabla} \cdot \mathbf{u} = E_{ii}$

C.6 Phenomenological Models in Rheology

Rheology models are made by the assembling of elementary elements, springs and dashpots, which are associated:

- either in parallel with the following rules:
  - the imposed net stress is the sum of the stresses ($\mathbf{c} = \sum \mathbf{c}_b$) applied to each branch (subscript b),

---

[15] $\begin{pmatrix} c_1 \\ c_2 \\ c_3 \end{pmatrix} = \begin{pmatrix} C_{11} & C_{12} & C_{13} \\ C_{21} & C_{22} & C_{23} \\ C_{31} & C_{32} & C_{33} \end{pmatrix} \begin{pmatrix} \hat{n}_1 \\ \hat{n}_2 \\ \hat{n}_3 \end{pmatrix}.$

Table C.3. Basic rheology tests on viscoelastic materials (c: stress , e: strain, G: shear modulus).

	Static tests		Sinusoidal loading
	Creep	Relaxation	Harmonic shear
Input	c = cst.	e = cst.	$c = \hat{c}\exp\{\imath\omega t\}$
function	$\phi(t) = e(t)/c$	$\psi(t) = c(t)/e$	$e = \big(G'(\omega) - \imath G''(\omega)\big)\hat{c}\exp\{\imath\omega t\}$

- the undergone deformation is identical in each branch and equal to the net deformation ($e = e_b$);
- or in series with the following rules:
  - the imposed net stress is wholly borne by each element i: ($c = c_i$),
  - the net deformation is the sum of the deformation undergone by each element ($e = \sum e_i$).

The rheological features of the viscoelastic models are obtained using well-defined procedures (Table C.3).

Maxwell Model: is constituted by the association in series of a spring (subscript S) and a dashpot (subscript D):

$$\dot{u} = \dot{u}_S + \dot{u}_D = C\dot{c} + c/\mu \, ,$$

$$cc = c_S = c_D \, .$$

Viscoelastic materials are often assumed to behave like simple Maxwell models, with their material constants E and μ (associated with the spring and the dashpot, respectively; $C = 1/E$: elastic compliance), and the following rheological law:

$$(\mu/E)\dot{c} + c = \mu\dot{u} \, .$$

Constitutive laws of three-dimensional bodies are derived from the phenomenological one-dimensional model:

$$(\mu/E)\overset{\triangledown}{\mathbf{C}} + \mathbf{C} = 2\mu\mathbf{D}$$

where the upper convected time derivative of the stress tensor $\mathbf{C}$ is given by:

$$\overset{\triangledown}{\mathbf{C}} = D\mathbf{C}/Dt - (\nabla\mathbf{v})\mathbf{C} - \mathbf{C}(\nabla\mathbf{v})^T,$$

$D\cdot/Dt$ and $\nabla\mathbf{v}$ being the substantive derivative and the tensor of velocity derivatives, respectively.[16]

---

[16] The lower convected time derivative is:

$$\overset{\triangle}{\mathbf{C}} = \dot{\mathbf{C}} + \mathbf{C}(\nabla\mathbf{v}) + (\nabla\mathbf{v})^T\mathbf{C} \, .$$

Voigt or Kelvin-Voigt Model: is constituted by the association in parallel of a spring and a dashpot:

$$u_S = u_D = u\,,$$
$$c = c_S + c_D = (1/C)u + \mu\dot{u}\,.$$

Standard Linear Model (Boltzmann model): is constituted by the association in parallel of a spring and a Maxwell model (subscript M):

$$u_S = c_S/E,$$
$$\dot{u}_M = c_M/\mu + \dot{c}_M/E_M$$

Rheological data of these simple models are summarized in Tables C.4, C.5 and C.6.

Table C.4. Constitutive laws of the simplest rheology models for a stress step ($c(t) = c_0 h(t)$).

Model	Relation $c(e)$	Viscosity	Creep	Stress relaxation
Hooke	$c = Eu$	–	–	–
Newton	$c = \mu\dot{u}$	+	+	–
Maxwell	$\dfrac{c}{\mu} + \dfrac{\dot{c}}{E} = \dot{u}$	–	+	–
Voigt	$c = Eu + \mu\dot{u}$	+	–	–

Table C.5. Creep function for a unit stress step $f(t)$ in simple rheology models ($\cdot_M$: Maxwell model component of the Boltzmann model; Source: [249]).

Maxwell	$\left(C + \dfrac{t}{\mu}\right)f(t)$
Voigt	$C(1 - \exp\{-t/\mu C\})f(t)$
Boltzmann	$C\left(1 - \left(1 - \dfrac{\mu C_M}{\mu C(1 + C_M/C)}\right)\exp\{-t/(\mu C(1 + C_M/C))\}\right)f(t)$

Table C.6. Relaxation function for a unit strain step $u(t)$ in simple rheology models ($\cdot_M$: Maxwell model component of the Boltzmann model; Source: [249]).

Maxwell	$\exp\{-t/\mu C\}u(t)/C$
Voigt	$\mu\delta(t) + u(t)/C$
Boltzmann	$C^{-1}\left(1 - \left(1 - \dfrac{\mu C(1 + C_M/C)}{\mu C_M}\right)\exp\{-t/(\mu C(1 + C_M/C))\}\right)u(t)$

C.7 Wave Speed

The speed of the elastodynamic coupling wave is tightly linked to the vessel law, which relates the transmural pressure p to the cross-sectional area A. It is calculated by:

$$c^2 = \frac{A}{\rho \partial p / \partial A} = (\rho D)^{-1}.$$

Radial Propagation Mode (Young mode): In an elastic cylindrical tube of circular cross-section conveying an incompressible fluid, the "ideal" wave speed c_0 was given by T. Young in 1809 by:[17]

$$c_0^2 = \frac{Eh}{2\rho R} = \frac{Eh}{\rho d} \; ,$$

with the following conditions:
- incompressible inviscid fluid,
- infinitesimal wave amplitude and great wavelength,
- $h/R \ll 1$,
- purely elastic tube with small impedance,
- absence of wave reflection.

For a compressible fluid,[18] the wave speed is expressed by the Moens-Korteweg formula:

$$c^2 = (\frac{\rho}{K} + \frac{2\rho R}{Eh})^{-1} \; .$$

When E is equal to $10^6 \, Pa$ and $h/R = 0,1$, $c_0 \sim 7 \, \text{m/s}$.

In axially tethered cylindrical vessels (without axial displacement) of circular cross-section, of isotropic purely elastic wall which has a small thickness ($h \ll R$):

$$c^2 = \frac{Eh}{2(1 - \nu_P^2)\rho \bar{R}} \; .$$

Other expression can be found in the literature as the one provided by O. Frank (1899), for a cylindrical vessel of circular cross-section:

$$c^2 = \frac{R}{\rho} \frac{\Delta p}{\Delta R} = \frac{V}{\rho} \frac{\Delta p}{\Delta V} \; .$$

Bergel proposed the following formula for a tube with thick incompressible wall, filled with an inviscid fluid:

$$c^2 = \frac{E(R_e^2 - R_i^2)}{3\rho R_e^2} \; .$$

---

[17] The formula can be easily obtained, starting from $c_0^2 = (A/\rho)dp/dA = (R/2\rho)dp/dR$, with $p = c_\theta h/R$ and $c_\theta = E \, \Delta R/R$

[18] Most often the gas compressibility K is much lower than the respiratory conduit distensibility D.

Axial Propagation Mode (Lamb mode): The speed of axial waves of small frequencies has been given by H. Lamb (1898):

$$c_L^2 = \frac{E}{\rho_w(1 - \nu_P^2)} \, .$$

With the values used to calculate c_0, $c_L \sim 36\,\text{m/s}$ ($c_L/c_0 \sim 5$).
Torsional Wave: The speed of torsional wave is given by:

$$c_{\text{tors}}^2 = \frac{G}{\rho_w} \sim \frac{1}{\rho_w} \frac{E}{2(1 + \nu_P^2)} \, .$$

With the values used to calculate c_0, $c_{\text{tors}} \sim 20\,\text{m/s}$ ($c_{\text{tors}}/c_0 \sim 3$).

Wave Speed in Collapsible Tubes: The speed of the elasto-hydrodynamics coupling wave is computed from the tube law $\mathcal{A}(p)$. The speed c of propagation of small pressure wave in a collapsible tube is indeed given by:

$$c^2 = (A_i/\rho)(\partial p/\partial A_i) \, . \tag{C.5}$$

The pressure wave of small amplitude, which propagates with the speed c, depends thus on fluid inertia and wall compliance (*transverse propagation mode*). This simple relationship is valid under the following additional assumptions:

1. incompressible fluid of mass density ρ,
2. long straight collapsible tube,
3. constant geometry along the whole tube length of the reference configuration,
4. thin wall ($h_0/R_0 \ll 1$),
5. purely elastic wall,
6. uniform mechanical properties of the tube wall throughout the entire length in the unstressed configuration,
7. negligible wall inertia,
8. negligible viscous dissipation,
9. wavelet (infinitesimal amplitude),
10. one-dimensional motion of the fluid.

Experimental observations have shown that the wave speed reaches its minimum when contact occurs between the opposite walls [365]. In the slightly negative range of transmural pressures, both the tube cross-sectional area and high compliance entail a low wave speed, associated with a high fluid velocity. Critical conditions (superscript $*$) are reached when the local cross-sectional average fluid velocity $V_q(x)$ becomes equal to the wave speed c, as in trans-sonic air flows through rigid nozzles and open-channel flows at critical Froude number characterized by a hydraulic jump (Table C.7).

At each cross-section can be associated a critical flow rate q^*, which is related to the luminal area by $q^* = A_i c$. Due to the discontinuity induced

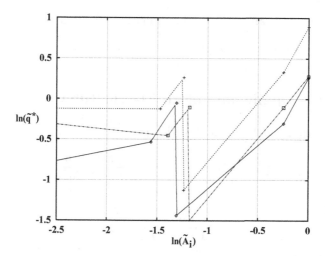

Figure C.2. $\ln(\tilde{A}_i)$ vs. $\ln(\tilde{q}^*)$ relationships for different tube ellipticities $k_0 = 1.005$ (continuous line), 2.8 (dotted line) and 10 (dashed line), with the characteristic values ($\tilde{A}_{i0}$: unstressed elliptical; $\tilde{A}_{ip}$: oval-shaped cross-section; $\tilde{A}_{ic}$: point contact, $\tilde{A}_{i\ell}$: line contact). When $p \leq p_\ell$, both the value and the sign of the slope of the relationship are affected by the tube ellipticity.

by the wall contact, the critical conditions must be different on the right (q_{c+}^*) and on the left (q_{c-}^*) side of the discontinuity (Fig. C.2). The slope of the q^* vs A_i curves depends on the tube law parameter $\mathcal{M}$:

$$dq^*/dA_i = (\mathcal{M}/2)(q^*/A_i), \qquad (C.6)$$

with $\mathcal{M} = 3 + \tilde{A}_i(\partial^2\tilde{A}_i/\partial\tilde{p}^2)/(\partial\tilde{A}_i/\partial\tilde{p})$.

In the pressure range $p \leq p_\ell$, for thin-walled collapsible tubes of unstressed elliptical cross-section, $\mathcal{M}[n(k_0,h_0)] = 2 - n$, i.e. $\mathcal{M}[n(1,0)] = 1/2$, $\mathcal{M}[n(2.8,0)] = 0$ and $\mathcal{M}[n(10,0)] = -1/4$. The slope of the relationship between p and q^* is positive when $\mathcal{M} > 0$ and is negative when $\mathcal{M} < 0$ [366].

Table C.7. Physical analogues. Flow subjected to small amplitude perturbations of large wavelength.

Wave type	Physical variable	State law	Wave speed
Elasto-hydrodynamics coupling wave	$A(\mathbf{x},t)$	$A = A(p)$	$c^2 = (A/\rho)(dp/dA)$
Sound wave (adiabatic conditions)	$\rho(\mathbf{x},t)$	$\rho = \rho(p_i)$	$c^2 = dp/d\rho$
Gravity wave	$h(\mathbf{x},t)$	$h = \dfrac{(p_i - p_{atm})}{\rho g}$	$c^2 = gh$

In the pressure range $p \leq p_\ell$, the similarity law gives:

$$(\tilde{c}/\tilde{c}_\ell)^2 = (\tilde{A}_i/\tilde{A}_{i\ell})^{-n} , \quad (\tilde{q}^*/\tilde{q}_\ell^*)^2 = (\tilde{A}_i/\tilde{A}_{i\ell})^{\mathcal{M}} . \tag{C.7}$$

C.8 Static Equilibrium of an Elastic Tube

Force equilibrium in the radial direction gives:

$$(p_i - p_e)R = pR = \tau_t h = \frac{Eh}{1 - \nu_P^2} \frac{\Delta R}{R_0} . \tag{C.8}$$

Therefore,

$$p = \frac{Eh}{1 - \nu_P^2} \left(\frac{1}{R_0} - \frac{1}{R} \right) . \tag{C.9}$$

When $(h/h_0)(R/R_0) \sim 1$, then

$$p = \frac{Eh_0}{1 - \nu_P^2} \frac{\Delta R}{R^2} . \tag{C.10}$$

C.9 Tube Deformation

C.9.1 Stresses and Strains

$$e_i = \frac{1}{E}(c_i - \nu_P(c_j + c_k)) , \quad i \neq j \neq k, \ i, j, k = r, \theta, x. \tag{C.11}$$

$$c_\theta = \frac{1}{R_e^2 - R_i^2} \left(p_i R_i^2 - p_e R_e^2 + p \left(\frac{R_e R_i}{R} \right)^2 \right) \tag{C.12}$$

$$c_r = \frac{1}{R_e^2 - R_i^2} \left(p_i R_i^2 - p_e R_e^2 - p \left(\frac{R_e R_i}{R} \right)^2 \right) \tag{C.13}$$

When p_i or $p_e = 0$, with $h/R \ll 1$, then $|c_r| \ll c_\theta$.
Generally $(p_i$ et $p_e \neq 0)$, $|c_r| \ll c_\theta$ if $p_i \ll p_e$ or $p_i \gg p_e$.

C.9.2 State Law

Free-End Tube

From (C.11) and (C.12),

$$p = Eh \left(\frac{1}{R_0} - \frac{1}{R} \right) . \tag{C.14}$$

Clamped-End Tube

$$p = \frac{Eh}{1-\nu_P^2}\left(\frac{1}{R_0} - \frac{1}{R}\right) = \frac{E}{1-\nu_P^2}\frac{h_0}{R_0}\left(\frac{R_0}{R} - \left(\frac{R_0}{R}\right)^2\right). \tag{C.15}$$

Clamped-End Tube with Small Pretension

$$p = \frac{E}{1-\nu_P^2}\frac{h_0}{\lambda_l R_0}\left(\frac{R_0}{R} - \left(\frac{R_0}{R}\right)^2(1 - \nu_P(\lambda_l - 1))\right). \tag{C.16}$$

Stress–Strain Relations

$$c_i - c_j = G(\lambda_i^2 - \lambda_j^2), \quad i \neq j, \ i,j = r,\theta,x, \tag{C.17}$$

with $G = E/(2(1+\nu_P))$ and $\lambda_k = L_k/L_{k0}$ ($\lambda_l = L/L_0$, $\lambda_r = R/R_0$, $\lambda_\theta = h/h_0$).

When the thin-walled tube ($h/R \ll 1$) has fixed ends,

$$c_\theta = pR/h = G(\lambda_\theta^2 - \lambda_r^2).$$

Therefore,

$$p = \frac{G}{\lambda_l}\frac{h_0}{R_0}\left(1 - \frac{1}{\lambda_l^2\lambda_r^4}\right), \tag{C.18}$$

An empirical correction factor κ_e has been proposed to take into account the wall thickness [843]:

$$\kappa_e = \frac{1 + h_0/(2R_{i0})}{(1 + h_0/R_{i0})^2}.$$

Distensible Tube

The cross-section is circular with unstressed radius R_0; the pressure force are balanced by tangential stresses. The deformed cross-section remains circular. Let $\{\lambda_k\}_{k=1}^3$ be the tension coefficients associated to the principal axes of the tube ($\lambda_1 = L/L_0$, $\lambda_2 = R/R_0$, $\lambda_3 = h/h_0$). The non-linear elastic behavior of the tube can be given by the following expression [844]:

$$p = G\kappa_e[h_0/(a_0\lambda_1^3)](\lambda_1^2 - (A_{i_0}/A_i)^2). \tag{C.19}$$

C.10 Collapsible Tubes

C.10.1 Tube Law

The pipe in its reference configuration is supposed to be stress free, in particular $p = 0$, i.e., unstressed configuration (subscript 0). The collapse is characterized by large variations in A_i under small variations of p before the contact configuration, when the opposite edges of the wetted perimeter touch (Fig. C.3). Consequently, the tube law exhibits a sigmoidal shape. Additionally, these huge changes in tube transverse configuration for slightly negative transmural pressure are observed in any compliant pipe, whether the unstressed cross-section is elliptical (e.g., [364]) or circular (e.g., [367]), whether the vessel has uniform homogeneous walls or is a composite material of nonuniform geometry [269], in vitro as well as in vivo [96].

Physiological vessels, susceptible to collapse, present noncircular unstressed cross-section. The unstressed cross-sectional shape is commonly assumed to be elliptic.[19] The tube collapse has been theoretically investigated in an infinitely long straight tube with a thin homogeneous isotropic and purely elastic wall and with uniform geometry and rheology [362, 364, 366, 845]. The transmural pressure is supposed to be uniformly distributed in every tube section and bending effects are assumed to be predominant. The floppy duct is subjected to a uniform transmural pressure p. The wall thickness, small relative to the wall mid-line curvature, is assumed to remain constant during the collapse. The neutral mid-surface is deformed without circumferential extension. The tube collapses then according to a bilobal collapsing mode (Fig. C.4).

When the unstressed cross-section is circular and p is slightly negative, the compliant tube keeps a circular cross-section down to the *buckling pressure* p_b. From this mechanical state, a small decrease in p produces a large change in A_i. Different modes of collapse can then be observed according to the number N of lobes, the lobes being the open part of the collapsed tube lumen which are associated with symmetry axes. Using computational thin-shell models of deformation in flexible tubes of infinite length and with a purely elastic wall, the buckling pressure is shown to be proportional to $N^2 - 1$ [362, 368]. Experimental evidence of such collapsing modes was obtained, the cross-section shape displaying usually four, three or two lobes either in tubes subjected to longitudinal bending effects [369] or in short and thin-walled pipes [367].

The tube law $\mathcal{A}(p)$, i.e., A vs. p relationship, depends on various factors: (1) tube geometry, (2) tube rheology, (3) prestresses, (4) vicinity loadings, (5) end effects, and (6) stresses developed (tension, bending).

Four characteristic transmural pressures are, at least, of interest in the collapse:

---

[19] The tube ellipticity k_0 is the ratio between the major semi-axis a_0 and the minor semi-axis of the neutral mid-line in the unstressed tube configuration.

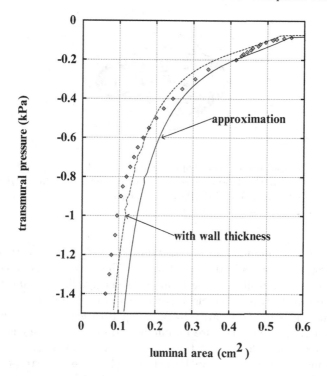

Figure C.3. Relation between the transmural pressure p (Pa) and the tube cross-sectional area A_i (cm^2). Comparison between simulations, using either the condition $z(0) = 0$ at contact (dashed line), or the condition $z(0) = h_0/2$ at contact (dotted line), and experimental results ($\diamond$) on a flexible pipe ($a_0 = 9.52$ mm, $k_0 = 1.337$, $h_0 = 0.43$ mm, $K = 24.47$ N/m^2).

1. *Ovalization pressure* p_p, for which the radius of curvature at mid-face becomes locally infinite (oval-shaped cross-section with parallel opposite edges).
2. *Stream division pressure* p_t, the greatest pressure associated with two lateral peak velocities within the cross-section.
3. *Point-contact pressure* p_c, at which the opposite faces touch for the first time.
4. *Line-contact pressure* p_ℓ, when the radius of curvature at the contact point becomes infinite.

Let s be the local curvilinear coordinate along the neutral line of any cross-section of the tube wall, with the origin at the material point at the mid-face $(0, a_0/k_0)$. The neutral line is oriented clockwise. A unit tangent $\mathbf{t}$ and a unit normal $\mathbf{n}$ are defined at each point M of the neutral line. The dimensionless local curvature $\tilde{\kappa}$ vs. dimensionless arclength $\tilde{s}$ relationships for

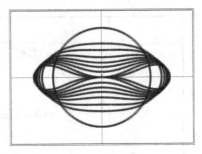

Figure C.4. Computational wall displacement during the collapse of a tube ($k_0 =$ 1.005, $h_0 = 0$) with the following set of dimensionless quantities ($\tilde{p}, \tilde{A}_i$): $(0, 1)$, $(-1.60, 0.89)$, $(-1.70, 0.79)$, $(-1.85, 0.67)$, $(-2.00, 0.57)$, $(-2.15, 0.48)$, $(-2.35, 0.38)$, $(p_c = -2.64, 0.27)$, $(p_\ell = -5.20, 0.21)$

the investigated configurations are shown in Fig. C.5. In the unstressed cross-section XS_0, $\tilde{\kappa}$ is constant and equal to -1. The curvature is indeed defined at the point of the wall mid-line from the angle gradient $d\theta/d\tilde{s}$, where θ is the angle between $\hat{e}_x$ and the clockwise oriented unit tangent.[20] At mid-face ($\tilde{s} = 0$), the curvature is negative in XS_q, equal to zero in XS_ℓ, and positive in XS_t and XS_c. The curvature minimum is observed at the edge whatever XS; the maximum is located at the mid-face, except in XS_ℓ.

Three modes of collapse can be defined. (1) **Mode 1** corresponds to the collapse before contact ($p_c < p \leq 0$), characterized by a high tube compliance. The transversal density of the distributed external force **f** induced by the pressure load is given by $\mathbf{f} = p\mathbf{n}$. The stress resultant acting from one part of the wall to the other $\mathbf{c}(s)$ is continuous everywhere. (2) **Mode 2** is characterized by a contact at a single point ($p_\ell < p \leq p_c$). The curvature at the contact point decreases from a finite value down to zero. A contact reaction appears (see below) and the resultant stress is discontinuous at the contact point ($s = 0$). The contact generates a local reaction $\mathbf{r}_c$ at the contact point ($s = 0$), which increases when p decreases from p_c ($\mathbf{r}_c = \mathbf{r}_{c0}(p_c)$) down to p_ℓ ($\mathbf{r}_c = \mathbf{r}_{c0}(p_\ell)$) (Fig. C.6). As soon as p undergoes an infinitesimal decrease, say $p = p_{\ell-}$, the reaction initiates its splitting into two components $\mathbf{r}_\ell(p_{\ell-}) = \frac{1}{2}\mathbf{r}_{c0}(p_\ell)$. When $p < p_\ell$, the reaction is distributed along the contact segment of length $2\,s_c$, with maxima $\mathbf{r}_\ell(p)$ located at both ends of the contact segment ($s = \pm s_c$). These points associated to concentrated force migrate laterally when p continues to decrease, whereas the reaction amplitude exerted on the line of contact, which spreads out, decreases. (3) **Mode 3** is defined by a contact on a line segment ($p \leq p_\ell$). The contact segment appears and length-

---

[20] Along the selected quarter of the wall, ($\tilde{x} \geq 0$ and $\tilde{y} \geq 0$), θ decreases from zero to $-\pi/2$, whereas $\tilde{s}$ rises from zero to $\pi/2$; $\theta = -\tilde{s}$; therefore, $d\theta/d\tilde{s} = -1$. The unit normal points toward the tube outside (and not toward the tube lumen as in direct Frenet basis, i.e., not toward the curvature center, because of negative applied resulting pressure).

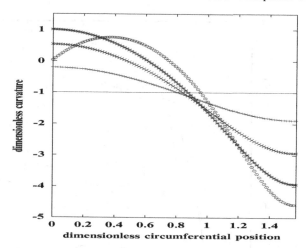

Figure C.5. Relationships between $\tilde{\kappa}$ and $\tilde{s}$ for the unstressed XS_0 (solid line) and four collapsed bidimensional configurations: XS_q (+), XS_t (×), XS_c (*), and XS_ℓ (○).

ens, while the transmural pressure continue to decrease. Besides, the contact phenomenon at $p = p_\ell$ is displayed by a slight change in direction of the $\mathcal{A}(p)$

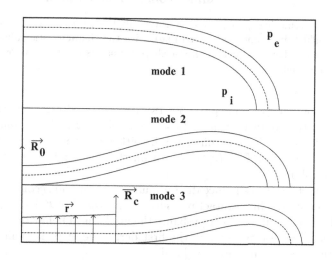

modes of collapse

Figure C.6. The three collapse modes in the case $k_0 = 1.6$ and $h_0/a_0 = 0.1$: (**top**) $p_c < p \leq 0$, (**mid**) $p_\ell \leq p \leq p_c$ and (**bottom**) $p < p_\ell$. Reaction loading at contact between opposite walls of the flexible pipe ($\mathbf{R}_0 \equiv \mathbf{r}_{c0}$, $\mathbf{R}_c \equiv \mathbf{r}_\ell$, $\mathbf{r} \equiv \mathbf{r}_d$):

$$\mathbf{r}_{c0}(p_c) \longrightarrow \mathbf{r}_{c0}(p_\ell) \xrightarrow{p<p_\ell} \begin{cases} \mathbf{r}_\ell(p) & s = s_c \\ \mathbf{r}_d(s,p) & 0 \leq s < s_c \end{cases}$$

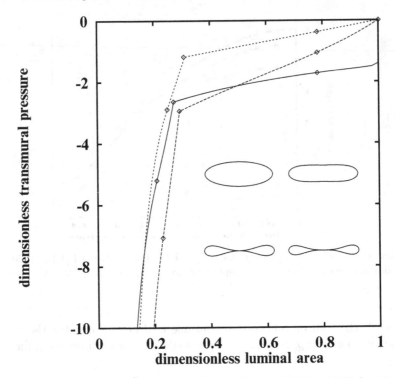

Figure C.7. Numerical $\tilde{p}(\tilde{A}_i)$ laws for three ellipticities $k_0 = 1.005$ (continuous line), 2.8 (dashed line), and 10 (dotted line), with the characteristic values ($\diamond$) corresponding to the displayed characteristic shapes (unstressed elliptical, oval-shaped, point- and line-contact).

curve. The contact reaction splits into two concentrated reactions applied at both ends of the contact segment ($s = \pm s_c$), where **c** is discontinuous. The transversal force density **f** is thus given by either $\mathbf{f} = p\mathbf{n}$ along the open part of the cross-section or $\mathbf{f} = (r_d(s) + p)\mathbf{n}$ on the contact line ($r_d(s)$: normal reaction distribution).

The contact reactions induce discontinuities in first and second derivatives of $\mathcal{A}(p)$ at point- and line-contact pressures $((\partial A_i/\partial p)|_{p_c}, (\partial^2 A_i/\partial p^2)|_{p_\ell})$ [363, 365]. Such discontinuities affect probably the mechanical behavior of the fluid-tube couple. The discontinuity in the first derivative at p_c is exhibited by a break in the slope of the tube law [364].

The tube law depends strongly on both tube geometry and rheology in the unstressed state (Fig. C.7). The main geometrical factors are the tube ellipticity k_0 [364], and the wall thickness h_0 [366] (Fig. C.3).

C.10.2 Normalization and Dimensionless Tube Law

The tube deformation, from rest to the line-contact pressure p_ℓ, is illustrated in Fig. C.7, using the dimensionless quantities $\tilde{p} = p/K$ and $\tilde{A}_i = A_i/A_{i_0}$. The bending stiffness K, used as the pressure scale, depends on the tube geometry and is proportional to the flexural rigidity $D = Eh_0^3/(12(1-\nu^2))$: $K = 2D/a_0^3$, where E and ν are the Young modulus and the Poisson coefficient, respectively. Fig. C.7 demonstrates that the tube law is affected by the unstressed elliticity. The speed scale $(K/\rho)^{1/2}$ is used to normalize the derived quantity c.

In the pressure range $p_\ell \leq p \leq 0$, any analytical expressions are very helpful in collapsible tube flow computations or experimental post-processings. Several expressions have been proposed, but they do not take into account the discontinuity of the tube law and its related physical phenomena. Analytical expressions must insure the continuity of the functions $c(p)$ and $q^\star(p)$, except at the contact condition $p = p_c$, the $q^\star(p)$ curve being piecewise fitted from numerical data. Algebraic expressions based on integration of the critical flow curve use nine coefficients ($\{\kappa_k\}_{k=1}^9$) [846]. Relationships $\tilde{p}(\tilde{A})$ have been given for the description of the tube law in the first three pressure intervals:

$$\tilde{p} = \kappa_1^2 \ln \tilde{A}_i - 2\kappa_1\kappa_2/\tilde{A}_i - \kappa_2^2/(2\tilde{A}_i^2 + \kappa_3), \qquad \tilde{p}_\ell \leq \tilde{p} \leq \tilde{p}_c$$

$$\tilde{p} = \exp\{2\kappa_4\}\tilde{A}_i^{2(\kappa_5-1)}/2(\kappa_5 - 1) + \kappa_6, \qquad \tilde{p}_c < \tilde{p} \leq \tilde{p}_p \qquad (C.20)$$

$$\tilde{p} = \exp\{2\kappa_7\}\tilde{A}_i^{2(\kappa_8-1)}/2(\kappa_8 - 1) + \kappa_9, \qquad \tilde{p}_p < \tilde{p} \leq 0$$

Variations with the tube ellipticity of the coefficients κ_i ($i = 1, \ldots, 3$ for the pressure range $[\tilde{p}_\ell, \tilde{p}_c]$ (collapse mode 2), $i = 4, \ldots, 6$ for $[\tilde{p}_c, \tilde{p}_p]$ (collapse mode 1), and $i = 7, \ldots, 9$ for $[\tilde{p}_p, 0]$ (collapse mode 1) of the analytical generalized tube law (C.20) are given in [361].

In the pressure range $\tilde{p} \leq \tilde{p}_\ell$, the similarity law can be used down to $\sim -30\tilde{p}_\ell$, when the tube shape remains self-similar with the twin-lobed characteristic configuration (for $\tilde{p} = \tilde{p}_\ell$). A specific mapping can always be determined between any collapsed cross-section of the investigated pipe, of unstressed ellipticity k_0, under transmural pressure $\tilde{p} < \tilde{p}_\ell$ ($p_\ell = p_\ell(k_0)$) and a reference collapsed cross-section of another tube of known properties, unstressed ellipticity $k_{\text{ref}.0}$, having the same shape at its own line-contact transmural pressure $p_{\text{ref}.\ell} = p_\ell(k_{\text{ref}.0})$.

The similarity has been used since 1972 for unstressed circular cross-section [362]. The pressure–area relationship for $p \leq p_\ell$ is generally expressed by the following equation, as verified by curve fittings:

$$\tilde{p} = -\mathcal{B}(k_0, h_0)\tilde{A}_i^{-n(k_0, h_0)} \qquad (C.21)$$

where $\mathcal{B}(k_0, h_0) = -\tilde{p}_\ell/\tilde{A}_{i\ell}$ [366]. The similarity law exponent n increases with k_0 for a given h_0. For instance for $h_0 = 0$, n rises from $3/2$ for $k_0 = 1$ up to a value of $9/4$ for $k_0 = 10$, being equal to 2 for $k_0 = 2.8$.

C.10.3 One-Dimensional Model of Flow in Collapsible Tubes

The one-dimensionality neglects the cross-distribution of physical quantities. It is thus assumed that: (1) the longitudinal area gradient is sufficiently small, (2) the tube axis remains straight during the test, (3) the velocity is uniformly distributed in the cross-section, and (4) the wall inertia and the fluid motion due to the collapsing process are negligible. Tube wall material is usually assumed to be homogeneous and purely elastic. The tube is subjected to an uniform and constant external pressure. The flowing fluid is assumed to be incompressible; the wave speed depends mainly on the tube compliance.

Mass and momentum conservation in the case of the one-dimensional unsteady flow give the following system of equations:

$$\frac{\partial A_i}{\partial t} + \frac{\partial A_i U}{\partial x} = 0 \, ,$$
$$\frac{\partial A_i U}{\partial t} + \frac{\partial A_i U^2}{\partial x} = -\frac{A_i}{\rho}\left(\frac{\partial p}{\partial x} + f_v(x, q)\right) . \tag{C.22}$$

The 1D model predicts not only the occurrence of a critical section but also the existence of critical uniform pipe segments ($dA_i/dx = 0$ and $S = 1$). The critical conditions are reached at once either in the upstream uniform segment or at the downstream end, whether the inlet cross-sectional area A_{i1} is greater than A_{i2} or not and whether $\mathcal{M}$ is positive or not.

C.10.3.1 Similarity conditions

The similarity conditions between two tests are given by dimensionless governing parameters. The tube law must be of the same type, i.e., same k_0; the flow regime is either laminar or turbulent. A set of dimensionless parameters[21] can be actually defined: (1) an aspect ratio $\pi_1 = L/a_0$, which can also be considered as an axial viscous effect parameter; (2) a viscous effect parameter $\pi_2 = a_0(\rho K)^{1/2}/\mu$, which is a combination of elastic, inertial, and viscous forces;[22] and (3) for gravity-friction flows, a hydrostatic effect parameter $\pi_3 = \rho g L/K$. A flow-pattern dependent group of the two first quantities π_1/π_2^m ($m = 1$ or $m = 1/4$ whether the regime is laminar or turbulent) has been found to give a suitable similarity parameter in the description of $\Delta p - q$ curves. High values of π_1/π_2 indicate that, for given tube properties, the viscous forces affect flow significantly. When gravity and friction are combined, the hydrostatic similarity parameter $\pi_3 \sin\beta$ must be added. This parameter is needed to adjust the tube inclination angle to the space available to the set-up.

---

[21] Using the unstressed perimeter $\chi_0/4$ as the transverse characteristic length, the tube's aspect ratio is expressed by $\pi_1 = 4L/\chi_0$ and $\pi_1/\pi_2 = 16\mu L/\chi_0^2(\rho K)^{1/2}$.

[22] π_2 is the ratio between the square root of the product of inertia, elastic forces, and viscous forces.

C.10.3.2 Reynolds number

The Reynolds number Re can be expressed in dimensionless form by:[23]

$$\text{Re} = 4\pi\pi_2 k_0^{-1} \tilde{\chi}_i^{-1} \tilde{q}, \qquad (C.23)$$

In the range $p \leq p_\ell$, the fluid domain can be deconstructed into two fluid regions: (1) an area of very small velocity magnitude near the contact point, where the viscous effects are predominant; and (2) an outer zone. One may suppose that the flow regime is mainly affected by this second region. Consequently, the usual cross-length scale, i.e., the hydraulic perimeter $d_h = 4A_i/\chi_i$, and the inertial effects might be underestimated. Each lobe of the line-contact cross-section can be modeled by a fluid domain of same area and same wetted perimeter, which includes an inner rectangular zone, of large aspect ratio, and an outer circular region. The radius of the latter is calculated such that the head loss through the equivalent cross-section is equal to the head loss through the actual one. The ratio between the Reynolds numbers in the actual and circular domains is found to be proportional to the ratio between the model radius and actual wetted perimeter. Assuming a critical Reynolds number of 2000 in the circular duct, a critical value in the collapsed tube is proposed to be equal to 1200 for the whole range $p \leq p_\ell$, when $k_0 = 1.005$ [847]. Only experiments should determine the transition regime, hence the critical Reynolds number.[24]

C.10.3.3 Head Losses and Shape Factor

The viscous head losses per unit length $f_v(q)$ in smooth-walled uniformly collapsed tubes are given by[25]

$$f_v = \frac{\Lambda}{d_h} \frac{\rho V^2}{2}, \qquad (C.24)$$

where Λ is the friction head loss coefficient : $\Lambda = f_{\text{lam}}/\text{Re}$ and $\Lambda = f_{\text{turb}}/\text{Re}^{1/4}$ (f_{lam} and f_{turb}: laminar and turbulent shape factors).[26] The shape factors depend on the transmural pressure p and on ellipticity k_0.

---

[23] The Reynolds number is expressed by $\text{Re} = 4\pi_2\tilde{q}/\tilde{\chi}_i$ when $\chi_0/4$ is used as the transverse length scale, and $\pi_2 = \chi_0(\rho K)^{1/2}/4\mu$.

   In a given tube conveying a given fluid at a given flow rate, when the tube wetted perimeter remains constant, i.e., for $p \geq p_\ell$, Re is constant along the whole collapsed tube length, but when the opposite walls are in contact over a line segment, the hydraulic perimeter and consequently Re vary along the tube length.

[24] The critical Reynolds number for laminar-turbulent transition was found experimentally to be equal to about 2800 for the approximately point-contact–shaped duct [847, Fig 9].

[25] Another usage gives the friction coefficient C_f, related to the wall shear stress τ_w by the formula $\tau_w = C_f\rho U^2/2$. Here, $f_v = (\chi_i/A_i)\tau_w$, which means that the dimensionless hydraulic loss factor $\Lambda = 4\,C_f$ for uniformly collapsed conduits.

[26] It is well known that the stronger the flow resistance induced by a given cross-section shape, the higher the shape factor: f_{lam} increases with the width to

Table C.8. Values of the laminar shape factor in the characteristic tube shapes for different ellipticities.

k_0	1	2	3	10
f_{L0}	64	67	71	77
f_{Lp}	71	75	79	86
f_{Lc}	50	49	48	47
$f_{L\ell}$	37	36	35	33

Laminar Pattern

The analytical solution is known for $\tilde{p} = 0$:

$$f_{\text{lam}_0} = 128\pi^2(1 + k_0^{-2})\tilde{\chi}_0^{-2} . \tag{C.25}$$

Besides, the laminar shape factor f_{lam} is computed for the following characteristic values of the transmural pressures $\tilde{p}_p$, $\tilde{p}_c$, $\tilde{p}_\ell$ [847] (Table C.8).
   In the range $\tilde{p} \leq \tilde{p}_\ell$:

$$f_{\text{lam}} = 39.0731 - 2.0929 k_0/\mathcal{A}_2 + 0.2484(k_0/\mathcal{A}_2)^2 - 0.0101(k_0/\mathcal{A}_2)^3 , \tag{C.26}$$

where $\mathcal{A}_2 = (\tilde{p}/\tilde{p}_\ell)^{(2n-3)/5n} = (\tilde{A}_i/\tilde{A}_{i\ell})^{(3-2n)/5}$. The theoretical values are close to experimental data in a tube of ellipticity $k_0 = 1.6$ in the range $[\tilde{p}_p, 0]$ (difference of about 2%); they are slightly overestimated in the range $[\tilde{p}_\ell, \tilde{p}_p]$ (difference of about 10%).

Turbulent Pattern

The value of the turbulent shape factor $f_{\text{turb}} = 0.26$ in the transmural pressure range $\tilde{p} \leq \tilde{p}_c$ was deduced from experimental data particularly in inclined collapsed tube in an uniform state [846]. During collapse mode 1, the classical value $f_{\text{turb}} = 0.316$ is used.

C.10.3.4 Inclined Collapsible Tube

Mass conservation gives:

$$\frac{d}{dt}(\rho A v) = \frac{\partial}{\partial t}(Av) + \frac{\partial}{\partial x}(Av) = 0 .$$

When the flow is steady:

$$\frac{\partial A}{\partial t} + \frac{\partial q}{\partial x} = 0 . \tag{C.27}$$

---

height ratio for a rigid duct of rectangular cross-section, with the ellipticity of rigid pipes of elliptical cross-section and with the internal to external diameter ratio of annuli between two rigid concentric cylinders.

Table C.9. Configurations of an inclined collapsible tube with respect to the respective values of hydrostatic and viscous forces on the one hand and flow regime on the other hand.

	$\rho g \sin \beta < f_v$	$\rho g \sin \beta = f_v$	$\rho g \sin \beta > f_v$
Subcritical $Ma < 1$	$dp/dx < 0$ Convergent tube	$dp/dx = 0$ Uniform tube	$dp/dx > 0$ Divergent tube
Transcritical $Ma = 1$	$dp/dx \to -\infty$ Throat	$0/0$	$dp/dx \to +\infty$ Jump
Supercritical $Ma > 1$	$dp/dx > 0$ Divergent tube	$dp/dx = 0$ Uniform tube	$dp/dx < 0$ Convergent tube

This principle can also be written as:

$$q(\mathbf{x}, t) = A(\mathbf{x}, t)v(\mathbf{x}, t) = \text{cst.} \implies dq = A\,dv + v\,dA = 0$$
$$\implies \frac{dA}{A} + \frac{dv}{v} = 0 . \qquad (C.28)$$

Momentum conservation gives:

$$\rho A d\,x\,v \frac{dv}{dx} = pA - (p + dp)(A + dA) - \chi_i\,dx\,\tau_w + \rho A\,dx\,g \sin \beta$$

where β is the tube inclination angle, $\chi_i = 4A/d_h$ the tube wetted perimeter (d_h: hydraulic diameter) and $\tau_w = -\mu(dv/dn)|_{r=R}$ (n: wall local normal):

$$\rho v \frac{dv}{dx} = -\frac{dp}{dx} - \frac{\chi_i}{A}\tau_w + \rho g \sin \beta = -\frac{dp}{dx} - f_v + \rho g \sin \beta .$$

Introducing the wave speed in the left hand side of the equation,

$$(1 - Ma^2)\frac{dp}{dx} = \rho g \sin \beta - f_v . \qquad (C.29)$$

Replacing p by A,

$$(1 - Ma^2)\frac{dA}{dx} = \frac{A}{\rho c^2}(\rho g \sin \beta - f_v) . \qquad (C.30)$$

The 1D model relates the tube shape to the existing forces and flow regime (Table C.9).

C.10.3.5 Collapsible Tube Connected to a Finite Reservoir, Subjected to a Pressure Ramp

The momentum conservation states that:

$$\frac{\partial v}{\partial t} + v\frac{\partial v}{\partial x} = -\frac{1}{\rho}\left(\frac{\partial p_i}{\partial x} + f_v\right). \tag{C.31}$$

Multiplying by A and applying the mass conservation:

$$\frac{\partial q}{\partial t} = -\left(2v\frac{\partial q}{\partial x} + \frac{A}{\rho}\left((1 - Ma^2)\frac{\partial p}{\partial x} + f_v\right)\right). \tag{C.32}$$

Such an equation gives the sign of $\partial q/\partial t$ with respect to the signs and relative magnitudes of the terms of the equation right-hand side. When the tube collapses, the first term (Tr1) $\partial A/\partial t < 0$, so, $2v\partial q/\partial x > 0$. According to $Ma <, =$ or > 1, the second term (Tr2) $(A/\rho)(1 - Ma^2)\partial p/\partial x <, =$ or > 0. The third term (Tr3) has the sign of $f_v > 0$,

Case 1 $Ma < 1$:

$$\partial q/\partial t = -[(\text{Tr1} > 0) + (\text{Tr2} < 0) + (\text{Tr3} > 0)].$$

the first term is negligible in the collapse beginning. Therefore, $\partial q/\partial t > 0$
Case 2 $Ma > 1$:

$$\partial q/\partial t = -[(\text{Tr1} > 0) + (\text{Tr2} > 0) + (\text{Tr3} > 0)].$$

$\partial q/\partial t < 0$
Case 3 $Ma = 1$: When the previously subcritical flow $\partial q/\partial t > 0$ becomes transcritical $\partial q/\partial t < 0$, the flow rate reaches a maximum $\partial q/\partial t = 0$. When such a phenomenon occurs during the acceleration stage (Tr1 et Tr3 are still negligible), then $Ma = 1$.

C.10.3.6 Limitations of the One-Dimensional Model

As soon as the opposite walls come into contact, the flow splits up into two separate streams. The upstream and downstream effects of the split on the flow field, as well as transverse velocities generated during wall motion (although the additional flow rate can represent a small, but substantial part of the total flow rate [848]) are neglected. Moreover, the 1D assumption breaks in the tube segment where an elastic jump or a shock wave-like transition occurs [849]. Also, longitudinal physical factors of the deformation are neglected. In particular, the longitudinal tension developed during wall deformation modifies the local tube law, which in turn affects wave-speed value and flow behavior. Stability analysis requires at least a bidimensional model [850–852]. Above all, the axial change in cross-sectional area can be experimentally large. In particular, end effects due to the mounting of the compliant tube on rigid ducts influence the shape and area of the cross-section. Downstream from the constriction designed by the collapsed tube, the induced divergent segment elicits jet and flow separation, especially in the case of wall contact. Flow separation affects strongly the pressure distribution in collapsed tube conveying steady flows [853].

C.10.4 Three-Dimensional Flow in a Collapsed Tube

Few 3D investigations have been carried out, although time-dependent velocity profiles have been measured by laser Doppler velocimetry in an oscillating collapsible tube [854]. The 3D velocity field of a steady flow was computed in a frozen collapsed configuration with a downstream contact segment designed from ultrasound echographic measurements of a Starling resistor conveying a critical flow[27] [371]. Between the two side jets emerging from the contact zone and immediateley behind the contact zone, as well as along diverging walls, swirls occur. Side jets run centrally and partially merge in the proximal segment of the rigid attachment duct. Very low steady Re flow has been studied in a collapsing elastic tube, without opposite wall contact [370].

C.11 Dispersion in Fluid Flows

Consider the brief injection (δ-input) of a tracer in a vessel. Although the tracer concentration is uniform at the entrance of the test section, the convected tracer concentration axially stretches, when it is observed short times after the tracer bolus injection. The concentration spreading in the channel is due to the development of the velocity profile. The tracer bolus spreads as an evolving Gaussian curve with the traveled distance, with a peak concentration $c_{\max} \propto \mathrm{Pe}^{-1/2}$. The dispersion induced by the convection is damped by transverse molecular diffusion which homogenizes the tracer distribution, a blunt distribution being observed at long time intervals. This molecular diffusion involves an axial dispersion-related diffusivity $\mathcal{D}_d$:[28]

$$c_t = \mathcal{D}_d \, c_{zz}.$$

The axial mixing, the so-called *Taylor dispersion*, which involves convection, then differs from pure diffusion. The higher the diffusivity, the lower the dispersion.

C.12 Porous Medium

Different liquids can mix in a porous solid. Conversely, a porous medium can separate mixtures (filtration). Particles bigger than the pore size or molecules that bind to the solid walls are stopped. Additionally, velocity fluctuations are damped by the wall friction in the porous medium.

Consider the brief injection of a tracer into a homogeneous porous solid. The transit times of the different tracer molecules across the porous solid

---

[27] The critical reference state is defined by a localized throat near the tube outlet with a local speed equal to 20 times the entrance cross-sectional average velocity.

[28] Usual molecular diffusion refers to any kind of random processes.

differ; the concentration-time curve is Gaussian. Longitudinal and the transverse *dispersions* are not equal (anisotropy). Tracer dispersion is associated with molecular diffusion. The *Taylor dispersion*, associated with the coupling of molecular diffusion to convection, is not involved in porous media. Indeed in a porous body, the speed of a given molecule in the pore center is relatively high, but the speed can be reduced in the following pore, the molecule coming into contact with the wall or close to it, and conversely.

Low-speed flow through a rigid porous medium is described by the Darcy law $\nabla p = -(\mu/\mathcal{P}_D)\breve{v}$ ($\mathcal{P}$: Darcy permeability) and high-speed flow by the law $\nabla p = -(\mu/\mathcal{P}_D)\breve{v} - \kappa\rho v^2$, which both relate the pressure (p) to the volume-averaged velocity ($\breve{v}$). Flow through a deformable porous body deforms the material, and the deformation affects the flow.

C.13 Turbulence

Turbulent flow is characterized by:

1. a three-dimensional nature;
2. a random process with statistical independency between spatially and temporally distant observations;
3. a nonlinear phenomenon with an important convection inertia;
4. a high Reynolds number (strong inertia);
5. a convective mixing;
6. a strong energy dissipation;
7. a velocity field which can be often decomposed into a mean- and a fluctuating component;
8. an energy transfer between large and small flow structures and vice versa;
9. a vorticity field with intense fluctuations of the velocity curl and creation and destruction of vortices;
10. an additional momentum transport associated with the Reynolds stress which requires a turbulent dynamic viscosity μ_T.

C.14 Pressure Units

In the international unit system (SI), meter, kilogram, and second are the units of length, mass, and time, respectively. However, old pressure units are still used in physiology. Pressure conversion factors are given in Table C.10.

Table C.10. Conversion table for pressure units.

	Pascal (Pa)
1 cm of water	98.04
1 mm of Hg	133
1 mb	102
1 dyn/cm^2	0.1

D

Numerical Simulations

> *Les mathématiciens n'étudient pas des objets mais les relations entre ces objets.[Mathematicians do not study objects, but relationships between these objects.]* (H Poincaré)

The three basic natural sciences - biology, chemistry and physics - are involved in investigation of the cardiovascular system. These three sciences interact with mathematics to understand the functioning of the blood flow. However, the solutions of complex problems become more qualitative than quantitative, as pointed out by H. Poincaré.

> " *In the natural sciences, we no longer find these conditons: homogeneity, relative independence of remote parts, simplicity of the elementary facts, and this is why naturalists need to resort to other methods of generalization.* " (H Poincaré) [855]

From the ancient times, people have invented means to compute (Babylonian charts, abaci, etc.). Computers are required tools for numerical calculations. The computer was first used in a heuristic (tentatively used to discover; ερίναζω: help to fecundate) manner by J. von Neumann and S. Ulam for pure mathematics problems. Numerical simulations have been carried out since the mid-twentieth century.

Numerical simulations yield evolution of the physical quantities in time and space with a good resolution that cannot be obtained by measurements. The numerical model thus allows a better understanding of the implicated phenomena of the investigated physical problem. The computational model also allows one to study the role played by influence parameters, all other variables being constant.

" *Dans le cas très fréquent où le nombre des variables en présence est considérable... la règle ca-
pitale à suivre... est de s'astreindre à laisser sys-
tématiquement invariable dans chaque expérience
tous les facteurs sauf un seul...[In the very frequent
case with a huge number of involved quantities...
the chief rule to follow... is to force oneself in every case to keep constant all factors except a single
one...]* " (Le Chatelier)

D.1 Numerical Model

The qualitative description of the physical problem of interest relies on a
mathematical model usually defined by partial differential equations that link
the relevant variables. The set of partial differential equations is associated
with initial and boundary conditions. Normally, boundary conditions are required for a unique solution and initial conditions for the distribution of the
involved quantities at the initial time. The mathematical model is determined
using assumptions and approximations of physical reality.

When the equation set cannot be solved analytically, approximation
schemes are used. The discretization of the partial differential equations leads
to a linear algebraic system solved by a suitable method [628, 856–859]. In
the case of the finite element method, the continuum is discretized into a set
of polygons on its surface and polyhedra in its volume.

Numerical modeling includes several stages. The mathematical model associated with the physical problem corresponds to a boundary-value problem:

$$\mathcal{L}\mathbf{u} = \mathbf{f},$$

where $\mathcal{L}$ is a differential operator, $\mathbf{u}$ is the real-value solution of the problem
and $\mathbf{f}$ a given function. In continuum mechanics, the partial differential equations result from equilibrium of fluxes and forces in an infinitesimal control
volume. The problem is well-conditionned when a small relative variation in
a quantity causes only a slight relative change in the results.

The mathematical framework is analyzed by functional analysis. The existence, uniqueness, and essential properties of the solution are provided at least
for a simplified model. The numerical model builds and analyzes a discrete
approximation:

$$\mathbf{A}\mathbf{u}_h = \mathbf{f}_h,$$

where $\mathbf{A}$ is the approximation matrix, $\mathbf{u}_h$ the vector of N unknowns or degrees of freedom associated with the solution representation after the domain
discretization. Subscript h refers to the mesh, h being the characteristic space
step. $\mathbf{f}_h$ is the vector containing the discretization of $\mathbf{f}$ and the boundary
conditions. Although linear systems can be solved by direct methods, the resolution of the discret problem is most often based on iterative procedures

(Jacobi, Gauss-Seidel, GMRES, and Newton methods). A sequence $\{\mathbf{u}_h\}_1^N$ of estimates of $\mathbf{u}_h$ is thus generated. There is a convergence of iterations when:

$$\lim_{n \to \infty} \mathbf{u}_h^n = \mathbf{u}_h.$$

D.2 Approximation Methods

The numerical analysis yields an approximate solution of the problem with a given precision and finite number of elementary operations. Various approximation methods include: (1) deterministic techniques, such as finite difference methods, finite volume methods, finite element methods, spectral methods,[1] particle methods,[2] cellular automata;[3] and (2) stochastic methods, such as Monte-Carlo algorithms.[4] The choice depends on the problem's nature and domain geometry. Whatever the selected method, approximation convergence must be ensured:

$$\lim_{h \to 0} \mathbf{u}_h = \mathbf{u}.$$

Numerous sources of error occur during the different modeling steps: (1) truncature and round-off errors,[5] (2) iteration errors, (3) approximation errors, and (4) modeling errors (errors on values of input data, determination of equations, hypotheses, those due to mesh, etc.):

---

[1] The basis functions of the spectral methods are global, whereas they are local in the finite element methods, being associated with mesh nodes.

[2] The studied system is considered as a set of moving particles, the motion of which is described by their position labeled with respect to a frame, velocity, mutual interaction force, and possible momentum exchange. In each mesh element, the averaged particle density is taken into account, and not the large number of particles.

[3] Cellular automata were introduced by J. von Neumann and S. Ulam to model auto-organization processes in biological systems. This deterministic technique links each mesh node to a discrete state, with takes only few values, sometimes two (states either excited or at rest). They have been used to simulate certain hydrodynamics processes.

[4] Monte-Carlo methods have been proposed by S. Ulam and N. Metropolis. Computations generate elementary random processes and select the phenomena with sufficient physical realizability, particularly ensuring conservation principles.

[5] Any number is represented by a finite bit (information binary unit) number. Computational writing approximates numbers with a fixed precision of significative digit. Moreover, computations of certain functions yield approximate results.

$$u_{calc} - u_{reel} = \begin{cases} u_{calc} - u_h^n & (1) \\ + \\ u_h^n - u_h & (2) \\ + \\ u_h - u & (3) \\ + \\ u - u_{reel} & (4) \end{cases}$$

Successive steps of modeling and simulation propagate errors. Conditioning estimates the sensitivity with respect to the variations of a given variable. Bad conditioning can occur for certain variation interval of the variable. The stability defines the sensitivity of the numerical procedure with respect to round-off errors. A numerical scheme is stable in the absence of error amplification.

D.3 Basic Techniques in Discretization

The approximate solution u_h in a finite-dimensional space is computed rather than the prediction u of the behavior of the explored complicated medium. The small parameter h (mesh length scale) is assumed to tend toward zero when the computational domain tends toward the continuum. The computation is based on a set of notions and numerical concepts: (1) linkers, such as *interpolations*, between the finite-dimensional space and the continuum; (2) *consistency* which states that the approximate state is closer to the real state when the characteristic grid spacing h tends to zero; (3) *stability* which states that u_h remains in a bounded set when h tends toward zero, and is associated with error damping when numerical computation proceeds; and (4) *convergence* which states that the numerical procedure produces a solution that approaches the exact solution as the grid spacing h is reduced (u_h is close enough to u). Several methods can be used.

D.3.1 Finite Difference Method

The differentiation operators are replaced by difference operators derived from truncations of Taylor series. A finite difference discrete system is determined as a function of linear combinations of translations of the unknown function. The finite difference method substitutes the continuum by a mesh of regularly spaced nodes.[6] A finite difference discrete system is determined as a function of linear combinations of translations of the unknown function.

The derivatives are discretized using the Taylor formulas:

$$u_{i+1,j,\ell} = u_{i,j,\ell} + \Delta x \frac{\partial u}{\partial x_i} + \frac{(\Delta x)^2}{2} \frac{\partial^2 u}{\partial x_i^2} + \mathcal{O}[(\Delta x)^3],$$

---

[6] The mesh is such that $x_i = i\Delta x$, $y_j = j\Delta y$, and $z_\ell = \ell\Delta z$ for node (x_i, y_j, z_ℓ).

$$u_{i-1,j,\ell} = u_{i,j,\ell} - \Delta x \frac{\partial u}{\partial x_i} + \frac{(\Delta x)^2}{2} \frac{\partial^2 u}{\partial x_i^2} + \mathcal{O}[(\Delta x)^3],$$

The difference between these two formulas leads to the *central difference*:

$$\frac{\partial u}{\partial x_i} = (u_{i+1,j,\ell} - u_{i-1,j,\ell})/2\Delta x + \mathcal{O}[(\Delta x)^2],$$

for an approximation of order 2. The sum of the two Taylor formulas leads to:

$$\frac{\partial^2 u}{\partial x_i^2} = (u_{i+1,j,\ell} - 2u_{i,j,\ell} + u_{i-1,j,\ell})/(\Delta x)^2 + \mathcal{O}[(\Delta x)^2].$$

The first Taylor formula can give the *forward difference*:

$$\frac{\partial u}{\partial x_i} = (u_{i+1,j,\ell} - u_{i,j,\ell})/\Delta x + \mathcal{O}[(\Delta x)],$$

the approximation of order 1 is not as good. The second Taylor formula can give the *backward difference*:

$$\frac{\partial u}{\partial x_i} = (u_{i,j,\ell} - u_{i-1,j,\ell})/\Delta x + \mathcal{O}[(\Delta x)].$$

Many finite difference schemes cannot be applied to the computation of discontinuous problems. Furthermore, the mapping of the computational domain onto the physical one must be regular for sufficiently accurate truncated Taylor formulas.

D.3.2 Finite Volume Method

The finite volume method introduces the notion of fluxes across the cells which compose the computational domain. Several techniques have been implemented (cell-centered technique, vertex-centered technique) whether the primal mesh provides the partition or the flux derivation is obtained on a dual partition.

A typical element of a 2D staggered mesh is displayed in Fig. D.1. The finite volume method guarantees conservation of fluid variables in each control volume.[7] Numerical schemes which have *conservativeness* ensure global conservation of fluid variables (throughout the entire computational domain), by means of consistent expressions for fluxes of these variables across the cell edges (2D space) or faces (3D space) of adjacent control volumes. Handling of the relative magnitude of convection and diffusion allows *transportiveness*.

---

[7] The conservation of a fluid variables, such as the velocity, within a control volume is ensured by the balance between the processes, which increase and decrease this variable. The time rate of the variable change in the control volume is equal to the sum of the net flux of the variable due to convection across the control volume, the diffusion transport, and the net rate of creation or vanishing inside the control volume.

In a orthogonal curvilinear coordinate system, $\{x_i\}_{i=1}^3$ (x_3 : axial coordinate), $\{h_i\}_{i=1}^3$ the variable scaling factors allows to map the Cartesian coordinates to the curvilinear coordinates ($h_i = |\partial \mathbf{r}/\partial x_i|$). The curvature radii of of lines $x_1 = cst$ et $x_2 = cst$ are given by $1/R_{c_j} = (1/(h_1 h_2))\partial h_i/\partial h_j$.

The continuity equation is written as:

$$\frac{1}{h_1 h_2}\frac{\partial}{\partial x_1}(\rho h_2 u_1) + \frac{1}{h_1 h_2}\frac{\partial}{\partial x_2}(\rho h_1 u_2) + \frac{\partial}{\partial x_3}(\rho u_3) = 0.$$

The transport equation of the physical quantity $g(\mathbf{x})$ is given by:

$$\frac{1}{h_1 h_2}\frac{\partial}{\partial x_1}(\rho h_2 u_1 g) + \frac{1}{h_1 h_2}\frac{\partial}{\partial x_2}(\rho h_1 u_2 g) + \frac{\partial}{\partial x_3}(\rho u_3 g) =$$

$$\frac{1}{h_1 h_2}\frac{\partial}{\partial x_1}\left(\gamma_g \frac{h_2}{h_1}\frac{\partial g}{\partial x_1}\right) + \frac{1}{h_1 h_2}\frac{\partial}{\partial x_2}\left(\gamma_g \frac{h_1}{h_2}\frac{\partial g}{\partial x_2}\right) + \frac{\partial}{\partial x_3}\left(\gamma_g \frac{\partial g}{\partial x_3}\right) + S_g,$$

with γ_g corresponds to diffusion coefficient and S_g to the source term for g (Table D.1).

These equations have been solved for bend flows using hexahedra. The discrete formulation of the equations is obtained by integration over the cell.

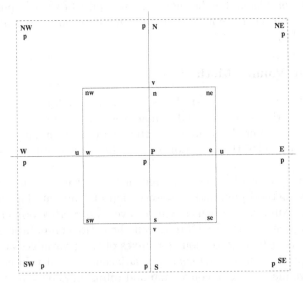

Figure D.1. Two-dimensional control volume in finite volume method. Upper case and lower case letters stands for nodes at which pressure (p) and velocity components (u, v) are stored, respectively. The "velocity nodes" are staggered with respect to the storage locations of all other physical variables. The pressure is stored at the cell center and each velocity component at corresponding mid-edge (normal to the velocity component).

The continuity equation is given by (U,u: upstream nodes, D,d: downstream nodes, W,w,E,e,N,n,S,s: lateral nodes; Fig. D.1):

$$\left(\rho u_1 \Delta x_2 \Delta x_3\right)_e - \left(\rho u_1 \Delta x_2 \Delta x_3\right)_w + \left(\rho u_2 \Delta x_1 \Delta x_3\right)_n - \left(\rho u_2 \Delta x_1 \Delta x_3\right)_s$$

$$+ \left(\rho u_3 \Delta x_1 \Delta x_2\right)_d - \left(\rho u_3 \Delta x_1 \Delta x_2\right)_u = 0,$$

and the transport equation by:

$$\left[\left(\rho u_1 \Delta x_2 \Delta x_3\, g\right)_e - \left(\rho u_1 \Delta x_2 \Delta x_3\, g\right)_w\right]$$

$$+ \left[\left(\rho u_2 \Delta x_1 \Delta x_3\, g\right)_n - \left(\rho u_2 \Delta x_1 \Delta x_3\, g\right)_s\right]$$

$$+ \left[\left(\rho u_3 \Delta x_1 \Delta x_2\, g\right)_d - \left(\rho u_3 \Delta x_1 \Delta x_2\, g\right)_u\right] =$$

Table D.1. Source terms of transport equation in the finite volume method for the three velocity components.

g	S_g
u_1	$-\dfrac{1}{h_1}\dfrac{\partial p}{\partial x_1} - \dfrac{\rho u_1 u_2}{R_{c1}} + \dfrac{\rho u_2^2}{R_{c2}} + \dfrac{1}{h_1 h_2}\dfrac{\partial}{\partial x_1}\left[h_2\mu\left(\dfrac{1}{h_1}\dfrac{\partial u_1}{\partial x_1} + \dfrac{2u_2}{R_{c1}}\right)\right]$ $+\dfrac{1}{h_1 h_2}\dfrac{\partial}{\partial x_2}\left[h_1\mu\left(\dfrac{1}{h_1}\dfrac{\partial u_2}{\partial x_1} - \dfrac{u_2}{R_{c2}} - \dfrac{u_1}{R_{c1}}\right)\right]$ $+\dfrac{\mu}{R_{c1}}\left(\dfrac{1}{h_1}\dfrac{\partial u_2}{\partial x_1} + \dfrac{1}{h_2}\dfrac{\partial u_1}{\partial x_2} - \dfrac{u_2}{R_{c2}} - \dfrac{u_1}{R_{c1}}\right)$ $-2\dfrac{\mu}{R_{c2}}\left(\dfrac{1}{h_2}\dfrac{\partial u_2}{\partial x_2} + \dfrac{u_1}{R_{c2}}\right) + \dfrac{\partial}{\partial x_3}\left(\dfrac{\mu}{h_1}\dfrac{\partial u_3}{\partial x_1}\right).$
u_2	$-\dfrac{1}{h_2}\dfrac{\partial p}{\partial x_2} - \dfrac{\rho u_1 u_2}{R_{c2}} + \dfrac{\rho u_1^2}{R_{c1}} + \dfrac{1}{h_1 h_2}\dfrac{\partial}{\partial x_2}\left[h_1\mu\left(\dfrac{1}{h_2}\dfrac{\partial u_2}{\partial x_2} + \dfrac{2u_1}{R_{c2}}\right)\right]$ $+\dfrac{1}{h_1 h_2}\dfrac{\partial}{\partial x_1}\left[h_2\mu\left(\dfrac{1}{h_2}\dfrac{\partial u_1}{\partial x_2} - \dfrac{u_2}{R_{c2}} - \dfrac{u_1}{R_{c1}}\right)\right]$ $+\dfrac{\mu}{R_{c2}}\left(\dfrac{1}{h_1}\dfrac{\partial u_2}{\partial x_1} + \dfrac{1}{h_2}\dfrac{\partial u_1}{\partial x_2} - \dfrac{u_2}{R_{c2}} - \dfrac{u_1}{R_{c1}}\right)$ $-2\dfrac{\mu}{R_{c1}}\left(\dfrac{1}{h_1}\dfrac{\partial u_1}{\partial x_1} + \dfrac{u_2}{R_{c1}}\right) + \dfrac{\partial}{\partial x_3}\left(\dfrac{\mu}{h_2}\dfrac{\partial u_3}{\partial x_2}\right).$
u_3	$\dfrac{\partial p}{\partial x_3} - \dfrac{\partial}{\partial x_3}\left(\mu\dfrac{\partial u_3}{\partial x_3}\right) + \dfrac{1}{h_1 h_2}\dfrac{\partial}{\partial x_1}\left(h_2\mu\dfrac{\partial u_1}{\partial x_3}\right) + \dfrac{1}{h_1 h_2}\dfrac{\partial}{\partial x_2}\left(h_1\mu\dfrac{\partial u_2}{\partial x_3}\right).$

$$\left[\left(\gamma_g \frac{\Delta x_2 \Delta x_3}{\Delta x_1}\right)_e (g_E - g_P) - \left(\gamma_g \frac{\Delta x_2 \Delta x_3}{\Delta x_1}\right)_w (g_P - g_W)\right]$$

$$+\left[\left(\gamma_g \frac{\Delta x_1 \Delta x_3}{\Delta x_2}\right)_n (g_N - g_P) - \left(\gamma_g \frac{\Delta x_1 \Delta x_3}{\Delta x_2}\right)_s (g_P - g_S)\right]$$

$$+\left[\left(\gamma_g \frac{\Delta x_1 \Delta x_2}{\Delta x_3}\right)_d (g_D - g_P) - \left(\gamma_g \frac{\Delta x_1 \Delta x_2}{\Delta x_3}\right)_u (g_P - g_U)\right]$$

$$+\left(S_g \, \Delta x_1 \Delta x_2 \Delta x_3\right)_P.$$

$\Delta x_{i(m)}$ being the distance between node P and adjacent cell center (m) in the direction x_i. When $g \equiv u_i$, S^g contains a pressure gradient term:

$$\left(\Delta x_i \Delta x_j\right)_P \left(p_{in(k)} - p_{out(k)}\right),$$

where subscripts $in(k)$ et $out(k)$ correspond to values at mesh nodes in the entry and exit cross-sections associated with velocity component u_k.

A linear interpolation assesses the physical quantity at the required position using a spatial weighting factor, which for node (m) on coordinate axis x_i between nodes P et M, is given by:

$$f_{(m)} = \left|\frac{x_{i(m)} - x_{iP}}{x_{i(M)} - x_{iP}}\right|.$$

D.3.3 Vortex-in-Cell Lagrangian Method

The vortex-in-cell Lagrangian method can be employed to compute nonzero-mean sinusoidal flow in stenoses, planar bends, and bifurcations [318]. The method is based on the stream function-vorticity formulation of the 2D Navier-Stokes equations:

$$\Delta \Psi = -\omega, \quad \omega_t + \Psi_y \omega_x - \Psi_x \omega_y = \nu \nabla^2 \omega.$$

The first step consists in transforming the flow field (x, y) in a rectangular domain (x', y'):

$$\nabla^2 \Psi = -J\omega, \quad J\omega_t + \Psi_{y'} \omega_{x'} - \Psi_{x'} \omega_{y'} = \nu \nabla^2 \omega,$$

where J is the jacobian function of the transformation.

The Poisson equation can be solved on a uniform grid at each time step in direction x' using fast Fourier transform.[8] The circulation Γ of a moving pinpoint vortex k in a given mesh cell is projected on the four cell vertices

---

[8] The normal direction is heterogeneously discretized due to the boundary layer.

with weighting by areas of corresponding sectors among the four delineated zones within the cell. The vorticity distribution is then given by:

$$\omega_i = \Gamma_k A_i / A^2 , \quad 1 \le i \le 4.$$

Weighting ensures circulation and momentum conservation. A finite difference scheme leads, after FFT, to solve Ψ along each y'-constant grid line and the velocity components are computed. The second equation is split into a viscous term $\omega_{t_{diff}} = \nu \nabla^2 \omega$ and a convective term $\omega_{t_{conv}} = -(\mathbf{u} \cdot \nabla)\omega$. The problem is solved assuming a uniform injection velocity.

D.3.4 Finite Element Method

The finite element method is an approximation procedure suitable for complicated domain geometries, such as the vasculature (Table D.2). It solves a system of partial differential equations in appropriate functional spaces. The functional spaces V defined for the continuum are approximated by finite-dimension functional spaces V_h:

$$V_h \equiv V_h^n[k, r] : \phi(\mathbf{x}) \in C^r(\overline{\Omega}),$$

V_h being the space of polynomials of degree $\le k$ in $\mathbb{R}^n$, defined in domain Ω_h, with continuous derivatives of order r.[9] The finite element method is based on a representation by interpolation functions, the coefficients of which are obtained from the set of nodal values.

The field of a given variable $\mathbf{u}(\mathbf{x})$ of the physical problem is represented by approximate values at Nn mesh nodes such that the differences $e(\mathbf{x}) = \mathbf{u}_h(\mathbf{x}) - \mathbf{u}(\mathbf{x})$ are small. In each element K, the field of the unknown $\mathbf{u}$ is approximated $(\mathbf{u}_h)$ by interpolation function $\{\varphi_i\}$ defined by nodal values $\{\mathbf{u}^{(i)}\}$, which are specified with the type of finite element adequate to the problem, at nodes either uniquely at the element boundary, or both at the

Table D.2. Some features of the finite difference (FDM) and finite element (FEM) methods.

	FDM	FEM
Formulation	Differential	Integral
Domain geometry	Simple	Complicated
Discret domain	Node network	Element set
Approximation	Pointwise	Piecewise

---

[9] The space V_h is such that for any sampled element v_h, v_h is continuous over the finite element K, v_h is piecewise continuous over the discrete domain Ω_h, and v_h satisfies the boundary conditions of the problem. In mathematical notation, $\forall v_h \in V_h, v_h \in C^r(\Omega), v_h|_K \in P^k, \forall K \in \mathcal{T}_h, v_h|_{\Gamma_h} = v_{\Gamma}.$

element boundary and inside it. Assembling then leads to the solution on the entire domain.[10]

The equations are written using the integral formulation[11] in the volume of element K generated by discretization T_h of the continuum domain. The approximate value of the unknown $\mathbf{u}_h$ is expressed by a linear combination (sum of products) of interpolation functions $\{ \varphi_i(\mathbf{x}) \}_{i=1}^{Nn}$ of the basis of space V_h, and nodal values $\mathbf{u}_h^{(i)} = \mathbf{u}_h(\mathbf{x}^{(i)})$ at a given instant:

$$\mathbf{u}_h(\mathbf{x}, t) = \sum_{i=1}^{Nn} \varphi_i(\mathbf{x}) \mathbf{u}^{(i)} \Big|_t .$$

The interpolation functions satisfy the following conditions: (1) $\varphi_i(\mathbf{x}^{(j)}) = \delta_{ij}$, (2) its degree depends on the node number, (3) its nature depends on the number and on the type of nodal unknowns, and (4) continuity on the element boundary.

The variable $\mathbf{u}_h$ is then defined on the whole domain Ω_h. The simplest interpolation uses piecewise continuous affine basis functions. The continuity is ensured because, for a neighboring element, $\mathbf{u}^{k'}$ is defined by the same value on the shared nodes, or edges or faces: $\mathbf{u}^k = \mathbf{u}^{k'}$.

The finite element is an affine set with interpolation operator: $\{K, P_T, \Sigma_T\}$, where $K \in T_h$ (T_h: domain mesh) can be the image of the reference finite element $\widehat{K}$, P_T the approximation polynomial space of dimension N_T ($P_T = \{v_h\big|_T, v_h \in V_h\}$, $\dim(P_T) = n + k$), Σ_T a set of N_T degrees of freedom

(N_T linear forms $\{\varphi_i\} \in P_T$, such that $\forall u \in P_T$, $u(\mathbf{x}) = \sum_{i=1}^{N_T} u^{(i)} \varphi_i(\mathbf{x})$.

The degrees of freedom are either the variables at the mesh nodes (Lagrange finite element) or the variables and their derivatives (Hermite finite element). The test functions (interpolation functions) can be derived from barycentric coordinates (Fig. D.2).

The domain discretization leads to a set of connected polyhedra with the following rules: (1) the element interior is not empty and (2) the intersection of two elements is void, a common node, edge, or face. Different estimators assess the mesh quality.

The equation formulation is based on projection (with the scalar product meaning) the unknown $\mathbf{u}$ on a functional space, i.e., to multiply by a test function $\mathbf{v}$ and to integrate. The integral formulation of the problem can be

---

[10] The finite element is characterized by several features: (1) its geometrical type, (2) the node number, (3) the types of nodal variables, and (4) the type of interpolation functions.

[11] The multiplication by test functions and integration of the equations over the domain, followed by bypart integration leads to a weak formulation, the derivability of the unknown u being reduced and the one of the test function v being augmented.

obtained either by the weighted residual method,[12] or the variational formulation.[13] The integration by parts leads to a weak formulation of the equations, because of the reduction in conditions on $\mathbf{u}_h$ in V_h, but the definition criteria of the functional space are strong. Such a method deals with natural boundary conditions[14] and discontinuities.

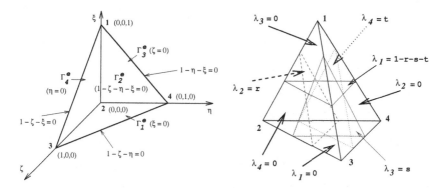

Figure D.2. Barycentric coordinates in the tetrahadron

---

[12] Let $\mathbf{u}_h$ be the approximate solution of $\mathcal{L}\mathbf{u} = \mathbf{f}$. The test functions $\mathbf{v}_h$, or weighting functions, are interpolation functions. $\mathcal{L}\mathbf{u}_h - \mathbf{f}$ defines the residual. The equation terms are residual-weighted integrals by test functions $\mathbf{v}_h$.

[13] Certain physical problems can be defined by a functional

$$\pi = \int F\left(\mathbf{u}(\mathbf{x},t), \nabla\mathbf{u}(\mathbf{x},t), \mathbf{x}, t\right),$$ to be minimized. Because equations are equivalent, the integral formulation is called variational in the absence of minimization problem. Consider a simple case, the Stokes problem, i.e., the steady flow of an incompressible Newtonian fluid in a domain subjected to body forces $\mathbf{f}$, inertia forces being negligible. The test function $\mathbf{v}$ satisfies the incompressibility condition $\nabla \cdot \mathbf{v} = 0$ in Ω and the boundary condition $\mathbf{v}|_\Gamma = 0$. The integration in Ω, followed by processing using Green formulas, leads to the following variational formulation:

$$a(\mathbf{u}, \mathbf{v}) = L(\mathbf{v}), \quad \forall \mathbf{v} \in \mathcal{V},$$

with $a(\mathbf{u}, \mathbf{v}) = \mu \sum_{1}^{3} \int_\Omega u_{,i} v_{,i} \, d\Omega$ and $L(\mathbf{v}) = \int_\Omega \mathbf{f} \cdot \mathbf{v} \, d\Omega.$

[14] There are two kinds of boundary conditions: the essential BCs, which must be explicitly satisfied and the natural BCs implicitly imposed. The latter appear in the equation and are automatically verified.

References

Introduction

1. Bachelard G (1934 [1968]) Le nouvel esprit scientifique [The new scientific spirit]. Presse Universitaire de France, Paris
2. Bachelard G (1938 [1969]) La psychanalyse du feu [The Psychoanalysis of fire]. Gallimard, Paris

Chap. 1. Anatomy of the Cardiovascular System

3. Gray H (1995) Gray's anatomy: anatomy descriptive and surgical. Barnes & Noble
4. Rouvière H, Delmas A (2002) Anatomie humaine descriptive, topographique et fonctionnelle[Descriptive human anatomy, topographical and functional] I–IV. Masson, Paris
5. Iacobellis G, Corradi D, Sharma AM (2005) Epicardial adipose tissue: anatomic, biomolecular and clinical relationships with the heart. Nature Clinical Practice Cardiovascular Medicine 2:536–543.
6. Kozlowski D, Owerczuk A, Piwko G, Kozlowska M, Bigus K, Grzybiak M (2002) The topography of the subthebesian fossa in relation to neighbouring structures within the right atrium. Folia Morphologica 62:65–70
7. Stradins P, Lacis R, Ozolanta I, Purina B, Ose V, Feldmane L, Kasyanov V (2004) Comparison of biomechanical and structural properties between human aortic and pulmonary valve. European Journal of Cardio-thoracic Surgery 26:634–639
8. Swanson WM, Clark RE (1974) Dimensions and geometric relationships of the human aortic valve as a function of pressure. Circulation Research 35:871–882
9. Grande KJ, Kunzelman KS, Cochran RP, David TE, Verrier ED (1993) Porcine aortic leaflet arrangement may contribute to clinical xenograft failure. ASAIO Journal 39:918–922
10. Valsalva AM (1740) Opera. Venice
11. Bellhouse BJ (1969) Velocity and pressure distributions in the aortic valve. Journal of Fluid Mechanics 37:587–600

12. Jatene MB, Monteiro R, Guimaraes MH, Veronezi SC, Koike MK, Jatene FB, Jatene AD (1999) Aortic valve assessment. Anatomical study of 100 healthy human hearts. Arquivos Brasileiros de Cardiologia 73:81-86

13. Anderson RH, Razavi R, Taylor AM (2004) Cardiac anatomy revisited. Journal of Anatomy 205:159–177

14. Anderson RH, Webb S, Brown NA (1999) Clinical anatomy of the atrial septum with reference to its developmental components. Clinical Anatomy 12:362–374

15. Gray AL, Johnson TA, Ardell JL, Massari VJ (2004) Parasympathetic control of the heart. II. A novel interganglionic intrinsic cardiac circuit mediates neural control of heart rate. Journal of Applied Physiology 96:2273–2278

16. Chambers R, Zweifach BW (1994) Topography and function of the mesenteric capillary circulation. The American Journal of Anatomy 75:175–205

17. Lee JS (2000) Biomechanics of the microcirculation, an integrative and therapeutic perspective. Annals of Biomedical Engineering 28:1–13

18. Hilgers RHP, Schiffers PMH, Aartsen WM, Fazzi GE, Smits JFM, De Mey JGR (2004) Tissue angiotensin-converting enzyme in imposed and physiological flow-related arterial remodeling in mice. Arteriosclerosis, Thrombosis, and Vascular Biology 24:892–897

19. Mori D, Yamaguchi T (2002) Computational fluid dynamics modeling and analysis of the effect of 3-D distortion of the human aortic arch. Computer Methods in Biomechanics and Biomedical Engineering 5:249–260

20. Cebral JR (2005) http://www. scs. gmu. edu/~jcebral/

Chap. 2. Physiology of the Cardiovascular System

21. Poon CS, Merrill CK (1997) Decrease of cardiac chaos in congestive heart failure, Nature 389:492–495

22. Penney D (2003) Cardiac cycle, http://www. coheadquarters. com/PennLibr/MyPhysiology/

23. Robinson TF, Factor SM, Sonnenblick EH (1986) The heart as a suction pump. Scientific American 6:62-69

24. Takahashi E, Asano K (2002) Mitochondrial respiratory control can compensate for intracellular O2 gradients in cardiomyocytes at low PO2. American Journal of Physiology – Heart and Circulatory Physiology 283:H871–H878

25. Guyton AC, Hall JE (1996) Textbook of medical physiology. Saunders

26. Milnor WR (1982) Haemodynamics. Williams & Wilkins, Baltimore

27. Silbernagl S, Despopoulos A (2001) Atlas de poche de physiologie[Pocket atlas of physiology]. Flammarion, Paris

28. Davies JI, Struthers AD (2003) Pulse wave analysis and pulse wave velocity: a critical review of their strengths and weaknesses. Journal of Hypertension 21:463–472

29. Meaney, E, Alva F, Moguel R, Meaney A, Alva J, Webel R (2000) Formula and nomogram for the sphygmomanometric calculation of the mean arterial pressure. Heart 84:64

30. Learoyd BM, Taylor MG (1996) Alterations with age in the viscoelastic properties of human arterial walls. Circulation Research 18:278–292

31. Mills CJ, Gabe IT, Gault JH, Mason DT, Ross J Jr, Braunwald E, Shilling-ford JP (1970) Pressure-flow relationships and vascular impedance in man. Cardiovascular Research 4:405–417

32. Anliker M et al (1977) Non-invasive measurement of blood flow, In: Hwang NHC, Normann NA (eds) Cardiovascular flow dynamics and measurements. University Park Press, Baltimore

33. Kalmanson D, Veyrat C (1978) Clinical aspects of venous return: a veloci-metric approach to a new system dynamics concept. In: Baan J, Noordegraaf A, Raines J (eds) Cardiovascular System Dynamics. MIT Press, Cambridge

34. Moreno AH (1978) Dynamics of pressure in the central veins. In: Baan J, Noordegraaf A, Raines J (eds) Cardiovascular System Dynamics. MIT Press, Cambridge

35. Hoffman JI, Spaan JA (1990) Pressure-flow relations in coronary circulation. Physiological Reviews 70:331–390

36. Scaramucci J (1695) De motu cordis, theorema sexton. Theoremata familiaria viros eruditos consulentia de variis physico medicis lucubrationibus iucta leges mecanicas. Urbino, Italy: Apud Joannem Baptistam Bustum, 70–81

37. Mori H, Tanaka E, Hyodo K, Mohammed MU, Sekka T, Ito K, Shinozaki Y, Tanaka A, Nakazawa H, Abe S, Handa S, Kubota M, Tanioka K, Umetani K, Ando M (1999) Synchrotron microangiography reveals configurational changes and to-and-fro flow in intramyocardial vessels. American Journal of Physiology. Heart Circulation Physiology 276:H429–H437

38. Lenègre J, Blondeau M, Bourdarias JP, Gerbaux A, Himbert J, Maurice P (1973) Cœur et circulation[Heart and circulation]. Pathologie médicale 3. Flammarion Médecine-Sciences, Paris

39. West JB (1974) Respiratory Physiology. Williams & Wilkins, Baltimore

40. Michel CC, Curry FE (1999) Microvascular permeability. Physiological Reviews 79:703–761

41. Rabiet M-J, Plantier J-L, Rival Y, Genoux Y, Lampugnani M-G, Dejana E (1996) Thrombin-induced increase in endothelial permeability is associated with changes in cell-to-cell junction organization. Arteriosclerosis, Thrombosis, and Vascular Biology 16:488–496

42. Weinbaum S, Curry FE (1995) Modelling the structural pathways for tran-scapillary exchange. Symposia of the Society for Experimental Biology 49:323–345

43. Agre P, Brown D, Nielsen S (1995) Aquaporin water channels: unanswered questions and unresolved controversies. Current Opinion in Cell Biology 7:472–483

44. Tarbell JM, Demaio L, Zaw MM (1999) Effect of pressure on hydraulic con-ductivity of endothelial monolayers: role of endothelial cleft shear stress. Jour-nal of Applied Physiology 87:261–268

45. Chen SC, Liu KM, Wagner RC (1998) Three-dimensional analysis of vacuoles and surface invaginations of capillary endothelia in the eel rete mirabile. The Anatomical Record 252:546–553

46. Davidson AJ, London B, Block GD, Menaker M (2005) Cardiovascular tissues contain independent circadian clocks. Clinical and Experimental Hyperten-sion 27:307–311

47. Curtis AM, Cheng Y, Kapoor S, Reilly D, Price TS, FitzGerald GA (2007) Circadian variation of blood pressure and the vascular response to asyn-

chronous stress. Proceedings of the National Academy of Sciences of the United States of America 104:3450–3455

48. McNamara P, Seo SP, Rudic RD, Sehgal A, Chakravarti D, FitzGerald GA (2001) Regulation of CLOCK and MOP4 by nuclear hormone receptors in the vasculature: a humoral mechanism to reset a peripheral clock. Cell 105:877–889

49. Damiola F, Le Minh N, Preitner N, Kornmann B, Fleury-Olela F, Schibler U (2000) Restricted feeding uncouples circadian oscillators in peripheral tissues from the central pacemaker in the suprachiasmatic nucleus. Genes & Development 14:2950–2961

50. Balsalobre A, Brown SA, Marcacci L, Tronche F, Kellendonk C, Reichardt HM, Schutz G, Schibler U (2000) Resetting of circadian time in peripheral tissues by glucocorticoid signaling. Science 289:2344–2347

51. Lemmer B (1992) Cardiovascular chronobiology and chronopharmacology. In: Touitou Y, Haus E (eds) Biologic Rhythms in Clinical and Laboratory Medicine, 418–427. Springer-Verlag, Berlin.

52. Pennes HH (1948) Analysis of tissue and arterial blood temperatures in the resting human forearm. Journal of Applied Physiology 1:93–122

53. Valvano JW, Bioheat transfer. users.ece.utexas.edu/~valvano/research/jwv.pdf

54. Werner J, Brinck H (2001) A three-dimensional vascular model and its applicatoion to the determination of the spatial variations in the arterial, venous, and tissue temperature distribution, In: Leondes C (ed) Biofluid Methods in Vascular and Pulmonary Systems. CRC Press, Boca Raton

55. Wissler EH (1998) Pennes' 1948 paper revisited. Journal of Applied Physiology 85:35–41

56. Arkin H, Xu LX, Holmes KR (1994) Recent developments in modeling heat transfer in blood perfused tissues. IEEE Transactions on Biomedical Engineering 41:97–107

57. Tedgui A, Lévy B (1994) Biologie de la paroi artérielle[Biology of the arterial wall]. Masson, Paris

58. Harder DR, Roman RJ, Gebremedhin D, Birks EK, Lange AR (1998) A common pathway for regulation of nutritive blood flow to the brain: arterial muscle membrane potential and cytochrome P450 metabolites. Acta Physiologica Scandinavica 164:527–532

59. Fischell TA, Bausback KN, McDonald TV (1990) Evidence for altered epicardial coronary artery autoregulation as a cause of distal coronary vasoconstriction after successful percutaneous transluminal coronary angioplasty. The Journal of Clinical Investigation 86:575–584

60. Guyenet PG (2006) The sympathetic control of blood pressure. Nature Reviews – Neuroscience 7:335–346

61. Ieda M, Kanazawa H, Kimura K, Hattori F, Ieda Y, Taniguchi M, Lee JK, Matsumura K, Tomita Y, Miyoshi S, Shimoda K, Makino S, Sano M, Kodama I, Ogawa S, Fukuda K (2007) Sema3a maintains normal heart rhythm through sympathetic innervation patterning. Nature Medicine 13:604–612

62. Koba S, Gao Z, Xing J, Sinoway LI, Li J (2006) Sympathetic responses to exercise in myocardial infarction rats: a role of central command. American Journal of Physiology – Heart and Circulatory Physiology 291, H2735-H2742

63. Crowley SD, Gurley SB, Herrera MJ, Ruiz P, Griffiths R, Kumar AP, Kim HS, Smithies O, Le TH, Coffman TM (2006) Angiotensin II causes hypertension

and cardiac hypertrophy through its receptors in the kidney. Proceedings of the National Academy of Sciences of the United States of America 103:17985–17990

64. Smith OA, Astley CA, Spelman FA, Golanov EV, Bowden DM, Chesney MA, Chalyan V (2000) Cardiovascular responses in anticipation of changes in posture and locomotion. Brain Research Bulletin 53:69–76

65. Taylor JA, Eckberg DL (1996) Fundamental relations between short-term RR interval and arterial pressure oscillations in humans. Circulation 93:1527–1532

66. Zhang R, Khoo MSC, Wu Y, Yang Y, Grueter CE, Ni G, Price EE, Thiel W, Guatimosim S, Song LS, Madu EC, Shah AN, Vishnivetskaya TA, Atkinson JB, Gurevich VV, Salama G, Lederer WJ, Colbran RJ, Anderson ME (2005) Calmodulin kinase II inhibition protects against structural heart disease. Nature Medicine 11:409–417

67. Goyal RK (1989) Muscarinic receptor subtypes: physiology and clinical implications. The New England Journal of Medicine 321:1022–1029

68. Lymperopoulos A, Rengo G, Funakoshi H, Eckhart AE, Koch WJ (2007) Adrenal GRK2 upregulation mediates sympathetic overdrive in heart failure. Nature Medicine 13:315–323

69. Klabunde RE (2004) Cardiovascular Physiology Concepts. Lippincott Williams & Wilkins (http://cvphysiology. com/).

70. Kraske S, Cunningham JT, Hajduczok G, Chapleau MW, Abboud FM, Wachtel RE (1998) Mechanosensitive ion channels in putative aortic baroreceptor neurons. American Journal of Physiology – Heart and Circulatory Physiology 275:1497–1501

71. Krauhs JM (1979) Structure of rat aortic baroreceptors and their relationship to connective tissue. Journal of Neurocytology 8:401–414

72. Monti A, Médigue C, Sorine M (2002) Short-term modelling of the controlled cardiovascular system. ESAIM: Proceedings 12:115–128

73. Hammer PE, Saul JP (2005) Resonance in a mathematical model of baroreflex control: arterial blood pressure waves accompanying postural stress. American Journal of Physiology – Regulatory, Integrative and Comparative Physiology 288:R1637–R1648.

74. Degtyarenko AM, Kaufman MP (2005) MLR-induced inhibition of barosensory cells in the NTS. American Journal of Physiology – Heart and Circulatory Physiology. To appear

75. Kondoh G, Tojo H, Nakatani N, N Komazawa, Murata C, Yamagata K, Maeda Y, Kinoshita T, Okabe M, Taguchi R, Takeda J (2005) Angiotensin-converting enzyme is a GPI-anchored protein releasing factor crucial for fertilization. Nature Medicine 11:160–166

76. Cantin M, Genest J (1986) Le cœur est une glande endocrine[The heart is an endocrine gland]. Pour la Science 102:43–49

77. Lombes M, Farman N, Oblin ME, Baulieu EE, Bonvalet JP, Erlanger BF, Gasc J (1990) Immunohistochemical localization of renal mineralocorticoid receptor by using an anti-idiotypic antibody that is an internal image of aldosterone. Proceedings of the National Academy of Sciences of the United States of America 87:1086–1088

78. Sasano H, Fukushima K, Sasaki I, Matsuno S, Nagura H, Krozowski ZS (1992) Immunolocalization of mineralocorticoid receptor in human kidney, pancreas, salivary, mammary and sweat glands: a light and electron microscopic immunohistochemical study. Journal of Endocrinology 132:305–310

79. Kurbel S, Dodig K, Radic R (2002) The osmotic gradient in kidney medulla: a retold story. Advances in Physiology Education 26:278–281

80. Lalioti MD, Zhang J, Volkman HM, Kahle KT, Hoffmann KE, Toka HR, Nelson-Williams C, Ellison DE, Flavell R, Booth CJ, Lu Y, Geller DS, Lifton RP (2006) Wnk4 controls blood pressure and potassium homeostasis via regulation of mass and activity of the distal convoluted tubule. Nature Genetics 38:1124–1132

81. Coffman TM (2006) A WNK in the kidney controls blood pressure. Nature Genetics 38:1105–1106

82. Jentsch TJ, Hübner CA, Fuhrmann JC (2004) Ion channels: Function unravelled by dysfunction. Nature Cell Biology 6:1039–1047 (2004)

83. Abboud FM, Floras JS, Aylward PE, Guo GB, Gupta BN, Schmid PG. Role of vasopressin in cardiovascular and blood pressure regulation. Blood Vessels 27:106–115

84. Koshimizu TA, Nasa Y, Tanoue A, Oikawa R, Kawahara Y, Kiyono Y, Adachi T, Tanaka T, Kuwaki T, Mori T, Takeo S, Okamura H, Tsujimoto G (2006) V1a vasopressin receptors maintain normal blood pressure by regulating circulating blood volume and baroreflex sensitivity. Proceedings of the National Academy of Sciences of the United States of America 103:7807–7812

85. de Bold AJ (1985) Atrial natriuretic factor: a hormone produced by the heart. Science 230:767–770

86. Tsujita Y, Muraski J, Shiraishi I, Kato T, Kajstura J, Anversa P, Sussman MA (2006) Nuclear targeting of Akt antagonizes aspects of cardiomyocyte hypertrophy. Proceedings of the National Academy of Sciences of the United States of America 103:11946-11951

87. Suga SI, Itoh H, Komatsu Y, Ishida H, Igaki T, Yamashita J, Doi K, Chun TH, Yoshimasa T, Tanaka I, Nakao K (1998) Regulation of endothelial production of C-type natriuretic peptide by interaction between endothelial cells and macrophages. Endocrinology 139:1920–1926

Chap. 3. Images, Signals and Measurements

88. Cebral JR, Löhner R (2001) From medical images to anatomically accurate finite element grids. International Journal for Numerical Methods in Engineering 51:985–1008

89. Thiriet M, Brugières P, Bittoun J, Gaston A (2001) Computational flow models in cerebral congenital aneurisms I: Steady flow. Revue Mécanique et Industries 2:107–118

90. Salmon S, Thiriet M, Gerbeau J-F (2003) Medical image-based computational model of pulsatile flow in saccular aneurisms. Mathematical Modelling and Numerical Analysis 37:663–679

91. Milner JS, Moore JA, Rutt BK, Steinman DA (1998) Hemodynamics of human carotid artery bifurcations: computational studies with models reconstructed from magnetic resonance imaging of normal subjects. Journal of Vascular Surgery 1998:143–156

92. Moore JA, Rutt BK, Karlik SJ, Yin K, Ethier CR (1999) Computational blood flow modeling based on in vivo measurements. Annals of Biomedical Engineering 27:627–640

93. Ladak HM, Milner JS, Steinman, DA (2000) Rapid 3D segmentation of the carotid bifurcation from serial MR images. Journal of Biomechanical Engineering 122:96–99

94. Papaharilaou Y, Doorly DJ, Sherwin SJ, Peiró J, Griffith C, Chesire N, Zervas V, Anderson J, Sanghera B, Watkins N, Caro CG (2002) Combined MRI and computational fluid dynamics detailed investigation of flow in idealised and realistic arterial bypass graft models. Biorheology 39:525–532

95. Gill JD, Ladak HM, Steinman DA, Fenster A (2000) Accuracy and variability assessment of a semiautomatic technique for segmentation of the carotid arteries from 3D ultrasound images. Medical Physics 27:1333–1342

96. Thiriet M, Maarek JM, Chartrand DA, Delpuech C, Davis L, Hatzfeld C, Chang HK (1989) Transverse images of the human thoracic trachea during forced expiration. Journal of Applied Physiology 67:1032–1040

97. Bachelard G (1940) La philosophie du non : essai d'une philosophie du nouvel esprit scientifique[The philosophy of no: a philosophy of the new scientific spirit]. Presse Universitaire de France, Paris

98. Bittoun J (1998) Basic principles of magnetic resonance imaging. In: Cerdan S, Haase A, Terrier F (eds) Spectroscopy and clinical MRI. Springer, New York

99. Singer JR (1959) Blood flow rates by nuclear magnetic resonance measurements. Science 130:1652–1653

100. McCready VR, Leach M, Ell PJ (1987) Functional studies using NMR. Springer, New York

101. Zerhouni EA, Parish DM, Rogers WJ, Yang A, Shapiro EP (1988) Human heart: tagging with MR imaging – a method for noninvasive assessment of myocardial motion. Radiology 169:59–63

102. Axel L, Dougherty L (1989) Heart wall motion: improved method of spatial modulation of magnetization for MR imaging. Radiology, 172:349–350

103. Mosher TJ, Smith MB (1990) A DANTE tagging sequence for the evaluation of translational sample motion. Magnetic Resonance in Medicine 15:334–339

104. Fischer SE, McKinnon GC, Maier SE, Boesiger P (1993) Improved myocardial tagging contrast. Magnetic Resonance in Medicine 30:191–200

105. McVeigh ER (1996) MRI of myocardial function: motion tracking techniques. Magnetic Resonance Imaging 14:137-150

106. Basser PJ, Mattiello J, LeBihan D (1994) MR diffusion tensor spectroscopy and imaging. Biophysical Journal 66:259–267

107. Hsu EW, Muzikant AL, Matulevicius SA, Penland RC, Henriquez CS (1998) Magnetic resonance myocardial fiber-orientation mapping with direct histological correlation. American Journal of Physiology – Heart and Circulatory Physiology 274:H1627–H1634

108. Scollan DF, Holmes A, Winslow R, Forder J (1998) Histological validation of reconstructed myocardial microstructure obtained from diffusion tensor magnetic resonance imaging. American Journal of Physiology – Heart and Circulatory Physiology 275 44:H2308–H2318

109. Helm PA, Tseng HJ, Younes L, McVeigh ER, Winslow RL (2005) Ex vivo 3D diffusion tensor imaging and quantification of cardiac laminar structure. Magnetic Resonance in Medicine 54:850-859

110. Winslow RL, Scollan DF, Holmes A, Yung CK, Zhang J, Jafri MS (2000) Electrophysiological modeling of cardiac ventricular function: from cell to organ. Annual Review of Biomedical Engineering 2:119–155

111. Papademetris X, Sinusas AJ, Dione DP, Constable RT, Duncan JS (2002) Estimation of 3-D left ventricular deformation from medical images using biomechanical models. IEEE Transactions on Medical Imaging 21:786–800

112. Sinusas AJ, Papademetris X, Constable RT, Dione DP, Slade MD, Shi P, Duncan JS (2001) Quantification of 3-D regional myocardial deformation: shape-based analysis of magnetic resonance images. American Journal of Physiology – Heart and Circulatory Physiology 281:698-714

113. Declerck J, Ayache N, McVeigh ER (1999) Use of a 4D planispheric transformation for the tracking and the analysis of LV motion with tagged MR images. In: Chen C-T, Clough AV (eds) Medical imaging 1999: Physiology and Function from Multidimensional Images. SPIE, Bellingham

114. Kozerke S, Scheidegger MB, Pedersen EM, Boesiger P (1999) Heart motion adapted cine phase-contrast flow measurements through the aortic valve. Magnetic Resonance in Medicine 42:970–978

115. Gleich B, Weizenecker J (2005) Tomographic imaging using the nonlinear response of magnetic particles. Nature 435:1214–1217

116. Schaar JA, de Korte CL, Mastik F, Baldewsing R, Regar E, de Feyter P, Slager CJ, van der Steen AF, Serruys PW (2003) Intravascular palpography for high-risk vulnerable plaque assessment. Herz 28:488–495

117. Kanai H, Hasegawa H, Chubachi N, Koiwa Y, Tanaka M (1997) Noninvasive evaluation of local myocardial thickening and its color-coded imaging. IEEE Transactions on Ultrasonics, Ferroelectrics, and Frequency Control 44:752–768

118. Moore JA, Steinman DA, Ethier CR (1998) Computational blood flow modelling: errors associated with reconstructing finite element models from magnetic resonance images. Journal of Biomechanics 31:179–184

119. Boissonnat J-D (1988) Shape reconstruction from planar cross-sections. Computer Vision, Graphics, and Image Processing 44:1–29

120. Boissonnat J-D, Chaine R, Frey P, Malandain G, Salmon S, Saltel E, Thiriet M (2005) From arteriographies to computational flow in saccular aneurisms: the INRIA experience. Medical Image Analysis 9:133–143

121. Boissonnat, J-D, Cazals F (2002) Smooth surface reconstruction via natural neighbour interpolation of distance functions. Computational Geometry 22:185–203

122. Osher S, Sethian JA (1988) Fronts propagating with curvature-dependent speed: algorithms based on Hamilton–Jacobi formulations. Journal of Computational Physics 79:12–49

123. Sethian, JA (1996) Level Set Methods: Evolving Interfaces in Geometry, Fluid Mechanics, Computer Vision, and Materials Science. Cambridge University Press, Cambridge

124. Lorensen, WE, Cline HE (1987) Marching cubes: a high resolution 3D surface construction algorithm. Computer Graphics 21:163–169

125. Delingette H, Hébert M, Ikeuchi K (1992) Shape representation and image segmentation using deformable surfaces. Image and Vision Computing 10:132–144

126. Taylor CA, Hughes TJR, Zarins CK (1998) Finite element modeling of blood flow in arteries. Computer Methods in Applied Mechanics and Engineering 158:155–196

127. Peiró J, Giordana S, Griffith C, Sherwin SJ (2002) High-order algorithms for vascular flow modelling. International Journal for Numerical Methods in Fluids 40:137–151
128. Sherwin SJ, Peiró J (2002) Mesh generation in curvilinear domains using high-order elements, International Journal for Numerical Methods in Engineering 53:207–223
129. Giachetti A, Tuveri M, Zanetti G (2003) Reconstruction and web distribution of measurable arterial models. Medical Image Analysis 7:79–93
130. Cohen LD (1991) On active contour models and balloons. Computer Vision, Graphics, and Image Processing 53:211–218
131. Xu C, Prince JL (1998) Snakes, shapes and gradient vector flow. IEEE Transactions on Image Processing 7:359–369
132. McInerney T, Terzopoulos D (1995) Topologically adaptable snakes. In: Proceedings of the fifth international conference on computer vision. IEEE
133. Krissian K, Malandain G, Ayache N (1998) Model based multiscale detection and reconstruction of 3D vessels. INRIA Research Report RR-3442
134. Fetita C, Prêteux F, Beigelman-Aubry C, Grenier P (2004) Pulmonary airways: 3D reconstruction from multi-slice CT and clinical investigation, IEEE Transactions on Medical Imaging 23:1353–1364
135. Perchet D, Fetita CI, Vial L, Prêteux F, Sbiêrlea-Apiou G, Thiriet M (2004) Virtual investigation of pulmonary airways in volumetric computed tomography, Computer Animation and Virtual Worlds 15:361–376
136. George P-L, Hecht H, Saltel E (1990) Fully automatic mesh generator for 3D domains of any shape. Impact of Computing in Science and Engineering 2:187–218
137. George P-L (1990) Génération automatique de maillages[Automatic Mesh Generation]. Masson, Paris
138. George P-L, Borouchaki H (1997) Triangulation de Delaunay et maillage [Delaunay triangulation and mesh]. Hermès, Paris
139. Frey PJ, George P-L (1999) Maillages[Meshes]. Hermès, Paris
140. George P-L, Hecht F (1999), Nonisotropic grid. In: Thompson JF, Soni BK, Weatherill NP (eds) Handbook of Grid Generation. CRC Press, Boca Raton, FL
141. Mohammadi B, George P-L, Hecht F, Saltel, E (2000) 3D Mesh adaptation by metric control for CFD. Revue européenne des éléments finis 9:439–449
142. Habashi WG, Dompierre J, Bourgault Y, Fortin M, Vallet M-G (1998) Certifiable computational fluid mechanics through mesh optimization. AIAA Journal, 1998, 36:703–711
143. Fortin, M (2000) Anisotropic mesh adaptation through hierarchical error estimators. In: Minev, P, Yanping L (eds) Scientific Computing and Applications. Nova Science Publishers, Commack
144. Dompierre J, Vellet M-G, Bourgault Y, Fortin M, Habashi WG (2002) Anisotropic mesh adaptation: towards user-independent, mesh-independent and solver-independent CFD III: Unstructured meshes. International Journal for Numerical Methods in Fluids 39:675–702
145. Fortin A, Bertrand F, Fortin M, Boulanger-Nadeau PE, El Maliki A, Najeh N (2004) Adaptive remeshing strategy for shear-thinning fluid flow simulations. Computers and Chemical Engineering 28:2363–2375

146. Cebral JR, Lohner R (2001) From medical images to anatomically accurate finite element grids. International Journal for Numerical Methods in Engineering 51:985–1008

147. Taubin G (1995) Curve and surface smoothing without shrinkage. In: Proceedings of the fifth international conference on computer vision. IEEE

148. Frey PJ, Borouchaki H (1998) Geometric surface mesh optimization. Computing and Visualization in Science 1:113–121

149. Thiriet M, Graham JMR, Issa RI (1992) A pulsatile developing flow in a bend. Journal de Physique III 2:995–1013

150. Frey PJ, Borouchaki H (2003) Surface meshing using a geometric error estimate. International Journal for Numerical Methods in Engineering 58:227–245

151. Belhamadia Y, Fortin A, Chamberland É (2004) Anisotropic mesh adaptation for the solution of the Stefan problem. Journal of Computational Physics 194:233–255

152. Belhamadia Y, Fortin A, Chamberland É (2004) Three-dimensional anisotropic mesh adaptation for phase change problems, Journal of Computational Physics 201:753–770

153. McGraw-Hill encyclopedia of science and technology (1960) McGraw-Hill, New York

154. Conrad WA, McQueen DM, Yellin EL (1980) Steady pressure flow relations in compressed arteries: Possible origin of Korotkoff sounds. Medical & Biological Engineering & Computing 18:419–426

155. Risacher F (1995) Étude de la propagation de l'onde de pouls par pléthysmographie d'impédance électrique[Study of the propagation of pulse waves by electric impedance plethysmography]. PhD Thesis, Université Claude Bernard, Lyon

156. O'Rourke MF, Avolio AP, Kelly RP (1992) The Arterial Pulse. Lea & Febiger, Baltimore

157. Penaz J (1992) Criteria for set point estimation in the volume clamp method of blood pressure measurement. Physiological Research 41:5–10

158. Omboni S, Parati G, Frattola A, Mutti E, Di Rienzo M, Castiglioni P, Mancia G (1993) Spectral and sequence analysis of finger blood pressure variability. Comparison with analysis of intra-arterial recordings. Hypertension 22:26–33

159. Novak V, Novak P, Schondorf R (1994) Accuracy of beat-to-beat noninvasive measurement of finger arterial pressure using the Finapres: a spectral analysis approach. Journal of Clinical Monitoring 10:118–126

160. Imholz BP, Wieling W, van Montfrans GA, Wesseling KH (1998) Fifteen years experience with finger arterial pressure monitoring: assessment of the technology. Cardiovascular Research 38:605–616

161. Coron J-M, Crépeau E (2003) Exact boundary controllability of a nonlinear KdV equation with critical lengths. INRIA Research Report RR-5000

162. Whitman GB (1999) Linear and Nonlinear Waves. Wiley-Interscience, New York

163. Crépeau E, Sorine M (2005) personal communication

164. Tasu J-P, Mousseaux E, Colin P, Slama MS, Jolivet O, Bittoun J (2002) Estimation of left ventricle performance through temporal pressure variations measured by MR velocity and acceleration mappings. Journal of Magnetic Resonance Imaging 16:246–252

165. Kitney RI, Giddens DP (1983) Analysis of blood velocity waveforms by phase shift averaging and autoregressive spectral estimation Journal of Biomechanical Engineering 105:398–401

166. Thiriet M, Cybulski G, Darrow RD, Doorly DJ, Dumoulin C, Tarnawski M, Caro CG (1997) Apports et limitations de la vélocimétrie par résonance magnétique nucléaire en biomécanique. Mesures dans un embranchement plan symétrique[Contributions and limitations of the nuclear magnetic resonance velocimetry in biomechanics. Measures in a two plane symmetrical bifurcation]. Journal de Physique III7:771–787

167. Durand E, Guillot G, Darrasse L, Tastevin G, Nacher PJ, Vignaud A, Vattolo D, Bittoun J (2002) CPMG measurements and ultrafast imaging in human lungs with hyperpolarized helium-3 at low field (0. 1 T). Magnetic Resonance in Medicine 47:75–81

168. Dumoulin CL, Hart HR Jr (1986) Magnetic resonance angiography. Radiology 161 717–720

169. Dumoulin CL, Souza SP, Hart HR (1987) Rapid scan magnetic resonance angiography. Magnetic Resonance in Medicine 5:238–245

170. Dumoulin CL, Souza SP, Walker MF, Yoshitome E (1988) Time-resolved magnetic resonance angiography. Magnetic Resonance in Medicine 6:275–286

171. Dumoulin CL, Souza SP, Walker MF, Wagle W (1989) Three-dimensional phase contrast angiography. Magnetic Resonance in Medicine 9:139–149

172. Dumoulin CL, Doorly DJ, Caro CG (1993) Quantitative measurement of velocity at multiple positions using comb excitation and Fourier velocity encoding. Magnetic Resonance in Medicine 29:44–52

173. Bittoun J, Jolivet O, Herment A, Itti E, Durand E, Mousseaux E, Tasu J-P (2000) Multidimensional MR mapping of multiple components of velocity and acceleration by Fourier phase encoding with a small number of encoding steps. Magnetic Resonance in Medicine 44:723–730

174. Durand E, Jolivet O, Itti E, Tasu J-P, Bittoun J (2001) Precision of magnetic resonance velocity and acceleration measurements: theoreticals issues and phantom experiments. Magnetic Resonance in Medicine 13:445–451

175. Liebman FM, Pearl J, Bagnol S (1962) The electrical conductance properties of blood in motion. Physics in Medicine and Biology 7:177–194

176. Geddes LA, Sadler C (1973) The specific resistance of blood at body temperature. Medical & Biological Engineering 11:336–339

177. Brody DA (1956) A theoretical analysis of intracavitary blood mass influence on the heart-lead relationship. Circulation Research 4:731–738

178. Gulrajani RM, Roberge FA, Mailloux GE (1989) The forward problem of electrocardiography. In Comprehensive Electrocardiology: Theory and Practice in Health and Disease. Macfarlane PW, Lawrie TDV (eds), 237–288, Pergamon Press, New York

179. Einthoven W (1895) Uber die Form des menschlichen Electrokardiograms[On the shape of the electrocardiogram in men]. Pflügers Archiv fur die gesamte Physiologie des Menschen und der Tiere60:101-123

180. Einthoven W (1908) Weiteres über das Elektrokardiogram[More on the electrocardiogram]. Pflügers Archiv fur die gesamte Physiologie des Menschen und der Tiere122:517-548

181. Einthoven W, Fahr G, de Waart A (1913) Uber die Richtung und die Manifeste Grösse der Potentialschwankungen im mennschlichen Herzen und über den

Einfluss der Herzlage auf die Form des Elektrokardiogramms[On the direction and manifest size of the variations of potential in the human heart and on the influence of the heart position in the shape of the electrocardiogram]. Pflügers Archiv fur die gesamte Physiologie des Menschen und der Tiere150:275-315

182. Goldberger E (1942) The aVL, aVR, and aVF leads. A simplification of standard lead electrocardiography. American Heart Journal 24:378–396

183. Einthoven W, Fahr G, de Waart A (1950) On the direction and manifest size of the variations of potential in the human heart and on the influence of the position of the heart on the form of the electrocardiogram. American Heart Journal 40:163–211

184. Boulakia M, Fernández MA, Gerbeau JF, Zenzemi N (2007) Numerical simulation of ECG, submitted to IEEE Transactions on Biomedical Engineering, also in Functional Imaging and Modeling of the Heart: FIMH07, Springer.

185. Frank E (1956) An accurate, clinically practical system for spatial vectorcardiography. Circulation 13: 737–749

186. Nousiainen J, Oja S, Malmivuo J (1994) Normal vector magnetocardiogram. I. Correlation with the normal vector ECG. Journal of Electrocardiology 27:221–231

187. Nousiainen J, Oja S, Malmivuo J (1994) Normal vector magnetocardiogram. II. Effect of constitutional variables. Journal of Electrocardiology 27:233–241

188. Baule GM, McFee R (1963) Detection of the magnetic field of the heart. American Heart Journal 66:95–96

189. van Oosterom A, Oostendorp TF, Huiskamp GJ, ter Brake HJ (1990) The magnetocardiogram as derived from electrocardiographic data. Circulation Research 67:1503–1509

190. Penney BC (1986) Theory and cardiac applications of electrical impedance measurements. CRC Critical Reviews in Biomedical Engineering 13:227-281

191. Aristote [Aristotle] (1979) Seconds analytiques[Posterior analytics]. Vrin, Paris

192. Ojha M, Cobbold RSC, Johnston KW, Hummel RL (1989) Pulsatile flow through constricted tubes: an experimental investigation using photochromic tracer methods. Journal of Fluid Mechanics 203:173–197

193. Lichtenstein O, Martinez-Val R, Mendez J, Castillo-Olivares JL (1986) Hydrogen bubble visualization of the flow past aortic prosthetic valves. Life Support Systems 4:141–149

194. Durst F, Melling A, Whitelaw JH (1981) Principles and practice of laser-Doppler anemometry. Academic Press, New York

195. Thiriet M, Graham JMR, Issa RI (1993) A computational model of wall shear and residence time of fluid particles conveyed by steady flow in a curved tube. Journal de Physique III 3:85–103

196. Thiriet M, Treiber J, Graham JMR (1988) Vélocimétrie par imagerie du déplacement de particules[Particle image velocimetry]. Archives internationales de physiologie et de biochimie 6:C92–C93

197. Wein O, Sobolik V (1987) Theory of direction sensitive probes for electrodiffusion measurement of wall velocity gradient. Collection of Czechoslovak Chemical Communications 52:2169–2180

198. Deslouis C, Gil O, Sobolik V (1990) Electrodiffusional probe for measurements of the wall shear rate vector. International Journal of Heat and Mass Transfer 33:1363–1366

199. Deslouis C, Gil O, Tribollet B (1990) Frequency response of electrochemical sensors to hydrodynamic fluctuations. Journal of Fluid Mechanics 215:85–100

200. Mao ZX, Hanratty, TJ (1986) Studies of the wall shear stress in a turbulent pulsating pipe flow. Journal of Fluid Mechanics 170:545–564

201. Chiang FP (2003) Evolution of white light speckle method and its application to micro/nanotechnology and heart mechanics. Optical Engineering 42:1288–1292

202. Thiriet M, Delpuech C, Piroird JM, Magnin I (1986) Banc d'analyse optique de la déformation de conduites souples[Optical bench analysis of the deformation of compliant vessels]. Innovation et technologie en biologie et médecine 8:99–107

203. Wu Z, Taylor LS, Rubens DJ, Parker KJ (2004) Sonoelastographic imaging of interference patterns for estimation of the shear velocity of homogeneous biomaterials. Physics in Medicine and Biology 49:911–922

204. Parker KJ, Taylor LS, Gracewski S, Rubens DJ (2005) A unified view of imaging the elastic properties of tissue. The Journal of the Acoustical Society of America 117:2705–2712

205. Bachelard G (1938 [1969]) La formation de l'esprit Scientifique. [The Formation of the Scientific Spirit]. Vrin, Paris

Chap. 4. Rheology

206. Bachelet C, Dantan P, Flaud P (2003) Indirect on-line determination of Newtonian fluid viscosity based on numerical flow simulations. The European Physical Journal – Applied Physics 21:67–73

207. Snabre P, Bitbol M, Mills P (1987) Cell disaggregation behavior in shear flow, Biophysical Journal 51:795–807

208. Chien S (1970) Shear dependence of effective cell volume as a determinant of blood viscosity. Science 168:977–978

209. Joly M, Lacombe C, Quemada D (1981) Application of the transient flow rheology to the study of abnormal human bloods. Biorheology 18:445–452

210. Easthope PL, Brooks DE (1980) A comparison of rheological constitutive functions for whole human blood. Biorheology 17:235–247

211. Chien S (1975), Biophysical behavior in suspensions. In: Surgenor D (ed) The red blood cell. Academic Press, New York

212. Bucherer C, Lacombe C, Lelièvre J-C (1998) Viscosité du sang humain[Viscosity of the human blood]. In: Jaffrin MY, Goubel F (eds) Biomécanique des fluides et des tissus[Fluid and tissue biomechanics], Masson, Paris

213. Thiriet M, Martin-Borret G, Hecht F (1996) Écoulement rhéofluidifiant dans un coude et une bifurcation plane symétrique. Application à l'écoulement sanguin dans la grande circulation[Shear-thinning flow in a bend and a symmetrical planar bifurcation. Application to the blood flow in large vessels]. Journal de Physique III 6:529–542

214. Rajagopal KR, Srinivasa AR (2000) A thermodynamic framework for rate-type fluid models. Journal of Non-Newtonian Fluid Mechanics 88:207–227

215. Anand M, Rajagopal KR (2002) A mathematical model to describe the change in the constitutive character of blood due to platelet activation. Comptes rendus de l'Académie des sciences – Série IIB : Mécanique 330:557–562

216. Grashow JS, Sacks MS, Liao J, Yoganathan AP (2006) Planar biaxial creep and stress relaxation of the mitral valve anterior leaflet. Annals of Biomedical Engineering 34:1509–1518

217. Kim H, Lu J, Sacks MS, Chandran KB (2006) Dynamic simulation pericardial bioprosthetic heart valve function. Journal of Biomechanical Engineering 128:717–724

218. Jennings LM, Butterfield M, Booth C, Watterson KG, Fisher J (2002) The pulmonary bioprosthetic heart valve: its unsuitability for use as an aortic valve replacement. The Journal of Heart Valve Disease 11:668–678

219. Pinto JG, Fung YC (1973) Mechanical properties of the heart muscle in the passive state, Journal of Biomechanics 6:597–616

220. Peskin CS (1975) Mathematical aspects of heart physiology. Courant Institute of Mathematical Sciences.

221. Ohayon J, Chadwick RS (1988) Effects of collagen microstructure on the mechanics of the left ventricle. Biophysical Journal 54:1077–1088

222. Ohayon J, Chadwick RS (1988) Theoretical analysis of the effects of a radial activation wave and twisting motion on the mechanics of the left ventricle. Biorheology 25:435–447

223. Oddou C, Ohayon J (1998) Mécanique de la structure cardiaque [Mechanics of the cardiac structure]. In: Jaffrin MY, Goubel F (eds) Biomécanique des fluides et des tissus [Fluid and tissue biomechanics]. Masson, Paris

224. Chuong CJ, Fung YC, (1986) On residual stresses in arteries, Journal of Biomechanical Engineering 108:189-192

225. Yang M, Taber LA, Clark EB (1994) A nonlinear poroelastic model for the trabecular embryonic heart. Journal of Biomechanical Engineering 116:213–223

226. Villarreal FJ, Lew WY, Waldman LK, Covell JW (1991) Transmural myocardial deformation in the ischemic canine left ventricle. Circulation Research 68:368–381

227. Sachse FB (2004) Computational Cardiology. Modeling of Anatomy, Electrophysiology, and Mechanics. Lecture Notes in Computer Science, 2966, Springer-Verlag, Berlin, Heidelberg

228. Bourdarias C, Gerbi S, Ohayon J (2003) A three dimensional finite element method for biological active soft tissue. Mathematical Modelling and Numerical Analysis 37:725–739

229. Ohayon J, Oddou C (1987) Analyse théorique de la performance cardiaque à partir d'un modèle de ventricule gauche tenant compte de la structure fibreuse du myocarde[Theoretical analysis of the cardiac performance using a left ventricular model with the myocardium fibrous structure]. Journal de Biophysique et de Biomécanique 11:145–154

230. Rygole AL, Ohayon J, Oddou C (1986) Modèles physiques en vue de l'estimation de la performance cardiaque[Physical models for the evaluation of cardiac performance]. Journal de Biophysique et de Biomécanique 10:31–39

231. Caillerie D, Mourad A, Raoult A (2003) Cell-to-muscle homogenization. Application to a constitutive law for the myocardium. Mathematical Modelling and Numerical Analysis 37:681–698

232. Mavrilas D, Sinouris EA, Vynios DH Papageorgakopoulou N (2005) Dynamic mechanical characteristics of intact and structurally modified bovine pericardial tissues. Journal of Biomechanics 38:761–768

233. Djerad SE, Du Burck F, Naili S, Oddou C (1992) Analyse du comportement rhéologique instationnaire d'un échantillon de muscle cardiaque[Analysis of the unsteady rheological behavior of a cardiac muscle sample]. Comptes rendus de l'Académie des sciences – Série II 315:1615–1621

234. Tsaturyan AK, Izacov VJ, Zhelamsky SV, Bykov BL (1984) Extracellular fluid filtration as the reason for the viscoelastic behaviour of the passive myocardium. Journal of Biomechanics 17:749–755

235. Terzaghi K (1925) Erdbaumechanik auf bodenphysikalischer Grundlage[Soil mechanics based on soil physics]. Deuticke, Leipzig

236. Biot MA (1935) Le problème de la consolidation des matières argileuses sous une charge[The problem of consolidation of clay matters under a load]. Annales de la société scientifique de Bruxelles – Série B55:110–113

237. Biot MA (1941) General theory of three-dimensional consolidation. Journal of Applied Physics 12:155–164

238. Biot MA (1955) Theory of elasticity and consolidation for a porous anisotropic solid. Journal of Applied Physics 26:182–185

239. Rice JR, Cleary MP (1976) Some basic stress diffusion solutions for fluid-saturated elastic porous media with compressible constituents. Reviews of Geophysics and Space Physics 14:227–241

240. Kenyon DE (1976) The theory of an incompressible solid-fluid mixture. Archive for Rational Mechanics and Analysis 62:131–147

241. Kenyon DE (1978), Consolidation in compressible mixture. Journal of Applied Physics 45:727–732

242. Parker KH, Mehta RV, Caro CG (1987) Steady flow in porous, elastically deformable materials. Journal of Applied Mechanics 54:794–800

243. Farrel DA, Greacen EL, Gurr CG (1966) Vapor transfer in soil due to air turbulence. Soil Science 102:305–313

244. Scotter DR, Thurtell GW, Raats PAC (1967) Dispersion resulting from sinusoidal gas flow in porous materials. Soil Science 104:306–308

245. Simon BR (1992) Multiphase poroelastic finite element models for soft tissue structures. Applied Mechanics Reviews 45:191–218

246. Kenyon DE (1979) A mathematical model of water flux through aortic tissue. Bulletin of Mathematical Biology 41:79–90

247. Sorek S, Sideman S (1986) A porous-medium approach for modeling heart mechanics I: Theory. Mathematical Bioscience, 81:1–14

248. Stergiopulos N, Tardy Y, Meister J-J (1993) Nonlinear separation of forward and backward running waves in elastic conduits. Journal of Biomechanics 26:201–209

249. Fung YC (1993) Biomechanics: mechanical properties of living tissues. Springer, New York

250. Hayashi K (1993) Experimental approaches on measuring the mechanical properties and constitutive laws of arterial walls. Journal of Biomechanical Engineering 115:481–488

251. Draney MT, Herfkens RJ, Hughes TJ, Pelc NJ, Wedding KL, Zarins CK, Taylor CA (2002) Quantification of vessel wall cyclic strain using cine phase contrast magnetic resonance imaging. Annals of Biomedical Engineering 30:1033–1045

252. Carew TE, Vaishnav RN, Patel DJ (1968) Compressibility of the arterial wall. Circulation Research 23:61–68

253. Bergel DH (1961) The static elastic properties of the arterial wall. The Journal of Physiology 156:445–457

254. Hudetz AG (1979) Incremental elastic modulus for orthotropic incompressible arteries. Journal of Biomechanics 12:651–655

255. Lu X, Pandit A, Kassab GS (2004) Biaxial incremental homeostatic elastic moduli of coronary artery: two-layer model, American Journal of Physiology – Heart and Circulatory Physiology 287:1663–1669

256. Fung YC, Liu SQ, Zhou JB (1993) Remodeling of the constitutive equation while a blood vessel remodels itself under stress. Journal of Biomechanical Engineering 115:453–459

257. Yu Q, Zhou J, Fung YC (1993) Neutral axis location in bending and Young's modulus of different layers of arterial wall. American Journal of Physiology – Heart and Circulatory Physiology 265:52–60

258. Xie J, Zhou J, Fung YC (1995) Bending of blood vessel wall: stress-strain laws of the intima-media and adventitial layers. Journal of Biomechanical Engineering 117:136–145

259. Stergiopulos N, Meister J-J (1995) Biomechanical and physiological aspects of arterial vasomotion. In: Jaffrin MY, Caro CG (eds) Biological Flows. Plenum Press, New York

260. Cabrera Fischer EI, Bia D, Camus JM, Zocalo Y, de Forteza E, Armentano RL (2006) Adventitia-dependent mechanical properties of brachiocephalic ovine arteries in in vivo and in vitro studies. Acta Physiologica 188:103-111

261. Fung YC, Fronek K, Patitucci P (1979) Pseudoelasticity of arteries and the choice of its mathematical expression. American Journal of Physiology – Heart and Circulatory Physiology 237:620–631

262. Flaud P, Quemada D (1988) A structural viscoelastic model of soft tissues. Biorheology 25:95–105

263. Pinto JG, Fung YC (1973) Mechanical properties of the stimulated papillary muscle in quick-release experiments. Journal of Biomechanics 6:617–630

264. Zulliger MA, Rachev A, Stergiopulos N (2004) A constitutive formulation of arterial mechanics including vascular smooth muscle tone. American Journal of Physiology – Heart and Circulatory Physiology 287:1335–1343

265. Holzapfel GA, Gasser TC (2000) A new constitutive framework for arterial wall mechanics and a comparative study of material models. Journal of Elasticity 61:1–48

266. Zulliger MA, Fridez P, Hayashi K, Stergiopulos N (2004) A strain energy function for arteries accounting for wall composition and structure. Journal of Biomechanics 37:989–1000

267. Humphrey JD, Rajagopal KR (2002) A constrained mixture model for growth and remodeling of soft tissues. Mathematical Models & Methods in Applied Sciences 12:407–430

268. Gleason RL, Humphrey JD (2004) A Mixture model of arterial growth and remodeling in hypertension: altered muscle tone and tissue turnover. Journal of Vascular Research 41:352–363

269. Begis D, Delpuech C, Le Tallec P, Loth L, Thiriet M, Vidrascu M (1988) A finite element model of tracheal collapse. Journal of Applied Physiology 64:1359–1368

270. Verdier C (2003) Rheological properties of living materials: from cells to tissues. Journal of Theoretical Medecine 5:67–91

271. Schmidt FG, Hinner B, Sackmann E, Tang JX (2000) Viscoelastic properties of semiflexible filamentous bacteriophage fd. Physical Review E 62:5509–5517

272. Evans EA (1973) New membrane concept applied to the analysis of fluid shear- and micropipette-deformed red blood cells. Biophysical Journal 13:941–954

273. Laurent V, Planus E, Isabey D, Lacombe C, Bucherer C (2000) Propriétés mécaniques de cellules endothéliales évaluées par micromanipulation cellulaire et magnétocytométrie[Mechanical properties of endothelial cells assessed by cell micromanipulation and magnetocytometry]. In: Ribreau C, Berthaud Y, Moreau M-R, Ratier L, Renaudaux JP, Thiriet M, Wendling S (eds) Mécanotransduction 2000. Tec&Doc, Paris

274. Lenormand G, Hénon S, Gallet F (2000) Détermination des modules élastiques du cytosquelette du globule rouge humain par pinces optiques[Determination of elastic moduli of the cytoskeleton of the human red blood cell by optical tweezers]. In: Ribreau C, Berthaud Y, Moreau M-R, Ratier L, Renaudaux JP, Thiriet M, Wendling S (eds) Mécanotransduction 2000. Tec&Doc, Paris

275. Yeung A, Evans E (1989) Cortical shell-liquid core model for passive flow of liquid-like spherical cells into micropipets. Biophysical Journal 56:139–149

276. Hochmuth RM (2000) Micropipette aspiration of living cells. Journal of Biomechanics 33:15–22

277. Binning G, Quate CF, Gerber C (1986) Atomic force microscope. Physical Review Letters 56:930–933

278. Canetta E (2004) Adhésion cellulaire. Application à la modélisation des tumeurs cancéreuses[Cellular adhesion. Application to the modeling of malignant tumors]. PhD thesis, University Joseph Fourier, Grenoble

279. Hénon S, Lenormand G, Richert A, Gallet F (1999) A new determination of the shear modulus of the human erythrocyte membrane using optical tweezers. Biophysical Journal 76:1145–1151

280. Moffitt JR, Chemla Y, Izhaky D, Bustamante C (2006) Differential detection of dual traps improves the spatial resolution of optical tweezers. Proceedings of the National Academy of Sciences of the United States of America 103:9006–9011

281. Caille N, Thoumine O, Tardy Y, Meister J-J (2002) Contribution of the nucleus to the mechanical properties of endothelial cells. Journal of Biomechanics 35:177–187

282. Geiger S (1994) Rhéologie des globules blancs humains. Application aux cellules leucémiques[Human white blood cell rheology. Application to leukemic cells]. PhD thesis, University Denis Diderot, Paris

Chap. 5. Hemodynamics

283. Obremski HJ, Fejer AA (1967) Transition in oscillating boundary layer flows. Journal of Fluid Mechanics 29:93–111

284. Kleinstreuer C, Lei M, Archie JP (2001) Hemodynamic simulations and optimal computer-aided desighs of branching blood. In: Leondes, C (ed) Biofluid Methods in Vascular and Pulmonary Systems. CRC Press, Boca Raton, FL

285. Comte A (1998) Cours de philosophie positive[Courses on positive philosophy]. Hermann, Paris

286. Idel'cik IE (1969) Mémento des pertes de charges singulières et de pertes de charges par frottement[Handbook of singular and friction head losses]. Eyrolles, Paris

287. Eustice J (1911) Experiments on stream-line motion in curved pipes. Proceedings of the Royal Society of London – Series A: Mathematical and Physical Sciences 85:119–131

288. Dean WR (1927) Note on the motion of fluid in a curved pipe. Philosophical Magazine 4:208–223

289. Dean WR (1928) The stream-line motion of fluid in a curved pipe. Philosophical Magazine 5:673–695

290. Dean WR (1928) Fluid motion in a curved channel. Proceedings of the Royal Society of London – Series A: Mathematical and Physical Sciences 121:402–420

291. Ito H (1959) Friction factor for turbulent flow in curved pipes. Transactions of the ASME – Journal of Basic Engineering 81:123–134

292. McConalogue DJ, Srivastava RS (1968) Motion of a fluid in a curved tube. Proceedings of the Royal Society of London – Series A: Mathematical and Physical Sciences 307:37–53

293. Bovendeerd PHM, van Steenhoven AA, van de Vosse FN, Vossers G (1987) Steady entry flow in a curved pipe. Journal of Fluid Mechanics 177:233–246

294. Yao LS, Berger SA (1975) Entry flow in a curved pipe. Journal of Fluid Mechanics 67:177–196

295. McConalogue DJ (1970) The effects of secondary flow on the laminar dispersion of an injected substance in a curved tube. Proceedings of the Royal Society of London – Series A: Mathematical and Physical Sciences 315:99–113

296. Smith FT (1976) Fluid flow into a curved pipe. Proceedings of the Royal Society of London – Series A: Mathematical and Physical Sciences 331:71–87

297. Ito H (1960) Pressure losses in smooth pipe bends, Transactions of the ASME – Journal of Basic Engineering 82:131–143

298. Lyne WH (1971) Unsteady viscous flow in a curved pipe. Journal of Fluid Mechanics 45:15–31

299. Smith FT (1975) Pulsatile flow in curved pipes. Journal of Fluid Mechanics 71:15–42

300. Talbot L, Gong KO (1983) Pulsatile entrance flow in a curved pipe. Journal of Fluid Mechanics 127:1–25

301. Chang LJ, Tarbell JM (1985) Numerical simulation of fully developed sinusoidal and pulsatile (physiological) flow in curved tubes. Journal of Fluid Mechanics 161:175–198

302. Hamakiotes CC, Berger SA (1990) Periodic flows through curved tubes: the effect of the frequency parameter. Journal of Fluid Mechanics 210:353–370

303. Tada S, Oshima S, Yamane R (1996) Classification of pulsating flow patterns in curved pipes. Journal of Biomechanical Engineering 118:311–317

304. Kang SG, Tarbell JM, The impedance of curved artery models. Journal of Biomechanical Engineering 105:275–282

305. Perktold K, Nerem RM, Peter RO (1991) A numerical calculation of flow in a curved tube model of the left main coronary artery. Journal of Biomechanics 1991, 24:175–189

306. Schilt S, Moore JE, Delfino A, Meister J-J (1996) The effects of time-varying curvature on velocity profiles in a model of the coronary arteries. Journal of Biomechanics 29:469–474

307. Waters SL, Pedley TJ (1999) Oscillatory flow in a tube of time-dependent curvature I: Perturbation to flow in a stationary curved tube. Journal of Fluid Mechanics 383:327–352

308. Thiriet M, Pares C, Saltel E, Hecht F (1992) Numerical model of steady flow in a model of the aortic bifurcation. Journal of Biomechanical Engineering 114:40–49

309. Bharadvaj BK, Mabon RF, Giddens DP (1982) Steady flow in a model of the human carotid bifurcation I: Flow visualization. Journal of Biomechanics 15:349–362

310. Bharadvaj BK, Mabon RF, Giddens DP (1982) Steady flow in a model of the human carotid bifurcation II: Laser-Doppler anemometer measurements, Journal of Biomechanics 15:363–378

311. Ku DN, Giddens DP, Phillips DJ, Strandness DE (1985) Hemodynamics of the normal human carotid bifurcation: in vitro and in vivo studies. Ultrasound in Medicine & Biology 11:13–26

312. Liepsch D, Poll A, Strigberger J, Sabbah HN, Stein PD (1989) Flow visualization studies in a mold of the normal human aorta and renal arteries. Journal of Biomechanical Engineering 111:222–227

313. Perktold K, Resch M, Florian H (1991) Pulsatile non-Newtonian flow characteristics in a three-dimensional human carotid bifurcation model. Journal of Biomechanical Engineering 113:464–475

314. Rieu R, Friggi A, Pélissier R (1985) Velocity distribution along an elastic model of human arterial tree. Journal of Biomechanics 18:703–715

315. Lutz RJ, Hsu L, Menawat A, Zrubek J, Edwards K (1983) Comparison of steady and pulsatile flow in a double branching arterial model. Journal of Biomechanics 16:753–766

316. Mark FF, Bargeron CB, Deters OJ, Friedman MH (1985) Nonquasi-steady character of pulsatile flow in human coronary arteries. Journal of Biomechanical Engineering 107:24–28

317. Sung HW, Yoganathan AP (1990) Axial flow velocity patterns in a normal human pulmonary artery model: pulsatile in vitro studies. Journal of Biomechanics 23:201–214

318. Graham JMR, Thiriet M (1990) Pulsatile flow through partially occluded ducts and bifurcations. In: Mosora F, Caro C, Krause E, Schmid-Schönbein H, Baquey C, Pélissier R (eds) Biomechanical Transport Processes. Plenum, New York

319. Ku DN, Liepsch D (1986) The effects of non-Newtonian viscoelasticity and wall elasticity on flow at a 90 degrees bifurcation. Biorheology 23:359–370

320. Lyne WH (1971) Unsteady viscous flow over a wavy wall. Journal of Fluid Mechanics 50:33–48

321. Ralph ME (1986) Oscillatory flows in wavy-walled tubes. Journal of Fluid Mechanics 168:515–540

322. Prandtl L (1905) Uber Flüssigkeitsbewegung bei sehr kleiner Reibung[On the motion of fluids with very little friction]. In: Krazer A (ed) Verhandlung des III internationalen Mathematiker-Kongresses. Teubner, Leipzig

323. Blasius H (1908) Grenzschichten in Flüssingkeiten mit kleiner Reibung [Boundary layers in fluids with small friction]. Zeitschrift für Mathematik und Physik 56:1–37

324. Gerrard JH, Hughes MD (1971) The flow due to an oscillating piston in a cylindrical tube: a comparison between experiment and a simple entrance flow theory. Journal of Fluid Mechanics 50:97–106

325. Atabek HB, Chang CC (1961) Oscillatory flow near the entry of a circular tube. Zeitschrift für Angewandte Mathematik und Physik 112:185–201

326. Chang CC, Atabek HB (1961) The inlet length for oscillatory flow and its effects on the determination of the rate of flow in arteries. Physics in Medicine and Biology 6:303–317

327. Mullin T, Greated CA (1980) Oscillatory flow in curved pipes I: The developing-flow case. Journal of Fluid Mechanics 98:383–396

328. Chandran KB, Yearwood TL (1981) Experimental study of physiological pulsatile flow in a curved tube. Journal of Fluid Mechanics 111:59–85

329. Valéry P (1932) Choses tues[Silent things]. Gallimard, Paris

330. Bernard C (1966) Introduction à l'étude de la médecine expérimentale[Introduction to the study of experimental medicine]. Garnier-Flammarion, Paris

331. Reynolds O (1883) On the experimental investigation of the circumstances which determine whether the motion of water shall be direct or sinuous, and the law of resistance in parallel channels. Philosophical Transactions of the Royal Society of London – Series A: Mathematical and Physical Sciences 174:935–982

332. Reynolds O (1895) On the dynamical theory of incompressible viscous fluids and the determination of the criterion, Philosophical Transactions of the Royal Society of London – Series A: Mathematical and Physical Sciences 186:123–164

333. Stettler JC, Fazle Hussain AKM (1986) On transition of the pulsatile pipe flow. Journal of Fluid Mechanics 170:169–197

334. Sarpkaya T (1966) Experimental determination of the critical Reynolds number for pulsating poiseuille flow, Transactions of the ASME – Journal of Basic Engineering 88:589–598

335. Akhavan R, Kamm RD, Shapiro AH (1991) An investigation of transition to turbulence in bounded oscillatory Stokes flows I: Experiments. Journal of Fluid Mechanics 225:395–422

336. Hino M, Sawamoto M, Takasu S (1976) Experiments on transition to turbulence in an oscillatory pipe flow. Journal of Fluid Mechanics 75:193–207

337. Ohmi M, Iguchi M, Kakehashi K, Masuda T (1982) Transition to turbulence and velocity distribution in an oscillating pipe flow. Bulletin of the Japan Society of Mechanical Engineers 25:365–371

338. Baldwin JT, Deutsch S, Petrie HL, Tarbell JM (1993) Determination of principal reynolds stresses in pulsatile flows after elliptical filtering of discrete velocity measurements. Journal of Biomechanical Engineering 115:396–403

339. Yellin EL (1966) Laminar-turbulent transition process in pulsatile flow. Circulation Research 19:791–804

340. Nerem RM, Seed WA (1972) An in vivo study of aortic flow disturbances. Cardiovascular Research 6:1–14

341. Clarion C, Pélissier R (1975) A theoretical and experimental study of the velocity distribution and transition to turbulence in free oscillatory flow. Journal of Fluid Mechanics 70:59–79

342. Taylor GI (1929) The criterion for turbulence in curved pipes, Proceedings of the Royal Society of London – Series A: Mathematical and Physical Sciences 124:243–249

343. Seminara G, Hall P (1976) Centrifugal instability of a Stokes layer: linear theory. Proceedings of the Royal Society of London – Series A: Mathematical and Physical Sciences 350:299–316

344. Papageorgiou D (1987), Stability of the unsteady viscous flow in a curved pipe. Journal of Fluid Mechanics 182:200–233

345. Shortis TA, Hall P (1999) On the nonlinear stability of the oscillatory viscous flow of an incompressible fluid in a curved pipe. Journal of Fluid Mechanics 379:145–163

346. Garg VK, Rouleau WT (1974) Stability of Poiseuille flow in a thin elastic tube. The Physics of Fluids 17:1103–1108

347. Stein PD, Walburn FJ, Blick EF (1980) Damping effect of distensible tubes on turbulent flow: implications in the cardiovascular system. Biorheology 17:275–281

348. Nerem RB, Seed WA, Wood WB (1972) An experimental study of the velocity distribution and transition to turbulence in the aorta. Journal of Fluid Mechanics 52:137–160

349. Lighthill J (1972), Physiological fluid dynamics: a survey. Journal of Fluid Mechanics 52:475–497

350. von Kerczek C, Davis SH (1974) Linear stability theory of oscillatory Stokes layers. Journal of Fluid Mechanics 62:753–773

351. Obremski HJ, Morkovin MV, Landahl M (1969) A portfolio of stability characteristics of incompressible boundary layers. AGARDograph 134, NATO, Paris

352. Hall P, Parker KH (1976) The stability of the decaying flow in a suddenly blocked channel. Journal of Fluid Mechanics 75:305–314

353. Dantan P, de Jouvenel F, Oddou C (1976) Stabilité d'un écoulement pulsé incompressible en tube cylindrique rigide[Stability of an incompressible pulsatile flow in a rigid cylindrical tube]. Journal de Physique 37:233

354. Bluestein D, Einav S (1994) Transition to turbulence in pulsatile flow through heart valves – A modified stability approach. Journal of Biomechanical Engineering 116:477–487

355. Despard RA, Miller JA (1971) Separation in oscillating laminar boundary layer flows. Journal of Fluid Mechanics 47:21–31

356. Sobey IJ (1985) Observation of waves during oscillatory channel flow. Journal of Fluid Mechanics 151:395–426

357. Uchida S, Aoki H (1977) Unsteady flows in a semi-infinite contracting or expanding pipe. Journal of Fluid Mechanics 82:371–387

358. Liepsch D, Moravec S (1984) Pulsatile flow of non-Newtonian fluid in distensible models of human arteries. Biorheology 21:571–586

359. Shapiro, AH (1977) Steady flow in collapsible tubes. Journal of Biomechanical Engineering 99:126–147

360. Ribreau C, Thiriet M (1998) Écoulements veineux[Venous flow]. In: Jaffrin MY, Goubel F (eds) Biomécanique des fluides et des tissus [Fluid and Tissue Biomechanics]. Masson, Paris

361. Thiriet M, Naili S, Langlet A, Ribreau C (2001) Flow in thin-walled collapsible tubes. In: Leondes C (ed) Biofluid Methods in Vascular and Pulmonary Systems. CRC Press, Boca Raton, FL

362. Flaherty JE, Keller JB, Rubinow S (1972) Post buckling behavior of elastic tubes and rings with opposite sides in contact. SIAM Journal on Applied Mathematics 23:446–455

363. Kresch E, Noordegraaf A (1972) Cross-sectional shape of collapsible tubes. Biophysical Journal 12:274–294

364. Bonis M, Ribreau C, Verchery G (1981) Étude théorique et expérimentale de l'aplatissement d'un tube élastique en dépression[Theoretical study of the collapse of an elastic tube]. Journal de Mécanique Appliquée 5:123–144

365. Bonis M, Ribreau C (1981) Wave speed in non circular collapsible ducts. Journal of Biomechanical Engineering 103:27–31

366. Ribreau C, Naili S, Bonis M, Langlet A (1993) Collapse of thin-walled elliptical tubes for high values of major-to-minor axis ratio. Journal of Biomechanical Engineering 115:432–440

367. Dion B, Naili S, Renaudeaux J-P, Ribreau C (1995) Buckling of elastic tubes: Study of highly compliant device. Medical & Biological Engineering & Computing 33:196–201

368. Tadjbakhsh I, Odeh F (1967) Equilibrium states of elastic rings. Journal of Mathematical Analysis and Applications 18:59–74

369. Palermo T, Flaud P (1987) Étude de l'effondrement à deux et trois lobes de tubes élastiques[Study of two- or three-lobe collapse of elastic tubes]. Journal de Biophysique et de Biomécanique 11:105–111

370. Heil M (1997) Stokes flow in collapsible tubes: computation and experiment. Journal of Fluid Mechanics 353:285–312

371. Thiriet M, Ribreau C (2000) Computational flow in a collapsed tube with wall contact. Revue Mécanique et Industries 1:349–364

372. Brown N (1993) Impedance matching at arterial bifurcations. Journal of Biomechanics 26:59–67

373. Olson RM (1968) Aortic blood pressure and velocity as a function of time and position. Journal of Applied Physiology 24:563–569

374. McDonald DA (1974) Blood Flow in Arteries. Edward Arnold, London

375. Parker KH, Jones CJ (1990) Forward and backward running waves in the arteries: analysis using the method of characteristics. Journal of Biomechanical Engineering 112:322–326

376. Platon [Plato] (1991) La république[The Republic]. Garnier-Flammarion, Paris

377. Batchelor GK (1967) An Introduction to Fluid Dynamics. Cambridge University Press, Cambridge

378. Thiriet M, Naili S, Ribreau C (2003) Entry length and wall shear stress in uniformly collapsed veins. Computer Modeling in Engineering & Sciences 4:473–488

379. Segré G, Silberberg A (1962) Behaviour of macroscopic rigid spheres in Poiseuille flow I: Determination of local concentration by statistical analysis of particle passages through crossed light beam. Journal of Fluid Mechanics 14:115–135

380. Segré G, Silberberg A (1962) Behavior of macroscopic rigid spheres in Poiseuille flow II: Experimental results and interpretation. Journal of Fluid Mechanics 14:136–157

381. Leighton D, Acrivos A (1987) The shear-induced migration of particles in concentrated suspensions. Journal of Fluid Mechanics 181:415–439

382. Koh CJ, Hookham P, Leal LG (1994) An experimental investigation of concentrated suspension flows in a rectangular channel. Journal of Fluid Mechanics 266:1–32

383. Karnis A, Goldsmith HL, Mason SG (1966) The flow of suspensions through tubes. Inertial effects. The Canadian Journal of Chemical Engineering 44:181–193

384. Karnis A, Goldsmith HL, Mason SG (1966) The kinetics of flowing dispersions I: Concentrated suspensions of rigid particles. Journal of Colloid and Interface Science 22:531–553

385. Kowalewski TA, (1980) Velocity profiles of suspension flowing through a tube. Archives of Mechanics 32:857–865

386. Feng J, Hu HH, Joseph DD (1994) Direct simulation of initial value problems for the motion of solid bodies in a Newtonian fluid II: Couette and Poiseuille flows. Journal of Fluid Mechanics 277:271–301

387. Takano M, Goldsmith HL, Mason SG (1968) The flow of suspensions through tubes VIII: Radial migration of particles in pulsatile flow. Journal of Colloid and Interface Science 27:253–267

388. Brenner H, Nadim A, Haber S (1987) Long-time molecular diffusion, sedimantation and Taylor dispersion of a fluctuating cluster of interacting Brownian particles. Journal of Fluid Mechanics 183:511–542

389. Jeffrey DJ, Acrivos A (1976) The rheological properties of suspensions of rigid particles. AIChE Journal 22:417–432

390. Ramanujan S, Pozrikidis C (1998) Deformation of liquid capsules enclosed by elastic membranes in simple shear flow: large deformations and the effect of fluid viscosities. Journal of Fluid Mechanics 361:117–143

391. Skalak R, Tozeren A, Zarda RP, Chien S (1973) Strain energy function of red blood cell membranes. Biophysical Journal 13:245–264

392. Barthes-Biesel D, Diaz A, Dhenin E (2002) Effect of constitutive laws for two-dimensional membranes on flow-induced capsule deformation. Journal of Fluid Mechanics 460:211–222

393. Barthes-Biesel D (1980) Motion of a spherical microcapsule freely suspended in a linear shear flow. Journal of Fluid Mechanics 100:831–853

394. Diaz A, Barthes-Biesel D (2002) Entrance of a bioartificial capsule in a pore. Computer Modeling in Engineering & Sciences 3:321–338

395. Dantan P (1985) Étude numérique et expérimentale de l'écoulement instationnaire d'un fluide visqueux incompressible dans une cavité de dimension variable. Modélisation de l'hémodynamique cardiaque[Unsteady flow of a viscous incompressible fluid through varying-size cavity : a numerical and experimental study; Modeling of heart hemodynamics]. PhD Thesis, Université Paris VII, Paris

396. Ebbers T, Wigstrom L, Bolger AF, Wranne B, Karlsson M (2002) Noninvasive measurement of time-varying three-dimensional relative pressure fields within the human heart. Journal of Biomechanical Engineering 124:288–293

397. Henderson Y, Johnson F (1912) Two modes of the closure of the heart valves. Heart 4:69–82

398. Bitbol M, Dantan P, Perrot P, Oddou C (1982) Collapsible tube model for the dynamics of closure of the mitral valve. Journal of Fluid Mechanics 114:187–211

399. Bellhouse BJ, Talbot L (1969) The fluid mechanics of the aortic valve. Journal of Fluid Mechanics 35:721–735

400. Lee CSF, Talbot L (1979) A fluid mechanical study on the closure of heart valves. Journal of Fluid Mechanics 91:41–63

401. van Steenhoven AA, Veenstra PC, Reneman RS (1982) The effect of some hemodynamic factors on the behaviour of the aortic valve. Journal of Biomechanics 15:941–950

402. van Steenhoven AA, van Dongen MEH (1979) Model studies of the closing behaviour of the aortic valve. Journal of Fluid Mechanics 90:21–32

403. Goldberger AL, Rigney DR, West BJ (1990) Chaos and fractals in human physiology. Scientific American 262:42–49

404. Baumgartner H, Schima H, Kuhn P (1991) Value and limitations of proximal jet dimensions for the quantitation of valvular regurgitation: an in vitro study using Doppler flow imaging. Journal of the American Society of Echocardiography 4:57–66

405. Miyake Y, Binder G (1970) Évolution d'un jet pulsant[Evolution of a pulsatile jet]. Comptes rendus hebdomadaires des séances de l'Académie des sciences – Série A 271:615–618

406. Oddou C, Dantan P, Flaud P, Geiger D (1979) Aspects of hydrodynamics in cardiovascular research. In: Hwang NHC, Gross DR, Patel DJ (eds) Quantitative Cardiovascular Studies, Clinical and Research Application of Engineering Principles. University Park Press, Baltimore

407. Fahraeus R, Lindqvist T (1931) The viscosity of the blood in narrow capillary tubes. The American Journal of Physiology 96:562–568

408. Desjardins C, Duling BR (1990) Heparinase treatment suggests a role for the endothelial cell glycocalyx in regulation of capillary hematocrit. American Journal of Physiology – Heart and Circulatory Physiology 258:647–654

409. Pries AR, Secomb TW, Jacobs H, Sperandio M, Osterloh K, Gaehtgens P (1997) Microvascular blood flow resistance: role of endothelial surface layer. American Journal of Physiology – Heart and Circulatory Physiology 273:2272–2279.

410. Feng J, Weinbaum S (2000) Lubrication theory in highly compressible porous media: the mechanics of skiing, from red cells to humans. Journal of Fluid Mechanics 422:281–317

Chap. 6. Numerical Simulations

411. Thom R. (1989) Paraboles et catastrophes[Parables and catastrophes]. Flammarion, Paris

412. Delhaas T, Arts T, Prinzen FW, Reneman RS (1993) Relation between regional electrical activation time and subepicardial fiber strain in the canine left ventricle. Pflügers Archiv 423:78–87

413. Faris OP, Evans FJ, Ennis DB, Helm PA, Taylor JL, Chesnick AS, Guttman MA, Ozturk C, ER. McVeigh ER (2003) Novel technique for cardiac electromechanical mapping with magnetic resonance imaging tagging and an epicardial electrode sock. Annals of Biomedical Engineering 31:430–440

414. Azhari H, Weiss JL, Rogers WJ, Siu CO, Zerhouni EA, Shapiro EP (1993) Noninvasive quantification of principal strains in normal canine hearts using tagged MRI images in 3-D. American Journal of Physiology – Heart and Circulatory Physiology 264:205–216

415. Bogaert J, Rademakers FE (2001) Regional nonuniformity of normal adult human left ventricle, American Journal of Physiology – Heart and Circulatory Physiology 280 (2):H610-620

416. Fogel MA, Weinberg PM, Hubbard A, Haselgrove J (2000) Diastolic biomechanics in normal infants utilizing MRI tissue tagging, Circulation 102:218–224

417. Hsu EW, Henriquez CS (2001) Myocardial fiber orientation mapping using reduced encoding diffusion tensor imaging. Journal of Cardiovascular Magnetic Resonance 3:339-347

418. Bestel J, Clément F, Sorine M (2001) A biomechanical model of muscle contraction. In: Niessen WJ, Viergever MA (eds), Medical Image Computing and Computer-Assisted Intervention. Springer, Berlin

419. Chapelle D, Clément F, Génot F, Le Tallec P, Sorine M, Urquiza JM (2001) A physiologically-based model for the active cardiac muscle. In: Katila T, Magnin IE, Clarysse P, Montagnat J, Nenonen J (eds) Functional Imaging and Modeling of the Heart. Springer, Berlin

420. Sermesant M, Forest C, Pennec X, Delingette H, Ayache N (2003) Deformable biomechanical models: application to 4D cardiac image analysis. Medical Image Analysis 7:475–488

421. van der Pol B, van der Mark J (1928) The heartbeat considered as a relaxation oscillation, and an electrical model of the heart. Philosophical Magazine 6:763–775

422. Hodgkin AL, Huxley AF (1952) A quantitative description of membrane current and its application to conduction and excitation in nerve. The Journal of Physiology 117:500-544

423. Noble D (1962) A modification of the Hodgkin-Huxley equations applicable to Purkinje fibre action and pacemaker potentials. The Journal of Physiology 160:317-352

424. Yanagihara K, Noma A, Irisawa H (1980) Reconstruction of sino-atrial node pacemaker potential based on the voltage clamp experiments. Japanese Journal of Physiology 30:841-857

425. Beeler GW, Reuter H (1977) Reconstruction of the action potential of ventricular myocardial fibres. The Journal of Physiology 268:177-210

426. Luo CH, Rudy Y (1991) A model of the ventricular cardiac action-potential: depolarization, repolarization, and their interaction. Circulation Research 68:1501–1526

427. FitzHugh R (1961) Impulses and physiological states in theoretical models of nerve membrane. Biophysical Journal 1:445–466

428. Nagumo J, Arimoto S, Yoshizawa S (1962) An active pulse transmission line simulating nerve axons. Proceedings of the IRE 50:2061–2070

429. Aliev RR, Panfilov AV (1996) A simple two-variable model of cardiac excitation. Chaos Solitons Fractals 7:293–301

430. Geselowitz DB, Miller WT (1983) A bidomain model for anisotropic cardiac muscle. Annals of Biomedical Engineering 11:191–206

431. Bourgault Y, Éthier M, LeBlanc VG (2003) Simulation of electrophysiological waves with an unstructured finite element method. Mathematical Modelling and Numerical Analysis 37:649–661

432. Mitchell CC, Schaeffer DG (2003) A two-current model for the dynamics of cardiac membrane. Bulletin of Mathematical Biology 65:767–793

433. Colli-Franzone P, Guerri L, Tentoni S (1990) Mathematical modeling of the excitation process in myocardial tissue: influence of fiber rotation on the wavefront propagation and potential field. Mathematical Biosciences 101:155–235

434. Sundnes J, Lines GT, Cai X, Nielsen BF, Mardal KA, Tveito A (2006) Computing the Electrical Activity in the Heart. Springer, Berlin

435. Saxberg BEH, Cohen RJ (1991) Cellular automata models of cardiac conduction. In: Glass L, Hunter P, McCulloch A (eds) Theory of Heart. Springer, Berlin

436. Siregar P, Sinteff JP, Chahine M, Lebeux P (1996) A cellular automata model of the heart and its coupling with a qualitative model. Computers and Biomedical Research 29:222-246

437. Siregar P, Sinteff JP, Julen N, Le Beux P (1998) An interactive 3D anisotropic cellular automata model of the heart. Computers and Biomedical Research 31:323–347

438. Pao YC, Robb RA, Ritman EL (1976) Plane-strain finite-element analysis of reconstructed diastolic left ventricular cross section. Annals of Biomedical Engineering 4:232–249

439. Neckyfarow CW, Perlman AB (1976) Deformation of the left ventrcle: material and geometric effect In: Saha S (ed) New England Bioengineering Conference. Karger, Basel

440. Heethaar RM, Pao YC, Ritman EL (1977) Computer aspects of three-dimensional finite element analysis of stresses and strains in the intact hear. Computers and Biomedical Research 10:271–285

441. Pao YC, Ritman EL (1998) Comparative characterization of the infarcted and reperfused ventricular wall muscles by finite element analysis and a myocardial muscle-blood composite model, Computers and Biomedical Research 31:18-31

442. Ghista DN, Chandran KB, Ray G, Reul H (1978) Optimal design of aortic leaflet prosthesis. Journal of Engineering Mechanics 104:97–117

443. Bestel J (2000) Modèle différentiel de la contraction musculaire contrôlée. Application au système cardio-vasculaire[Model of controlled muscle contraction. Application to the cardiovascular system]. PhD thesis, Université Paris Dauphine, Paris

444. Hunter PJ, McCulloch AD, ter Keurs HE (1998) Modelling the mechanical properties of cardiac muscle. Progress in Biophysics and Molecular Biology 69:289–331

445. Panerai RB (1980) A model of cardiac muscle mechanics and energetics. Journal of Biomechanics 13:929–940

446. Huxley AF (1957) Muscle structure and theories of contraction. Progress in Biophysics and Biophysical Chemistry 7:255–318

447. Zahalak GI (1981) A distribution-moment approximation for kinetic theories of muscular contraction. Mathematical Biosciences 114:55–89

448. Brookes PS, Yoon Y, Robotham JL, Anders MW, Sheu SS (2004) Calcium, ATP, and ROS: a mitochondrial love-hate triangle. American Journal of Physiology – Cell Physiology 287:817–833

449. Mirsky I, Parmley WW (1973) Assessment of passive elastic stiffness for isolated heart muscle and the intact heart. Circulation Research 33:233–243

450. Guccione J, McCulloch A (1991) Theory of heart: biomechanics, biophysics, and nonlinear dynamics of cardiac function. In: Glass L, Hunter P, McCulloch A (eds) Finite Element Modeling of Ventricular Mechanics. Springer, Berlin

451. Hunter PJ, Nash MP, Sands GB (1997) Computational electromechanics of the heart. In: Panfilov AV, Holden AV (eds) Computational Biology of the Heart. John Wiley & Sons

452. Wu JZ, Herzog W (1999) Modelling concentric contraction of muscle using an improved cross-bridge model. Journal of Biomechanics 32:837–848

453. Hill AV (1938) The heat of shortening and the dynamic constants in muscle. Proceedings of the Royal Society of London – Series B: Biological Sciences 126:136–195

454. Sainte-Marie J, Chapelle D, Sorine M (2003) Data assimilation for an electro-mechanical model of the myocardium, In: Bathe KJ (ed) Computational fluid and solid mechanics 2003. Elsevier, Amsterdam

455. Sainte-Marie J, Numerical simulations of a complete heart beat. http://www-rocq. inria. fr/MACS/Coeur/

456. Durrer D, van Dam RT, Freud GE, Janse MJ, Meijler FL, Arzbaecher RC (1970) Total excitation of the isolated human heart. Circulation 41:899–912

457. Stergiopulos N, Westerhof BE, Westerhof N (1999) Total arterial inertance as the fourth element of the windkessel model. American Journal of Physiology – Heart and Circulatory Physiology 276:81–88

458. Westerhof N, Bosman F, De Vries CJ Noordergraaf A (1969) Analog studies of the human systemic arterial tree. Journal of Biomechanics 2:121-143

459. Veronda DR, Westmann RA (1970) Mechanical characterization of skin. Finite deformation. Journal of Biomechanics 3:114–124

460. Pioletti DP, Rakotomanana, LR (2000) Nonlinear viscoelastic laws for soft biological tissues. European Journal of Mechanics A – Solids 19:749–759

461. McVeigh E, Faris O, Ennis D, Helm P, Evans F (2001) Measurement of ventricular wall motion, epicardial, electrical mapping, and myocardial fiber angles in the same heart. In: Katila T, Magnin IE, Clarysse P, Montagnat J, Nenonen J, (eds) Functional Imaging and Modeling of the Heart. Springer, Berlin

462. Burnes J, Taccardi B, Rudy Y (2000) A noninvasive imaging modality for cardiac arrhythmias. Circulation 102:2152–2158

463. Smith NP, Mulquiney PJ, Nash MP, Bradley CP, Nickerson DP, Hunter PJ (2002) Mathematical modelling of the heart: cell to organ. Chaos Solitons Fractals 13:1613–1621

464. Smith NP, Pullan AJ, Hunter PJ (2002) An anatomically based model of transient coronary blood flow in the heart. SIAM Journal on Applied Mathematics 62:990–1018

465. Cimrman R, Rohan E (2003) Modelling heart tissue using a composite muscle model with blood perfusion. In: Bathe KJ (ed) Computational Fluid and Solid Mechanics. Elsevier, Amsterdam

466. Huyghe JM, Arts T, van Campen DH (1992) Porous medium finite element model of the beating left ventricle. American Journal of Physiology – Heart and Circulatory Physiology 262:1256–1267

467. Delavaud E (2003) Couplage de l'écoulement entre le ventricule gauche et l'aorte[Coupling of the flow between the left ventricle and the aorta]. DESS, Université Paris VI, Paris

468. Olufsen MS (1998) Modeling the arterial system with reference to an anesthesia simulator. PhD thesis, Roskilde University

469. Watanabe H, Sugiura S, Kafuku H, Hisada T (2004) Multiphysics simulation of left ventricular filling dynamics using fluid-structure interaction finite element method. Biophysical Journal 87:2074–2085

470. Sun Y, Beshara M, Lucariello RJ, Chiaramida SA (1997) A comprehensive model for right-left heart interaction under the influence of pericardium and baroreflex. American Journal of Physiology. Heart and Circulatory Physiology 272:H1499–H1515

471. Olansen JB, Clark JW, Khoury D, Ghorbel F, Bidani A (2000) A closed-loop model of the canine cardiovascular system that includes ventricular interaction. Computers and Biomedical Research 33:260–295

472. Frank O (1899) Die Grundform des arteriellen Pulses[The basic shape of the arterial pulse]. Zeitschrift für Biologie 37:483–586

473. Stergiopulos N, Meister J-J, Westerhof N (1995) Evaluation of methods for estimation of total arterial compliance. American Journal of Physiology – Heart and Circulatory Physiology 268:1540–1548

474. Pontrelli G, Rossoni E (2003) Numerical modelling of the pressure wave propagation in the arterial flow. International Journal for Numerical Methods in Fluids 43:651–671

475. Calhoun D, LeVeque RJ (2000) A Cartesian grid finite-volume method for the advection-diffusion equation in irregular geometries. Journal of Computational Physics 157:143–180

476. Girault V, Raviart PA (1979) Finite element methods for Navier–Stokes equations. Springer, New York

477. Pironneau O (1988) Méthodes d'éléments finis pour les fluides[The finite element methods for fluids]. Masson, Paris

478. Conca C, Pares C, Pironneau O, Thiriet M (1995) A computational model of Navier–Stokes equations with imposed pressure and velocity fluxes. International Journal for Numerical Methods in Fluids 20:267–287

479. Pelletier D, Fortin A, Camarero R (1989) Are FEM Solutions of Incompressible Flows Really Incompressible? International Journal for Numerical Methods in Fluids 9:99–112

480. Frisch U, Hasslacher B, Pomeau Y (1986) Lattice-gas automata for the Navier-Stokes equation. Physical Review Letters 56:1505–1508

481. Chopard B, Luthi PO (1999) Lattice Boltzmann computations and application to physics. Theoretical Computer Science 217:115-130

482. Mei R, Luo LS, Shyy W (1999) An accurate curved boundary treatment in the lattice Boltzmann method. Journal of Computational Physics 155:307–330

483. Fang H, Wang Z, Lin Z, Liu M (2002) Lattice Boltzmann method for simulating the viscous flow in large distensible blood vessels. Physical Review E 65:051925

484. Harting J, Venturoli M, Coveney PV (2004) Large-scale grid-enabled lattice Boltzmann simulations of complex fluid flow in porous media and under shear. Philosophical Transactions of the Royal Society of London – Series A: Mathematical and Physical Sciences 362:1703–1722

485. Krafczyc M, Cerrilaza A, Schulz M, Rank E (1998) Analysis of 3D transient blood flow through an artificial aortic valve by lattice–Boltmann methods. Journal of Biomechanics 31:453–462

486. Glowinski R, Pan TW, Periaux J (1994) A fictitious domain method for external incompressible viscous flow modeled by Navier-Stokes. Computer Methods in Applied Mechanics and Engineering 111:283–303

487. Girault V, Glowinski R (1995) Error analysis of a fictitious domain method applied to a Dirichlet problem. Japan Journal of Industrial and Applied Mathematics 12:487–514

488. De Hart J, Baaijens FPT, Peters GWM, Schreurs PJG (2003) A computational fluid-structure interaction analysis of a fiber-reinforced stentless aortic valve. Journal of Biomechanics 36:699-712

489. Valéry P (1947) Mauvaises pensées et autres[Bad toughts and others]. Gallimard, Paris

490. Quarteroni A, Ragni S, Veneziani A (2001) Coupling between lumped and distributed models for blood flow problems. Computing and Visualization in Science 4:111–124

491. Olufsen MS, Peskin CS, Kim WY, Pedersen EM, Nadim A, Larsen J (2000) Numerical simulation and experimental validation of blood flow in arteries with structured-tree outflow conditions, Annals of Biomedical Engineering 28:1281–1299

492. Laganà K, Dubini G, Migliavacca F, Pietrabissa R, Pennati G, Veneziani A, Quarteroni A (2002) Multiscale modelling as a tool to prescribe realistic boundary conditions for the study of surgical procedures. Biorheology 39:359–364

493. Formaggia L, Gerbeau J-F, Nobile F, Quarteroni A (2002) Numerical treatment of defective boundary conditions for the Navier–Stokes equations. SIAM Journal on Numerical Analysis 40:376–401

494. Segers P, Dubois F, De Wachter D, Verdonck P (1998) Role and relevancy of a cardiovascular simulator. Journal of Cardiovascular Engineering 3:48–56.

495. Fernández M, Milišić V, Quarteroni A (2005) Analysis of a geometrical multiscale blood flow model based on the coupling of ODEs and hyperbolic PDEs. Multiscale Modeling & Simulation 4:215–236

496. Collins R, Maccario JA (1979) Blood flow in the lung. Journal of Biomechanics 12:373–395

497. Fung YC, Sobin SS (1969) Theory of sheet flow in lung alveoli. Journal of Applied Physiology 26:472–488

498. Formaggia L, Gerbeau J-F, Nobile F, Quarteroni A (2001) On the coupling of 3D and 1D Navier–Stokes equations for flow problems in compliant vessels. Computer Methods in Applied Mechanics and Engineering 191:561–582

499. Peiro J, Sherwin SJ, Parker KH, Franke V, Formaggia L, Lamponi D, Quarteroni A (2003) Numerical simulation of arterial pulse propagation using one-dimensional models. In Collins MW, Pontrelli G, Atherton MA (eds) Wall-Fluid Interactions in Physiological Flows, Advances in Computational Bioengineering, WIT Press, Ashurst

500. Sherwin SJ, Franke V, Peiro J, Parker KH (2003) One-dimensional modelling of a vascular network in space-time variables. Journal of Engineering Mathematics 47:217–250

501. Quarteroni A, Veneziani A (2003) Analysis of a geometrical multiscale model based on the coupling of PDE's and ODE's for blood flow simulations. Multiscale Modeling & Simulation 1:173–195

502. Kuhn T (1972) La structure des révolutions scientifiques. Flammarion, Paris [The structure of scientific revolutions (1962). University of Chicago Press, Chicago]

503. Maury B (1999) Direct simulations of 2D fluid-particle flows in biperiodic domains. Journal of Computational Physics 156:325–351

504. Maury B (2003) Fluid-particle shear flows. Mathematical Modelling and Numerical Analysis 37:699–708

505. Grandmont C, Maday Y (1998) Analyse et méthodes numériques pour la simulation de phénomènes d'interaction fluide-structure[Analysis and numerical methods for the simulation of fluid structure interaction phenomena]. In: Ahues M, Boukrouche M, Carasso C (eds), Actes du 29e congrès d'analyse numérique. Société de Mathématiques Appliquées et Industrielles, Paris

506. Grandmont C, Maday Y, Guimet V (1998) Results about some decoupling techniques for the approximation of the unsteady fluid-structure interaction. In: Bock HG, Kanschat G, Rannacher R, Brezzi F, Glowinski R, Kuznetsov YA, Périaux J (eds), ENUMATH 97. World Scientific Publishing, River Edge

507. Grandmont C, Maday Y (2000) Existence for an unsteady fluid-structure interaction problem. Mathematical Modelling and Numerical Analysis 34:609–636

508. Errate D, Esteban MJ, Maday Y (1994) Couplage fluide-structure. Un modèle simplifié en dimension 1[Fluid-structure interaction: a simplified model in dimension 1]. Comptes Rendus des Séances de l'Académie des Sciences – Série I – Mathématique318:275-281

509. Takahashi T, Tucsnak M (2004) Global strong solutions for the two dimensional motion of an infinite cylinder in a viscous fluid. Journal of Mathematical Fluid Mechanics 6:53–77

510. Desjardins B, Esteban MJ (2000) On weak solutions for fluid-rigid structure interaction: compressible and incompressible models. Communications in Partial Differential Equations 25:1399–1413

511. Conca C, San Martin JH, Tucsnak M (2000) Existence of solutions for the equations modelling the motion of a rigid body in a viscous fluid. Communications in Partial Differential Equations 25:1019–1042

512. Desjardins B, Esteban MJ, Grandmont C, Le Tallec P (2001) Weak solutions for a fluid-elastic structure interaction model. Revista Matemática Complutense 14:523–538

513. Causin P, Gerbeau J-F, Nobile F (2005) Added-mass effect in the design of partitioned algorithms for fluid-structure problems. Computer Methods in Applied Mechanics and Engineering 194:4506–4527

514. Čanić S, Mikelić A (2002) Effective equations describing the flow of a viscous incompressible fluid through a long elastic tube. Comptes Rendus Mécanique 330:661–666

515. Čanić S, Mikelić A (2003) Effective equations modeling the flow of a viscous incompressible fluid through a long elastic tube arising in the study of blood flow through small arteries. SIAM Journal on Applied Dynamical Systems 2:431–463

516. Farhat C, Lesoinne M, Maman, N (1995) Mixed explicit/implicit time integration of coupled aeroelastic problems: three field formulation, geometric conservation and distributed solution. International Journal for Numerical Methods in Fluids 21:807–835

517. Piperno S, Farhat C, Larrouturou B (1995) Partitioned procedures for the transient solution of coupled aeroelastic problems I: model problem, theory, and two-dimensional application. Computer Methods in Applied Mechanics and Engineering 124:79–112

518. Piperno S (1997) Explicit/implicit fluid/structure staggered procedures with a structural predictor and fluid subcycling for 2D inviscid aeroelastic simulations. International Journal for Numerical Methods in Fluids 25:1207–1226

519. Lanteri S (1996) Parallel solutions of compressible flows using overlapping and non-overlapping mesh partitioning strategies. Parallel Computing 22:943–968

520. Carré G, Lanteri S, Fournier L (2000) Parallel linear multigrid algorithms applied to the acceleration of compressible flows. Computer Methods in Applied Mechanics and Engineering 184:427–448

521. Joppich W, Mijalkovic S (1993) Multigrid Methods for Process Simulation. Springer, Berlin

522. Nobile F (2001) Numerical approximation of fluid-structure interaction problems with application to haemodynamics. PhD thesis, École Polytechnique Fédérale de Lausanne

523. Mouro J, Le Tallec P (2001) Fluid structure interaction with large structural displacements. Computer Methods in Applied Mechanics and Engineering 190:3039–3068

524. Deparis S, Fernández MA, Formaggia, L (2003) Acceleration of a fixed point algorithm for a fluid-structure interaction using transpiration condition. Mathematical Modelling and Numerical Analysis 37:601–616

525. Fernández MA, Moubachir M (2003) An exact Block-Newton algorithm for the solution of implicit time discretized coupled systems involved in fluid-structure interaction problems. In: Bathe KJ (ed) Computational Fluid and Solid Mechanics 2003. Elsevier, Amsterdam

526. Gerbeau J-F, Vidrascu M (2003) A quasi-Newton algorithm based on a reduced model for fluid-structure interaction problems in blood flows. Mathematical Modelling and Numerical Analysis 37:631–647

527. Chahboune B, Crolet J-M (1998) Numerical simulation of the blood-wall interaction in the human left ventricle. The European Physical Journal – Applied Physics 2:291–297

528. Brackbill JU, Pracht WE (1973) An implicit almost Lagrangian algorithm for magnetohydrodynamics. Journal of Computational Physics 13:455–482

529. Hughes TJR, Liu WK, Zimmermann TK (1981) Lagrangian–Eulerian finite element formulation for incompressible viscous flows. Computer Methods in Applied Mechanics and Engineering 29:329–349

530. Donea J (1983) Arbitrary Lagrangian–Eulerian finite element methods. In: Belytschko T, Hugues TJR (eds), Computers Methods for Transient Analysis. North-Holland, Amsterdam

531. Maury B (1996) Characteristics ALE method for the unsteady 3D Navier-Stokes equations with a free surface. International Journal of Computational Fluid Dynamics 6:175–188

532. van de Vosse FN, de Hart J, van Oijen CHGA, Bessems D, Gunther TWM, Segal A, BJBM Wolters BJBM, Stijnen JMA, FPT Baaijens FPT (2003) Finite-element-based computational methods for cardiovascular fluid-structure interaction. Journal of Engineering Mathematics 47:355–368

533. Quarteroni A, Formaggia L (2004) Mathematical modelling and numerical simulation of the cardiovascular system. In: Ayache N (ed), Handbook of numerical analysis. Volume XII. North-Holland, Amsterdam

534. Peskin CS (1977) Numerical analysis of blood flow in the heart. Journal of Computational Physics 25:220–252

535. Peskin CS, McQueen DM (1989) A three-dimensional computational method for blood flow in the heart I: immersed elastic fibers in a viscous incompressible fluid. Journal of Computational Physics 81:372-405

536. McQueen DM, Peskin CS (1989) A three-dimensional computational method for blood flow in the heart II: contractile fibers. Journal of Computational Physics 82:289-297

537. Peskin CS, McQueen DM (1993) Computational biofluid dynamics. In: Cheer AY, van Dam CP (eds) Fluid Dynamics in Biology. American Mathematical Society, Providence, RI

538. LeVeque RJ, Li Z (1997) Immersed interface methods for Stokes flow with elastic boundaries or surface tension. SIAM Journal on Scientific Computing 18:709-735

539. Li Z, Lai MC (2001) The immersed interface method for the Navier-Stokes equations with singular forces. Journal of Computational Physics 171:822–842

540. Bathe KJ, Iosilevich A, Chapelle D (2000) An evaluation of the MITC shell elements. Computers Structures 75:1–30

541. Bernardi C, Maday Y, Patera AT (1994) A new nonconforming approach to domain decomposition: the mortar element method. In: Brézis H, Lions J-L (eds), Nonlinear partial differential equations and their applications. Collège de France Seminar. Volume XI. Longman Scientific & Technical, Harlow

542. Gerbeau J-F, Vidrascu M, Frey P (2003) Fluid-structure interaction in blood flows on geometries based on medical imaging. Computers Structure 83:155–165

543. Lions J-L (1968) Contrôle optimal de systèmes gouvernés par des équations aux dérivées partielles[Optimal control of systems governed by partial differential equations]. Dunod, Paris

544. Delfour MC, Zolesio JP (2001) Shapes and Geometries: Analysis, Differential Calculus and Optimization. SIAM, Philadelphia

545. Cole JS, Wijesinghe LD, Watterson JK, Scott DJA (2002) Computational and experimental simulations for the haemodynamics a cuffed arterial bypas graft anastomoses. Proceedings of the Institution of Mechanical Engineers – Part H – Journal of Engineering in Medicine 216:135–143

546. Etave F, Finet G, Boivin M, Boyer, J-C, Riufol G, Thollet G (2001) Mechanical properties of coronary stents determined by using finite element analysis. Journal of Biomechanics 34:1065–1075

547. Quarteroni A, Rozza G (2003) Optimal control and shape optimization of aorto-coronaric bypass anastomoses. Mathematical Models & Methods in Applied Sciences 13:1801–1823

548. Quick CM, Young WL, Noordergraaf A (2001) Infinite number of solutions to the hemodynamic inverse problem. American Journal of Physiology – Heart and Circulatory Physiology 280:1472–1470

549. Stergiopulos M, Meister J-J, Westerhof N (1994) Simple and accurate way for estimating total and segmental arterial compliance: the pulse pressure method. Annals of Biomedical Engineering 22:392–397

550. Canty JMJ, Klocke FJ, Mates RE (1985) Pressure and tone dependence of coronary diastolic input impedance and capacitance. American Journal of Physiology – Heart and Circulatory Physiology 248:700–711

551. Goldwyn RM, Watt TBJ (1967) Arterial pressure contour analysis via a mathematical model for the clinical quantification of human vascular properties. IEEE Transactions on Biomedical Engineering 14:11–17

552. Lagrée P-Y (2000) An inverse technique to deduce the elasticity of a large artery. The European Physical Journal – Applied Physics 9:153–163

553. Fernández MA, Moubachir M (2002) Sensitivity analysis for an incompressible aeroelastic system. Mathematical Models & Methods in Applied Sciences 12:1109–1130

554. Moubachir M, Zolésio J-P (2002) Optimal control of fluid-structure interaction systems: the case of a rigid solid. INRIA report 4611

555. Muyl F, Dumas L, Herbert V (2004) Hybrid method for aerodynamic shape optimization in automotive industry, Computers & Fluids 33:849-858

556. Dumas L, El Alaoui L (2007) personal communication

557. Taylor GI (1953) Dispersion of soluble matter in solvent flowing slowly through a tube. Proceedings of the Royal Society of London – Series A: Mathematical and Physical Sciences 219:186–203

558. Taylor GI (1954) Conditions under which dispersion of a solute in a stream of solvent can be used to measure molecular diffusion. Proceedings of the Royal Society of London – Series A: Mathematical and Physical Sciences 225:473–477

559. Aris, R (1956) On the dispersion of a solute in a fluid through a tube. Proceedings of the Royal Society of London – Series A: Mathematical and Physical Sciences 235:67–77

560. Prosi M, Zunino P, Perktold K, Quarteroni A (2005) Mathematical and numerical models for transfer of low density lipoproteins through the arterial walls: a new methodology for the model set up with applications to the study of disturbed lumenal flow. Journal of Biomechanics 38:903–917

561. Manseau J (2002) Étude numérique d'un modèle de transport de macromolécules à travers la paroi artérielle[Numerical study of a macromolecule transport model through the arterial wall]. MSc thesis, École Polytechnique de Montréal

562. Curry FE (1984) Mechanism and thermodynamics of transcapillary exchange. In: Renkin EM, Michel CC (eds) Microcirculation, Handbook of Physiology. The Cardiovascular System, American Physiological Society, Bethesda, MD

563. Huang Y, Rumschitzki D, Chien S, Weinbaum S (1994) A fiber matrix model for the growth of macromolecule leakage spots in the arterial intima. Journal of Biomechanical Engineering 116:430–445

564. Caro CG, Lever MJ, Laver-Rudich Z, Meyer F, Liron N, Ebel W, Parker KH, Winlove CP (1980) Net albumin transport across the wall of the rabbit common carotid artery perfused in situ. Atherosclerosis 37:497–511

565. Tedgui A, Lever MJ (1985) The interaction of convection and diffusion in the transport of 131I- albumin within the media of the rabbit thoracic aorta. Circulation Research 57:856–863

566. Caro CG, Nerem RM (1973) Transport of C4-cholesterol between serum and wall in perfused dog common carotid artery. Circulation Research 32:187–205

567. Kaazempur-Mofrad MR, Ethier CR (2001) Mass transport in an anatomically realistic human right coronary artery. Annals of Biomedical Engineering 29:121–127

568. Tarbell JM (2003) Mass transport in arteries and the localization of atherosclerosis. Annual Review of Biomedical Engineering 5:79–118

569. Ethier CR (2002) Computational modeling of mass transfer and links to atherosclerosis. Annals of Biomedical Engineering 30:461–471

570. Moore JA, Ethier CR (1997) Oxygen mass transfer calculations in large arteries. Journal of Biomechanical Engineering 119:469–475

571. Fry DL (1987) Mass transport, atherogenesis, and risk. Arteriosclerosis 7:88–100

572. Tsay R, Weinbaum S, Pfeffer R (1989) A new model for capillary filtration based on recent electron microscopic studies of endothelial junctions. Chemical Engineering Communications 82:67–102

573. Garon A, Jullien S, Manseau J (2001) Numerical simulation of local mass transfer from an endovascular device. ASME-AIChE-AIAA 35th National Heat Transfer Conference, Anaheim, California

574. Grant J (2004) Modélisation du transport de macromolécules à travers la paroi artérielle[Modeling of the macromolecule transport through the arterial wall]. MSc thesis, École Polytechnique de Montréal

575. Rappitsch G, Perktold K, Pernkopf E (1997) Numerical modelling of shear-dependent mass transfer in large arteries. International Journal for Numerical Methods in Fluids 25:847–857

576. Anderson JL, Malone DM (1974) Mechanism of osmotic flow in porous membranes. Biophysical Journal 14:957–982

577. Wang DM, Tarbell JM (1995) Modeling interstitial flow in an artery wall allows estimation of wall shear stress on smooth muscle cells. Journal of Biomechanical Engineering 117:358–363

578. Huang ZJ, Tarbell JM (1997) Numerical simulation of mass transfer in porous media of blood vessel walls. American Journal of Physiology – Heart and Circulatory Physiology 273:464–477

579. Beard DA, Bassingthwaighte JB (2000) Advection and diffusion of substances in biological tissues with complex vascular networks. Annals of Biomedical Engineering 28:253–268.

580. Quarteroni A, Veneziani A, Zunino P (2002) Mathematical and numerical modeling of solute dynamics in blood flow and arterial walls. SIAM Journal on Numerical Analysis 39:1488–1511

581. Jäger W, Mikelić A (2000) On the interface boundary conditions by Beavers, Joseph and Saffman. SIAM Journal on Applied Mathematics 60:1111–1127.

582. Stangeby DK, Ethier CR (2002) Computational analysis of coupled blood-wall arterial LDL transport. Journal of Biomechanical Engineering 124:1–8

583. Quarteroni A, Veneziani A, Zunino P (2002) A domain decomposition method for advection-diffusion processes with application to blood solutes. SIAM Journal on Scientific Computing 23:1959–1980

584. Zunino P (2004) Multidimensional pharmacokinetic models applied to the design of drug eluting stents. Cardiovascular Engineering 4:181–191

Chap. 7. Cardiovascular Diseases

585. Mani A, Radhakrishnan J, Wang H, Mani A, Mani MA, Nelson-Williams C, Carew KS, Mane S, Najmabadi H, Wu D, Lifton RP (2007) LRP6 mutation in a family with early coronary disease and metabolic risk factors. Science 315:1278–1282

586. Durier S, Fassot C, Laurent S, Boutouyrie P, Couetil J-P, Fine E, Lacolley P, Dzau VJ, Pratt RE (2003) Physiological genomics of human arteries: quantitative relationship between gene expression and arterial stiffness. Circulation 108:1845–1851

587. Cowley AW (2006) The genetic dissection of essential hypertension. Nature Reviews – Genetics 7:829–840

588. Watkins H, Farrall M (2006) Genetic susceptibility to coronary artery disease: from promise to progress. Nature Reviews – Genetics 7:163–173

589. Helgadottir A, Manolescu A, Thorleifsson G, Gretarsdottir S, Jonsdottir H, Thorsteinsdottir U, Samani NJ, Gudmundsson G, Grant SF, Thorgeirsson G, Sveinbjornsdottir S, Valdimarsson EM, Matthiasson SE, Johannsson H, Gudmundsdottir O, Gurney ME, Sainz J, Thorhallsdottir M, Andresdottir M, Frigge ML, Topol EJ, Kong A, Gudnason V, Hakonarson H, Gulcher JR, Stefansson K (2004) The gene encoding 5-lipoxygenase activating protein confers risk of myocardial infarction and stroke. Nature Genetics 36:233–239

590. Wang X, Ria M, Kelmenson PM, Eriksson P, Higgins DC, Samnegård A, Petros C, Rollins J, Bennet AM, Wiman B, de Faire U, Wennberg C, Olsson PG, Ishii N, Sugamura K, Hamsten A, Forsman-Semb K, Lagercrantz J, Paigen B (2005) Positional identification of TNFSF4, encoding OX40 ligand, as a gene that influences atherosclerosis susceptibility. Nature Genetics 37:365–372

591. Gretarsdottir S, Thorleifsson G, Reynisdottir ST, Manolescu A, Jonsdottir S, Jonsdottir T, Gudmundsdottir T, Bjarnadottir SM, Einarsson OB, Gudjonsdottir HM, Hawkins M, Gudmundsson G, Gudmundsdottir H, Andrason H, Gudmundsdottir AS, Sigurdardottir M, Chou TT, Nahmias J, Goss S, Sveinbjörnsdottir S, Valdimarsson EM, Jakobsson F, Agnarsson U, Gudnason V, Thorgeirsson G, Fingerle J, Gurney M, Gudbjartsson D, Frigge ML, Kong A, Stefansson K, Gulcher JR (2003) The gene encoding phosphodiesterase 4D confers risk of ischemic stroke. Nature Genetics 35:131–138.

592. Shiffman D, Rowland CM, Louie JZ, Luke MM, Bare LA, Bolonick JI, Young BA, Catanese JJ, Stiggins CF, Pullinger CR, Topol EJ, Malloy MJ, Kane JP, Ellis SG, Devlin JJ (2006) Gene variants of VAMP8 and HNRPUL1 are associated with early-onset myocardial infarction. Arteriosclerosis, Thrombosis, and Vascular Biology 26:1613–1618

593. Frank D, Kuhn C, Katus HA, Frey N (2006) The sarcomeric Z-disc: a nodal point in signalling and disease. Journal of Molecular Medicine 84:446–468

594. Coucke PJ, Willaert A, Wessels MW, Callewaert B, Zoppi N, De Backer J, Fox JE, Mancini GMS, Kambouris M, Gardella R, Facchetti F, Willems PJ, Forsyth R, Dietz HC, Barlati S, Colombi M, Loeys B, De Paepe A (2006) Mutations in the facilitative glucose transporter GLUT10 alter angiogenesis and cause arterial tortuosity syndrome. Nature Genetics 38:452–457

595. Rodriguez-Viciana P, Tetsu O, Tidyman WE, Estep AL, Conger BA, Santa Cruz M, McCormick F, Rauen KA (2006) Germline mutations in genes within the MAPK pathway cause cardio-facio-cutaneous syndrome. Science 311:1287–1290

596. Ching YH, Ghosh TK, Cross SJ, Packham EA, Honeyman L, Loughna S, Robinson TE, Dearlove AM, Ribas G, Bonser AJ, Thomas NR, Scotter AJ, Caves LS, Tyrrell GP, Newbury-Ecob RA, Munnich A, Bonnet D, Brook JD (2005) Mutation in myosin heavy chain 6 causes atrial septal defect. Nature Genetics 37:423–428

597. Garg V, Muth AN, Ransom JF, Schluterman MK, Barnes R, King IN, Gross-feld PD, Srivastava D (2005) Mutations in NOTCH1 cause aortic valve disease. Nature 437:270–274

598. Griendling KK, Sorescu D, Ushio-Fukai M (2000) NAD(P)H oxidase: role in cardiovascular biology and disease. Circulation Research 86:494–501

599. Griendling KK, Sorescu D, Lassegue B, Ushio-Fukai M (2000) Modulation of protein kinase activity and gene expression by reactive oxygen species and their role in vascular physiology and pathophysiology. Arteriosclerosis, Thrombosis, and Vascular Biology 20:2175–2183

600. Yang B, Rizzo V (2007) TNFα potentiates protein-tyrosine nitration through activation of NADPH oxidase and eNOS localized in membrane rafts and caveolae of bovine aortic endothelial cells. American Journal of Physiology – Heart and Circulatory Physiology 292:H954–H962

601. Mehta PK, Griendling KK (2006) Angiotensin II cell signaling: physiological and pathological effects in the cardiovascular system. American Journal of Physiology – Cell Physiology 292:C82–C97

602. van Rooij E, Sutherland LB, Liu N, Williams AH, McAnally J, Gerard RD, Richardson JA, Olson EN (2006) A signature pattern of stress-responsive microRNAs that can evoke cardiac hypertrophy and heart failure. Proceedings of the National Academy of Sciences of the United States of America 103:18255–18260

603. Weyer GW, Jahromi BS, Aihara Y, Agbaje-Williams M, Nikitina E, Zhang ZD, Macdonald RL (2006) Expression and function of inwardly rectifying potassium channels after experimental subarachnoid hemorrhage. Journal of Cerebral Blood Flow and Metabolism 26:382–391

604. Adams JH, Duchen LW (eds) (1992) Greenfield's Neuropathology, 5th ed., Edward Arnold, London

605. He CM, Roach MR (1994) The composition and mechanical properties of abdominal aortic aneurysms. Journal of Vascular Surgery 20:6–13

606. Sobolewski K, Wolanska M, Bankowski E, Gacko M, Glowinski S (1995) Collagen, elastin and glycosaminoglycans in aortic aneurysms. Acta Biochimica Polonica 42:301–307

607. Dobrin PB, Mrkvicka R (1994) Failure of elastin or collagen as possible critical connective tissue alterations underlying aneurysmal dilatation. Cardiovascular Surgery 2:484–488

608. Pyo R, Lee JK, Shipley JM, Curci JA, Mao D, Ziporin SJ, Ennis TL, Shapiro SD, Senior RM, Thompson RW (2000) Targeted gene disruption of matrix metalloproteinase-9 (gelatinase B) suppresses development of experimental abdominal aortic aneurysms. The Journal of Clinical Investigation 105:1641–1649

609. van Laake LW, Vainas T, Dammers R, Kitslaar PJ, Hoeks AP, Schurink GW (2005) Systemic dilation diathesis in patients with abdominal aortic aneurysms: a role for matrix metalloproteinase-9. European Journal of Vascular and Endovascular Surgery 29:371–377

610. Thompson RW, Baxter BT (1999) MMP inhibition in abdominal aortic aneurysms: rationale for a prospective randomized clinical trial. Annals of the New York Academy of Sciences 878:159–178

611. Carmeliet P, Moons L, Lijnen R, Baes M, Lemaitre V, Tipping P, Drew A, Eeckhout Y, Shapiro S, Lupu F, Collen D (1997) Urokinase-generated plas-

min activates matrix metalloproteinases during aneurysm formation. Nature Genetics 17:439–444

612. Irizarry E, Newman KM, Gandhi RH, Nackman GB, Halpern V, Wishner S, Scholes JV, Tilson MD (1993) Demonstration of interstitial collagenase in abdominal aortic aneurysm disease. Journal of Surgical Research 54:571–574

613. Pagano MB, Bartoli MA, Ennis TL, Mao D, Simmons PM, Thompson RW, Pham CTN (2007) Critical role of dipeptidyl peptidase I in neutrophil recruitment during the development of experimental abdominal aortic aneurysms. Proceedings of the National Academy of Sciences of the United States of America 104:2855–2860

614. Tummala PE, Chen XL, Sundell CL, Laursen JB, Hammes CP, Alexander RW, Harrison DG, Medford RM (1999) Angiotensin II induces vascular cell adhesion molecule-1 expression in rat vasculature: A potential link between the renin-angiotensin system and atherosclerosis. Circulation 100:1223–1229

615. Boring L, Gosling J, Cleary M, Charo IF (1998) Decreased lesion formation in CCR2-/- mice reveals a role for chemokines in the initiation of atherosclerosis. Nature 394:894–897

616. Mancini GBJ, Henry GC, Macaya C, O'Neill BJ, Pucillo AL, Carere RG, Wargovich TJ, Mudra H, Luscher TF, Klibaner MI, Haber HE, Uprichard ACG, Pepine CJ, Pitt B (1996) Angiotensin-converting enzyme inhibition with quinapril improves endothelial vasomotor dysfunction in patients with coronary artery disease. The TREND (Trial on Reversing ENdothelial Dysfunction) Study. Circulation 94:258–265

617. Carmeliet P, Stassen JM, Schoonjans L, Ream B, van den Oord JJ, De Mol M, Mulligan RC, Collen D (1993) Plasminogen activator inhibitor-1 gene-deficient mice II: Effects on hemostasis, thrombosis, and thrombolysis. The Journal of Clinical Investigation 92:2756–2760

618. Yoshimura K, Aoki H, Ikeda Y, Fujii K, Akiyama N, Furutani A, Hoshii Y, Tanaka N, Ricci R, Ishihara T, Esato K, Hamano K, Matsuzaki M (2005) Regression of abdominal aortic aneurysm by inhibition of c-Jun N-terminal kinase. Nature Medicine 11:1330–1338

619. Li Z, Kleinstreuer C (2006) Effects of blood flow and vessel geometry on wall stress and rupture risk of abdominal aortic aneurysms. Journal of Medical Engineering & Technology 30:283–297

620. Raghavan ML, Vorp DA, Federle MP, Makaroun MS, Webster MW (2000) Wall stress distribution on three-dimensionally reconstructed models of human abdominal aortic aneurysm. Journal of Vascular Surgery 31:760–769

621. Salsac AV, Sparks SR, Lasheras JC (2004) Hemodynamic changes occurring during the progressive enlargement of abdominal aortic aneurysms. Annals of Vascular Surgery 18:14–21

622. Li Z, Kleinstreuer C (2006) Analysis of biomechanical factors affecting stent-graft migration in an abdominal aortic aneurysm model. Journal of Biomechanics 39:2264–2273

623. Alain [Chartier É] (1916) Éléments de philosophie[Elements of philosophy]. Gallimard, Paris

624. Koyré A (1971) Les étapes de la cosmologie scientifique[The stages of scientific cosmology]. Gallimard, Paris.

625. Arnold DN, Brezzi F, Fortin M (1984) A stable finite element for the Stokes equation. Calcolo 21:337–344

626. N'dri D, Garon A, Fortin A (2001) A new stable space-time formulation for two-dimensional and three-dimensional incompressible viscous flow. International Journal for Numerical Methods in Fluids 37:865–884

627. Pironneau O (1982) On the transport-diffusion algorithm and its application to the Navier–Stokes equations. Numerische Mathematik 38:309–332

628. Glowinski R (1984) Numerical methods for nonlinear variational problems. Springer, New York

629. Steinman DA, Milner JS, Norley CJ, Lownie SP, Holdsworth DW (2003) Image-based computational simulation of flow dynamics in a giant intracranial aneurysm. American Journal of Neuroradiology 24:559–566

630. Di Martino ES, Guadagni G, Fumero A, Ballerini G, Spirito R, Biglioli P, Redaelli A (2001) Fluid-structure interaction within realistic three-dimensional models of the aneurysmatic aorta as a guidance to assess the risk of rupture of the aneurysm. Medical Engineering & Physics 23:647–655

631. Engel FB, Hsieh PCH, Lee RT, Keating MT (2006) FGF1/p38 MAP kinase inhibitor therapy induces cardiomyocyte mitosis, reduces scarring, and rescues function after myocardial infarction. Proceedings of the National Academy of Sciences of the United States of America 103:15546–15551

632. Mieghem CA, Bruining N, Schaar JA, McFadden E, Mollet N, Cademartiri F, Mastik F, Ligthart JM, Granillo GA, Valgimigli M, Sianos G, Giessen WJ, Backx B, Morel MA, Es GA, Sawyer JD, Kaplow J, Zalewski A, Steen AF, Feyter P, Serruys PW (2005) Rationale and methods of the integrated biomarker and imaging study (IBIS): combining invasive and non-invasive imaging with biomarkers to detect subclinical atherosclerosis and assess coronary lesion biology. The International Journal of Cardiovascular Imaging 21:425–441

633. Barter P, Rye KA (2006) Are we lowering LDL cholesterol sufficiently? Nature Clinical Practice Cardiovascular Medicine 3:290–291

634. Choudhury RP, Fuster V, Fayad ZA (2004) Molecular, cellular and functional imaging of atherothrombosis. Nature Reviews – Drug Discovery 3:913–925

635. Amirbekian V, Lipinski MJ, Briley-Saebo KC, Amirbekian S, Aguinaldo JGS, Weinreb DB, Vucic E, Frias JC, Hyafil F, Mani V, Fisher EA, Fayad ZA (2007) Detecting and assessing macrophages in vivo to evaluate atherosclerosis noninvasively using molecular MRI. Proceedings of the National Academy of Sciences of the United States of America 104:961–966

636. Elmaleh DR, Fischman AJ, Tawakol A, Zhu A, Shoup TM, Hoffmann U, Brownell AL, Zamecnik PC (2006) Detection of inflamed atherosclerotic lesions with diadenosine-5', 5'''-P1, P4-tetraphosphate (Ap4A) and positron-emission tomography. Proceedings of the National Academy of Sciences of the United States of America 103:15992–15996

637. Jones GT, Jiang F, McCormick SP, Dusting GJ (2005) Elastic lamina defects are an early feature of aortic lesions in the apolipoprotein E knockout mouse. Journal of Vascular Research 42:237–246

638. Hansson GL, Libby P (2006) The immune response in atherosclerosis: a double-edged sword. Nature Reviews – Immunology 6:508-519

639. Mahley RW, Ji ZS (1999) Remnant lipoprotein metabolism: key pathways involving cell-surface heparan sulfate proteoglycans and apolipoprotein E. Journal of Lipid Research 40:1–16

640. Gargalovic PS, Imura M, Zhang B, Gharavi NM, Clark MJ, Pagnon J, Yang WP, He A, Truong A, Patel S, Nelson SF, Horvath S, BerlinerJA, Kirchgessner TG, Lusis AJ (2006) Identification of inflammatory gene modules based on variations of human endothelial cell responses to oxidized lipids. Proceedings of the National Academy of Sciences of the United States of America 103:12741–12746

641. Swirski FK, Pittet MJ, Kircher MF, Aikawa E, Jaffer FA, Libby P, Weissleder R (2006) Monocyte accumulation in mouse atherogenesis is progressive and proportional to extent of disease. Proceedings of the National Academy of Sciences of the United States of America 103:10340–10345

642. Wong CW, Christen T, Roth I, Chadjichristos CE, Derouette JP, Foglia BF, Chanson M, Goodenough DA, Kwak BR (2006) Connexin37 protects against atherosclerosis by regulating monocyte adhesion. Nature Medicine 12:950–954

643. Ait-Oufella H, Salomon BL, Potteaux S, Robertson AKL, Gourdy P, Zoll J, Merval R, Esposito B, Cohen JL, Fisson S, Flavell RA, Hansson GK, Klatzmann D, Tedgui A, Mallat Z (2006) Natural regulatory T cells control the development of atherosclerosis in mice. Nature Medicine 12:178–180

644. Hayashi T, Esaki T, Sumi D, Mukherjee T, Iguchi A, Chaudhuri G (2006) Modulating role of estradiol on arginase II expression in hyperlipidemic rabbits as an atheroprotective mechanism. Proceedings of the National Academy of Sciences of the United States of America 103:10485-10490

645. Park EM, Cho S, Frys KA, Glickstein SB, Zhou P, Anrather J, Ross ME, Iadecola C (2006) Inducible nitric oxide synthase contributes to gender differences in ischemic brain injury. Journal of Cerebral Blood Flow & Metabolism 26:392–401

646. Tiwari S, Zhang Y, Heller J, Abernethy DR, Soldatov NM (2006) Atherosclerosis-related molecular alteration of the human CaV1. 2 calcium channel α1C subunit. Proceedings of the National Academy of Sciences of the United States of America 103:17024–17029

647. Clarke MCH, Figg N, Maguire JJ, Davenport AP, Goddard M, Littlewood TD, Bennett MR (2006) Apoptosis of vascular smooth muscle cells induces features of plaque vulnerability in atherosclerosis. Nature Medicine 12:1075–1080

648. Li Y, Ge M, Ciani L, Kuriakose G, Westover EJ, Dura M, Covey DF, Freed JH, Maxfield FR, Lytton J, Tabas I (2004) Enrichment of endoplasmic reticulum with cholesterol inhibits sarcoplasmic-endoplasmic reticulum calcium ATPase-2b activity in parallel with increased order of membrane lipids: implications for depletion of endoplasmic reticulum calcium stores and apoptosis in cholesterol-loaded macrophages. Journal of Biological Chemistry 279:37030–37039

649. Feng B, Yao PM, Li Y, Devlin CM, Zhang D, Harding PH, Sweeney M, Rong JX, Kuriakose G, Fisher EA, Marks AR, Ron D, Tabas I (2003) The endoplasmic reticulum is the site of cholesterol-induced cytotoxicity in macrophages. Nature Cell Biology 5:781–792

650. Vengrenyuk Y, Carlier S, Xanthos S, Cardoso S, Ganatos P, Virmani R, Einav S, Gilchrist L, Weinbaum S (2006) A hypothesis for vulnerable plaque rupture due to stress-induced debonding around cellular microcalcifications in thin fibrous caps. Proceedings of the National Academy of Sciences of the United States of America 103:14678–14683

651. Ross R (1999) Atherosclerosis: an inflammatory disease. The New England Journal of Medecine 340:115–126

652. Eichner JE, Dunn ST, Perveen G, Thompson DM, Stewart KE, Stroehla BC (2002) Apolipoprotein E polymorphism and cardiovascular disease. American Journal of Epidemiology 155:487–495

653. Marsche G, Heller R, Fauler G, Kovacevic A, Nuszkowski A, Graier W, Sattler W, Malle E (2004) 2-chlorohexadecanal derived from hypochlorite-modified high-density lipoprotein-associated plasmalogen is a natural inhibitor of endothelial nitric oxide biosynthesis. Arteriosclerosis, Thrombosis, and Vascular Biology 24:2302–2306

654. Tanaka H, Sukhova GK, Swanson SJ, Cybulsky MI, Schoen FJ, Libby P (1994) Endothelial and smooth muscle cells express leukocyte adhesion molecules heterogeneously during acute rejection of rabbit cardiac allografts. American Journal of Pathology 144:938–951

655. Bevilacqua MP, Nelson RM, Mannori G, Cecconi O (1994) Endothelial-leukocyte adhesion molecules in human disease. Annual Review of Medicine 45:361–378

656. Altman R (2003) Risk factors in coronary atherosclerosis, athero-inflammation: the meeting point. Thrombosis Journal 1:4

657. Weill D, Wautier J-L, Dosquet C, Wautier M-P, Carreno MP, Boval B (1995) Monocyte modulation of endothelial leukocyte adhesion molecules. The Journal of Laboratory and Clinical Medicine 125:768–74

658. Henn V, Slupsky JR, Grafe M, Anagnostopoulos I, Forster R, Muller-Berghaus G, Kroczek RA (1998) CD40 ligand on activated platelets triggers an inflammatory reaction of endothelial cells. Nature 391:591–594

659. Cho N-H, Seong S-Y, Huh M-S, Kim N-H, Choi M-S, Kim I-S (2002) Induction of the gene encoding macrophage chemoattractant protein 1 by *orientia tsutsugamushi* in human endothelial cells involves activation of transcription factor activator protein 1. Infection and Immunity 70:4841–4850

660. Kai H, Ikeda H, Yasukawa H, Kai M, Seki Y, Kuwahara F, Ueno T, Sugi K, Imaizumi T (1998) Peripheral blood levels of matrix metalloproteases-2 and -9 are elevated in patients with acute coronary syndromes. Journal of the American College of Cardiology 32:368–372

661. Crawley JT, Goulding DA, Ferreira V, Severs NJ, Lupu F (2002) Expression and localization of tissue factor pathway inhibitor-2 in normal and atherosclerotic human vessels, Arteriosclerosis, Thrombosis, and Vascular Biology 22:218–224

662. Dichtl W, Nilsson L, Goncalves I, Ares MP, Banfi C, Calara F, Hamsten A, Eriksson P, Nilsson J (1999) Very low-density lipoprotein activated nuclear factor-κB in endothelial cells. Circulation Research 84:1085–1094

663. Orr AW, Sanders JM, Bevard M, Coleman E, Sarembock IJ, Schwartz MA (2005) The subendothelial extracellular matrix modulates NF-κB activation by flow: a potential role in atherosclerosis. The Journal of Cell Biology 169:191–202

664. Libby P, Ridker PM, Maseri A (2002) Inflammation and atherosclerosis. Circulation 105:1135-1143

665. Boyanovsky B, Karakashian A, King K, Giltyay NV, Nikolova-Karakashian M (2003) Ceramide-enriched low-density lipoproteins induce apoptosis in human microvascular endothelial cells. The Journal of Biological Chemistry 278:26992–26999

666. Cucina A, Borrelli V, Di Carlo A, Pagliei S, Corvino V, Santoro-D'Angelo L, Cavallaro A, Sterpetti AV (1999) Thrombin induces production of growth factors from aortic smooth muscle cells. The Journal of Surgical Research 82:61–66

667. Mallat Z, Besnard S, Duriez M, Mallat Z, Besnard S, Duriez M, Deleuze V, Emmanuel F, Bureau M. F, Soubrier F, Esposito B, Duez H, Fievet C, Staels B, Duverger N, Sherman D, Tedgui A (1999) Protective role of interleukin 10 in atherosclerosis. Circulation Research 85:E17–E24

668. Blankenberg S, Tiret L, Bickel C, Peetz D, Cambien F, Meyer J, Rupprecht HJ (2002) Interleukin-18 is a strong predictor of cardiovascular death in stable and unstable angina. Circulation 106:24–30

669. Ridker PM (2003) Clinical application of C-reactive protein for cardiovascular disease detection and prevention. Circulation 107:363–369

670. Ishikawa T, Hatakeyama K, Imamura T, Date H, Shibata Y, Hikichi Y, Asada Y, Eto T (2003) Involvement of C-reactive protein obtained by directional coronary atherotomy in plaque instability and developing restenosis in patients with stable or unstable angina pectoris. The American Journal of Cardiology 91:287–292

671. Sasu S, LaVerda D, Qureshi N, Golenbock DT, Beasley D (2001) Chlamydia pneumoniae and chlamydial heat shock protein 60 stimulate proliferation of human vascular smooth muscle cells via toll-like receptor 4 and p44/p42 mitogen activated protein kinase activation. Circulation Research 89:244–250

672. Chen KH, Guo X, Ma D, Guo Y, Li Q, Yang D, Li P, Qiu X, Wen S, Xiao R, Tang J (2004) Dysregulation of HSG triggers vascular proliferative disorders. Nature Cell Biology 6:872–883

673. Echtay KS, Roussel D, St-Pierre J, Jekabsons MB, Cadenas S, Stuart JA, Harper JA, Roebuck SJ, Morrison A, Pickering S, Clapham JC, Brand MD (2002) Superoxide activates mitochondrial uncoupling proteins. Nature 415:96–99.

674. Bernal-Mizrachi C, Gates AC, Weng S, Imamura T, Knutsen RH, DeSantis P, Coleman T, Townsend RR, Muglia LJ, Semenkovich CF (2005) Vascular respiratory uncoupling increases blood pressure and atherosclerosis. Nature 435:502–506.

675. Ohh M, Park CW, Ivan M, Hoffman MA, Kim TY, Huang LE, Pavletich N, Chau V, Kaelin WG (2000) Ubiquitination of hypoxia-inducible factor requires direct binding to the beta-domain of the von Hippel-Lindau protein. Nature Cell Biology 2:423–427

676. Rocnik E, Chow LH, Pickering JG (2000) Heat shock protein 47 is expressed in fibrous regions of human atheroma and is regulated by growth factors and oxidized low-density lipoprotein. Circulation 101:1237–1242.

677. Suzuki K, Sawa Y, Kaneda Y, Ichikawa H, Shirakura R, Matsuda H (1997) In vivo gene transfection with heat shock protein 70 enhances myocardial tolerance to ischemia-reperfusion injury in rat. The Journal of Clininical Investigation 99:1645–1650.

678. Martin JL, Mestril R, Hilal-Dandan R, Brunton LL, Dillmann WH (1997) Small heat shock proteins and protection against ischemic injury in cardiac myocytes. Circulation 96:4343–4348.

679. Tyson KL, Reynolds JL, McNair R, Zhang Q, Weissberg PL, Shanahan CM (2003) Osteo/chondrocytic transcription factors and their target genes exhibit

distinct patterns of expression in human arterial calcification. Arteriosclerosis, Thrombosis, and Vascular Biology 23:489–494

680. Bobryshev YV (2005) Transdifferentiation of smooth muscle cells into chondrocytes in atherosclerotic arteries in situ: implications for diffuse intimal calcification. The Journal of Pathology 205:641–650

681. Sata M, Saiura A, Kunusato A, Tojo A, Okada S, Tokuhisa T, Hirai H, Makuuchi M, Hirata Y, Nagai R (2002) Hematopoietic stem cells differentiate into vascular cells that participate in the pathogenesis of atherosclerosis. Nature Medicine 8:403–409

682. Linsel-Nitschke P, Tall AR (2005) HDL as a target in the treatment of atherosclerotic cardiovascular disease. Nature Reviews – Drug Discovery 4:193–205

683. Massaro M, Habib A, Lubrano L, Del Turco S, Lazzerini G, Bourcier T, Weksler BB, De Caterina R (2006) The omega-3 fatty acid docosahexaenoate attenuates endothelial cyclooxygenase-2 induction through both NADP(H) oxidase and PKCε inhibition. Proceedings of the National Academy of Sciences of the United States of America 103:15184–15189

684. Caro CG, Fitzgerald JM, Schroter RC (1971) Atherosclerosis and arterial wall shear: observations, correlation and proposal of a shear dependent mass transfer mechanism for atherogenesis. Proceedings of the Royal Society of London – Series B: Biological Sciences 177:109–159

685. Kumar RK, Balakrishnan KR (2005) Influence of lumen shape and vessel geometry on plaque stresses: possible role in the increased vulnerability of a remodelled vessel and the "shoulder" of a plaque. Heart 91:1459–1465

686. Andersson HI, Halden R, Glomsaker T (2000) Effects of surface irregularities on flow resistance in differently shaped arterial stenoses. Journal of Biomechanics 3:1257–1262

687. Lorthois S, Lagrée P-Y (2000) Flow in an axisymmetric convergent: evaluation of maximal wall shear stress. Comptes Rendus de l'Académie des Sciences – Série IIB – Mécanique 328:33–40

688. Hamacher-Brady A, Brady NR, Logue SE, Sayen MR, Jinno M, Kirshenbaum LA, Gottlieb RA, Gustafsson AB (2007) Response to myocardial ischemia/reperfusion injury involves Bnip3 and autophagy. Cell Death and Differentiation 14:146–157

689. Fielitz J, van Rooij E, Spencer JA, Shelton JM, Latif S, van der Nagel R, Bezprozvannaya S, de Windt L, Richardson JA, Bassel-Duby R, Olson EN (2007) Loss of muscle-specific RING-finger 3 predisposes the heart to cardiac rupture after myocardial infarction. Proceedings of the National Academy of Sciences of the United States of America 104:4377–4382

690. Gidday JM (2006) Cerebral preconditioning and ischaemic tolerance. Nature Reviews – Neuroscience 7:437–448

691. Doukas J, Wrasidlo W, Noronha G, Dneprovskaia E, Fine R, Weis S, Hood J, DeMaria A, Soll R, Cheresh D (2006) Phosphoinositide 3-kinase gamma/delta inhibition limits infarct size after myocardial ischemia/reperfusion injury. Proceedings of the National Academy of Sciences of the United States of America 103:19866–19871

692. Kawano T, Anrather J, Zhou P, Park L, Wang G, Frys KA, Kunz A, Cho S, Orio M, Iadecola C (2006) Prostaglandin E2 EP1 receptors: downstream effectors of COX-2 neurotoxicity. Nature Medicine 12, 225–229

693. Haudek SB, Xia Y, Huebener P, Lee JM, Carlson S, Crawford JR, Pilling D, Gomer RH, Trial J, Frangogiannis NG, Entman ML (2006) Bone marrow-derived fibroblast precursors mediate ischemic cardiomyopathy in mice. Proceedings of the National Academy of Sciences of the United States of America 103:18284–18289

694. Mirotsou M, Zhang Z, Deb A, Zhang L, Gnecchi M, Noiseux N, Mu H, Pachori A, Dzau V (2007) Secreted frizzled related protein 2 (Sfrp2) is the key Akt-mesenchymal stem cell-released paracrine factor mediating myocardial survival and repair. Proceedings of the National Academy of Sciences of the United States of America 104:1643–1648

695. Giddens DP, Kitney RI (1985) Blood flow disturbances and spectral analysis. In: Bernstein EF (ed) Noninvasive Diagnostic Techniques in Vascular Disease. Mosby, St Louis

696. Deshpande MD, Giddens DP (1980) Turbulence measurements in a constricted tube. Journal of Fluid Mechanics 97:65–89

697. Siegel JM, Markou CP, Ku DN, Hanson SR (1994) A scaling law for wall shear rate through an arterial stenosis. Journal of Biomechanical Engineering 116:446–451

698. Deplano V, Siouffi M (1999) Experimental and numerical study of pulsatile flows through stenosis: wall shear stress analysis. Journal of Biomechanics 32:1081–1090

699. Ahmed SA, Giddens DP (1984) Pulsatile poststenotic flow studies with laser Doppler anemometry. Journal of Biomechanics 17:695–705

700. Siouffi M, Deplano V, Pélissier R. (1998) Experimental analysis of unsteady flows through a stenosis. Journal of Biomechanics 31:11–19

701. Steinman DA, Poepping TL, Tambasco M, Rankin RN, Holdsworth DW (2000) Flow patterns at the stenosed carotid bifurcation: effect of concentric versus eccentric stenosis. Annals of Biomedical Engineering 28:415–423

702. Gandhi RH, Irizarry E, Nackman GB, Halpern VJ, Mulcare RJ, Tilson MD (1993) Analysis of the connective tissue matrix and proteolytic activity of primary varicose veins. Journal of Vascular Surgery 18:814–820

703. Sansilvestri-Morel P, Rupin A, Jaisson S, Fabiani JN, Verbeuren TJ, Vanhoutte PM (2002) Synthesis of collagen is dysregulated in cultured fibroblasts derived from skin of subjects with varicose veins as it is in venous smooth muscle cells. Circulation 106:479–483

704. Sansilvestri-Morel P, Rupin A, Badier-Commander C, Kern P, Fabiani JN, Verbeuren TJ, Vanhoutte PM (2001) Imbalance in the synthesis of collagen type I and collagen type III in smooth muscle cells derived from human varicose veins. Journal of Vascular Research 38:560–568

705. de Simone JG (1991) The ostial valve of the junction of the internal saphenous vein and the wall of the femoral vein. Phlébologie 44:427–459

706. Van Cleef JF (1997) Classification VCT (Valve, Cusp, Tributary) et endoscopie veineuse[V. C. T. (Valve, Cusp, Tributary) classification and venous endoscopy]. Journal des Maladies Vasculaires 22:101–104

707. Zhong W, Mao S, Tobis S, Angelis E, Jordan MC, Roos KP, Fishbein MC, de Alboran IM, MacLellan WR (2006) Hypertrophic growth in cardiac myocytes is mediated by Myc through a cyclin D2-dependent pathway. EMBO Journal 25:3869–3879

708. Hilfiker-Kleiner D, Kaminski K, Podewski E, Bonda T, Schaefer A, Sliwa K, Forster O, Quint A, Landmesser U, Doerries C, Luchtefeld M, Poli V,

Schneider MD, Balligand JL, Desjardins F, Ansari A, Struman I, Nguyen NQ, Zschemisch NH, Klein G, Heusch G, Schulz R, Hilfiker A, Drexler H (2007) A cathepsin D-cleaved 16 kDa form of prolactin mediates postpartum cardiomyopathy. Cell 128:589–600

709. Arking DE, Pfeufer A, Post W, Kao WHL, Newton-Cheh C, Ikeda M, West K, Kashuk C, Akyo M, Perz S, Jalilzadeh S, Illig T, Gieger C, Guo CY, Larson MG, Wichmann HE, Marbán E, O'Donnell CJ, Hirschhorn JN, Kääb S, Spooner PM, Meitinger T, Chakravarti A (2006) A common genetic variant in the NOS1 regulator NOS1AP modulates cardiac repolarization. Nature Genetics 38:644–651

710. Winfree AT, Strogatz SH (1984) Organizing centres for three-dimensional chemical waves. Nature 311:611–615

711. Panfilov AV, Pertsov AM (1984) Vortex ring in a 3-dimensional active medium described by reaction-diffusion equations (Russian). Doklady Akademii Nauk SSSR 274:1500–1503

712. Panfilov AV, Keener JP (1993) Generation of reentry in anisotropic myocardium. Journal of Cardiovascular Electrophysiology 4:412–421

713. Panfilov AV, Kerkhof PL (2004) Quantifying ventricular fibrillation: in silico research and clinical implications. IEEE Transactions on Biomedical Engineering 51:195–196; Erratum. 558

714. Nash MP, Panfilov AV (2004) Electromechanical model of excitable tissue to study reentrant cardiac arrhythmias. Progress in Biophysics and Molecular Biology 85:501–522

715. Wyman BT, Hunter WC, Prinzen FW, McVeigh ER (1999) Mapping propagation of mechanical activation in the paced heart with MRI tagging. American Journal of Physiology – Heart and Circulatory Physiology 276:881–891

716. Sermesant M, Rhode KS, Anjorin A, Hegde S, Sanchez-Ortiz G, Rueckert D, Lambiase P, Bucknall C, Hill D, Razavi R (2004) Simulation of the elctromechanical activity of the heart using XMR interventional imaging. In : Barillot C, Haynor DR, Hellier P (eds) Medical Image Computing and Computer-assisted Intervention – MICCAI 2004. Springer, Berlin

717. Nattel S (2002) New ideas about atrial fibrillation 50 years on. Nature 415:219–226

718. Sasano T, McDonald AD, Kikuchi K, Donahue JK (2006) Molecular ablation of ventricular tachycardia after myocardial infarction. Nature Medicine 12:1256–1258

719. Moe GK, Rheinboldt WC, Abildskov JA (1964) A computer model of atrial fibrillation. American Heart Journal 67:200–220

720. Gray RA, Pertsov AM, Jalife J (1998) Spatial and temporal organization during cardiac fibrillation. Nature 392:75–78

721. Witkowski FX, Leon LJ, Penkoske PA, Giles WR, Spano ML, Ditto WL, Winfree AT (1998) Spatiotemporal evolution of ventricular fibrillation. Nature 392:78–82

722. Winfree AT (1989) Electrical instability in cardiac muscle: phase singularities and rotors. Journal of Theoretical Biology 138:353–405

723. Panfilov AV, (1998) Spiral breakup as a model of ventricular fibrillation. Chaos 8:57–64

724. Samie FH, Berenfeld O, Anumonwo J, Mironov SF, Udassi S, Beaumont J, Taffet S, Pertsov AM, Jalife J (2001) Rectification of the background potas-

sium current: a determinant of rotor dynamics in ventricular fibrillation. Circulation Research 89:1216–1223

725. Choi BR, Liu T, Salama G (2001) The distribution of refractory periods influences the dynamics of ventricular fibrillation. Circulation Research 88: e49–e58

726. Sanguinetti MC, Tristani-Firouzi M (2006) hERG potassium channels and cardiac arrhythmia. Nature 440:463–469

727. Sidhu J, RobertsR (2003) Genetic basis and pathogenesis of familial WPW syndrome. Indian Pacing and Electrophysiology Journal 3:197–201

728. Xie M, Zhan D, Dyck JRB, Li Y, Zhang H, Morishima M, Mann DL, Taffet GE, Baldini A, Khoury DS, Schneider MD (2006) A pivotal role for endogenous TGF-beta-activated kinase-1 in the LKB1/AMP-activated protein kinase energy-sensor pathway. Proceedings of the National Academy of Sciences of the United States of America 103:17378–17383

729. Sheu TWH, Tsai SF, Hwang WS, Chang TM (1999) A finite element study of the blood flow in total cavopulmonary connection. Computers & Fluids 28:19–39

730. Tsai SF, Sheu TWH, Chang TM (2001) Lung effect on the hemodynamics in pulmonary artery. International Journal for Numerical Methods in Fluids 36:249–263

731. Dubini G, de Leval MR, Pietrabissa R, Montevecchi FM, Fumero R (1996) A numerical fluid mechanical study of repaired congenital heart defects. Application to the total cavopulmonary connection. Journal of Biomechanics 29:111–121

Chap. 8. Treatments of Cardiovascular Diseases

732. Marescaux J, Leroy J, Gagner M, Rubino F, Mutter D, Vix M, Butner SE, M. K. Smith MK (2001) Transatlantic robot-assisted telesurgery. Nature 413:379–380

733. Krupa A, Gangloff J, Doignon C, M. de Mathelin M, Morel G, Leroy J, Soler L, Marescaux J (2003) Autonomous 3D positioning of surgical instruments in robotized laparoscopic surgery using visual serving. IEEE Transactions on Robotics Automation 19:842–853

734. Coste-Manière E, Adhami L, Mourgues F, Bantiche O (2004) Optimal planning of robotically assisted heart surgery: transfer precision in the operating room. The International Journal of Robotics Research 23:539–548

735. Borst C (2000) Operating on a beating heart. Scientific American 283:58–63

736. Cattin P, Dave H, Grunenfelder J, Szekely G, Turina M, Zuend G (2004) Trajectory of coronary motion and its significance in robotic motion cancellation. European Journal of Cardio-Thoracic Surgery 25:786–790

737. Garrigue S, Pepin JL, Defaye P, Murgatroyd F, Poezevara Y, Clementy J, Levy P (2007) High prevalence of sleep apnea syndrome in patients With long-term pacing. The European multicenter polysomnographic study. Circulation

738. Falk V, Mourgues F, Adhami L, Jacobs S, Thiele H, Nitzsche S, Mohr F, Coste-Manière E (2005) Cardio navigation: planning, simulation and augmented reality to robotic assisted endoscopic bypass grafting. The Annals of Thoracic Surgery 79:2040–2047

739. Soler L, Delingette H, Malandain G, Montagnat J, Ayache N, Koehl C, Dour-the O, Malassagne B, Smith M, Mutter D, Marescaux J (2001) Fully automatic anatomical, pathological, and functional segmentation from CT scans for hepatic surgery. Computer Aided Surgery 6:131–142

740. Ochando JC, Homma C, Yan Y, Hidalgo A, Garin A, Tacke F, Angeli V, Li Y, Boros P, Ding Y, Jessberger R, Trinchieri G, Lira SA, Randolph GJ, Bromberg JA (2006) Alloantigen-presenting plasmacytoid dendritic cells mediate tolerance to vascularized grafts. Nature Immunology 7:652–662

741. Lei M, Kleinstreuer C, Archie JP (1997) Hemodynamic simulations and computer-aided designs of graft-artery junctions. Journal of Biomechanical Engineering 119:343–348

742. Giordana S, Sherwin SJ, Peiro J, Doorly DJ, Papaharilaou Y, Caro CG, Watkins N, Cheshire N, Jackson M, Bicknall C, Zervas V (2005) Automated classification of peripheral distal by-pass geometries reconstructed from medical data. Journal of Biomechanics 38:47–62

743. Agoshkov V, Quarteroni A, Rozza G (2006) Shape design in aorto-coronaric bypass anastomoses using perturbation theory. SIAM Journal of Numerical Analysis 44:367–384

744. Rozza G (2005) Real-time reduced-basis techniques in arterial bypass geometries. In "Computational Fluid and Solid Mechanics, 1283–1287, Bathe KJ (ed), Elsevier, Amsterdam

745. Johnston SC, Wilson CB, Halbach VV, Higashida RT, Dowd CF, McDermott MW, Applebury CB, Farley TL, Gress DR (2000) Endovascular and surgical treatment of unruptured cerebral aneurysms: comparison of risks. Annals of Neurology 48:11–19

746. Burgreen GW, Antaki JF, Wu ZJ, Holmes AJ (2001) Computational fluid dynamics as a development tool for rotary blood pumps. Artificial Organs 25:336–340

747. Jarvik RK (1981) The total artificial heart. Scientific American 244:74–80

748. Grandmont C, Maday Y, Métier P (2002) Existence of a solution for a unsteady elasticity problem in large displacement and small perturbation. Comptes Rendus Mathématique. Académie des Sciences. Paris 334:521–526

749. Métier P (2003) Modélisation, analyse mathématique et applications numériques de problèmes d'interaction fluide-structure instationnaires[Modeling, mathematical analysis and numerical applications of unsteady fluid structure interaction problems]. PhD thesis, University Pierre & Marie Curie, Paris

750. Doyle MG, Tavoularis S, Bourgault Y (2004) Computation of blood flow in a diaphragme-type ventricular assist device. In: Chen S, McIllwain S (eds) Proceedings of the 12th Annual Conference of the CFD Society of Canada

751. Doyle MG, Tavoularis S, Bourgault Y (2005) Closed-loop simulation of a ventricular assist device coupled with a circulatory system model. In: Bathe KJ (ed) Computational Fluid and Solid Mechanics 2005. Elsevier, Amsterdam

752. Farinas MI, Garon A (2004) Application of DOE for optimal turbomachinery design. AIAA-2004-2139 34th AIAA Fluid Dynamics Conference and Exhibit, Portland, Oregon

753. Garon A, Farinas, MI (2004) Fast three-dimensional numerical hemolysis approximation. Artificial Organs 28:1016–1025

754. Comolet R, Miraoui M (1987) Performance de valves à feuillets onguiformes[Performance of nail-shaped membrane valves]. Journal de Biophysique et de Biomécanique 11:55–61

755. Bruss KH, Reul H, van Gilse J, Knott E (1983) Pressure drop and velocity fields at four mechanical heart valve prostheses: Bjork–Shiley Standard, Bjork–Shiley Concave-Convex, Hall–Kaster and St. Jude Medical. Life Support Systems 1:3–22

756. Woo YR, Sung HW, Williams FP, Yoganathan AP (1986) In vitro fluid dynamic characteristics of aortic bioprostheses: old versus new. Life Support Systems 4:63–85

757. van Steenhoven AA, van Duppen TJ, Cauwenberg JW, van Renterghem RJ (1982) In vitro closing behaviour of Bjork-Shiley, St Jude and Hancock heart valve prostheses in relation to the in vivo recorded aortic valve closure. Journal of Biomechanics 15:841–848

758. Chandran KB, Cabell GN, Khalighi B, Chen CJ (1984) Pulsatile flow past aortic valve bioprostheses in a model human aorta. Journal of Biomechanics 17:609–619

759. Garitey V (1994) Étude expérimentale de l'écoulement intraventriculaire en aval de différent types de prothèses valvulaires[Experimental study of the intraventricular flow downstream from different types of valvular prothesis]. PhD thesis, University Aix–Marseille II

760. Yamaji S, Imai S, Saito F, Yagi H, Kushiro T, Uchiyama T (2006) Does high-power computed tomography scanning equipment affect the operation of pacemakers? Circulation Journal 70:190–197

761. Chen KH, Guo X, Ma D, Guo Y, Li Q, Yang D, Li P, Qiu X, Wen S, Xiao RP, Tang J (2004) Dysregulation of HSG triggers vascular proliferative disorders. Nature Cell Biology 6:872–883

762. Mercurius KO, Morla AO (1998) Inhibition of vascular smooth muscle cell growth by inhibition of fibronectin matrix assembly. Circulation Research 82:548–556

763. Dumoulin C, Cochelin B (2000) Mechanical behaviour modelling of balloon-expandable stents. Journal of Biomechanics 33:1461–1470

764. Delfour MC, Garon A, Longo V (2005) Modeling and design of coated stents to optimize the effect of the dose. SIAM Journal on Applied Mathematics 65:858-881

765. Cribier A, Savin T, Saoudi N, Behar P, Rocha P, Mechmeche R, Berland J, Letac B (1986) Percutaneous transluminal aortic valvuloplasty using a balloon catheter. A new therapeutic option in aortic stenosis in the elderly. Archives des Maladies du Coeur et des Vaisseaux 79:1678–1686

766. Cribier A, Eltchaninoff H, Tron C, Bauer F, Agatiello C, Sebagh L, Bash A, Nusimovici D, Litzler PY, Bessou JP, Leon MB (2004) Early experience with percutaneous transcatheter implantation of heart valve prosthesis for the treatment of end-stage inoperable patients with calcific aortic stenosis. Journal of the American College of Cardiology 43:698–703

767. Boudjemline Y, Bonhoeffer P (2002) Steps towards percutaneous aortic valve replacement. Circulation 105:775–778

768. Ferrari M, Figulla HR, Schlosser M, Tenner I, Damm C, Guyenot V, Werner GS, Hellige G (2004) Transarterial aortic valve replacement with a self expanding stent in pigs. Heart 90:1326–1331

769. Liou TM, Liou SN, Shu KL (2004) Intra-anevrismal flow with helix and mesh stent placement across side-wall aneurysm pore of a straight parent vessel. Journal of Biomechanical Engineering 126:36–43

770. Lohner R, Castro M, Cebral JR, Putman C (2005) Clinical applications of patient-specific vascular CFD. Workshop of the CRM spring school, Montréal

771. Turjman F, Massoud TF, Sayre J, Vinuela F (1998) Predictors of aneurysmal occlusion in the period immediately after endovascular treatment with detachable coils: a multivariate analysis. American Journal of Neuroradiology 19:1645–1651

772. Tsai SF, Rani HP, Sheu TWH, Thiriet M (2008) Study of pulsatile blood and injected medicine fluid flow in TACE therapy (revised for Computer Methods in Biomechanics and Biomedical Engineering)

773. Kim KS, Lei Y, Stolz DB, Liu D (2007) Bifunctional compounds for targeted hepatic gene delivery. Gene Therapy 14:704–708

774. Huang K, Voss B, Kumar D, Hamm HE, Harth E (2007) Dendritic molecular transporters provide control of delivery to intracellular compartments. Bioconjugate Chemistry 18:403–409

775. Patri AK, Kukowska-Latallo JF, Baker JR (2005) Targeted drug delivery with dendrimers: comparison of the release kinetics of covalently conjugated drug and non-covalent drug inclusion complex. Advanced Drug Delivery Reviews 57:2203–2214

776. Simberg D, Duza T, Park JH, Essler M, Pilch J, Zhang L, Derfus AM, Yang M, Hoffman RM, Bhatia S, Sailor MJ, Ruoslahti E (2007) Biomimetic amplification of nanoparticle homing to tumors. Proceedings of the National Academy of Sciences of the United States of America 104:932–936

777. Dayton PA, Zhao S, Bloch SH, Schumann P, Penrose K, Matsunaga TO, Zutshi R, Doinikov A, Ferrara KW (2006) Application of ultrasound to selectively localize nanodroplets for targeted imaging and therapy. Molecular Imaging 5:160–174

778. Crowder KC, Hughes MS, Marsh JN, Barbieri AM, Fuhrhop RW, Lanza GM, Wickline SA (2005) Sonic activation of molecularly-targeted nanoparticles accelerates transmembrane lipid delivery to cancer cells through contact-mediated mechanisms: implications for enhanced local drug delivery. Ultrasound in Medicine & Biology 31:1693–1700

779. Rapoport N, Gao Z, Kennedy A (2007) Multifunctional nanoparticles for combining ultrasonic tumor imaging and targeted chemotherapy. Journal of the National Cancer Institute 99:1095–1106

780. Aalto S, Haarala C, Brück A, Sipilä H, Hämäläinen H, Rinne JO (2006) Mobile phone affects cerebral blood flow in humans. Journal of Cerebral Blood Flow & Metabolism 26:885–890

781. Cotin S, Delingette H, Clément J-M, Marescaux J, Ayache N (1997) Simulation active de chirurgie endoscopique[Active simulation of endoscopic surgery]. Revue Européenne de Technologie Biomédicale19:167–172

782. Marescaux J, Clément J-M, Tassetti V, Koehl C, Cotin S, Russier Y, Mutter D, Delingette H, Ayache N (1998) Virtual reality applied to hepatic surgery simulation: the next revolution. Annals of Surgery 228:627–634

Conclusion

783. Pascal B (1670) Pensées. [Thoughts] Gallimard, Paris

784. Pic de la Mirandole J (1993) Œuvres philosophiques [Philosophical works]. Presse Universitaire de France, Paris

App. A. Anatomical, Biological, Medical Glossaries

785. Shibata R, Sato K, Pimentel DR, Takemura Y, Kihara S, Ohashi K, Funahashi T, Ouchi N, Walsch K (2005) Adiponectin protects against myocardial ischemia-reperfusion injury through AMPK- and COX-2-dependent mechanisms. Nature Medicine 11:1096–1103

786. Ai D, Fu Y, Guo D, Tanaka H, Wang N, Tang C, Hammock BD, Shyy JYJ, Zhu Y (2007) Angiotensin II up-regulates soluble epoxide hydrolase in vascular endothelium in vitro and in vivo. Proceedings of the National Academy of Sciences of the United States of America 104:9018–9023

787. Ress A, Moelling K (2005) Bcr is a negative regulator of the Wnt signalling pathway. EMBO Reports 6:1095–1100

788. Rothblat GH, de la Llera-Moya M, Atger V, Kellner-Weibel G, Williams DL, Phillips, MC (1999) Cell cholesterol efflux: integration of old and new observations provides new insights. Journal of Lipid Research 40:781–796

789. van Meer G (2005) Cellular lipidomics. The EMBO Journal 24:3159–3165

790. Pufahl RA, Singer CP, Peariso KL, Lin SJ, Schmidt PJ, Fahrni CJ, Culotta VC, Penner-Hahn JE, O'Halloran TV (1997) Metal ion chaperone function of the soluble Cu(I) receptor Atx1. Science 278:853–856

791. Buck E, Wells JA (2005) Disulfide trapping to localize small-molecule agonists and antagonists for a G-protein-coupled receptor. Proceedings of the National Academy of Sciences of the United States of America 102:2719–2724

792. Haslbeck M, Franzmann T, Weinfurtner D, Buchner J (2005) Some like it hot: the structure and function of small heat-shock proteins. Nature Structural Molecular Biology 12:842–846

793. Kasper LH, Boussouar F, Boyd K, Xu W, Biesen M, Rehg J, Baudino TA, Cleveland JL, Brindle PK (2005) Two transactivation mechanisms cooperate for the bulk of HIF-1-responsive gene expression. EMBO Journal 24:3846–3858

794. Liu ST, Chan GK, Hittle JC, Fujii G, Lees E, Yen TJ (2003) Human MPS1 kinase is required for mitotic arrest induced by the loss of CENP-E from kinetochores. Molecular Biology of the Cell 14:1638–1651

795. Glass CK, Ogawa S (2006) Combinatorial roles of nuclear receptors in inflammation and immunity. Nature Reviews – Immunology 6:44–55

796. Hochedlinger K, Jaenisch R (2006) Nuclear reprogramming and pluripotency. Nature 441:1061–1067

797. Shinohara ML, Lu L, Bu J, Werneck MBF, Kobayashi KS, Glimcher LH, Cantor H (2006) Osteopontin expression is essential for interferon-alpha production by plasmacytoid dendritic cells. Nature Immunology 7:498–506

798. Tian T, Harding A, Inder K, Plowman S, Parton RG, Hancock JF (2007) Plasma membrane nanoswitches generate high-fidelity Ras signal transduction. Nature Cell Biology 9:905–914

799. Harding A, Tian T, Westbury E, Frische E, Hancock JF (2005) Subcellular localization determines MAP kinase signal output. Current Biology 15:869–873

800. Trinchieri G, Sher A (2007) Cooperation of Toll-like receptor signals in innate immune defence. Nature Reviews – Immunology 7:179–190

801. Wenk MR, De Camilli P (2004) Protein-lipid interactions and phosphoinositide metabolism in membrane traffic: Insights from vesicle recycling in nerve

terminals. Proceedings of the National Academy of Sciences of the United States of America 101:8262–8269

802. Wieland T, Mittmann C (2003) Regulators of G-protein signalling: multifunctional proteins with impact on signalling in the cardiovascular system. Pharmacology & Therapeutics 97:95–115

803. Thiébot M-H, Hamon M (1996) Un agent multiple : la sérotonine[A multiple agent: the serotonin]. Pour la Science 221:82–89

804. O'Neill LAJ, Bowie AG (2007) The family of five: TIR-domain-containing adaptors in Toll-like receptor signalling. Nature Reviews – Immunology 7:353–364

805. Weis SM, Cheresh DA (2005) Pathophysiological consequences of VEGF-induced vascular permeability. Nature 437:497–504

806. Takenawa T, Suetsugu S (2007) The WASP-WAVE protein network: connecting the membrane to the cytoskeleton. Nature Reviews Molecular Cell Biology 8:37–48

807. Davidson G, Wu W, Shen J, Bilic J, Fenger U, Stannek P, Glinka A, Niehrs C (2005) Casein kinase 1 gamma couples Wnt receptor activation to cytoplasmic signal transduction. Nature 438:867–872

808. Karin M, Greten FR (2005) NF-κB: linking inflammation and immunity to cancer development and progression. Nature Reviews Immunology 5:749–759

809. De Coppi P, Bartsch G, Siddiqui MM, Xu T, Santos CC, Perin L, Mostoslavsky G, Serre AC, Snyder EY, Yoo JJ, Furth ME, Soker S, Atala A (2007) Isolation of amniotic stem cell lines with potential for therapy. Nature Biotechnology 25:100–106

App. B. Adipocyte

810. Bastelica D, Morange P, Berthet B, Borghi H, Lacroix O, Grino M, Juhan-Vague I, Alessi MC (2002) Stromal cells are the main plasminogen activator inhibitor-1-producing cells in human fat: evidence of differences between visceral and subcutaneous deposits. Arteriosclerosis, Thrombosis, and Vascular Biology 22:173–178

811. Van Gaal LF, Mertens IL, De Block CE (2006) Mechanisms linking obesity with cardiovascular disease. Nature 444:875–880

812. Rosen ED, Spiegelman BM (2006) Adipocytes as regulators of energy balance and glucose homeostasis. Nature 444:847–853

813. Tilg H, Moschen AR (2006) Adipocytokines: mediators linking adipose tissue, inflammation and immunity. Nature Reviews Immunology 6:772–783

814. Alexander SPH, Mathie A, Peters JA (2007) Guide to receptors and channels. British Journal of Pharmacology 150:S1–S168

815. Kumada M, Kihara S, Ouchi N, Kobayashi H, Okamoto Y, Ohashi K, Maeda K, Nagaretani H, Kishida K, Maeda N, Nagasawa A, Funahashi T, Matsuzawa Y (2004) Adiponectin specifically increased tissue inhibitor of metalloproteinase-1 through interleukin-10 expression in human macrophages. Circulation 109:2046–2049

816. Kobayashi H, Ouchi N, Kihara S, Walsh K, Kumada M, Abe Y, Funahashi T, Matsuzawa Y (2004) Selective suppression of endothelial cell apoptosis by the high molecular weight form of adiponectin. Circulation Research 94:e27-31

817. Oda A, Taniguchi T, Yokoyama M (2001) Leptin stimulates rat aortic smooth muscle cell proliferation and migration. Kobe Journal of Medical Sciences 47:141–150

818. Sierra-Honigmann MR, Nath AK, Murakami C, García-Cardena G, Papapetropoulos A, Sessa WC, Madge LA, Schechner JS, Schwabb MB, Polverini PJ, Flores-Riveros JR (1998) Biological action of leptin as an angiogenic factor. Science 281:1683–1686

819. Fukuhara A, Matsuda M, Nishizawa M, Segawa K, Tanaka M, Kishimoto K, Matsuki Y, Murakami M, Ichisaka T, Murakami H, Watanabe E, Takagi T, Akiyoshi M, Ohtsubo T, Kihara S, Yamashita S, Makishima M, Funahashi T, Yamanaka S, Hiramatsu R, Matsuzawa Y, Shimomura I (2005) Visfatin: a protein secreted by visceral fat that mimics the effects of insulin. Science 307:426–430

App. C. Basic Aspects in Mechanics

820. Pedley TJ (1976) Viscous boundary layers in reversing flow. Journal of Fluid Mechanics 74:59-79

821. Jan DL, Shapiro AH, Kamm RD (1989) Some features of oscillatory flow in a model bifurcation. Journal of Applied Physiology 67:147–159

822. Poiseuille J-LM (1840) Recherches expérimentales sur le mouvement des liquides dans les tubes de très petits diamètres [Experimental studies on the movement of liquids in tubes of very small diameter]. Comptes Rendus de l'Académie des Sciences 11:961–967; 1041–1048

823. Womersley J (1955) Method for the calculation of velocity, rate of flow and viscous drag in arteries when the pressure gradient is known. The Journal of Physiology 127:553–563

824. Boussinesq JV (1891) Calcul de la moindre longueur que doit avoir un tube circulaire, évasé à son entrée, pour qu'un régime sensiblement uniforme s'y établisse, et de la dépense de charge qu'y entraîne l'établissement de ce régime[Calculation of the least length that a circular tube, flared at its entrance, in order to convey a fairly uniform flow regime and the pressure drop that establishes this flow regime]. Comptes Rendus de l'Académie des Sciences 113:49–51

825. Comolet R (1976) Mécanique expérimentale des fluides [Experimental Fluid Mechanics], Volume 2. Masson, Paris

826. Janna WS (1998) Internal incompressible viscous flow. In: Jonhson RW (ed) Handbook of Fluid Dynamics. CRC Press, Boca Raton, FL

827. Padet J (1991) Fluides en écoulement[Flowing Fluids]. Masson, Paris

828. Schiller L (1922) Die Entwicklung der laminaren Gesschwindigkeitsverteilung und ihre Bedeutung fur Zähigkeitsmessungen[The evolution of the laminar velocity distribution and its significance for the measurement of viscosity]. Zeitschrift für Angewandte Mathematik und Mechanik 2:96–106

829. Goldstein S (1938) Modern Developments in Fluid Dynamics. Oxford University Press, London

830. Langhaar HL (1942) Steady flow in the transition length of a straight tube. Journal of Applied Mechanics 9:A55–A58

831. Fediaevski C, Voitkounski I, Faddeev Y (1974) Mécanique des fluides[Fluid Mechanics]. Mir, Moscow

832. Caro CGC, Pedley TJ, Schroter RC, Seed WA (1978) The Mechanics of the Circulation. Oxford University Press, London

833. Kay JM, Nedderman RM, (1985) Fluid Mechanics and Transfer Processes. Cambridge University Press, Cambridge

834. Tritton DJ (1988) Physical Fluid Dynamics. Oxford University Press, London

835. Targ SM (1951) Basic problems of the theory of laminar flows (Russian). Moscow

836. Lew HS, Fung YC (1970) Entry flow into blood vessels at arbitrary Reynolds number. Journal of Biomechanics 3:23–38

837. Schlichting H (1955) Boundary-Layer Theory. McGraw Hill, New York

838. Brodkey RS (1995) The Phenomena of Fluid Motion. Dover, New York

839. Joulié R (1998) Mécanique des fluides appliquée[Applied Fluid Mechanics]. Ellipses, Paris

840. Shah RK, London AL (1978) Laminar Flow Forced Convection in Ducts. Academic Press, New York

841. Liepsch D (1986) Flow in émentotubes and arteries. Biorheology 23:395–433

842. Atkinson B, Brocklebank MP, Card CCH, Smith JM (1969) Low Reynolds number developing flows. AIChE Journal 15:548–553

843. Taylor LA, Gerrard JH (1977) Pressure radius relationships for elastic tubes and their application to arteries. Medical & Biology Engineering & Computing 15:11–21

844. Ribreau C, Bonis M (1978) Propagations et écoulements dans les tubes collabables. Contributions à l'étude des vaisseaux[Propagations and flows in collapsible tubes. Contributions to the blood vessel studies]. Journal Français de Biophysique et Médecine Nucléaire 3:153-158

845. Kresch E (1979) Compliance of flexible tubes. Journal of Biomechanics 12:825-839

846. Ribreau C (1991) Sur la loi d'état, la loi de perte de charge, et la nature de l'écoulement permanent en conduite collabable inclinée[On the tube-law, the head-loss, and the properties of steady flow in inclined collapsible ducts]. PhD thesis, Université Paris XII

847. Ribreau C, Naili S, Langlet A (1994) Head losses in smooth pipes obtained from collapsed tubes. Journal of Fluids and Structures 8:183–200

848. Thiriet M, Bonis M, Adedjouma AS, Yvon JP (1989) A numerical model of expired flow in a monoalveolar lung model subjected to pressure ramps, Journal of Biomechanical Engineering 111:9–16

849. Kececioglu I, McClurken ME, Kamm RD, Shapiro AH (1981) Steady supercritical flow in collapsible tube 1: experimental observations. Journal of Fluid Mechanics 109:367–389

850. Weaver DS, Paidoussis MP, (1977) On collapse and flutter phenomena in thin tubes conveying fluid. Journal of Sound and Vibration 50:117–132

851. Matsuzaki Y, Fung YC (1977) Stability analysis od striaght and buckled two-dimensional channels conveying an incompressible flow. Journal of Applied Mechanics 99:548–552

852. Grotberg JB, Gravely N (1989) Flutter in collapsible tubes: a theoretical model of wheezes, Journal of Applied Physiology 66:2262–2273

853. Matsuzaki Y, Ikeda T, Kitegawa T, Sakada S (1994) Analysis of flow in a two-dimensional collapsible channel using universal tube law. Journal of Biomechanical Engineering 116:469–476
854. Ohba K, Sakurai A, Oka J (1989) Self-excited oscillations in collapsible tube IV: laser Doppler measurement of local flow field. Technological report. Kansai University

App. D. Numerical Simulations

855. Poincaré H (1946) The foundations of Sciences. Science Press, Lancaster, PA
856. Girault V, Raviart P-A (1986) Finite Element Methods for Navier–Stokes Equation. Theory and Algorithms. Springer Series Computational Mathematics 5. Springer, Berlin
857. Temam R (1977) Navier–Stokes Equations. Theory and Numerical Analysis. North-Holland, Amsterdam
858. Quarteroni A, Valli A (1994) Numerical approximation of partial differential equations. Springer Series in Computational Mathematics 23, Springer, Berlin
859. Quarteroni A, Sacco R, Saleri F (2000) Numerical mathematics. Texts in Applied Mathematics 37, Springer, New York

Notations

A

$\mathcal{A}(p)$: area-pressure relation
A: Almansi strain tensor
A: cross sectional area
A: actin binding site
a: acceleration
a: major semi-axis
AA: arachidonic acid
AAA: abdominal aortic aneurism
Aaa: ATPase associated with diverse
    cellular activities
AAAP: aneurism-associated antigenic
    protein
AAK: adaptin-associated kinase
ABC: ATP binding cassette transporter
ABP: actin binding protein
AC: atrial contraction
ACase: adenyl cyclase
ACAT: acyl CoA-cholesterol acyltrans-
    ferase
ACC: acetyl coenzyme-A carboxylase
ACE: angiotensin converting enzyme
ACh: acetylcholine
ACTH: adrenocorticotropic hormone
Ad: adrenaline
ADAM: a disintegrin and metallopro-
    tease
ADAMTS: a disintegrin and metallo-
    protease with thrombosspondin
ADP: adenosine diphosphate
AF: atrial fibrillation
AGF: autocrine growth factor
Aip: actin interacting protein

AKAP: A-kinase anchoring protein
ALE: arbitrary Eulerian Lagrangian
AMPK: AMP-activated protein kinase
AmyR: amylin receptor
Ang: angiopoietin
Ank: ankyrin
ANP: atrial natriuretic peptide
ANS: autonomic nervous system
ANT: adenine nucleotide transporter
AOC: amine oxidase copper-containing
    protein
AoV: aortic valve
AP: activator protein
AP: activating enhancer-binding protein
APC: adenomatous polyposis coli
    protein
APl: action potential
Apn: adiponectin
Apo: apolipoprotein
Aqp: aquaporin
AR: adrenergic receptor
AR: area ratio
Arf: ADP-ribosylation factor
ARNO: Arf nucleotide site opener
ARP: absolute refractory period
ARP: actin-related protein
Artn: artemin
ARVD: arrythmogenic right ventricular
    dystrophy
ASAP: artery-specific antigenic protein
ASP: actin-severing protein
ASK: apoptosis signal-regulating kinase
AT: antithrombin

ATF: activating transcription factor
ATG: autophagy-related gene
ATMK: ataxia telangiectasia mutated
    kinase
ATn: angiotensine
ATng: angiotensinogen
ATP: adenosine triphosphate
ATR: angiotensin receptor
ATRK: ATM and Rad3-related kinase
AVN: atrioventricular node
AVV: atrioventricular valves
AW: analysis window

B

B: Biot-Finger strain tensor
B: bulk modulus
$\mathcal{B}$: bilinear form
b: minor semi-axis
b: body force
$\hat{\mathbf{b}}$: unit binormal
BBB: blood-brain barrier
BC: boundary condition
BCL: B-cell leukemia/lymphoma
Bdk: bradykinin
BEM: boundary element method
bFGF: basic fibroblast growth factor
BFU-E: burst forming unit-erythroid
BM: basement membrane
BNP: B-type natriuretic peptide
BOC: brother of CDO
$B\varphi$: basophil

C

C: stress tensor
$\mathcal{C}$: compliance
C: chronotropy
C_D: drag coefficient
C_f: friction coefficient
C_L: lift coefficient
c: stress vector
c_τ: shear
c_w: wall shear stress
c: concentration
c_p: wave speed
CA: computed angiography
Ca: calcium
CABG: coronary artery bypass grafting

Cam: calmodulin
CamK: calmodulin-dependent kinase
cAMP: cyclic adenosine monophosphate
CaR: calcium-sensing receptor
CAS: Crk-associated substrate
Cav: caveolin
CBF: coronary blood flow
CBF: core binding factor
CBP: CREB binding protein
CD: cluster determinant protein
Cdc42: cell-division cycle-42
CdK: cyclin-dependent kinase
Cdm: caldesmon
CDO: cell adhesion molecule-
    related/downregulated by
    oncogenes
CETP: cholesterol ester transfer protein
CFD: computational fluid dynamics
cFos: cellular Finkel Biskis Jinkins
    murine osteosarcoma virus
    sarcoma oncogene
CFU: colony-forming unit
cGMP: cyclic guanosine monophosphate
CGRP: calcitonin gene-related peptide
CI: cardiac index
Cin: chronophin
cJun: avian sarcoma virus-17 oncogene
CK: creatine kinase
CK: casein kinase
CRLR: calcitonin receptor-like receptor
CMC: cardiomyocyte
cMyb: myeloblastosis oncogene
cMyc: myelocytomatosis oncogene
Cn: collagen
CnF: collagen fiber
CNGC: cyclic nucleotide-gated channel
CNS: central nervous system
CO: cardiac output
COx: cyclooxygenase
CoP: coat protein
Cr: creatine
CRAC: Ca^{++}-release–activated Ca^{++}
    current
CREB: cAMP responsive element-
    binding protein
cRel: reticuloendotheliosis oncogene
CRH: corticotropin-releasing hormone
Crk: chicken tumor virus regulator of
    kinase

CRP: C-reactive protein
Cs: cholesterol
CsE: cholesteryl esters
CSF: cerebrospinal fluid
CSK: C-terminal Src kinase
Csk: cytoskeleton
Csq: calsequestrin
CT: computed tomography
CTL: cytotoxic T lymphocyte
CtR: calcitonin receptor
CVI: chronic venous insufficiency
CVLM: caudal ventrolateral medulla
CVP: central venous pressure
CVS: cardiovascular system
cyCK: cytosolic creatine kinase

D

D: dromotropy
D: vessel distensibility
$\mathcal{D}$: diffusion coefficient
D: deformation rate tensor
d: displacement vector
D: flexural rigidity
d: duration
DAG: diacylglycerol
DCA: directional coronary atherectomy
DCT: distal convoluted tubule
De: Dean number
DH: Dbl homology
Dhh: desert hedgehog
DICOM: digital imaging and communication for medicine
DISC: death-inducing signaling complex
DNA: deoxyribonucleic acid
DUS: Doppler ultrasound

E

E: strain tensor
E: elastic modulus
E: elastance
$\mathcal{E}$: energy
$\{\hat{\mathbf{e}}_i\}_{i=1}^{3}$: basis
e: strain vector
EBCT: electron beam CT
EC: endothelial cell
ECA: external carotide artery
ECF: extracellular fluid

ECG: electrocardiogram
ECM: extracellular matrix
EDGR: endothelial differentiation gene receptor
EDV: end-diastolic volume
EEL: external elastic lamina
EGF: epidermal growth factor
eIF: eukaryotic translation initiation factor
ELAM: endothelial–leukocyte adhesion molecules
ELCA: excimer laser coronary angioplasty
En: elastin
EnF: elastin fiber
EPDC: epicardial derived cell
Epo: erythropoietin
ER: endoplasmic reticulum
ERGIC: ER Golgi intermediate compartment
ERK: extracellular signal-regulated protein kinase
ERM: ezrin–radixin–moesin
ERP: effective refractory period
ESCRT: endosomal sorting complexes required for transport
ESV: end-systolic volume
ET: endothelin
ETR: endothelin receptor
EVAR: endovascular aneurism repair
Eφ: eosinophil

F

F: transformation gradient tensor
f: surface force
$\hat{\mathbf{f}}$: fiber direction unit vector
f: binding frequency
f_c: cardiac frequency
f: friction shape factor
f_v: head loss per unit length
FAD: flavine adenine dinucleotide
FADD: Fas receptor associated death domain
FAK: focal adhesion kinase
FB: fibroblast
FC: fibrocyte
FDM: finite difference method
FEM: finite element method

FGF: fibroblast growth factor
FHL: four and a half LIM-only protein
FlIP: flice-inhibitory protein
FN: fibronectin
Fn: fibrin
Fng: fibrinogen
FoxO: forkhead transcription factor
FR: flow ratio
FSH: follicle-stimulating hormone
FSI: fluid-structure interaction
FVM: finite volume method

G

G: Green-Lagrange strain tensor
G: shear modulus
G': storage modulus
G'': loss modulus
G: conductance
G_p: pressure gradient
G_h: hydraulic conductivity
g: gravity acceleration
g: physical quantity
g: detachement frequency
G protein: guanine nucleotide-binding
    protein
Gab: Grb2-associated binder
GAG: glycosaminoglycan
GAK: cyclin G-associated kinase
Gal: galanin
GAP: GTPase-activating protein
GCAP: guanylyl cyclase-activating
    protein
GDP: guanosine diphosphate
GDI: guanine nucleotide-dissociation
    inhibitor
GEF: guanine nucleotide-exchange
    factor
GF: growth factor
GFP: geodesic front propagation
GH: growth hormone
GHRH: growth hormone-releasing
    hormone
GIT: GPCR kinase-interacting protein
GKAP: guanylate kinase-associated
    protein
GluT: glucose transporter
GnRH: gonadotropin-releasing hormone
GP: glycoprotein

GPI: glycosyl phosphatidylinositol
    protein
GPCR: G-protein–coupled receptor
GPx: glutathione peroxidase
Grb: growth factor receptor-bound
    protein
GRK: GPCR kinase
GSK: glycogen synthase kinase
GTP: guanosine triphosphate

H

H: height
$\mathcal{H}$: history function
h: thickness
HAT: histone acetyltransferase
Hb: hemoglobin
HCT: helical CT
HDAC: histone deacetylase
HDL: high density lipoprotein
hERG: human ether-a-go-go related
    gene
HES: hairy enhancer of split
hFABP: heart fatty acid binding protein
HGF: hepatocyte growth factor
HIF: hypoxia-inducible factor
His: histamine
HMWK: high molecular weight
    kininogen
Hrt: Hairy-related transcription factor
HS: heparan sulfate
HSC: hematopoietic stem cell
Hsp: heat shock protein
HSPG: heparan sulfate proteoglycan
Ht: hematocrit

I

I: identity tensor
i: current
I: inotropy
IC: isovolumetric contraction
ICA: internal carotide artery
ICAM: intercellular adhesion molecule
ICF: intracellular fluid
ICliP: intramembrane-cleaving protease
IEL: internal elastic lamina
Ifn: interferon
Ig: immunoglobulin

IGF: insulin-like growth factor
IH: intimal hyperplasia
Ihh: indian hedgehog
IL: interleukin
IP: inositol phosphate
IP3: inositol triphosphate
IR: isovolumetric relaxation
ISA: intracranial saccular aneurism
ISG: interferon stimulated gene product
IVC: inferior vena cava
IVUS: intravascular US

J

J: flux
J_m: cell surface current density
JAM: junctional adhesion molecule
JaK: Janus kinase
JNK: c-Jun N-terminal kinase

K

$\mathbf{K}$: conductivity tensor
K: bending stiffness
K: reflection coefficient
K_m: compressibility
k: cross section ellipticity
k_c: spring stiffness
k_{ATP}: myosin ATPasic rate
KHC: kinesin heavy chain
Kk: kallikrein
KLC: kinesin light chain
KlF: Kruppel-like factor

L

$\mathbf{L}$: velocity gradient tensor
L: inertance
L: length
LA: left atrium
LCA: left coronary artery
LCAT: lysolecithin cholesterol
    acyltransferase
LCC: left coronary cusp
LDL: low density lipoproteins
LDV: laser Doppler velocimetry
Le: entry length
LH: luteinizing hormone
LKlF: lung Kruppel-like factor

Lkt: leukotriene
Ln: laminin
LMR: laser myocardial revascularization
LP: lipoprotein
LPase: lipoprotein lipase
LPS: lipopolysaccharide
LQTS: long-QT syndrome
LRP: LDL receptor-related protein
LSK: Lin-SCA1+ KIT+ cell
LSV: long saphenous vein
LTBP: latent TGFβ-binding protein
LV: left ventricle
LVAD: left ventricular assist device
LXR: liver X receptor
$L\varphi$: lymphocyte

M

Ma: Mach number
MAP: microtubule-associated protein
mAP: mean arterial pressure
MAPK: mitogen-activated protein
    kinase
MARK: microtubule affinity regulating
    kinase
MBP: myosin binding protein
MCP: monocyte chemoattractant
    protein
MEJ: myoendothelial junction
MGP: matrix Gla protein
MHC: major histocompatibility
    complex
miCK: mitochondrial creatine kinase
MIS: mini-invasive surgery
MIT: mini-invasive therapy
MiV: mitral valve
MKP: mitogen-activated protein kinase
    phosphatase
MLC: myosin light chain
MLCK: myosin light chain kinase
MLCP: myosin light chain phosphatase
MLP: muscle LIM protein
MMP: matrix metalloproteinase
mtMMP: membrane-type MMP
Mo: monocyte
MPO: median preoptic nucleus
Mpo: myeloperoxidase
MRI: magnetic resonance imaging

MRTF: myocardin-related transcription factor
MSSCT: multi-slice spiral CT
mTOR: mammalian target of rapamycin
mTORc: mTOR complex
MVO2: myocardial oxygen consumption
MWSS: maximal wall shear stress
mmCK: myofibrillar creatine kinase
MuRF: muscle-specific ring finger
MyHC: myosin heavy chain
$M\varphi$: macrophage

N

N: sarcomere number
$\hat{n}$: unit normal vector
n: PAM density with elongation x
n: myosin head density
NAD: nicotine adenine dinucleotide
NAd: noradrenaline
NCC: non-coronary cusp
NCS: neuronal calcium sensor
NCX: Na^+–Ca^{++} exchanger
NF: nuclear factor
NFAT: nuclear factor of activated T cells
NHE: sodium–hydrogen exchanger
NHERF: NHE regulatory factor
NIK: NFκB-inducing kinase
NIP: neointimal proliferation
NmU: neuromedin-U
NO: nitric oxide
NOS: nitric oxide synthase
NOx: NAD(P)H oxidase
NPC: Niemann-Pick disease C protein
NpY: neuropeptide-Y
NRSTK: non-receptor serine/threonine kinase
NRTK: non-receptor tyrosine kinase
NSF: N-ethylmaleimide-sensitive factor
NST: nucleus of the solitary tract
$N\varphi$: neutrophil

O

OSI: oscillatory shear index
OVLT: organum vasculosum lamina terminalis

P

$\mathcal{P}$: permeability
P: power
p: pressure
PAF: platelet activating factor
PAFAH: platelet-activating factor acetylhydrolase
PAI: plasminogene activator inhibitor
PAK: p21-activated kinase
PAR: partitioning defective protein
Pax: paxillin
PC: protein C
PCMRV: phase-contrast MR velocimetry
PCr: phosphocreatine
PCT: proximal convoluted tubule
PDE: phosphodiesterase
PDE: partial differential equation
PDGF: platelet derived growth factor
PDK: phosphoinositide-dependent kinase
Pe: Péclet number
PE: pulmonary embolism
PECAM: platelet endothelial cell adhesion molecule
PEO: proepicardial organ
PET: positron emission tomography
PF: platelet factor
PG: prostaglandin
pGC: particulate guanylyl cyclase
PGC: PPARγ coactivator
PGI2: prostacyclin
PGF: paracrine growth factor
PH: pleckstrin homology
PI: phosphoinositide
PI3K: phosphatidylinositol 3-kinase
PIP2: phosphatidyl inositol diphosphate
PIV: particle image velocimetry
PIX: p21-interacting exchange factor
PK: protein kinase
PKL: paxillin kinase linker
Pkp: plakophilin
PL: phospholipase
PLb: phospholamban
PLd: phospholipid
PLTP: phospholipid transfer protein
PMCA: plasma membrane Ca-ATPase

PMR: percutaneous (laser) myocardial revascularization
Pn: plasmin
Png: plasminogen
PoG: proteoglycan
Pon: paraoxonase
PPAR: peroxisome proliferator–activated receptor
preKk: prekallikrein
PS: protein S
PSC: pluripotent stem cell
PSEF: pseudo-strain energy function
PSer: phosphatidylserine
PTA: plasma thromboplastin antecedent
PTCA: percutaneous transluminal coronary angioplasty
PTCRA: PTC rotational burr atherectomy
PTEN: phosphatase and tensin homologue deleted on chromosome 10
PTFE: polytetrafluoroethylene
PTH: parathyroid hormone
PTHRP: parathyroid hormone-related protein
PTP: protein tyrosine phosphatase
PTK: protein tyrosine kinase
PuV: pulmonary valve
PVNH: paraventricular nucleus of hypothalamus
PWS: pulse wave speed
PYK: proline-rich tyrosine kinase
P2X: purinergic ligand-gated channel

Q

q: flow rate

R

R: resistance
R_h: hydraulic radius
R_g: gas constant
r: radial coordinate
RA: right atrium
RAAS: renin–angiotensin–aldosterone system
Rab: Ras from brain

RACC: receptor-activated cation channel
RACK: receptor for activated C-kinase
RAMP: receptor activity–modifying protein
Ran: Ras-related nuclear protein
Rap: Ras-related protein
RAR: retinoic acid receptor
Ras: (superfamily of small GTPases/genes)
RBC: red blood cell
RC: ryanodine calcium channel
RCA: right coranary artery
RCC: right coronary cusp
Re: Reynolds number
RFA: radiofrequency ablation
Rheb: Ras homolog enriched in brain
RHS: equation right hand side
Rho: Ras homology
RIAM: Rap1-GTP-interacting adapter
RIP: receptor-interacting protein
RKIP: Raf kinase inhibitor protein
RNA: ribonucleic acid
dsRNA: double-stranded RNA
mRNA: messenger RNA
miRNA: microRNA
rRNA: ribosomal RNA
siRNA: small interfering RNA
tRNA: transfer RNA
RNABP: RNA binding protein
RNP: ribonucleoprotein
snoRNP: small nucleolar ribonucleoprotein
Robo: roundabout
ROI: region of interest
ROK: Rho kinase
ROS: reactive oxygen species
RPTP: receptor protein tyrosine phosphatase
RSE: rapid systolic ejection
RSMCS: robot-supported medical and surgical system
RSK: ribosomal S6 kinase
RSTK: receptor serine/threonine kinase
RTK: receptor tyrosine kinase
Runx: Runt-related transcription factor
RV: right ventricle
RVF: rapid ventricular filling
RVLM: rostral ventrolateral medulla

RVMM: rostral ventromedial medulla
RXR: retinoid X receptor

S

S: Cauchy-Green deformation tensor
s: sarcomere length
SAA: serum amyloid A
SAC: stretch-activated channel
sAC: soluble adenylyl cyclase
SAH: subarachnoid hemorrhage
SAN: sinoatrial node
SAPK: stress-activated MAPK
Sc: Schmidt number
SCA: stem cell antigen
SCF: stem cell factor
SDF: stromal cell-derived factor
SE: systolic ejection
SEF: strain-energy function
SERCA: SR Ca ATPase
SFK: Src-family kinase
SFO: subfornical organ
sGC: soluble guanylyl cyclase
SGK: serum- and glucocorticoid-induced
    kinase
SH: Src homology
Shc: Src-homologous and collagen-like
    substrate
Shc: Src homology 2 domain containing
    transforming protein
Shh: sonic hedgehog
SHIP: SH-containing inositol phos-
    phatase
SHP: SH-containing protein tyrosine
    phosphatase
SIP: steroid receptor coactivator–
    interacting protein
SKIP: sphingosine kinase-1 interacting
    protein
SLAM: signaling lymphocytic activation
    molecule
SMC: smooth muscle cell
SNAP: soluble N-ethylmaleimide-
    sensitive factor-attachment
    protein
SNARE: SNAP receptor
SOD: superoxide dismutase
Sos: Son-of-sevenless
SPECT: single photon emission CT

Sph: sphingosine
SPN: supernormal period
SQTS: short-QT syndrome
SR: sarcoplasmic reticulum
SR: scavenger receptor
SRF: serum response factor
SSAC: shear stress-activated channel
SSE: slow systolic ejection
Ssh: slingshot protein
SSV: short saphenous vein
St: Strouhal number
STAT: signal transducer and activator
    of transduction
STIM: stromal interaction molecule
Sto: Stokes number
SUMO: small ubiquitin-related modifier
SV: stroke volume
SVC: superior vena cava
SVF: slow ventricular filling
SVR: systemic vascular resistance
SW: stroke work
S1P: sphingosine 1-phosphate

T

T: extrastress tensor
T: temperature
$\hat{\mathbf{t}}$: unit tangent
t: time
TACE: tumour necrosis factorα-
    converting enzyme
TACE: transarterial chemoembolization
TAK: TGFβ-activated kinase
TC: thrombocyte
TCF: T-cell factor
TcR: T-cell receptor
TCA: tricarboxylic acid
TEA: transluminal extraction atherec-
    tomy
TEM: transendothelial migration
Ten: tenascin
TFPI: tissue factor pathway inhibitor
TG: triglyceride
TGF: transforming growth factor
TGN: trans-Golgi network
TJ: tight junction
TKR: tyrosine kinase receptor
TLR: Toll-like receptors
TLT: TREM-like transcript

TM: thrombomodulin
TMC: twisting magnetocytometry
TMy: tropomyosin
TN: troponin
Tn: thrombin
TNF: tumor necrosis factor
TNFR: tumor necrosis factor receptor
TORC: transducer of regulated CREB
tPA: tissue thromboplastin activator
Tpo: thrombopoietin
TRADD: tumor necrosis factor-receptor
    associated death domain
TRAF: tumor necrosis factor-receptor
    associated factor
TREM: triggering receptor expressed
    on myeloid cells
TRH: thyrotropin-releasing hormone
TRPC: transient receptor potential
    channel
TrV: tricuspid valve
Trx: thioredoxin
TSH: thyroid-stimulating hormone
Tsp: thrombospondin
TxnIP: thioredoxin-interacting protein

U

$\mathbf{U}$: right stretch tensor
$\mathbf{u}$: displacement vector
$\mathbf{u}$: electrochemical command
Ub: ubiquitin
UCP: uncoupling protein
UDP: uridine diphosphate-glucose
UK: urokinase
Unc: uncoordinated receptor
US: ultrasound
USC: unipotential stem cell
USI: ultrasound imaging
UTP: uridine triphosphate

V

$\mathbf{V}$: left stretch tensor
V: volume
V_q: cross-sectional average velocity
$\mathbf{v}$: velocity vector
$\mathbf{v}$: recovery variable
VAMP: vesicle-associated membrane
    protein

VAV: ventriculoarterial valve
VCAM: vascular cell adhesion molecule
VCt: vasoconstriction
VDC: voltage-dependent channel
VDt: vasodilation
VEGF: vascular endothelial growth
    factor
VF: ventricular filling
VGC: voltage-gated channel
VIP: vasoactive intestinal peptide
VLDL: very low density lipoprotein
VN: vitronectin
VR: venous return
VVO: vesiculo–vacuolar organelle
vWF: von Willenbrand factor

W

$\mathbf{W}$: vorticity tensor
$\mathcal{W}$: strain energy density
W: work, deformation energy
$\mathbf{w}$: grid velocity
WBC: white blood cell
WSS: wall shear stress
WSSTG: WSS transverse gradient

X

X: reactance
$\mathbf{x}$: position vector
$\{x, y, z\}$: Cartesian coordinates

Y

Y: admittance coefficient

Z

Z: impedance
ZO: zonula occludens

Miscellaneous

3DR: three-dimensional reconstruction
5HT: serotonin

Σc: sympathetic
$p\Sigma c$: parasympathetic

Greek Letters

α: volumic fraction
α: convergence/divergence angle

α: attenuation coefficient
α_k: kinetic energy coefficient
α_m: momentum coefficient
β: inclination angle
$\{\beta_i\}_1^2$: myocyte parameters
Γ: domain boundary
Γ_L: local reflection coefficient
Γ_G: global reflection coefficient
γ: activation factor
γ_g: amplitude ratio of $\mathbf{g}$
$\dot{\gamma}$: shear rate
δ: boundary layer thickness
ϵ: strain
ε: small quantity
ζ: singular head loss coefficient
ζ: transmural coordinate
$\{\zeta_j\}_1^3$: local coordinate
η: azimuthal spheroidal coordinate
θ: circonferential polar coordinate
θ: $(\hat{\mathbf{e}}_x, \hat{\mathbf{t}})$ angle
κ: wall curvature
κ_c: curvature ratio
κ_d: drag coefficient
κ_h: hindrance coefficient
κ_o: osmotic coefficient
κ_s: size ratio
$\{\kappa_k\}_{k=1}^9$: tube law coefficients
κ_e: correction factor
Λ: head loss coefficient
λ_L: Lamé coefficient
λ: stretch ratio
λ: wavelength
μ: dynamic viscosity
μ_L: Lamé coefficient
ν: kinematic viscosity
ν_c: cardiac frequency
ν_P: Poisson ratio
Π: osmotic pressure
ρ: mass density
τ: time constant
Φ: potential
$\phi(t)$: creep function
φ: phase
χ_i: wetted perimeter
$\psi(t)$: relaxation function
Ψ: porosity
ω: angular frequency
Ω: computational domain

Subscripts

A: atrial (alveolar)
Ao: aortic
a: arterial
app: apparent
b: blood
c: contractile
c: center
$\cdot_c$: point-contact
D: Darcy (filtration)
d: diastolic
dyn: dynamic
e: external
e: extremum
eff: effective
f: fluid
g: grid
i: internal
inc: incremental
l: limit
$\cdot_\ell$: line-contact
max: maximum
m: muscular
met: metabolic
P: pulmonary
p: parallel
p: particule
$\cdot_q$: quasi-ovalisation
r: radial
rel: relative
S: systemic
s: solute
s: serial
s: systolic
$\cdot_t$: stream division
T: total
t: time derivative of order 1
tt: time derivative of order 2
tis: tissue
V: ventricular
v: veinous
w: wall
w: water (solvent)
$\cdot_\Gamma$: boundary
θ: azimuthal
+: positive command
−: negative command
$\bullet_*$: at interface

0: reference state ($\cdot_0$: unstressed or low
   shear rate)
$\cdot_\infty$: high shear rate

Superscripts

^{a}: active state
^{e}: elastic
^{f}: fluid
^{h}: hypertensive
^{n}: normotensive
^{p}: passive state
p: power
^{s}: solid
^{T}: transpose
^{v}: viscoelastic
*: scale
*: complex variable

Mathematical Notations

$\Delta\bullet$: difference

$\delta\bullet$: incriment
$d\bullet/dt$: time gradient
∇: gradient operator
$\nabla\cdot$: divergence operator
∇^2: Laplace operator
$|\ |_+$: positive part
$|\ |_-$: negative part
$\dot{\bullet}$: time derivative
$\bar{\bullet}$: time mean
$\breve{\bullet}$: space averaged
$\langle\bullet\rangle$: ensemble averaged
$\tilde{\bullet}$: dimensionless
$\bullet^+$: normalized ($\in [0,1]$)
$\hat{\bullet}$: peak value
$\bullet_\sim$: modulation amplitude
$\det(\bullet)$: determinant
$\text{cof}(\bullet)$: cofactor
$\text{tr}(\bullet)$: trace

Chemical Notations

$[\bullet]$: concentration

Index